T0211087

Lecture Notes in Computer Science   **13952**

Founding Editors

Gerhard Goos
Juris Hartmanis

The series Lecture Notes in Computer Science (LNCS), including its subseries Lecture Notes in Artificial Intelligence (LNAI) and Lecture Notes in Bioinformatics (LNBI), has established itself as a medium for the publication of new developments in computer science and information technology research, teaching, and education.

LNCS enjoys close cooperation with the computer science R & D community, the series counts many renowned academics among its volume editors and paper authors, and collaborates with prestigious societies. Its mission is to serve this international community by providing an invaluable service, mainly focused on the publication of conference and workshop proceedings and postproceedings. LNCS commenced publication in 1973.

Ivan Lirkov · Svetozar Margenov
Editors

Large-Scale
Scientific Computations

14th International Conference, LSSC 2023
Sozopol, Bulgaria, June 5–9, 2023
Revised Selected Papers

 Springer

Editors
Ivan Lirkov (iD)
Institute of Information and Communication
Technologies
Sofia, Bulgaria

Svetozar Margenov (iD)
Institute of Information and Communication
Technologies
Sofia, Bulgaria

ISSN 0302-9743   ISSN 1611-3349 (electronic)
Lecture Notes in Computer Science
ISBN 978-3-031-56207-5   ISBN 978-3-031-56208-2 (eBook)
https://doi.org/10.1007/978-3-031-56208-2

Preface

The 14th International Conference on Large-Scale Scientific Computations (LSSC 2023) was held in Sozopol, Bulgaria, June 5–9, 2023. The conference was organized by the Institute of Information and Communication Technologies at the Bulgarian Academy of Sciences and MIV Consult Ltd. in cooperation with Sozopol municipality.

The plenary invited speakers and lectures were:

- G. Pichot, Flow Simulations in Large-Scale Fractured Media
- D. Boffi, Reduced Order Methods for Parametric Eigenvalue Problems
- B. Alexandrov, SmartTensors: A New AI Platform for Big-Data Analytics and Feature Extractions
- S. Caucao, T. Li, I. Yotov, Mixed Finite Element Methods for Fluid-Poroelastic Structure Interaction
- M. Neytcheva, The Power of Block Preconditioners
- J. Schöberl, Flexible Multigrid Methods within the Finite Element Library NGSolve

The success of the conference and the present volume are the outcome of the joint efforts of many partners from various institutions and organizations. First, we would like to thank all the members of the Scientific Committee for their valuable contribution forming the scientific face of the conference, as well as for their help in reviewing contributed papers. We especially thank the organizers of the special sessions. We are also grateful to the staff involved in the local organization.

Traditionally, the purpose of the conference is to bring together scientists working with large-scale computational models in natural sciences and environmental and industrial applications, as well as specialists in the field of numerical methods and algorithms for modern high-performance computers. The invited lectures reviewed some of the most advanced achievements in the field of numerical methods and their efficient applications. The conference talks were presented by researchers from academic institutions and practical industry engineers including applied mathematicians, numerical analysts, and computer experts. The general theme for LSSC 2023 was "Large-Scale Scientific Computing" with a particular focus on the organized special sessions.

The special sessions and organizers were:

- Preconditioning and Multilevel Methods, In memory of Owe Axelsson — M. Neytcheva and P. Vassilevski
- Fractures and Mixed Dimensional Modeling: Discretizations, Solvers, and Methodology — J. Adler, X. Hu, G. Pichot, and L. Zikatanov
- Modeling, Discretization and Solvers for Interface-Coupled Multiphysics Problems — A. Budisa, M. Kuchta, K.-A. Mardal, and L. Zikatanov
- Machine Learning and Model Order Reduction for Large Scale Predictive Simulations — B. Haasdonk, O. Iliev, and M. Ohlberger
- Fractional Differential Problems: Theoretical Aspects, Algorithms and Applications — L. Aceto and S. Harizanov

- Variational Analysis and Optimal Control — M. Krastanov and V. Veliov
- Stochastic Optimal Control and Numerical Methods in Economics and Finance — D. Ghilli, M. Leocata, and G. Livieri
- Tensor Methods for Big Data Analytics and Low-Rank Approximations of PDEs Solutions — B. Alexandrov, H. Djidjev, and G. Manzini
- Applications of Metaheuristics to Large-Scale Problems — S. Fidanova and G. Luque
- Large-Scale Models: Numerical Methods, Parallel Computations and Applications — K. Georgiev and Z. Zlatev
- HPC and HPDA: Algorithms and Applications — A. Karaivanova, T. Gurov, and E. Atanassov

About 150 participants from all over the world attended the conference representing some of the strongest research groups in the field of advanced large-scale scientific computing. For the conference proceedings in total 61 papers were received. The organizers of the special sessions managed the reviewing process in their sessions. Following an initial single-blind review phase by reviewers, the special session organizers provided a recommendation for each paper. Following a final revision of the papers the editors of the volume made their final decision about the acceptance of the papers. This process resulted in publication of 50 papers by authors from 19 countries.

The next international conference on LSSC will be organized in June 2025.

February 2024

Ivan Lirkov
Svetozar Margenov

Organization

Program Committee Chairs

Svetozar Margenov   Institute of Information and Communication Technologies, BAS, Bulgaria

Ivan Lirkov   Institute of Information and Communication Technologies, BAS, Sofia, Bulgaria

Scientific Committee

Lidia Aceto   Università del Piemonte Orientale, Italy

James Adler   Tufts University, USA

Boian Alexandrov   Los Alamos National Laboratory, USA

Emanouil Atanassov   Institute of Information and Communication Technologies, BAS, Bulgaria

Daniele Boffi   KAUST, Saudi Arabia

Ana Budisa   Simula Research Laboratory, Norway

Ivan Dimov   Institute of Information and Communication Technologies, BAS, Bulgaria

Stefka Dimova   Sofia University, Bulgaria

Hristo Djidjev   Los Alamos National Laboratory, USA

Stefka Fidanova   Institute of Information and Communication Technologies, BAS, Bulgaria

Krassimir Georgiev   Institute of Information and Communication Technologies, BAS, Bulgaria

Daria Ghilli   University of Pavia, Italy

Todor Gurov   Institute of Information and Communication Technologies, BAS, Bulgaria

Bernard Haasdonk   University of Stuttgart, Germany

Stanislav Harizanov   Institute of Information and Communication Technologies, BAS, Bulgaria

Xiaozhe Hu   Tufts University, USA

Oleg Iliev   ITWM, Germany

Aneta Karaivanova   Institute of Information and Communication Technologies, BAS, Bulgaria

Mikhail Krastanov   Institute of Mathematics and Informatics, BAS, Bulgaria

Johannes Kraus   University of Duisburg-Essen, Germany

Contents

**Fractional Differential Problems: Theoretical Aspects, Algorithms
and Applications**

Variational Analysis and Optimal Control

Stochastic Optimal Control and Numerical Methods in Economics
and Finance

Tensor Methods for Big Data Analytics and Low-Rank
Approximations of PDEs Solutions

Applications of Metaheuristics to Large-Scale Problems

Large-Scale Models: Numerical Methods, Parallel Computations and Applications

HPC and HPDA: Algorithms and Applications

Contributed Papers

Invited Papers

An Implementation of a Coarse-Fine Mesh Stabilized Schwarz Method for a Three-Space Dimensional PDE-Problem

Owe Axelsson[1,2], Roman Kohut[1], and Maya Neytcheva[2(✉)]

[1] The Czech Academy of Sciences, Institute of Geonics, Ostrava, Czech Republic
`roman.kohut@ugn.cas.cz`
[2] Department of Information Technology, Uppsala University, Uppsala, Sweden
`maya.neytcheva@it.uu.se`

Abstract. When solving very large scale problems on parallel computer platforms, we consider the advantages of domain decomposition in strips or layers, compared to general domain decomposition splitting techniques. The layer sub-domains are grouped in pairs, ordered as odd-even respectively even-odd and solved by a Schwarz alternating iteration method, where the solution at the middle interfaces of the odd-even groups is used as Dirichlet boundary conditions for the even-odd ordered groups and vice versa. To stabilize the method the commonly used coarse mesh method can be replaced by a coarse-fine mesh method. A component analysis of the arising eigenvectors demonstrates that this solution framework leads to very few Schwarz iterations. The resulting coarse-fine mesh method entails a coarse mesh of a somewhat large size. In this study it is solved by two methods, a modified Cholesky factorization of the whole coarse mesh matrix and a block-diagonal preconditioner, based on the coarse mesh points and the inner node points. Extensive numerical tests show that the latter method, being also computationally cheaper, needs very few iterations, in particular when the domain has been divided in many layers and the coarse to fine mesh size ratio is not too large.

Keywords: Preconditioning · Domain decomposition in layers · Schwarz method · Coarse-fine mesh stabilization

1 Introduction

The numerical solution of three-dimensional (3D) partial differential equation (PDE) problems leads to very large scale algebraic systems for which domain decomposition methods are useful as they can enable an efficient usage of parallel computer platforms, see e.g. [1–6]. In many practical problems, such as in

This work has been initiated by Owe Axelsson before he passed away. The other two authors acknowledge his contributions and pay their respect by finalizing and bringing the idea to the interested readers' audience.

porous media, it can be advantageous to partition the domain in layers based on the different material properties. In an earlier work [7], for two-dimensional (2D) problems, a special construction of Schwarz alternating iteration method has been proposed for such layered domains. Namely, an alternating Schwarz iteration method is used, which interlaces between odd-numbered and even-numbered layers with maximum possible overlap between the subdomains. It has been demonstrated that this framework leads to very few Schwarz iterations when it is stabilized by a particular coarse-fine mesh method. Thereby the solution at the middle layers of the first group are used as Dirichlet boundary conditions for the iterations on the second group. As it is well known, Schwarz iteration methods without stabilization converge very slowly, which is due to the fact that the eigenvector modes in directions orthogonal to the layer interfaces are very slowly approximated, in fact, the iterative process needs $O(m^2)$ Schwarz iterations for full convergence, where m is the number of layers.

In this paper we consider three-dimensional (3D) PDE problems described by a differential operator with discontinuous coefficients between the different layers. We remark that the use of Schwarz alternating iterations with strong (maximal) overlap, can be particularly advantageous also when solving problems with narrowly oscillating coefficients or even with randomly defined coefficients, i.e. subject to measurements with stochastic variations. It is then important to use a very fine mesh in the subdomains. For the coarse mesh one can use a harmonic or arithmetic average of the coefficients on each subdomain and a much coarser mesh. In the present paper the latter method is used.

For 3D problems the coarse-mesh system becomes large scale. For its solution, two methods are compared, namely a modified Cholesky factorization and an inner preconditioned iteration method where the preconditioner is block-diagonal containing a block corresponding to the coarse node points and a block corresponding to the inner node points, which are themselves uncoupled in the preconditioner, an idea, related to the two-level grid preconditioning methods, see e.g. [8–12].

Employing overlapping subdomains avoids the commonly used conforming boundary conditions on intersecting lines [1, 2], i.e. faces in 3D, between neighboring subdomains, that is of the type $\frac{\partial u}{\partial n}|_{\Gamma_-} + \frac{\partial u}{\partial n}|_{\Gamma_+} = 0$, where Γ is an intersecting plane. Instead in the above implementation of Schwarz iteration method, only Dirichlet boundary values at interfaces, computed during the iterations, are used, which clearly simplifies the implementation of the method and makes it more stable.

Section 2 of the paper describes the general properties of the Schwarz method, and its stabilization. The methods, used for the coarse-fine mesh stabilized method for 3D problems, are discussed in Sect. 3. Extensive numerical test results are presented in Sect. 4 and the paper ends with some concluding remarks.

2   The Schwarz Iteration Method and Its Stabilization

The classical Schwarz alternating iterations method [5] iterates between two subdomains with some overlap. In this paper we use subdomains with full overlap.

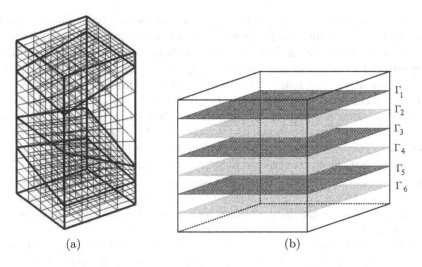

(a)   (b)

Fig. 1. A 3D domain subdivided in layers

2.1 Schwarz Alternating Iteration Method with Maximal Overlap

Consider the solution of a given elliptic PDE problem

$$\mathcal{L}u \equiv -\sum_{i,j=1}^{3} \frac{\partial}{\partial x_i}\left(\rho_{ij}\frac{\partial u}{\partial x_j}\right) = f(x) \tag{1}$$

in a bounded domain $\Omega \in R^3$, discretized by some appropriate discretization method, in this work, standard piece-wise linear finite elements on hexahedral elements. We assume that the coefficient matrix $[\rho_{ij}]$ is symmetric positive definite and that the physical domain Ω exhibits a layered structure, where the coefficients ρ_{ij} may vary strongly between these layers, see Fig. 1 for an illustration. In Fig. 1(a) the interfaces between the layers are marked with bold lines and, as the figure indicates, the layers may have irregular shape. In Fig. 1(b) we illustrate the geometry used in the current numerical expaeriment, with interfaces marked by filled surfaces.

As a solution framework for the discrete algebraic system of equations we use domain decomposition and the Schwarz alternating method. We associate subdomains with the layers in Ω and use a domain decomposition in $m+1$ strips, ordered consecutively and separated by interfaces Γ_i, $i = 1, 2, \cdots, m$. Assume for simplicity that m is odd.

The classical Schwarz alternating method is an iteration method that succeedingly in turns solves the systems corresponding to two large subdomains, the union of which covers the whole domain, with some overlap. To apply the Schwarz iteration method, in this paper we constitute the two subdomains so as the first contains all odd-numbered layers and the second - all even-numbered layers. The overlap between the two so-defined subdomains consists of all interfaces between the layers.

Starting with some initial guess of the solution on the even-numbered interfaces, for instance computed by interpolation of the already known solution on a coarse mesh, we solve the local problems with Dirichlet boundary conditions given on the two adjacent subdomains $\Gamma_{i-2}, \Gamma_i, i = 2, 4, \cdots, m - 1$. The computed values at $\Gamma_i, i = 1, 3, \cdots, m$ are then used as Dirichlet boundary values for the local subdomains between $(\Gamma_1, \Gamma_3), (\Gamma_3, \Gamma_5), \cdots, (\Gamma_{m-2}, \Gamma_m)$. The latter two solutions form one step of the full Schwarz alternating method between the first and second group of subdomains. It corresponds to an overlapping domain decomposition with maximal overlap of neighboring subdomains.

We note that the so-constructed Schwarz alternating method avoids the need to use some special interface conditions, such as $\frac{\partial u}{\partial n}\big|_{\Gamma_i}^{-} + \frac{\partial u}{\partial n}\big|_{\Gamma_i}^{+} = 0$, where the superscripts $\pm$ stand for values taken on the upper and lower sides of the interface. Such interface conditions are often used in the standard domain decomposition methods.

As it is well known, the so-described alternating solution procedure can be repeated a number of times but it can not approximate the globally spread components of the eigenfunctions of the solution, so it needs some enhancement, for instance, by adding a coarse mesh stabilization. In the current solution framework we suggest two ways to this, described in Sect. 2.2 and referred in the sequel to as '*coarse*' and '*coarce-fine*' mesh stabilizations.

2.2   Coarse and Coarse-Fine Mesh Stabilization

To avoid a large number of Schwarz iterations one must couple the method with some stabilization method, involving globally coupled node-points. Such combined methods are referred to as two-level methods.

Coarse Mesh Stabilization. Consider first the use of a standard coarse mesh stabilization, i.e., let $\mathcal{A} = \begin{bmatrix} A_{11} & A_{12} \\ A_{21} & A_C \end{bmatrix}$ be a splitting of the global matrix corresponding to the use of a finite element method for Eq. (1) and a splitting in fine and coarse mesh nodes. Here A_{11} corresponds to the fine mesh matrix formed by fine mesh finite element basis functions, A_C to the coarse mesh basis functions and A_{12}, A_{21}, to the coupling between these sets of mesh nodes. In the striped subdomain method we let all coarse mesh nodes be located on the interfaces between the subdomains. One can use the block-diagonal part of $\mathcal{A}$, i.e. $\mathcal{B} = \begin{bmatrix} A_{11} & 0 \\ 0 & A_C \end{bmatrix}$, as a preconditioner. Then as it is known, see e.g. [8,11,12], the condition number of $\mathcal{B}^{-1}\mathcal{A}$ equals $\kappa(\mathcal{B}^{-1}\mathcal{A}) = \frac{1+\gamma}{1-\gamma}$, where $\gamma \leq \sqrt{\frac{m_0^2 - 1}{m_0^2}} = \sqrt{1 - (\frac{h}{H})^2}$ and $m_0 = H/h$, H being the width of the layers, h is the fine mesh width, and γ is the constant in the strengthened Cauchy-Bunyakowski-Schwarz inequality corresponding to the bilinear product,

$$a(u, v) = \int_\Omega \mathcal{L}uv\, dx = \int_\Omega \sum \rho_{ij} \frac{\partial u}{\partial x_i} \frac{\partial u}{\partial x_j}\, d\Omega.$$

Here, for notational simplicity and without loss of generality, we assume homogeneous boundary conditions on the outer boundary $\partial\Omega$. This bound on γ holds for orthogonal grids in both 2D and 3D problems if $[\rho_{ij}]$ is a diagonal matrix.

Based on the above it follows that the number of iterations, such as in a preconditioned conjugate gradient, PCG method, increases as $O(\kappa(\mathcal{B}^{-1}\mathcal{A})^{1/2}) = O(m_0)$, where $\kappa(\ )$ denotes the spectral condition number. Hence, the method is not of optimal order with respect to the mesh ratio H/h but in practice this ratio can be controlled.

Coarse-Fine Mesh Stabilization. The two-level method can be significantly improved by replacing the coarse mesh with a coarse-fine mesh which can be constructed by connecting the fine mesh nodes on the interfaces with the fine mesh nodes on the two nearest interfaces. In this way we form a thin long triangular mesh in 2D problem or correspondingly in a 3D problem, and a coarse-fine matrix computed from all fine mesh node-points on the interfaces. There is no need to use interpolation on the intersection lines as they are already given with maximal accuracy. However, to reduce the solution cost of the arising coarse-fine mesh equations, in particular for 3D problems, one can go a middle way, that is not include all fine mesh node points, and interpolate the corresponding solution to the mesh-nodes not used in the fine-coarse stabilization. For instance, let

$$x_{i,j}^{(c)} = ((i-1)H, (j-1)H), \qquad i,j = 1,2,\cdots, H^{-1}+1$$

be the coarse mesh nodes and

$$x_{i,j}^{(f)} = ((i-1)h, (j-1)h), \qquad i,j = 1,2,\cdots, h^{-1}+1$$

be the nodes of the fine mesh.

If we use only each second ($\alpha = 2$), fourth ($\alpha = 4$), or eighth ($\alpha = 8$), where for simplicity we assume that h^{-1} is divisible by α, then the coarse-fine mesh nodes used are

$$x_{i,j}^{(c,f)} = (iH^{-1}, jh^{-1}), \quad i = 1,2,\cdots, H^{-1}+1, \quad j = \alpha h^{-1}, 2\alpha h^{-1}, \cdots, h^{-1}+1.$$

Either way, now the initial values for the Schwarz iteration method are more accurate, which indicates that there will be much fewer Schwarz iterations needed to satisfy a relative required accuracy tolerance of the residuals. This is also clearly demonstrated in the numerical tests.

2.3   Implementation of the Stabilized Method

The stabilized method can be formulated as a defect-correction method as follows. After constructing the finite element mesh, select the coarse-fine mesh points on the even numbered interfaces and all, each second, or say each fourth node points on those edges and correspondingly on faces in 3D problems, and form the corresponding finite element matrix $\mathcal{A}_{CF}$. Then, given an initial approximation $u^{(0)}$, say $u^{(0)} = 0$, for $k = 0, 1, \cdots$ until convergence,

(i) Compute the residual $\delta f^{(k)} = f - \mathcal{A}u^{(k)}$ and solve the coarse-fine mesh system
$$\mathcal{A}_{CF}\delta u^{(k)} = \delta f^{(k)}$$
restricted to the coarse-fine mesh points.

(ii) When applicable, interpolate $\delta u^{(k)}$ to the remaining node points, i.e. node points not already included, on the even numbered interfaces.

(iii) Use the Schwarz alternating method once, or more times, to approximately solve $\mathcal{A}\delta u^{(k)} = \delta f^{(k)}$ as described in Subsect. 2.1.

(iv) Update the solution, $u^{(k+1)} = u^{(k)} + \delta u^{(k)}$ and Repeat.

In this stabilization method, when solving the coarse-fine system, the corresponding errors on the intersecting lines or planes will be small due to the finer mesh used there. In general, there could be larger error components along the directions orthogonal to the interfaces. However, these are going to be damped out by the so-stabilized global alternating Schwarz method. Hence, in the combined Schwarz iterations with coarse-fine mesh stabilization, all eigenvector components of the solution are rapidly well approximated.

Depending on the width H of the subdomains one can use a direct solution for the arising sub-systems in the Schwarz method if H is small, i.e. H/h is not very large. However, in general H is not small and it would be less costly to instead use an iterative solution method, such as the incomplete Cholesky factorization method, or its rowsum modified version, MIC, see [9,10], as we now describe.

2.4 Modified Incomplete Factorization Methods

In order to get sparse factors, in incomplete factorization methods one computes a sparse factorization of a given, normally sparse matrix, typically by deleting entries arising in the factorization which falls outside an a priori chosen sparsity pattern. Since deleted entries have mostly negative signs, this makes the factorized matrix used as a preconditioner "too heavy" so the eigenvalues of the preconditioned matrix corresponding to the smoother solution modes, which normally dominates in the solution, get then very small, which causes delays in the rate of convergence of the iteration method.

As it has been shown in [8,9], see also [13–17], one way to avoid this is to make small perturbations of the given matrix before it is factorized. The methods can be described as a form of a generalized SSOR method, see [8,9], i.e. as approximate factorization where only a diagonal matrix need to be computed.

Consider first the method where no a priori perturbation is used. Let $A = (a_{ij})$ be a real symmetric $n \times n$ matrix,

$$A = D - L - L^T \tag{2}$$

where D is the diagonal and $(-L)$ is the strictly lower triangular part of A. Let

$$B = (X - L)X^{-1}(X - L^T) \tag{3}$$

be an approximate factorization of A where $X = \text{diag}(x_1, \ldots, x_n)$ is a diagonal matrix determined by the condition of equal rowsums:

$$Be = Ae \qquad e = (1, \ldots, 1)^T \in \mathbb{R}^n. \tag{4}$$

This factorization is referred to as MIC(0)* or *generalized SSOR factorization* in [8, 13, 16].

For the purpose of preconditioning, we adhere to the case where X and thus B, are positive definite, which we call *stable* MIC(0)* factorization. It is readily proven that if

$$L \geq 0, \quad Ae \geq 0 \quad \text{and} \quad Ae + L^T e > 0,$$

then A is weakly diagonally dominant matrix and $D - L$ is strictly diagonally dominant. Further, since $B - A = X + LX^{-1}L^T - D$, it follows that

$$x_i = a_{ii} - \sum_{k=1}^{i-1} \frac{a_{ik}}{x_k} \sum_{j=k+1}^{n} a_{kj} \tag{5}$$

which gives positive values x_i and the diagonal matrix $X = \text{diag}(x_1, \ldots, x_n)$ defines a stable MIC(0)* factorization of A.

Consider now application of small perturbations of A before the factorization is applied. Thereby we follow [14, 15] and [17].

Let $A = (a_{ij})$ be a regular symmetric weakly diagonally dominant $n \times n$ matrix which has nonpositive off-diagonal entries. Moreover, let $\widetilde{D} = \widetilde{D}(\xi) = \text{diag}(\widetilde{d}_1, \ldots, \widetilde{d}_h)$,

$$\widetilde{d} = \begin{cases} \xi a_{ii} & \text{if } a_{ii} \geq 2 \sum_{j>i} -a_{ij} \\ \xi^{1/2} a_{ii} & \text{if } a_{ii} < 2 \sum_{j>i} -a_{ij} \end{cases} \tag{6}$$

where $0 < \xi < 1$ is a constant. Note that $\widetilde{D}(\xi) = \xi D_1 + \xi^{-1/2} D_2$ where D_1, D_2 are two diagonal matrices such that $D_1 D_2 = 0$ and $D_1 + D_2 = D$ is the diagonal of A.

Then there is the stable MIC(0)* factorization C of the matrix $A' = A + \widetilde{D}$ and it holds that

$$\langle Bu, u \rangle \leq \langle A'u, u \rangle \leq (1 + \xi^{-1/2}) \langle Bu, u \rangle \qquad \forall u \in \mathbb{R}^n.$$

Furthermore

$$cond(B^{-1}A') \leq (1 + \mu_\xi)(1 + \xi^{-1/2})$$

where

$$\mu_\xi = \max_{u \neq 0} \frac{\langle \widetilde{D}(\xi)u, u \rangle}{\langle Au, u \rangle} \leq (\xi + \xi^{1/2}) \max_{u \neq 0} \frac{\langle Du, u \rangle}{\langle Au, u \rangle}. \tag{7}$$

As it has been shown in [8, 14, 16], for second order elliptic problems, where the second perturbation in Eq. (6) is only applied in a fraction $(O(\xi^{1/2}))$ of the node points, one can let $\xi = O(h^2)$ and the spectral condition number of

$C^{-1}A$ becomes then bounded above by $O(h^{-1})$. In practice, when the method is accelerated by a conjugate gradient method, since the number of iterations then grow as $O(h^{-1/2})$, this doesn't lead to many iterations.

As it has been shown in [13,17] the method is applicable also for certain elasticity problems.

An efficient implementation

Since both the preconditioner and the given matrix are symmetric and positive definite it is advisable to utilize this to write also the preconditioned matrix in a symmetric and positive definite form.

Given $B^{-1}(Ay - f) = 0$, scale here both B and A with $X^{-1/2}$ from both sides, to get

$$\widehat{B}^{-1}(\widehat{A}\widehat{y} - \widehat{f}) = 0, \tag{8}$$

where $\widehat{A} = X^{-1/2}AX^{-1/2}$, $\widehat{f} = X^{-1/2}f$, $\widehat{y} = X^{1/2}y$ and $\widehat{B} = (I - \widehat{L})(I - \widehat{L}^T)$, $\widehat{L} = X^{-1/2}LX^{-1/2}$. Here Eq. (8) is rewritten in the form

$$(I - \widehat{L})^{-1}\widehat{A}(I - \widehat{L}^T)^{-1}z = (I - \widehat{L})^{-1}\widehat{f}, \tag{9}$$

where $z = (I - \widehat{L}^T)\widehat{y}$. It follows that z can be computed efficiently by use of the classical, one term version of the conjugate gradient method, which simplyfies the implementation. Besides a multiplication with $\widehat{A}$, each iteration step requires only one solution with each of the triangular systems $I - \widehat{L}$ respectively $I - \widehat{L}^T$, where no divisions but only simple recursions are used, so the total expense is not much larger than matrix vector multiplication with $\widehat{A}$. For the computation of $\widehat{y} = (I - \widehat{L}^T)^{-1}z$ only one additional cheap triangular matrix solve is needed.

Since the coarse-fine matrix can also be relatively large it can be efficient to use this method for those systems also, which avoids fill-in entries in the matrix, or one can use a version MIC(2) (see [9,10]) which permits fill-in of the order of the number of non-zero entries in the given matrix. It has been shown in [9] that MIC(2) is a good compromise between obtaining a faster convergent method without any need to include too much fill-in. The MIC(2) method has been used in [13] for the solution of Navier's equations in elasticity. However, in this paper we use just MIC(0), i.e. without any fill-in.

The MIC method is not of optimal order but nearly so since the number of iterations for smooth coefficients grow only as $O(h^{-1})$, and as $O(h^{-1/2})$ when one uses the preconditioner with the MIC method in a CG method. Sometimes it converges even faster, see [9,10].

Extensive numerical tests have been reported for discontinuous coefficient problems in [7], for the solution of an elliptic problems $\mathcal{L}u = f$ in 2D problems, where $\mathcal{L} = -\nabla \cdot (\varrho\nabla)$, with a variable positive coefficient ϱ. For simplicity ϱ was assumed to be constant in each subdomain. The domain is a unit square and a standard obtuse triangulation with piecewise linear finite element basis functions is used. For illustrations, in Table 1 the number of Schwarz iteration and computer times for different values of H and a relative stopping criterion ε in each subdomain, are shown for such a 2D problem. It can be seen the number of

Table 1. 2D, piecewise constant coefficients: Total number of Schwarz iterations and computing times for $h = 1/128$ and a variable number of strips

No. strips (ratio H/h)	Rel. stop.crit. ε on Ω_i	Method					
		(a) no stab.		(b) coarse mesh stab.		(c) fine-coarse mesh stab	
		iter	time (s)	iter	time (s)	iter	time (s)
8 (16)	10^{-2}	50	6.3	34	4.9	8	1.6
	10^{-4}	50	7.5	32	6.9	4	1.5
	10^{-6}	50	8.5	32	9.3	4	2.2
16 (8)	10^{-2}	195	13.8	58	5.4	6	1.1
	10^{-4}	195	16.8	49	6.6	4	1.2
	10^{-6}	195	18.1	49	9.1	4	1.7
32 (4)	10^{-2}	761	43.3	57	4.4	6	1.9
	10^{-4}	761	46.2	27	3.6	4	1.8
	10^{-6}	761	49.8	30	5.7	4	2.2

Schwarz iterations is fixed, i.e. in particular does not depend on h for the coarse-fine mesh method, but clearly the computer time depends somewhat on H. It is also seen that the number of iterations depends little on the inner solution tolerance used for the Schwarz solutions on each subdomain Ω_i, which shows that one can lower the elapsed computer time even further using weaker tolerances. To illustrate the latter on a simple two-dimensional version of Eq. (1) and constant ρ, in Table 1 results are given for various values of ε. Here, method (a) refers to a pure Schwarz method, method (b) to a coarse-mesh stabilized Schwarz method and method (c) to a coarse-fine mesh stabilized Schwarz method. (For these tests a computer with two Intel Xeon CPU, 3.40 GHz and 4-GB RAM is used.) It is seen that the coarse-fine mesh stabilization method strongly outperforms the pure coarse mesh stabilized method. The results in [7] also indicate that it is most efficient to stabilize after each complete Schwarz alternating iteration step, that is, one does not gain anything by stabilizing only after, say, each second Schwarz iteration.

In Sect. 4 such advantageous results are shown to hold also for 3D problems. The major difference between the method when applied to 2D domains and applied to 3D domains is that the coarse-fine mesh matrix can have very larger sizes and special solution methods to solve the arising linear systems are needed. This is discussed in Sect. 3, where also some implementation details are presented.

3 An Approach to Implement Schwarz Alternating Iteration Method and Coarse-Fine Mesh Stabilization for 3D Domains

Consider the 3D elliptic problem Eq. (1) in a domain $\Omega \subset \mathbb{R}^3$, $x = [x_1, x_2, x_3]$ and where $[\rho_{ij}]^3$ is a positive definite matrix. As before, for notational simplicity, assume that the boundary conditions, Dirichlet or Neumann, are homogeneous. The domain is divided in subdomains as illustrated in Fig. 1.

3.1 Two-Level Layered Subdivisions

We discuss first some implementational issues related to the use of a two-level layer subdomain division.

The data for each subdomain can be stored on its own parallel processor. However, assume that each processor consists itself of a number of parallel processors or cores. We can then use a modification of the coarse-fine mesh stabilized Schwarz method. First, to limit the size of the matrix, the fine-coarse matrix shall not contain all node points on the intersecting surfaces. Instead, when forming the coarse-fine mesh matrix, we take only, say each 4'th, each 8'th or each 16'th of the mesh points. To avoid much fill-in due to the now somewhat irregular sparsity pattern we do not form an incomplete factorization of the matrix but the corresponding linear algebraic systems can be solved by use of an algebraic multilevel method, see e.g. [11,12]. Thereby one starts with a very coarse mesh, say containing only each 32'nd mesh point, computes the corresponding solution and interpolates it to the next finer mesh, consisting of each 16'th mesh points, which is solved by iteration based on the interpolated values as initial approximations.

This can continue to the next matrix, formed by each 8'th mesh point, and so on. This forms a sort of V-cycle. One can also go back one level each time, which will then form a W-cycle.

At any rate, in this way one can solve the global coarse-fine mesh matrix system to a sufficient accuracy which suffices for the next step of the Schwarz alternating iteration method. In a similar way as in Sect. 2, this is constructed from the subdomains bounded by the odd numbered surfaces and even numbered surface, respectively. Now, however, each pair of subdomains is in general of too large size, so the corresponding systems must be solved by an inner level of coarse-fine stabilized Schwarz method. Then each such pair of subdomains can be partitioned in smaller subdomains, as it is illustrated in Fig. 2. As for the global coarse-fine matrix, here the coarse-fine matrix can be constructed from only a subset of the interface surface between the so split subdomains. In this way, the demand of memory to store the matrices, now on the subprocessors of the global processor, can be acceptable. The solution method for the so split pair of subdomains is then similar to the method presented in Sect. 2. In this way one has a double two-level form of coarse-fine Schwarz method. Clearly, the number of inner-most subdomain solvers multiply up. Based on the results in Table 1 we can expect a total number of such subdomain solvers being the square of the number for the 2D domain problem, that is $4^2 = 16$, $5^2 = 25$ or $6^2 = 36$ subdomain solvers can be expected. Because of the large scale of the global matrix, this is still acceptable.

More detailed discussion on the topic is found in [19].

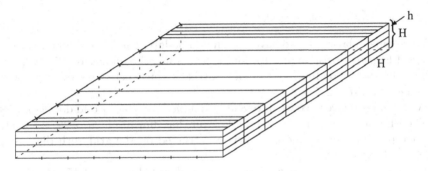

Fig. 2. Division of subdomains in layers

3.2 Solving the Problems Without Use of a Double Layered Subdomain Division

The use of two-level layered subdomain divisions can imply some more involved implementation. Therefore we consider now a method without use of it. The coarse-fine mesh method is then a very large scale and for its solution we compare two methods, one based on incomplete Cholesky factorization and one - using an inner iterative method with a block-diagonal preconditioner as shown in Sect. 2.2. In addition we sparsify the blocks by deleting all couplings between the inner line or face nodes, in this way obtaining an easily parallelizable solution method for the inner node part of the matrix. The coarse mesh matrix need, however, still a somewhat heavy computation at least for larger values of H. Further details and a description of the corresponding tests are found in the next section.

4 Numerical Tests

Consider a 3D elliptic problem $-\nabla(\rho\nabla u) = f$ in Ω with Dirichlet BC $u = 0$ on $\delta\Omega$, where $f \equiv 1$ and Ω is the unit cube. A rectangular uniform grid with $257 \times 257 \times 257$ nodes is used, which represents $l = 256$ FE layers in the z direction. If nz is the number of nodes in the z direction, then $l = nz - 1$. We use Schwarz alternating iteration method with coarse-fine mesh stabilization. The domain Ω is divided into $2m$ horizontal layers, so each layer contains $(nz-1)/2m$ FE layers. The alternating Schwarz method is working on m "even" ordered double-layers and $m - 1$ "odd" ordered double-layers. We solve the homogeneous coefficient case and several heterogeneous cases where we have $\rho = 1$ in the odd layers and sequentially $\rho = 1, 5, 10, 20, 50, 100$ in the even layers. We use coarse-fine grid with the number of nodes in z-direction corresponding to the number of layers (for $2m$ layers the coarse-fine grid has $2m + 1$ nodes in the z direction).

The computations are performed on Super Micro computer (symmetric multiprocessor) with 8x octa-core processor Intel Xeon E7-8837. The parallel part of programming used the OpenMP paradigm.

We solve first five basic tasks:

1. We assume 8 layers, so we have 4 even double-layers and 3 odd double layers. The FE grid corresponding to double-layers has the dimension $257 \times 257 \times 65$. The corresponding coarse-fine matrix is constructed on the grid with dimension $257 \times 257 \times 9$.
2. We assume 16 layers, so we have 8 even double-layers and 7 odd double layers. The FE grid corresponding to double-layers has dimension $257 \times 257 \times 33$. The corresponding coarse-fine matrix is constructed on the grid with dimension $257 \times 257 \times 17$.
3. We assume 32 layers, so we have 16 even double-layers and 15 odd double layers. The FE grid corresponding to double-layers has the dimension $257 \times 257 \times 17$. The corresponding coarse-fine matrix is constructed on the grid with dimension $257 \times 257 \times 33$.
4. We assume 64 layers, so we have 32 even double-layers and 31 odd double layers. The FE grid corresponding to double-layers has dimension $257 \times 257 \times 9$. The corresponding coarse-fine matrix is constructed on the grid with dimension $257 \times 257 \times 65$.
5. We assume 128 layers, so we have 64 even double-layers and 63 odd double layers. The FE grid corresponding to double-layers has the dimension $257 \times 257 \times 5$. The corresponding coarse-fine matrix is constructed on the grid with dimension $257 \times 257 \times 129$.

All five tasks are solved for the following six values of ρ in odd layers (1, 5, 10, 20, 50, 100). Thus, we solve 30 numerical examples.

The relative stopping criteria is fixed to $\varepsilon_S = 10^{-6}$ for the alternating Schwarz, $\varepsilon_C = 10^{-9}$ for the coarse grid solution and $\varepsilon_S = 10^{-9}$ for double-layer grid solution. (The chosen rather small tolerance for the Schwarz solutions on each subdomain ensures unified stopping criterion when comparing the results when varying heterogeneity and doublelayer sizes.)

Table 2 shows the number of iterations for all 30 numerical examples. Here **itS** denotes the number of iterations for the two level alternating Schwarz method, **itC** denotes the averaged number of iterations for each coarse-fine grid solution, and **itDL** presents the averaged number of iterations for each double-layer grid solution. It is seen that the method is stable with respect to the discontinuities of the coefficients. The Schwarz alternating iteration steps are very few. The number of coarse grid solutions varies somewhat with the mesh size. This has also been mentioned in Sect. 3.

The coarse-fine grid contains a very large number of nodes in the xy-index plane, the FE elements are very slim in the xy-direction and a very long in the z-direction. We include a comparison with a more standard choice of a regular coarse grid, which is uniform.

1. For task 1 (8 layers) we use a coarse grid with dimension $9 \times 9 \times 9$. The dimension is too small and the coarse approximation is very inaccurate for strong heterogeneity.
2. For task 2 (16 layers) we use a coarse grid with dimension $17 \times 17 \times 17$.

Table 2. Performance results for varying numbers of layers and size of discontinuities

number of layers	$\rho = 1$			$\rho = 5$			$\rho = 10$		
(H/h)	itS	itC	itDL	itS	itC	itDL	itS	itC	itDL
8(32)	4	25	31	4	30	33	4	31	33
16(16)	4	24	21	4	27	22	4	28	22
32(8)	4	23	14	4	24	14	3	31	15
64(4)	3	24	9	3	27	9	3	29	9
128(2)	3	24	6	3	42	6	3	46	5
	$\rho = 20$			$\rho = 50$			$\rho = 100$		
	itS	itC	itDL	itS	itC	itDL	itS	itC	itDL
8(32)	4	32	33	4	32	32	4	31	31
16(16)	4	30	21	4	31	20	3	37	22
32(8)	3	33	15	3	34	15	3	35	14
64(4)	3	32	10	3	33	10	3	35	10
128(2)	3	53	5	3	64	5	2	90	6

3. For task 3 (32 layers) we use a coarse grid with dimension $33 \times 33 \times 33$.
4. For task 4 (64 layers) we use a coarse grid with dimension $65 \times 65 \times 65$.
5. For task 5 (128 layers) we use a coarse grid with dimension $129 \times 129 \times 129$.

The results are presented in Table 3. It is seen that when using coarse mesh stabilization, the number of Schwarz alternating iteration steps is approximately twice as large as when using coarse-fine mesh stabilization. So even the calculated time for the parallel solution of tasks on double-layers is approximately double. The total number of iterations for solving coarse-fine and coarse tasks is approximately the same, but the matrix for the coarse task is much smaller, so the time will also be much smaller.

Table 3. Iteration counts for various number of layers and fine-coarse grid and coarse grid; $\rho = 1$ for odd layers, $\rho = 100$ for even layers

n.l.	fine-coarse				coarse			
(H/h)	itS	itC	itDL	dimension	itS	itC	itDL	dimension
8(32)	4	31	31	$257 \times 257 \times 9$				$9 \times 9 \times 9$
16(16)	3	37	22	$257 \times 257 \times 17$	6	12	25	$17 \times 17 \times 17$
32(8)	3	35	14	$257 \times 257 \times 33$	6	17	16	$33 \times 33 \times 33$
64(4)	3	35	10	$257 \times 257 \times 65$	5	23	11	$65 \times 65 \times 65$
128(2)	2	90	6	$257 \times 257 \times 129$	5	29	6	$129 \times 129 \times 129$

Table 4. Iteration counts for the coarse-fine stabilization systems for various number of layers and with coefficient $\rho = 1$ and $\rho = 100$ for adjacent layers

number of layers	Block preconditioner			Cholesky IF preconditioner		
(H/h)	itS	itC	itDL	itS	itC	itDL
8(32)	4	16	31	4	31	31
16(16)	3	16	22	3	37	22
32(8)	3	13	14	3	35	14
64(4)	3	9	9	3	35	10
128(2)	2	7	6	2	90	6

For clarity, the dimensions of the coarse grids are given. We note that $h = 1/512$ leads to a too large problem for the available memory on the computer used.

As a general conclusion, it is seen that the number of alternating Schwarz iterations are very few and essentially independent of the number of layers and coefficient discontinuities. Furthermore, these iterations decrease to about half when a coarse-fine mesh stabilization is used compared to the coarse mesh stabilization. However, the number of iterations for the solution of the coarse-fine mesh stabilization matrix system for the chosen solvers is rather high and much larger than for the standard coarse mesh stabilization. The number of iterations to solve the double-layer system is also high and clearly increases with number of layers. It is about half when the fine-coarse stabilization is used compared to those for the usual coarse mesh stabilization.

As shown in [7] one can further reduce the computational cost by use of variable stopping tolerance, that is, rough tolerances for the first iterations where the iterative error is large anyway and smaller tolerances when one approaches the exact solution.

Since the solution cost of the coarse-fine mesh matrix systems for 3D problems by the incomplete Cholesky factorization method is very high, we consider an alternative iteration method with a block-diagonal preconditioner. Here one block consists of the coarse matrix and the other blocks are formed by the inner node-points ordered in the direction vertical to the interface planes. Furthermore the coupling between these lines are deleted, in this way forming diagonal block matrices with very small bandwidth, which are easy to solve.

As Table 4 shows, the number of iterations needed to solve the coarse-fine matrix systems is now significantly reduced in particular if we use many layers and do not use a higher value of the ratio H/h than, say 8. For instance, compared to the incomplete Cholesky method, the computational cost is reduced by a factor of about $1/12$ for $H/h = 2$, by a factor of about $1/4$ for $H/h = 4$ and by a factor of about $1/3$ for $H/h = 8$. The number of Schwarz alternating iterations is the same and, hence, the number of double-layer solutions is also essentially the same. This shows that the coarse-fine matrix stabilization can outperform the coarse matrix stabilization not only for 2D problems but also for

Table 5. Computing time for various number of processors

Number of processors	computing time (seconds)	speedup
1	509	
2	496	1.026
4	305	1.669
8	197	2.584
16	144	3.535
32	115	4.426

layer divided subdomains in 3D even when one takes the total computational cost into account. By use of more sophisticated preconditioners this can likely be even further improved.

For the presented numerical tests we use the OpenMP shared-memory parallel programming model and parallelization only for the solution on double-layers (see Fig. 2). Here we use the preconditioned CG method, where the preconditioning is the additive overlapping one-level Schwarz method with subproblems replaced by incomplete Cholesky factorization. The computing times are presented in Table 5. The speedup is small because parallel computing is used in a part of the code only.

We note that the coarse-fine and the coarse matrices are created using the procedure described in [18].

5 Concluding Remarks

It has been shown that a properly stabilized Schwarz alternating iteration method based on use of two sets of subdomains with strong (maximal) overlap, leads to very few Schwarz iterations. Furthermore, all subdomain problems can be solved fully in parallel, that is, on a parallel computer they will be computationally cheap with respect to elapsed computer time.

The previously successful use of a coarse-fine mesh stabilized alternating Schwarz method for use with double layer pairs of 2D subdomains, reported in [7], has been shown to work well also for 3D problems. Here the solution cost of the coarse-fine mesh matrix systems can be rather high but it has been shown that utilizing a certain block-diagonal preconditioner this cost can be reduced to an acceptable level.

A possible further improvement is to use a two-level mesh subdomain division, that is, dividing also each 3D subdomain in layers can be an efficient method to use in a distributed memory computer platforms with more heterogeneous processing units.

Acknowledgements. The research has been supported by European Union's Horizon 2020 research and innovation programme under grant agreement number 847593 and by The Czech Radioactive Waste Repository Authority (SÚRAO) under grant agreement number SO2020-017.

References

1. Chan, T.F., Mathew, T.: Domain decomposition algorithms. Acta Numer. **3**, 61–143 (1994)
2. Smith, B.F., Bjørstad, P.E., Gropp, W.D.: Domain decomposition: parallel multilevel methods for elliptic partial differential equations. Cambridge University Press, New York, NY, USA (1996). ISBN 0-521-49589-X
3. Quarteroni, A., Valli, A.: Domain Decomposition Methods for Partial Differential Equations. Oxford University Press, Oxford, UK (1999). ISBN 9780198501787
4. Xu, J.: Iterative methods by space decomposition and subspace correction. SIAM Rev. **34**(4), 581–613 (1992)
5. Toselli, A., Widlund, O.B.: Domain Decomposition Methods - Algorithms and Theory. Volume 34 of Springer Series of Computational Mathematics: Springer, Berlin, Heidelberg (2005). https://doi.org/10.1007/b137868. ISBN 978-3-540-20696-5
6. Blaheta, R.: Space decomposition preconditioners and parallel solvers. In: Feistauer, M., Dolejsi, V., Knobloch, P., Najzar, K. (eds.) Numerical Mathematics and Advanced Applications, LNCS, pp. 20–38. Springer, Berlin, Heidelberg (2004). https://doi.org/10.1007/978-3-642-18775-9_2
7. Axelsson, O., Gustafsson, I.: A coarse-fine mesh stabilization for an alternating Schwarz domain decomposition method. Numer. Linear Algebra Appl. **26**, 1–19 (2019)
8. Axelsson, O., Barker, A.: Finite Element Solution of Boundary Value Problems: Theory and Computation. Academic Press, Inc., Cambridge (1984). Reprinted in SIAM Classics in Applied Mathematics; 2001. ISBN 978-0-89871-499-9
9. Gustafsson, I.: A class of first order factorization methods. BIT **18**, 142–156 (1978)
10. Axelsson, O.: Iterative Solution Methods. Cambridge University Press, Cambridge (1994). ISBN 9780521555692
11. Axelsson, O., Vassilevski, P.S.: Algebraic multilevel preconditioning methods II. SIAM J. Numer. Anal. **27**, 1569–1590 (1990)
12. Vassilevski, P.S.: Multilevel Block Factorization Preconditioners. Springer-Verlag, New York (2008). https://doi.org/10.1007/978-0-387-71564-3, ISBN 978-0-387-71563-6
13. Axelsson, O., Gustafsson, I.: Iterative methods for the solution of the Navier equations of elasticity. Comput. Methods Appl. Mech. Eng. **15**, 241–258 (1978)
14. Gustafsson, I.: Modified incomplete Cholesky (MIC) methods. In: Evans, D.J. (ed.), Preconditioning Methods, Theory and Applications, pp. 265–293 (1983)
15. Axelsson, O.: A survey of preconditioned iterative methods for linear systems of algebraic equations. BIT **25**, 165–187 (1985)
16. Axelsson, O.: On iterative solution of elliptic difference equations on a mesh-connected array of processors. Int. J. High Speed Comput. **1**, 165–183 (1989)
17. Blaheta, R.: Displacement decomposition - incomplete factorization preconditioning techniques for linear elasticity problems. Numer. Linear Algebra Appl. **1**, 107–128 (1994)
18. Kohut, R.: Parallel solution of elasticity problems using overlapping aggregations. Appl. Math. **63**, 603–628 (2018)
19. Axelsson, O.: Extensions of a coarse-fine mesh stabilized Schwarz alternating iteration domain decomposition method. J. Comput. Appl. Math. **364**, 112341 (2020)

Mixed Finite Element Methods
for the Navier–Stokes–Biot Model

Sergio Caucao[1,2], Aashi Dalal[3], Tongtong Li[4], and Ivan Yotov[3(✉)]

[1] Departamento de Matemática y Física Aplicadas,
Universidad Católica de la Santísima Concepción, Concepción, Chile
`scaucao@ucsc.cl`
[2] Grupo de Investigación en Análisis Numérico y Cálculo Científico,
GIANuC[2], Concepción, Chile
[3] Department of Mathematics, University of Pittsburgh, Pittsburgh, USA
`aad100@pitt.edu, yotov@math.pitt.edu`
[4] Department of Mathematics, Dartmouth College, Hanover, USA
`tongtong.li@dartmouth.edu`

Abstract. We present two mixed finite element methods for the quasistatic Navier–Stokes–Biot model. The methods are based on a fully-mixed formulation, using pseudostress, velocity, and vorticity for Navier–Stokes, stress, displacement, and rotation for elasticity, and velocity and pressure for Darcy flow. To handle the advective term, augmenting Galerkin-type terms are introduced in the Navier–Stokes model in the first method, while the second method uses a Banach space formulation. We discuss the relative advantages and disadvantages of the two methods and illustrate their behavior with numerical experiments.

Keywords: Navier–Stokes–Biot · Mixed finite elements · Banach space · Augmented formulation

1 Introduction

Fluid-poroelastic structure interaction (FPSI) refers to the interaction between a free viscous fluid and the flow in adjacent poroelastic medium. Such physical phenomenon occurs in a wide range of applications, including flows in fractured poroelastic media, surface–subsurface hydrological systems, and cardiovascular flows. The free fluid flow is modeled by the Stokes or the Navier–Stokes equations, while the flow in the poroelastic medium is modeled by the Biot system of poroelasticity. The two regions are coupled through dynamic and kinematic interface conditions, including continuity of normal velocity, balance of fluid force, conservation of momentum, and the Beavers–Joseph–Saffman slip with friction condition. Most of the previous theoretical work on the mathematical model and its numerical approximation has been on the Stokes–Biot model, see e.g. [2,5,7,8,12,14,16,17,20]. The Navier–Stokes–Biot model is better suitable for fast flows, such as blood flows or flows through industrial filters. This is a

I. Lirkov and S. Margenov (Eds.): LSSC 2023, LNCS 13952, pp. 19–31, 2024.
https://doi.org/10.1007/978-3-031-56208-2_2

more challenging problem, due to the presence of the nonlinear advective term in the fluid momentum equation. The well-posedness of the model with a non-mixed Darcy formulation is established in [13]. Computational results for the Navier–Stokes–Biot model can be found in [3,6,7,11]. Recently, an augmented fully-mixed formulation and its mixed finite element approximation have been developed and analyzed in [15].

In this paper we present two mixed finite element methods for the quasistatic Navier–Stokes–Biot model, which are based on a fully-mixed formulation. The Navier–Stokes equations are reformulated in terms of three variables: weakly symmetric nonlinear pseudostress, velocity, and vorticity. The elasticity formulation involves weakly symmetric poroelastic stress, displacement, and rotation, while Darcy flow is modeled by the classical dual mixed velocity-pressure formulation. The weak stress symmetry allows for the use of simpler finite elements. The two methods we consider use the same mixed finite element method for Biot, but differ in the Navier–Stokes discretization. In particular, they differ in the handling of the nonlinear advective term in the Navier–Stokes momentum equation and the related regularity of the fluid velocity. The first method is the augmented method from [15], where redundant Galerkin-type terms arising from the equilibrium and constitutive equations are introduced. As a result, the pseudostress is in H(div) and the fluid velocity is in H^1. The second method is related to the one introduced in [11]. It is based on a Banach space formulation where the divergence of the pseudostress is in $L^{4/3}$ and the fluid velocity is in L^4. While the augmented method is analyzed in [15], the Banach space method has not been analyzed. We present both semidiscrete methods as degenerate evolution problems in a mixed form, indicating how the techniques from [15] can be applied for the analysis of the Banach space method. We further discuss important features of the two methods, including local mass and momentum conservation, accurate approximations for the stresses and the Darcy velocity, locking-free behavior, and robustness with respect to the physical parameters. We further discuss the relative advantages and disadvantages of the two methods and illustrate their behavior with numerical experiments.

We end this section with some definitions and notation. For a domain $\mathcal{O} \subseteq \mathbb{R}^n$, $n \in \{2,3\}$, and $p \in [1, +\infty]$, we denote by $L^p(\mathcal{O})$ and $W^{s,p}(\mathcal{O})$ the usual Lebesgue and Sobolev spaces. If $p = 2$ we write $H^s(\mathcal{O})$ in place of $W^{s,2}(\mathcal{O})$. Let $(\cdot, \cdot)_\mathcal{O}$ be the $L^2(\mathcal{O})$-inner product. For $\Gamma \subset \partial\mathcal{O}$, let $\langle \cdot, \cdot \rangle_\Gamma$ be the $L^2(\Gamma)$ inner product or duality pairing. By $\mathbf{M}$ and $\mathbb{M}$ we will denote the vectorial and tensorial counterparts of the generic scalar functional space M. For vector fields $\mathbf{v} = (v_i)_{i=1:n}$ and $\mathbf{w} = (w_i)_{i=1:n}$, we set the gradient, divergence, and tensor-product operators, as $\nabla\mathbf{v} := \left(\frac{\partial v_i}{\partial x_j}\right)_{i,j=1:n}$, $\mathrm{div}(\mathbf{v}) := \sum_{j=1}^n \frac{\partial v_j}{\partial x_j}$, and $\mathbf{v} \otimes \mathbf{w} := (v_i w_j)_{i,j=1:n}$. For any tensor field $\boldsymbol{\tau} := (\tau_{ij})_{i,j=1:n}$, its deviatoric part is defined as $\boldsymbol{\tau}^{\mathrm{d}} := \boldsymbol{\tau} - \frac{1}{n}\mathrm{tr}(\boldsymbol{\tau})\,\mathbf{I}$, where $\mathbf{I}$ is the identity matrix in $\mathbb{R}^{n \times n}$ and $\mathrm{tr}(\boldsymbol{\tau}) := \sum_i^n \tau_{ii}$. We also recall the space $\mathbf{H}(\mathrm{div}; \mathcal{O}) := \left\{\mathbf{w} \in \mathbf{L}^2(\mathcal{O}) : \mathrm{div}(\mathbf{w}) \in L^2(\mathcal{O})\right\}$. The space of matrix valued functions with rows in $\mathbf{H}(\mathrm{div}; \mathcal{O})$ is denoted by $\mathbb{H}(\mathbf{div}; \mathcal{O})$.

2 Model Problem

Let $\Omega \subset \mathrm{R}^n$, $n \in \{2,3\}$ be a Lipschitz domain with polytopal boundary, which is subdivided into two non-overlapping and possibly non-connected regions: a fluid region Ω_f and a poroelastic region Ω_p. Let $\Gamma_{fp} = \partial\Omega_f \cap \partial\Omega_p$ denote the (nonempty) interface between these regions and let $\Gamma_f = \partial\Omega_f \setminus \Gamma_{fp}$ and $\Gamma_p = \partial\Omega_p \setminus \Gamma_{fp}$ denote the external parts of the boundary $\partial\Omega$. We denote by $\mathbf{n}_f$ and $\mathbf{n}_p$ the unit normal vectors which point outward from $\partial\Omega_f$ and $\partial\Omega_p$, respectively, noting that $\mathbf{n}_f = -\mathbf{n}_p$ on Γ_{fp}. Let $(\mathbf{u}_\star, p_\star)$ be the velocity-pressure pair in $\Omega_\star$ with $\star \in \{f,p\}$, and let $\boldsymbol{\eta}_p$ be the displacement in Ω_p. Let $\mu > 0$ be the fluid viscosity, let ρ be the density, let $\mathbf{f}_\star$ be the body force terms, which do not depend on time, and let q_p be external source or sink term. The flow in Ω_f is governed by the Navier–Stokes equations:

$$\rho\,(\nabla\mathbf{u}_f)\,\mathbf{u}_f - \mathbf{div}(\boldsymbol{\sigma}_f) = \mathbf{f}_f, \quad \mathrm{div}(\mathbf{u}_f) = 0 \quad \text{in } \Omega_f \times (0,T], \tag{1a}$$

$$(\boldsymbol{\sigma}_f - \rho\,(\mathbf{u}_f \otimes \mathbf{u}_f))\,\mathbf{n}_f = \mathbf{0} \quad \text{on } \Gamma_f^{\mathrm{N}} \times (0,T], \quad \mathbf{u}_f = \mathbf{0} \quad \text{on } \Gamma_f^{\mathrm{D}} \times (0,T], \tag{1b}$$

where $\boldsymbol{\sigma}_f := -p_f\,\mathbf{I} + 2\,\mu\,\mathbf{e}(\mathbf{u}_f)$ denotes the stress tensor, $\mathbf{e}(\mathbf{v}) := \dfrac{1}{2}(\nabla\mathbf{v} + (\nabla\mathbf{v})^{\mathrm{t}})$, and $\Gamma_f = \Gamma_f^{\mathrm{N}} \cup \Gamma_f^{\mathrm{D}}$. While the standard Navier–Stokes equations are presented above to describe the behavior of the fluid in Ω_f, in this work we make use of an equivalent version of (1) based on the introduction of a pseudostress tensor combining the stress tensor $\boldsymbol{\sigma}_f$ with the convective term. More precisely, analogously to [10] and [9], we introduce the nonlinear-pseudostress tensor

$$\mathbf{T}_f := \boldsymbol{\sigma}_f - \rho\,(\mathbf{u}_f \otimes \mathbf{u}_f) = -p_f\,\mathbf{I} + 2\,\mu\,\mathbf{e}(\mathbf{u}_f) - \rho\,(\mathbf{u}_f \otimes \mathbf{u}_f) \quad \text{in } \Omega_f \times (0,T]. \tag{2}$$

In this way, applying the matrix trace to the tensor $\mathbf{T}_f$, and utilizing the incompressibility condition $\mathrm{div}(\mathbf{u}_f) = 0$ in $\Omega_f \times (0,T]$, one arrives at

$$p_f = -\frac{1}{n}\,(\mathrm{tr}(\mathbf{T}_f) + \rho\,\mathrm{tr}(\mathbf{u}_f \otimes \mathbf{u}_f)) \quad \text{in } \Omega_f \times (0,T]. \tag{3}$$

Hence, replacing back (3) into (2), and using the definition of the deviatoric operator, we obtain $\mathbf{T}_f^{\mathrm{d}} = 2\,\mu\,\mathbf{e}(\mathbf{u}_f) - \rho\,(\mathbf{u}_f \otimes \mathbf{u}_f)^{\mathrm{d}}$. Therefore (1) can be rewritten, equivalently, as the set of equations with unknowns $\mathbf{T}_f$ and $\mathbf{u}_f$, given by

$$\frac{1}{2\,\mu}\,\mathbf{T}_f^{\mathrm{d}} = \nabla\mathbf{u}_f - \boldsymbol{\gamma}_f(\mathbf{u}_f) - \frac{\rho}{2\,\mu}\,(\mathbf{u}_f \otimes \mathbf{u}_f)^{\mathrm{d}} \quad \text{in } \Omega_f \times (0,T], \tag{4a}$$

$$-\mathbf{div}(\mathbf{T}_f) = \mathbf{f}_f, \quad \mathbf{T}_f = \mathbf{T}_f^{\mathrm{t}} \quad \text{in } \Omega_f \times (0,T], \tag{4b}$$

$$\mathbf{T}_f\mathbf{n}_f = \mathbf{0} \quad \text{on } \Gamma_f^{\mathrm{N}} \times (0,T], \quad \mathbf{u}_f = \mathbf{0} \quad \text{on } \Gamma_f^{\mathrm{D}} \times (0,T], \tag{4c}$$

where $\boldsymbol{\gamma}_f(\mathbf{u}_f) := \dfrac{1}{2}\left(\nabla\mathbf{u}_f - (\nabla\mathbf{u}_f)^{\mathrm{t}}\right)$ is the vorticity. Notice that, as suggested by (3), p_f is eliminated from the present formulation and can be computed afterwards in terms of $\mathbf{T}_f$ and $\mathbf{u}_f$. In addition, the fluid stress $\boldsymbol{\sigma}_f$ can be recovered from (2).

In turn, let $\boldsymbol{\sigma}_e$ and $\boldsymbol{\sigma}_p$ be the elastic and poroelastic stress tensors, respectively:

$$A(\boldsymbol{\sigma}_e) = \mathbf{e}(\boldsymbol{\eta}_p) \text{ and } \boldsymbol{\sigma}_p := \boldsymbol{\sigma}_e - \alpha_p\, p_p\, \mathbf{I} \text{ in } \Omega_p \times (0,T], \tag{5}$$

where $0 < \alpha_p \le 1$ is the Biot–Willis constant, and A is the symmetric and positive definite compliance tensor, satisfying, for some $0 < a_{\min} \le a_{\max} < \infty$,

$$\forall\, \boldsymbol{\tau} \in \mathrm{R}^{n\times n}, \quad a_{\min}\, \boldsymbol{\tau}:\boldsymbol{\tau} \le A(\boldsymbol{\tau}):\boldsymbol{\tau} \le a_{\max}\, \boldsymbol{\tau}:\boldsymbol{\tau} \quad \forall\, \mathbf{x} \in \Omega_p.$$

In the isotropic case A has the form, for all symmetric tensors $\boldsymbol{\tau}$,

$$A(\boldsymbol{\tau}) := \frac{1}{2\mu_p}\left(\boldsymbol{\tau} - \frac{\lambda_p}{2\mu_p + n\lambda_p}\mathrm{tr}(\boldsymbol{\tau})\mathbf{I}\right), \text{ with } A^{-1}(\boldsymbol{\tau}) = 2\mu_p\boldsymbol{\tau} + \lambda_p\mathrm{tr}(\boldsymbol{\tau})\mathbf{I},$$

where $0 < \lambda_{\min} \le \lambda_p(\mathbf{x}) \le \lambda_{\max}$ and $0 < \mu_{\min} \le \mu_p(\mathbf{x}) \le \mu_{\max}$ are the Lamé parameters. In this case, $\boldsymbol{\sigma}_e := \lambda_p\mathrm{div}(\boldsymbol{\eta}_p)\mathbf{I} + 2\mu_p\mathbf{e}(\boldsymbol{\eta}_p)$, $a_{\min} = \dfrac{1}{2\mu_{\max} + n\lambda_{\max}}$, and $a_{\max} = \dfrac{1}{2\mu_{\min}}$. The poroelasticity region Ω_p is governed by the quasistatic Biot system [4]:

$$-\mathbf{div}(\boldsymbol{\sigma}_p) = \mathbf{f}_p, \quad \mu\, \mathbf{K}^{-1}\, \mathbf{u}_p + \nabla p_p = \mathbf{0} \text{ in } \Omega_p \times (0,T], \tag{6a}$$

$$\partial_t\left(s_0\, p_p + \alpha_p\, \mathrm{div}(\boldsymbol{\eta}_p)\right) + \mathrm{div}(\mathbf{u}_p) = q_p \text{ in } \Omega_p \times (0,T], \tag{6b}$$

$$\mathbf{u}_p \cdot \mathbf{n}_p = 0 \text{ on } \Gamma_p^N \times (0,T], \quad p_p = 0 \text{ on } \Gamma_p^D \times (0,T], \tag{6c}$$

$$\boldsymbol{\sigma}_p\mathbf{n}_p = \mathbf{0} \text{ on } \tilde{\Gamma}_p^N \times (0,T], \quad \boldsymbol{\eta}_p = \mathbf{0} \text{ on } \tilde{\Gamma}_p^D \times (0,T], \tag{6d}$$

where $\Gamma_p = \Gamma_p^N \cup \Gamma_p^D = \tilde{\Gamma}_p^N \cup \tilde{\Gamma}_p^D$, $0 < s_{0,\min} \le s_0 \le s_{0,\max}$ is a constant storage coefficient and $\mathbf{K}$ the symmetric and uniformly positive definite rock permeability tensor, satisfying, for some constants $0 < k_{\min} \le k_{\max}$,

$$\forall\, \mathbf{w} \in \mathrm{R}^n \quad k_{\min}\, \mathbf{w}\cdot\mathbf{w} \le (\mathbf{Kw})\cdot\mathbf{w} \le k_{\max}\, \mathbf{w}\cdot\mathbf{w} \quad \forall\, \mathbf{x} \in \Omega_p.$$

Next, we introduce the transmission conditions on the interface Γ_{fp}:

$$\mathbf{u}_f \cdot \mathbf{n}_f + \left(\partial_t\boldsymbol{\eta}_p + \mathbf{u}_p\right)\cdot\mathbf{n}_p = 0, \quad \boldsymbol{\sigma}_f\mathbf{n}_f + \boldsymbol{\sigma}_p\mathbf{n}_p = \mathbf{0} \text{ on } \Gamma_{fp} \times (0,T], \tag{7a}$$

$$\boldsymbol{\sigma}_f\mathbf{n}_f + \mu\, \alpha_{\mathrm{BJS}}\sqrt{\mathbf{K}_\mathbf{t}^{-1}}\left((\mathbf{u}_f - \partial_t\boldsymbol{\eta}_p)\cdot\mathbf{t}_f\right)\mathbf{t}_f = -p_p\mathbf{n}_f \text{ on } \Gamma_{fp} \times (0,T]. \tag{7b}$$

For simplicity we have adopted notation for a two-dimensional domain in (7b), where $\mathbf{t}_f$ is a unit tangent vector on Γ_{fp}, $\mathbf{K}_\mathbf{t} = (\mathbf{K}\,\mathbf{t}_f)\cdot\mathbf{t}_f$, and $\alpha_{\mathrm{BJS}} \ge 0$ is an experimentally determined friction coefficient. The equations in (7a) correspond to mass conservation and conservation of momentum on Γ_{fp}, respectively, whereas (7b) can be decomposed into its normal and tangential components representing the balance of force and the Beavers–Joseph–Saffman (BJS) slip with friction condition, respectively. We observe that the second equations in (7a) and (7b) can be rewritten in terms of tensor $\mathbf{T}_f$ as follows:

$$\mathbf{T}_f\mathbf{n}_f + \rho(\mathbf{u}_f \otimes \mathbf{u}_f)\mathbf{n}_f + \boldsymbol{\sigma}_p\mathbf{n}_p = \mathbf{0}, \tag{8a}$$

$$\mathbf{T}_f\mathbf{n}_f + \rho(\mathbf{u}_f \otimes \mathbf{u}_f)\mathbf{n}_f + \mu\, \alpha_{\mathrm{BJS}}\sqrt{\mathbf{K}_\mathbf{t}^{-1}}\left((\mathbf{u}_f - \partial_t\boldsymbol{\eta}_p)\cdot\mathbf{t}_f\right)\mathbf{t}_f = -p_p\mathbf{n}_f. \tag{8b}$$

The system of Eqs. (4)–(8) is complemented by the initial condition $p_p(\mathbf{x}, 0) = p_{p,0}(\mathbf{x})$ in Ω_p, which is used to construct suitable compatible initial data for the rest of the variables in a way that all equations in the system, except for the unsteady conservation of mass equation in (6b), hold at $t = 0$.

3   Weak Formulations

In this section we present two mixed variational formulations associated to the coupled problem (4)–(8).

3.1   Augmented Mixed Formulation

First, we follow an augmented approach developed in [15]. More precisely, we introduce the Hilbert spaces associated to the Navier–Stokes unknowns

$$\mathbb{X}_f := \left\{ \mathbf{R}_f \in \mathbb{H}(\mathbf{div}; \Omega_f) : \quad \mathbf{R}_f \mathbf{n}_f = \mathbf{0} \text{ on } \Gamma_f^{\mathrm{N}} \right\},$$

$$\mathbf{V}_f := \left\{ \mathbf{v}_f \in \mathbf{H}^1(\Omega_f) : \quad \mathbf{v}_f = \mathbf{0} \text{ on } \Gamma_f^{\mathrm{D}} \right\}. \tag{9}$$

In turn, in order to deal with the unknowns in the Biot region we introduce the Hilbert spaces:

$$\mathbb{X}_p := \left\{ \boldsymbol{\tau}_p \in \mathbb{H}(\mathbf{div}; \Omega_p) : \quad \boldsymbol{\tau}_p \mathbf{n}_p = \mathbf{0} \text{ on } \tilde{\Gamma}_p^{\mathrm{N}} \right\},$$

$$\mathbf{V}_s := \mathbf{L}^2(\Omega_p), \quad \mathbb{Q}_p := \left\{ \boldsymbol{\chi}_p \in \mathbb{L}^2(\Omega_p) : \quad \boldsymbol{\chi}_p^{\mathrm{t}} = -\boldsymbol{\chi}_p \right\},$$

$$\mathbf{V}_p := \left\{ \mathbf{v}_p \in \mathbf{H}(\mathrm{div}; \Omega_p) : \quad \mathbf{v}_p \cdot \mathbf{n} = 0 \text{ on } \Gamma_p^{\mathrm{N}} \right\}, \quad W_p := L^2(\Omega_p). \tag{10}$$

Next, consider the spaces of traces $\Lambda_p := H^{1/2}(\Gamma_{fp})$ and $\boldsymbol{\Lambda}_s := [H^{1/2}(\Gamma_{fp})]^n$. Inspired by [1], we include the structure velocity $\mathbf{u}_s := \partial_t \boldsymbol{\eta}_p \in \mathbf{V}_s$ and the Lagrange multipliers $\boldsymbol{\theta} := \mathbf{u}_s|_{\Gamma_{fp}} \in \boldsymbol{\Lambda}_s$ and $\lambda := p_p|_{\Gamma_{fp}} \in \Lambda_p$ as additional unknowns. In addition, we employ a mixed elasticity formulation with weak stress symmetry, introducing as a new unknown the structure rotation operator $\boldsymbol{\gamma}_p := \frac{1}{2} (\nabla \mathbf{u}_s - (\nabla \mathbf{u}_s)^{\mathrm{t}}) \in \mathbb{Q}_p$. In turn, from the definition of the elastic and poroelastic stress tensors $\boldsymbol{\sigma}_e, \boldsymbol{\sigma}_p$, see (5), we deduce the identities

$$\mathrm{div}(\boldsymbol{\eta}_p) = A(\alpha_p\, p_p\, \mathbf{I}) : \mathbf{I} + A(\boldsymbol{\sigma}_p) : \mathbf{I} \tag{11}$$

$$\text{and} \quad \partial_t A(\boldsymbol{\sigma}_p) = \nabla \mathbf{u}_s - \boldsymbol{\gamma}_p - \partial_t A (\alpha_p\, p_p\, \mathbf{I}). \tag{12}$$

We group spaces, unknowns and test functions as follows:

$$\mathbf{Q} := \mathbb{X}_p \times W_p \times \mathbf{V}_p \times \mathbb{X}_f \times \mathbf{V}_f \times \boldsymbol{\Lambda}_s, \quad \mathbf{S} := \Lambda_p \times \mathbf{V}_s \times \mathbb{Q}_p,$$

$$\mathbf{p} := (\boldsymbol{\sigma}_p, p_p, \mathbf{u}_p, \mathbf{T}_f, \mathbf{u}_f, \boldsymbol{\theta}) \in \mathbf{Q}, \quad \mathbf{r} := (\lambda, \mathbf{u}_s, \boldsymbol{\gamma}_p) \in \mathbf{S},$$

$$\mathbf{q} := (\boldsymbol{\tau}_p, w_p, \mathbf{v}_p, \mathbf{R}_f, \mathbf{v}_f, \boldsymbol{\phi}) \in \mathbf{Q}, \quad \mathbf{s} := (\xi, \mathbf{v}_s, \boldsymbol{\chi}_p) \in \mathbf{S}.$$

We test the first equation of (4a), the second equation of (6a), and (12) with arbitrary $\mathbf{R}_f \in \mathbb{X}_f, \mathbf{v}_p \in \mathbf{V}_p$, and $\boldsymbol{\tau}_p \in \mathbb{X}_p$, integrate by parts, and utilize the fact that $\mathbf{T}_f^d : \mathbf{R}_f = \mathbf{T}_f^d : \mathbf{R}_f^d$. We further test (6b) with $w_p \in \mathrm{W}_p$ employing (11), and impose the remaining equations weakly, as well as the symmetry of $\mathbf{T}_f, \boldsymbol{\sigma}_p$, and the transmission conditions in the first equation of (7a) and (8). Then, we can write the augmented weak formulation associated to (4)–(8) in an operator notation as a degenerate evolution problem in a mixed form (see [15, Section 3] for details): Given $\mathbf{f}_f \in \mathbf{L}^2(\Omega_f), \mathbf{f}_p \in \mathbf{L}^2(\Omega_p)$, and $q_p : [0, T] \to \mathrm{L}^2(\Omega_p)$, find $(\mathbf{p}, \mathbf{r}) : [0, T] \to \mathbf{Q} \times \mathbf{S}$ such that, for a.e. $t \in (0, T)$,

$$
\begin{aligned}
\partial_t \, \mathcal{E}(\mathbf{p}(t)) + (\mathcal{A} + \mathcal{K}_{\mathbf{u}_f(t)})(\mathbf{p}(t)) + \mathcal{B}'(\mathbf{r}(t)) &= \mathcal{F}(t) \quad \text{in } \mathbf{Q}', \\
-\mathcal{B}(\mathbf{p}(t)) \qquad\qquad\qquad\qquad\qquad\quad &= \mathcal{G} \quad \text{in } \mathbf{S}',
\end{aligned}
\tag{13}
$$

where, given $\mathbf{w}_f \in \mathbf{V}_f$, the operators $\mathcal{A} : \mathbf{Q} \to \mathbf{Q}'$, $\mathcal{K}_{\mathbf{w}_f} : \mathbf{Q} \to \mathbf{Q}'$, and $\mathcal{B} : \mathbf{Q} \to \mathbf{S}'$, are defined by

$$
\begin{aligned}
\mathcal{A}(\mathbf{p})(\mathbf{q}) &:= \mu \, (\mathbf{K}^{-1} \mathbf{u}_p, \mathbf{v}_p)_{\Omega_p} + \frac{1}{2\,\mu} \, (\mathbf{T}_f^d, \mathbf{R}_f^d)_{\Omega_f} + \kappa_1 \, (\mathbf{div}(\mathbf{T}_f), \mathbf{div}(\mathbf{R}_f))_{\Omega_f} \\
&+ \kappa_2 \left(\mathbf{e}(\mathbf{u}_f) - \frac{1}{2\,\mu} \mathbf{T}_f^d, \mathbf{e}(\mathbf{v}_f) \right)_{\Omega_f} + (\mathbf{u}_f, \mathbf{div}(\mathbf{R}_f))_{\Omega_f} - (\mathbf{v}_f, \mathbf{div}(\mathbf{T}_f))_{\Omega_f} \\
&+ (\boldsymbol{\gamma}_f(\mathbf{u}_f), \mathbf{R}_f)_{\Omega_f} - (\mathbf{T}_f, \boldsymbol{\gamma}_f(\mathbf{v}_f))_{\Omega_f} + \langle \mathbf{T}_f \mathbf{n}_f, \mathbf{v}_f \rangle_{\Gamma_{fp}} - \langle \mathbf{R}_f \mathbf{n}_f, \mathbf{u}_f \rangle_{\Gamma_{fp}} \\
&+ \mu \, \alpha_{\mathrm{BJS}} \, \langle \sqrt{\mathbf{K}_\mathbf{t}^{-1}} (\mathbf{u}_f - \boldsymbol{\theta}) \cdot \mathbf{t}_f, (\mathbf{v}_f - \boldsymbol{\phi}) \cdot \mathbf{t}_f \rangle_{\Gamma_{fp}} \\
&- (p_p, \mathrm{div}(\mathbf{v}_p))_{\Omega_p} + (w_p, \mathrm{div}(\mathbf{u}_p))_{\Omega_p} + \langle \boldsymbol{\sigma}_p \mathbf{n}_p, \boldsymbol{\phi} \rangle_{\Gamma_{fp}} - \langle \boldsymbol{\tau}_p \mathbf{n}_p, \boldsymbol{\theta} \rangle_{\Gamma_{fp}},
\end{aligned}
\tag{14}
$$

$$
\mathcal{K}_{\mathbf{w}_f}(\mathbf{p})(\mathbf{q}) := \frac{\rho}{2\,\mu} ((\mathbf{u}_f \otimes \mathbf{w}_f)^d, \mathbf{R}_f - \kappa_2 \, \mathbf{e}(\mathbf{v}_f))_{\Omega_f} + \rho \langle \mathbf{w}_f \cdot \mathbf{n}_f, \mathbf{u}_f \cdot \mathbf{v}_f \rangle_{\Gamma_{fp}},
$$

with $\kappa_1, \kappa_2 > 0$ defined in [15, Lemma 3.3] (see also [15, Remark 3.4]), and

$$
\mathcal{B}(\mathbf{q})(\mathbf{s}) := (\mathbf{v}_s, \mathbf{div}(\boldsymbol{\tau}_p))_{\Omega_p} + (\boldsymbol{\chi}_p, \boldsymbol{\tau}_p)_{\Omega_p} + \langle \mathbf{v}_f \cdot \mathbf{n}_f + (\boldsymbol{\phi} + \mathbf{v}_p) \cdot \mathbf{n}_p, \xi \rangle_{\Gamma_{fp}},
$$

whereas the operator $\mathcal{E} : \mathbf{Q} \to \mathbf{Q}'$ is given by

$$
\mathcal{E}(\mathbf{p})(\mathbf{q}) := s_0(p_p, w_p)_{\Omega_p} + (A(\boldsymbol{\sigma}_p + \alpha_p p_p \mathbf{I}), \boldsymbol{\tau}_p + \alpha_p w_p \mathbf{I})_{\Omega_p},
\tag{15}
$$

and the functionals $\mathcal{F} \in \mathbf{Q}'$, $\mathcal{G} \in \mathbf{S}'$ are defined as

$$
\mathcal{F}(\mathbf{q}) := (q_p, w_p)_{\Omega_p} + (\mathbf{f}_f, \mathbf{v}_f - \kappa_1 \, \mathbf{div}(\mathbf{R}_f))_{\Omega_f} \quad \text{and} \quad \mathcal{G}(\mathbf{s}) := (\mathbf{f}_p, \mathbf{v}_s)_{\Omega_p}.
$$

The κ_1 and κ_2 terms are redundant consistent Galerkin-type terms arising from the equilibrium and constitutive equations, respectively. Notice that the symmetry of the tensor $\mathbf{T}_f$ is imposed in an ultra-weak sense (see the seventh and eighth terms in (14) and [15, eq. (3.6)] for details).

3.2 Banach Space Mixed Formulation

Alternatively to (13), we now proceed analogously to [9, 11] (see also [1, Section 3]) and derive a Banach space weak formulation to deal with the nonlinear pseudostress tensor. To that end, we introduce the spaces:

$$\widetilde{\mathbb{X}}_f := \Big\{ \mathbf{R}_f \in \mathbb{L}^2(\Omega_f) : \quad \mathbf{div}(\mathbf{R}_f) \in \mathbf{L}^{4/3}(\Omega_f) \text{ and } \mathbf{R}_f \mathbf{n}_f = \mathbf{0} \text{ on } \Gamma_f^{\mathrm{N}} \Big\},$$

$$\widetilde{\mathbf{V}}_f := \mathbf{L}^4(\Omega_f), \quad \mathbb{Q}_f := \Big\{ \boldsymbol{\chi}_f \in \mathbb{L}^2(\Omega_f) : \quad \boldsymbol{\chi}_f^{\mathrm{t}} = -\boldsymbol{\chi}_f \Big\}. \tag{16}$$

Notice that for this formulation the vorticity $\boldsymbol{\gamma}_f := \boldsymbol{\gamma}_f(\mathbf{u}_f) \in \mathbb{Q}_f$ as well as the trace of the velocity on the interface $\boldsymbol{\varphi} := \mathbf{u}_f|_{\Gamma_{fp}} \in \Lambda_f := [\mathrm{H}^{1/2}(\Gamma_{fp})]^n$ are included as additional unknowns. For the Biot unknowns we proceed as in (13).

Now, we group the spaces and test functions as follows:

$$\widetilde{\mathbf{Q}} := \big(\widetilde{\mathbb{X}}_f \times \mathbf{V}_p \times \mathbb{X}_p \times W_p \big) \times \big(\Lambda_f \times \Lambda_s \times \Lambda_p \big), \quad \widetilde{\mathbf{S}} := \widetilde{\mathbf{V}}_f \times \mathbf{V}_s \times \mathbb{Q}_f \times \mathbb{Q}_p,$$

$$\widetilde{\mathbf{p}} := (\mathbf{p}_1, \mathbf{p}_2) = \big((\mathbf{T}_f, \mathbf{u}_p, \boldsymbol{\sigma}_p, p_p), (\boldsymbol{\varphi}, \boldsymbol{\theta}, \lambda) \big) \in \widetilde{\mathbf{Q}}, \quad \widetilde{\mathbf{r}} := (\mathbf{u}_f, \mathbf{u}_s, \boldsymbol{\gamma}_f, \boldsymbol{\gamma}_p) \in \widetilde{\mathbf{S}},$$

$$\widetilde{\mathbf{q}} := (\mathbf{q}_1, \mathbf{q}_2) = \big((\mathbf{R}_f, \mathbf{v}_p, \boldsymbol{\tau}_p, w_p), (\boldsymbol{\psi}, \boldsymbol{\phi}, \xi) \big) \in \widetilde{\mathbf{Q}}, \quad \widetilde{\mathbf{s}} := (\mathbf{v}_f, \mathbf{v}_s, \boldsymbol{\chi}_f, \boldsymbol{\chi}_p) \in \widetilde{\mathbf{S}}.$$

The Banach spaces-based weak formulation is: Given $\mathbf{f}_f \in \mathbf{L}^2(\Omega_f)$, $\mathbf{f}_p \in \mathbf{L}^2(\Omega_p)$, and $q_p : [0, T] \to \mathrm{L}^2(\Omega_p)$, find $(\widetilde{\mathbf{p}}, \widetilde{\mathbf{r}}) : [0, T] \to \widetilde{\mathbf{Q}} \times \widetilde{\mathbf{S}}$ such that, for a.e. $t \in (0, T)$,

$$\begin{aligned} \partial_t \widetilde{\mathcal{E}}(\widetilde{\mathbf{p}}(t)) + (\widetilde{\mathcal{A}} + \mathcal{C}_{\boldsymbol{\varphi}(t)})(\widetilde{\mathbf{p}}(t)) + (\widetilde{\mathcal{B}}' + \widetilde{\mathcal{K}}_{\mathbf{u}_f(t)})(\widetilde{\mathbf{r}}(t)) &= \widetilde{\mathcal{F}}(t) \quad \text{in } \widetilde{\mathbf{Q}}', \\ - \widetilde{\mathcal{B}}(\widetilde{\mathbf{p}}(t)) &= \widetilde{\mathcal{G}} \quad \text{in } \widetilde{\mathbf{S}}', \end{aligned} \tag{17}$$

where, the operator $\widetilde{\mathcal{E}} : \widetilde{\mathbf{Q}} \to \widetilde{\mathbf{Q}}'$ has the same formula as the one defined in (15). In addition, given $\boldsymbol{\zeta} \in \Lambda_f$ and $\mathbf{w}_f \in \mathbf{V}_f$, the operators $\widetilde{\mathcal{A}} : \widetilde{\mathbf{Q}} \to \widetilde{\mathbf{Q}}'$, $\mathcal{C}_{\boldsymbol{\zeta}} : \widetilde{\mathbf{Q}} \to \widetilde{\mathbf{Q}}'$, $\widetilde{\mathcal{K}}_{\mathbf{w}_f} : \widetilde{\mathbf{Q}} \to \widetilde{\mathbf{S}}'$, and $\widetilde{\mathcal{B}} : \widetilde{\mathbf{Q}} \to \widetilde{\mathbf{S}}'$, are defined by

$$\widetilde{\mathcal{A}}(\widetilde{\mathbf{p}})(\widetilde{\mathbf{q}}) := \mathbf{a}(\mathbf{p}_1)(\mathbf{q}_1) + \mathbf{b}(\mathbf{p}_2)(\mathbf{q}_1) + \mathbf{b}(\mathbf{q}_2)(\mathbf{p}_1) - \mathbf{c}(\mathbf{p}_2)(\mathbf{q}_2),$$

$$\mathbf{a}(\mathbf{p}_1)(\mathbf{q}_1) := \frac{1}{2\mu}(\mathbf{T}_f^{\mathrm{d}}, \mathbf{R}_f^{\mathrm{d}})_{\Omega_f} + \mu(\mathbf{K}^{-1}\mathbf{u}_p, \mathbf{v}_p)_{\Omega_p}$$

$$+ (w_p, \mathrm{div}(\mathbf{u}_p))_{\Omega_p} - (p_p, \mathrm{div}(\mathbf{v}_p))_{\Omega_p},$$

$$\mathbf{b}(\mathbf{q}_2)(\mathbf{q}_1) := -\langle \mathbf{R}_f \mathbf{n}_f, \boldsymbol{\psi} \rangle_{\Gamma_{fp}} - \langle \boldsymbol{\tau}_p \mathbf{n}_p, \boldsymbol{\phi} \rangle_{\Gamma_{fp}} + \langle \mathbf{v}_p \cdot \mathbf{n}_p, \xi \rangle_{\Gamma_{fp}},$$

$$\mathbf{c}(\mathbf{p}_2)(\mathbf{q}_2) := \mu\, \alpha_{\mathrm{BJS}} \langle \sqrt{\mathbf{K}_{\mathbf{t}}^{-1}}(\boldsymbol{\varphi} - \boldsymbol{\theta}) \cdot \mathbf{t}_f, (\boldsymbol{\psi} - \boldsymbol{\phi}) \cdot \mathbf{t}_f \rangle_{\Gamma_{fp}}$$

$$+ \langle \boldsymbol{\psi} \cdot \mathbf{n}_f, \lambda \rangle_{\Gamma_{fp}} + \langle \boldsymbol{\phi} \cdot \mathbf{n}_p, \lambda \rangle_{\Gamma_{fp}} - \langle \boldsymbol{\varphi} \cdot \mathbf{n}_f, \xi \rangle_{\Gamma_{fp}} - \langle \boldsymbol{\theta} \cdot \mathbf{n}_p, \xi \rangle_{\Gamma_{fp}},$$

$$\mathcal{C}_{\boldsymbol{\zeta}}(\widetilde{\mathbf{p}})(\widetilde{\mathbf{q}}) := \rho \langle \boldsymbol{\zeta} \cdot \mathbf{n}_f, \boldsymbol{\varphi} \cdot \boldsymbol{\psi} \rangle_{\Gamma_{fp}},$$

$$\widetilde{\mathcal{K}}_{\mathbf{w}_f}(\widetilde{\mathbf{r}})(\widetilde{\mathbf{q}}) := \frac{\rho}{2\mu}((\mathbf{u}_f \otimes \mathbf{w}_f)^{\mathrm{d}}, \mathbf{R}_f)_{\Omega_f},$$

$$\widetilde{\mathcal{B}}(\widetilde{\mathbf{q}})(\widetilde{\mathbf{s}}) := (\mathbf{v}_f, \mathbf{div}(\mathbf{R}_f))_{\Omega_f} + (\mathbf{v}_s, \mathbf{div}(\boldsymbol{\tau}_p))_{\Omega_p} + (\mathbf{R}_f, \boldsymbol{\chi}_f)_{\Omega_f} + (\boldsymbol{\tau}_p, \boldsymbol{\chi}_p)_{\Omega_p},$$

and the functionals $\widetilde{\mathcal{F}} \in \widetilde{\mathbf{Q}}'$, $\widetilde{\mathcal{G}} \in \widetilde{\mathbf{S}}'$ are defined as

$$\widetilde{\mathcal{F}}(\widetilde{\mathbf{q}}) := (q_p, w_p)_{\Omega_p} \quad \text{and} \quad \widetilde{\mathcal{G}}(\widetilde{\mathbf{s}}) := (\mathbf{f}_f, \mathbf{v}_f)_{\Omega_f} + (\mathbf{f}_p, \mathbf{v}_s)_{\Omega_p}.$$

4 Numerical Methods

For the space discretization we consider mixed finite element approximations for both weak formulations (13) and (17). Let $\mathcal{T}_h^f$ and $\mathcal{T}_h^p$ be affine finite element partitions of Ω_f and Ω_p, respectively, which may be non-matching along the interface Γ_{fp}. We denote by $\mathbf{BDM}_1$ and $\mathrm{P}_1^{\mathrm{dc}}$ the lowest order Brezzi-Douglas-Marini space and discontinuous piecewise linear polynomials, respectively. Then, for both formulations we consider the discrete spaces $\mathbb{X}_{ph} \times \mathbf{V}_{sh} \times \mathbb{Q}_{ph} = \mathbb{BDM}_1 - \mathbf{P}_0 - \mathbb{P}_0$ for stress-displacement-rotation in elasticity, $\mathbf{V}_{ph} \times \mathrm{W}_{ph} = \mathbf{BDM}_1 - \mathrm{P}_0$ for Darcy velocity-pressure, and $\Lambda_{sh} \times \Lambda_{ph} = \mathbf{P}_1^{\mathrm{dc}} - \mathrm{P}_1^{\mathrm{dc}}$ for the Lagrange multipliers, whereas for the Navier–Stokes unknowns we distinct the approaches:

- **Augmented approach:** $\mathbb{X}_{fh} \times \mathbf{V}_{fh} = \mathbb{BDM}_1 - \mathbf{P}_1$ for the pseudostress and velocity.
- **Banach approach:** $\widetilde{\mathbb{X}}_{fh} \times \widetilde{\mathbf{V}}_{fh} \times \mathbb{Q}_{fh} = \mathbb{BDM}_1 - \mathbf{P}_0 - \mathbb{P}_0$ for the pseudostress, velocity and vorticity, and $\Lambda_{fh} = \mathbf{P}_1^{\mathrm{dc}}$ for the additional Lagrange multiplier.

We note that weak stress symmetry allows for the use of the $\mathbb{BDM}_1$ space to approximate the pseudostress and the poroelastic stress. Higher order spaces may also be used [15], but for simplicity we focus here on the lowest order case. The well-posedness of the discrete version of the augmented formulation (13) considering the above discrete spaces, which makes use of a fixed-point strategy along with the theory of degenerate parabolic systems [19], is established in [15, Theorem 5.6], see also [15, Theorem 5.8] for the error analysis. For the analysis of the Banach space method, one can proceed as in [15], due to the similar structure of (13) and (17) as degenerate evolution problems in a mixed form, but now considering arguments similar to [9]. This is a topic of ongoing research.

Both methods exhibit local momentum conservation for the poroelastic stress, local mass conservation for the Darcy fluid, accurate approximations for the stresses and the Darcy velocity, locking-free behavior, and robustness with respect to the physical parameters. Regarding the comparison between the two approaches we comment that for the augmented formulation neither the vorticity γ_f nor the Lagrange multiplier φ are included as additional unknowns since $\mathbf{u}_f$ is sought in $\mathbf{H}^1(\Omega_f)$ instead of $\mathbf{L}^4(\Omega_f)$. In addition, it provides smooth approximations for both the pseudostress, in $\mathbb{H}(\mathbf{div}; \Omega_f)$, and the fluid velocity, in $\mathbf{H}^1(\Omega_f)$. On the other hand, the Banach space formulation, although more expensive in terms of number of variables, utilizes the natural spaces arising from the application of the Cauchy–Schwarz and Hölder inequalities to the terms resulting from the testing and integration by parts of the equations of the model. One consequence is that it exhibits local momentum conservation for the pseudostress, due to the $\mathbb{BDM}_1 - \mathbf{P}_0$ pseudostress-velocity pair. Furthermore, it can be implemented as a multipoint stress-flux mixed finite element method, which allows for local elimination of the stresses and the Darcy velocity, as well as the vorticity and rotation in the case of $\mathbb{P}_1$ elements for these two variables [11].

5   Numerical Results

We present two numerical experiments to illustrate the convergence and applicability of the methods, using the backward Euler method for time discretization.

Example 1: Convergence test. In this example we study numerically the convergence in space, using unstructured triangular grids. The total simulation time is $T = 0.01$ and the time step is $\Delta t = 10^{-3}$, which is sufficiently small, so that the time discretization error does not affect the convergence rates. The domain consists of $\Omega_f = (0, 1) \times (0, 1)$, $\Gamma_{fp} = (0, 1) \times \{0\}$, and $\Omega_p = (0, 1) \times (-1, 0)$. We take $\Gamma_f^D = (0, 1) \times \{1\}$, $\Gamma_p^D = (0, 1) \times \{-1\}$, and $\tilde{\Gamma}_p^D = \Gamma_p$. The true solution in the Navier–Stokes region is

$$\mathbf{u}_f = \exp(t) \begin{pmatrix} \sin(\pi x)\cos(\pi y) \\ -\sin(\pi y)\cos(\pi x) \end{pmatrix}, \quad p_f = \exp(t)\sin(\pi x)\cos\left(\frac{\pi y}{2}\right) + 2\pi\cos(\pi t).$$

The Biot solution is chosen accordingly to satisfy the interface conditions (7):

$$p_p = \exp(t)\sin(\pi x)\cos\left(\frac{\pi y}{2}\right), \quad \mathbf{u}_p = -\frac{1}{\mu}\mathbf{K}\nabla p_p, \quad \boldsymbol{\eta}_p = \sin(\pi t)\begin{pmatrix} \cos(y) - 3x \\ y + 1 \end{pmatrix}.$$

The model parameters are

$$\mu = 1, \quad \rho = 1, \quad \lambda_p = 1, \quad \mu_p = 1, \quad s_0 = 1, \quad \mathbf{K} = \mathbf{I}, \quad \alpha_p = 1, \quad \alpha_{\text{BJS}} = 1.$$

We run a sequence of mesh refinements with non-matching grids along Γ_{fp}. For the sake of space we display the results only for the Banach formulation. Nevertheless, we stress that both schemes (13) and (17) converge optimally, see [15] for convergence studies of the augmented method. The results are reported on Table 1. We note that the fluid pressure and displacement at t_n are recovered by the discrete version of (3) and the formula $\boldsymbol{\eta}_p^n = \Delta t\, \mathbf{u}_s^n + \boldsymbol{\eta}_p^{n-1}$, respectively. We observe at least first order convergence for all subdomain variables in their natural norms. The Lagrange multiplier variables, which are approximated in $\mathbf{P}_1^{\text{dc}} - \mathbf{P}_1^{\text{dc}} - \mathbf{P}_1^{\text{dc}}$, exhibit second order convergence in the L^2-norm on Γ_{fp}, which is consistent with the order of approximation.

Example 2: Air flow through a filter. The setting in this example is similar to the one presented in [18], where the Navier–Stokes–Darcy model is considered. The domain is a two-dimensional rectangular channel with length 2.5 m and width 0.25 m, which on the bottom center is partially blocked by a square poroelastic filter of length and width 0.2 m, see Fig. 1 (top). The model parameters are set as

$$\mu = 1.81 \times 10^{-8} \text{ kPa s}, \quad \rho = 1.225 \times 10^{-3} \text{ Mg/m}^3, \quad s_0 = 7 \times 10^{-2} \text{ kPa}^{-1},$$
$$\mathbf{K} = [0.505, -0.495; -0.495, 0.505] \times 10^{-6} \text{ m}^2, \quad \alpha_{\text{BJS}} = 1.0, \quad \alpha = 1.0.$$

Note that μ and ρ are parameters for air. The permeability tensor $\mathbf{K}$ is obtained by rotating the identity tensor by a $45°$ rotation angle in order to consider the effect of material anisotropy on the flow. We further consider a stiff material in the poroelastic region with parameters: $\lambda_p = 1 \times 10^4$ kPa and $\mu_p = 1 \times 10^5$ kPa.

Table 1. [Example 1] Average number of Newton iterations, mesh sizes, errors, rates of convergences.

iter	h_f	$\|\mathbf{e}_{\mathbf{T}_f}\|_{\ell^2(0,T;\mathbb{X}_f)}$		$\|\mathbf{e}_{\mathbf{u}_f}\|_{\ell^2(0,T;\mathbf{V}_f)}$		$\|\mathbf{e}_{\gamma_f}\|_{\ell^2(0,T;\mathbb{Q}_f)}$		$\|\mathbf{e}_{p_f}\|_{\ell^2(0,T;L^2(\Omega_f))}$	
		error	rate	error	rate	error	rate	error	rate
2.2	0.1964	1.6E-01	–	1.2E-02	–	2.7E-02	–	7.2E-03	–
2.2	0.0997	7.9E-02	1.0059	5.6E-03	1.0686	1.3E-02	1.0960	2.5E-03	1.5343
2.2	0.0487	3.9E-02	1.0028	2.7E-03	1.0358	6.5E-03	0.9605	1.1E-03	1.2087
2.2	0.0250	1.9E-02	1.0323	1.3E-03	1.0249	3.3E-03	1.0319	5.1E-04	1.1042
2.2	0.0136	9.7E-03	1.1425	6.7E-04	1.1440	1.6E-03	1.1575	2.5E-04	1.2087
2.2	0.0072	4.8E-03	1.1040	3.3E-04	1.1152	8.1E-04	1.0939	1.2E-04	1.1099

h_p	$\|\mathbf{e}_{\sigma_p}\|_{\ell^\infty(0,T;\mathbb{X}_p)}$		$\|\mathbf{e}_{\mathbf{u}_s}\|_{\ell^2(0,T;\mathbf{V}_s)}$		$\|\mathbf{e}_{\gamma_p}\|_{\ell^2(0,T;\mathbb{Q}_p)}$		$\|\mathbf{e}_{\mathbf{u}_p}\|_{\ell^2(0,T;\mathbf{V}_p)}$		$\|\mathbf{e}_{p_p}\|_{\ell^\infty(0,T;\mathrm{W}_p)}$	
	error	rate	error	rate	error	rate	error	rate	error	rate
0.2828	2.7E-01	–	4.3E-02	–	8.6E-02	–	1.0E-01	–	7.5E-02	–
0.1646	1.4E-01	1.2732	2.2E-02	1.2275	2.2E-02	2.5658	5.0E-02	1.3529	3.8E-02	1.2480
0.0779	6.7E-02	0.9650	1.1E-02	0.9618	4.9E-03	1.9684	2.4E-02	0.9896	1.9E-02	0.9328
0.0434	3.4E-02	1.1690	5.4E-03	1.1866	1.5E-03	2.0132	1.2E-02	1.2361	9.4E-03	1.2150
0.0227	1.7E-02	1.0634	2.7E-03	1.0668	5.8E-04	1.4877	5.9E-03	1.0739	4.7E-03	1.0658
0.0124	8.4E-03	1.1462	1.4E-03	1.1456	2.7E-05	1.2917	2.9E-03	1.1452	2.4E-03	1.1429

$\|\mathbf{e}_{\eta_p}\|_{\ell^2(0,T;L^2(\Omega_p))}$		h_{tf}	$\|\mathbf{e}_{\varphi}\|_{\ell^2(0,T;\mathbf{L}^2(\Gamma_{fp}))}$		h_{tp}	$\|\mathbf{e}_{\theta}\|_{\ell^2(0,T;\mathbf{L}^2(\Gamma_{fp}))}$		$\|\mathbf{e}_{\lambda}\|_{\ell^2(0,T;\mathbf{L}^2(\Gamma_{fp}))}$	
error	rate		error	rate		error	rate	error	rate
2.7E-04	–	1/8	6.5E-03	–	1/5	9.3E-03	–	1.1E-03	–
1.4E-04	1.2253	1/16	1.5E-03	2.0936	1/10	2.8E-03	1.7323	2.7E-04	2.0019
6.7E-05	0.9615	1/32	3.7E-04	2.0245	1/20	6.8E-04	2.0386	6.7E-05	1.9977
3.4E-05	1.1865	1/64	9.4E-05	1.9893	1/40	1.7E-04	1.9889	1.7E-05	1.9941
1.7E-05	1.0668	1/128	2.4E-05	1.9613	1/80	4.3E-05	1.9925	4.3E-06	1.9826
8.4E-06	1.1456	1/256	5.6E-06	2.0985	1/160	1.1E-05	2.0094	1.1E-06	1.9632

The top and bottom of the domain are rigid, impermeable walls. The flow is driven by a pressure difference $\Delta p = 2 \times 10^{-6}$ kPa between the left and right boundary, see Fig. 1 (bottom) for the boundary conditions. The body force terms $\mathbf{f}_f$ and $\mathbf{f}_p$ and external source q_p are set to zero. For the initial conditions, we consider

$$p_{p,0} = 100 \text{ kPa}, \quad \sigma_{p,0} = -\alpha_p \, p_{p,0} \, \mathbf{I}, \quad \mathbf{u}_{f,0} = \mathbf{0} \text{ m/s}.$$

The computational matching grid along Γ_{fp} has a characteristic parameter $h = \max\{h_f, h_p\} = 0.018$. The total simulation time is $T = 400$ s with $\Delta t = 1$ s.

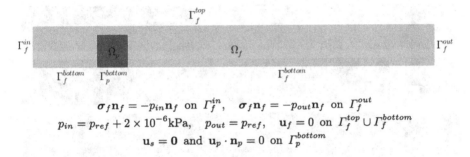

$$\boldsymbol{\sigma}_f \mathbf{n}_f = -p_{in}\mathbf{n}_f \text{ on } \Gamma_f^{in}, \quad \boldsymbol{\sigma}_f \mathbf{n}_f = -p_{out}\mathbf{n}_f \text{ on } \Gamma_f^{out}$$
$$p_{in} = p_{ref} + 2 \times 10^{-6}\text{kPa}, \quad p_{out} = p_{ref}, \quad \mathbf{u}_f = 0 \text{ on } \Gamma_f^{top} \cup \Gamma_f^{bottom}$$
$$\mathbf{u}_s = 0 \text{ and } \mathbf{u}_p \cdot \mathbf{n}_p = 0 \text{ on } \Gamma_p^{bottom}$$

Fig. 1. [Example 2] Top: computational domain and boundaries; channel Ω_f in gray, filter Ω_p in brown. Bottom: boundary conditions with $p_{ref} = 100\,\text{kPa}$. (Color figure online)

For the sake of space we display the results only for the augmented formulation. The computed magnitude of the velocities and pressures are displayed in Figs. 2 and 3, respectively. We observe high velocity in the narrow open channel above the filter. Vortices develop behind the obstacle, which travel with the fluid and are smoothed out at later times. A sharp pressure gradient is observed in the region above the filter, as well as within the filter, where the permeability anisotropy affects both the pressure and velocity fields. This example illustrates the ability of the mixed method to produce oscillation-free solution in a regime of challenging physical parameters, including small viscosity and permeability and large Lamé coefficients.

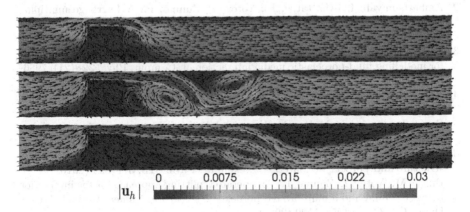

Fig. 2. [Example 2] Computed velocities $\mathbf{u}_{fh}$ and $\mathbf{u}_{ph}$ (arrows not scaled) and their magnitudes at times $t \in \{20, 80, 400\}$ (from top to bottom).

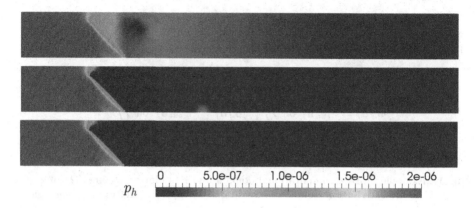

Fig. 3. [Example 2] Computed pressures $p_{fh} - p_{ref}$ and $p_{ph} - p_{ref}$ at times $t \in \{20, 80, 400\}$ (from top to bottom).

Acknowledgments. This work was supported in part by ANID-Chile through the projects CENTRO DE MODELAMIENTO MATEMÁTICO (FB210005) and Fondecyt 11220393; and by NSF grant DMS 2111129.

References

1. Ambartsumyan, I., Ervin, V.J., Nguyen, T., Yotov, I.: A nonlinear Stokes-Biot model for the interaction of a non-Newtonian fluid with poroelastic media. ESAIM Math. Model. Numer. Anal. **53**(6), 1915–1955 (2019)
2. Ambartsumyan, I., Khattatov, E., Yotov, I., Zunino, P.: A Lagrange multiplier method for a Stokes-Biot fluid-poroelastic structure interaction model. Numer. Math. **140**(2), 513–553 (2018)
3. Badia, S., Quaini, A., Quarteroni, A.: Coupling Biot and Navier-Stokes equations for modelling fluid-poroelastic media interaction. J. Comput. Phys. **228**(21), 7986–8014 (2009)
4. Biot, M.: General theory of three-dimensional consolidation. J. Appl. Phys. **12**, 155–164 (1941)
5. Bociu, L., Canic, S., Muha, B., Webster, J.T.: Multilayered poroelasticity interacting with Stokes flow. SIAM J. Math. Anal. **53**(6), 6243–6279 (2021)
6. Bukač, M.: A loosely-coupled scheme for the interaction between a fluid, elastic structure and poroelastic material. J. Comput. Phys. **313**, 377–399 (2016)
7. Bukač, M., Yotov, I., Zunino, P.: An operator splitting approach for the interaction between a fluid and a multilayered poroelastic structure. Numer. Methods Partial Differ. Eq. **31**(4), 1054–1100 (2015)
8. Bukač, M., Yotov, I., Zakerzadeh, R., Zunino, P.: Partitioning strategies for the interaction of a fluid with a poroelastic material based on a Nitsche's coupling approach. Comput. Methods Appl. Mech. Eng. **292**, 138–170 (2015)
9. Camaño, J., García, C., Oyarzúa, R.: Analysis of a momentum conservative mixed-FEM for the stationary Navier-Stokes problem. Numer. Methods Partial Differ. Eq. **37**(5), 2895–2923 (2021)

10. Camaño, J., Gatica, G.N., Oyarzúa, R., Tierra, G.: An augmented mixed finite element method for the Navier-Stokes equations with variable viscosity. SIAM J. Numer. Anal. **54**(2), 1069–1092 (2016)
11. Caucao, S., Li, T., Yotov, I.: A cell-centered finite volume method for the Navier–Stokes/Biot model. In: Klöfkorn, R., Keilegavlen, E., Radu, F.A., Fuhrmann, J. (eds.) FVCA 2020. SPMS, vol. 323, pp. 325–333. Springer, Cham (2020). https://doi.org/10.1007/978-3-030-43651-3_29
12. Caucao, S., Li, T., Yotov, I.: A multipoint stress-flux mixed finite element method for the Stokes-Biot model. Numer. Math. **152**(2), 411–473 (2022)
13. Cesmelioglu, A.: Analysis of the coupled Navier-Stokes/Biot problem. J. Math. Anal. Appl. **456**(2), 970–991 (2017)
14. Cesmelioglu, A., Chidyagwai, P.: Numerical analysis of the coupling of free fluid with a poroelastic material. Numer. Methods Partial Differ. Eq. **36**(3), 463–494 (2020)
15. Li, T., Caucao, S., Yotov, I.: An augmented fully mixed formulation for the quasistatic Navier–Stokes–Biot model. IMA J. Numer. Anal. drad036 (2023). https://doi.org/10.1093/imanum/drad036
16. Li, T., Yotov, I.: A mixed elasticity formulation for fluid-poroelastic structure interaction. ESAIM Math. Model. Numer. Anal. **56**(1), 01–40 (2022)
17. Ruiz-Baier, R., Taffetani, M., Westermeyer, H.D., Yotov, I.: The Biot-Stokes coupling using total pressure: formulation, analysis and application to interfacial flow in the eye. Comput. Methods Appl. Mech. Eng. **389**, 114384 (2022)
18. Schneider, M., Weishaupt, K., Gläser, D., Boon, W.M., Helmig, R.: Coupling staggered-grid and MPFA finite volume methods for free flow/porous-medium flow problems. J. Comput. Phys. **401**, 109012 (2020)
19. Showalter, R.E.: Monotone Operators in Banach Space and Nonlinear Partial Differential Equations. Mathematical Surveys and Monographs, p. 49. American Mathematical Society, Providence (1997)
20. Showalter, R.E.: Poroelastic filtration coupled to Stokes flow. In: Control Theory of Partial Differential Equations. Lecture Notes in Pure Applied Mathematics, vol. 242, pp. 229–241. Chapman & Hall/CRC, Boca Raton (2005)

Preconditioning and Multilevel Methods

Robust Iterative Solvers for Algebraic Systems Arising from Elliptic Optimal Control Problems

Ulrich Langer[1]([envelope]), Richard Löscher[2], Olaf Steinbach[2], and Huidong Yang[3]

[1] Institute for Computational Mathematics, Johannes Kepler University Linz, Altenbergerstr. 69, 4040 Linz, Austria
ulanger@numa.uni-linz.ac.at
[2] Institut für Angewandte Mathematik, Technische Universität Graz, Steyrergasse 30, 8010 Graz, Austria
loescher@math.tugraz.at, o.steinbach@tugraz.at
[3] Faculty of Mathematics, and Doppler Laboratory MaMS, University of Vienna, Oskar–Morgenstern–Platz 1, 1090 Wien, Austria
huidong.yang@univie.ac.at

Abstract. We consider tracking-type, distributed elliptic optimal control problems with standard L_2 and more general energy regularizations. We propose, analyze, and test new robust preconditioned iterative solvers for systems of linear algebraic equations arising from the finite element discretization of the reduced optimality systems defining the optimal solution in the case of the optimal choice of the regularization parameter. In particular, we study variable regularization parameters adapted to the local behavior of the mesh-size that can heavily change in the case of adaptive mesh refinements as required for discontinuous target functions.

Keywords: Elliptic optimal control problems · L_2 regularization · energy regularization · finite element discretization · iterative solvers

1 Introduction

In this paper, we consider elliptic optimal control problems of the form: find the state $y_\varrho \in Y = H_0^1(\Omega)$ and the control $u_\varrho \in U$ minimizing the cost functional

$$J(y_\varrho, u_\varrho) := \frac{1}{2} \|y_\varrho - y_d\|_{L_2(\Omega)}^2 + \frac{1}{2} \|\sqrt{\varrho}\, u_\varrho\|_U^2, \tag{1}$$

subject to the elliptic boundary value problem

$$B y_\varrho = u_\varrho \quad \text{in } U \subset Y^* = H^{-1}(\Omega), \tag{2}$$

where the computational domain $\Omega \subset \mathbb{R}^d$, with $d \in \{1, 2, 3\}$, is assumed to be bounded and Lipschitz. We use the standard notations for Lebesgue and Sobolev spaces. We emphasize that the regularization parameter $\varrho \in L_\infty(\Omega)$ can be a uniformly positive function. If $\varrho = \text{constant} > 0$, then $\frac{1}{2}\|\sqrt{\varrho}\, u_\varrho\|_U^2$ turns into

I. Lirkov and S. Margenov (Eds.): LSSC 2023, LNCS 13952, pp. 35–43, 2024.
https://doi.org/10.1007/978-3-031-56208-2_50

the usual Tikhonov-like regularization term $\frac{\varrho}{2}\|u_\varrho\|_U^2$. Besides the standard L_2 regularization where $U = L_2(\Omega)$, we also consider the energy regularization given by $U = Y^*$; see [5] for more details. We take the Laplace operator $-\Delta$ as model for the elliptic operator $B : Y \to Y^*$ defined by the variational identity

$$\langle By, z \rangle = (\nabla y, \nabla z)_{L_2(\Omega)} := \int_\Omega \nabla y \cdot \nabla z \, dx \quad \forall y, z \in Y, \tag{3}$$

where $\langle \cdot, \cdot \rangle : Y^* \times Y \to \mathbb{R}$ is the duality pairing. The unique solvability of the optimal control problem (1)–(2) follows from standard arguments; see, e.g., [8].

The optimal solution to (1)–(2) can be found by solving the first-order optimality system that is nothing but a system of coupled PDEs. The finite element (fe) discretization of the reduced (after elimination of the control u_ϱ) optimality system leads to a symmetric, but indefinite system of the form: find the fe nodal vectors $(\mathbf{p}_h, \mathbf{y}_h) \in \mathbb{R}^{N_h = n_h + n_h}$, representing the fe approximations to the adjoint p_ϱ and the state y_ϱ, such that

$$\begin{bmatrix} A_{\varrho h} & K_h \\ K_h^\top & -M_h \end{bmatrix} \begin{bmatrix} \mathbf{p}_h \\ \mathbf{y}_h \end{bmatrix} = \begin{bmatrix} \mathbf{0}_h \\ -\mathbf{y}_{dh} \end{bmatrix}, \tag{4}$$

where the stiffness matrix K_h, the mass matrix M_h, and the regularization matrix $A_{\varrho h}$ are symmetric and positive definite (spd), and $\mathbf{y}_{dh} \in \mathbb{R}^{n_h}$ is the fe load vector representing the desired state y_d. In the standard case of a constant positive regularization parameter ϱ, the regularization matrix $A_{\varrho h}$ equals $\varrho^{-1} M_h$ for the L_2 regularization and $\varrho^{-1} K_h$ for the energy regularization. In the case of a variable regularization parameter $\varrho(x)$, we have $A_{\varrho h} = M_{\varrho h}$ (L_2 regularization) and $A_{\varrho h} = K_{\varrho h}$ (energy regularization) defined by the variational identities

$$(M_{\varrho h} \mathbf{v}_h, \mathbf{w}_h) = (\varrho^{-1} v_h, w_h)_{L_2(\Omega)} \text{ and } (K_{\varrho h} \mathbf{v}_h, \mathbf{w}_h) = (\varrho^{-1} \nabla v_h, \nabla w_h)_{L_2(\Omega)} \tag{5}$$

for all $\mathbf{v}_h$ and $\mathbf{w}_h \in \mathbb{R}^{n_h}$, and the corresponding fe functions v_h and w_h from the fe space $P_h = Y_h = V_h \subset P = Y = V = H_0^1(\Omega)$ spanned by the fe basis functions $\varphi_{h1}, \ldots, \varphi_{hn_h}$, where $(\cdot, \cdot)$ denotes the Euclidean inner product. Here we only consider continuous, piecewise affine-linear basis functions on a shape-regular triangulation $\mathcal{T}_h$ of Ω, where h is a suitably chosen discretization parameter such that $n_h = O(h^{-d})$. We mention that the system (4) turns into an equivalent positive definite, but nonsymmetric (block-antisymmetric) system when the second block-row of (4) is multiplied by (-1). Further, eliminating the fe adjoint state $\mathbf{p}_h$ from (4), we arrive at the Schur-complement system: find the fe state $\mathbf{y}_h \in \mathbb{R}^{n_h}$ such that

$$S_{\varrho h} \mathbf{y}_h = \mathbf{y}_{dh} \tag{6}$$

with the spd Schur-complement matrix $S_{\varrho h} = K_h A_{\varrho h}^{-1} K_h + M_h$. Therefore, system (6) can be solved by the preconditioned conjugate gradient (PCG) method provided that a good preconditioner is available. The matrix-by-vector multiplication $S_{\varrho h} * \mathbf{y}_h$, which is the basic operation in the PCG, always requires the action of $A_{\varrho h}^{-1}$, i.e. the solution of a system with the matrix $A_{\varrho h}$ with high accuracy. This is a principle drawback of the Schur-complement approach. However, in the case of the constant energy regularization, $A_{\varrho h}^{-1} = \varrho K_h^{-1}$, and, therefore, $S_{\varrho h} = \varrho K_h + M_h$. A good preconditioner for $\varrho K_h + M_h$ will turn the PCG into a perfect solver. In general, when we want to avoid the action of $A_{\varrho h}^{-1}$, we can directly solve the saddle-point problem (4) by some Krylov subspace iteration method. There we only need preconditioners for $S_{\varrho h}$ and not the application of $S_{\varrho h}$. There is a huge amount of literature on efficient preconditioned solvers for saddle-point systems like (4) in general; see, e.g., [2,3,9]. Special h and ϱ robust preconditioned iterative solvers for discrete (reduced) optimality systems of the kind (4) in the case of the standard L_2 regularization with a constant regularization parameter ϱ were investigated, e.g., in [11,14]; see also [12] and [1] for handling control and state constraints, and the references therein.

In this paper, we consider the case of the optimal choice of the regularization parameter or function ϱ with respect to the best approximation of the desired state y_d by the computed fe state $y_{\varrho h}$ in the $L_2(\Omega)$ norm. The first step towards such estimates was made in [10] where $\|y_\varrho - y_d\|_{L_2(\Omega)}$ was estimated in terms of ϱ and the regularity of y_d without any discretization. Then these estimates have been used in [6] and [5,7] to show that the choices $\varrho = h^4$ and $\varrho = h^2$ lead to asymptotically optimal estimates of $\|y_{\varrho h} - y_d\|_{L_2(\Omega)}$ for the L_2 regularization and the energy regularization, respectively. These choices of the regularization parameter imply that the mass matrix M_h is spectrally equivalent to the Schur complement $S_{\varrho h}$ in both cases; see [6]. Therefore, its diagonal replacement $\text{diag}(M_h)$ or the lumped version $\text{lump}(M_h)$ can be used as simple preconditioners for $S_{\varrho h}$. The adaption of the regularization parameter ϱ to the local (element) mesh size h_τ for $\tau \in \mathcal{T}_h$ was studied in [5]. It was shown that $\varrho(x) = h_\tau^2$ for $x \in \tau$ again leads to best-balanced estimates of $\|y_{\varrho h} - y_d\|_{L^2(\Omega)}$ in the case of the energy regularization. In the same paper, numerical experiments supported that the mass matrix M_h is an efficient preconditioner for the Schur complement $S_{\varrho h}$, but without any rigorous analysis. We mention that the spectral analysis used in the case of a constant regularization parameter does not work here. In the next section, we will provide a new rigorous analysis of the spectral equivalence of the mass matrix M_h and the Schur complement $S_{\varrho h}$ in the case of the popular L_2 regularization. This result is the basis for the construction of efficient solvers for (4) or even (6) when $A_{\varrho h} = M_{\varrho h}$.

2   Solvers in the Case of Optimal L_2 Regularizations

Let us first recall the case of the L_2 regularization with constant $\varrho = h^4$ that is the optimal choice when aiming at the best approximation of the desired state y_d by the computed fe state $y_{\varrho h}$ with respect to the $L_2(\Omega)$ norm [6]. Then $A_{\varrho h} = M_{\varrho h} = \varrho^{-1} M_h$, and, therefore, $S_{\varrho h} = \varrho K_h M_h^{-1} K_h + M_h$. Expanding vectors $\mathbf{v}_h = \sum_{i=1}^{n_h} v_i^e \mathbf{e}_{hi}$ into the M_h-orthonormal eigenvector basis $\mathbf{e}_{h1}, \ldots, \mathbf{e}_{hn_h}$ provided by the generalized eigenvalue problem $K_h \mathbf{e}_{hi} = \lambda_{hi} M_h \mathbf{e}_{hi}$ with $(M_h \mathbf{e}_{hi}, \mathbf{e}_{hj}) = \delta_{ij}$ and eigenvalues $0 < \underline{c}_{\text{KM}} \leq \lambda_{h1} \leq \cdots \leq \lambda_{hn_h} \leq \bar{c}_{\text{KM}} h^{-2}$, we get

$$(S_{\varrho h} \mathbf{v}_h, \mathbf{v}_h) = ((\varrho K_h M_h^{-1} K_h + M_h)\mathbf{v}_h, \mathbf{v}_h) = \sum_{i=1}^{n_h} (\varrho \lambda_{hi}^2 + 1)(v_i^e)^2,$$

from which the spectral equivalence inequalities

$$\underline{c}_{SM} M_h \leq S_{\varrho h} \leq \overline{c}_{SM} M_h, \tag{7}$$

with $\underline{c}_{SM} = 1$ and $\overline{c}_{SM} = \overline{c}_{KM}^2 + 1$, immediately follow; see [6] for details.

Now let us consider the more interesting case of variable regularization parameters defined by

$$\varrho(x) = h_\tau^4 \quad \text{for all } x \in \tau \text{ and for all } \tau \in \mathcal{T}_h. \tag{8}$$

Then the regularization matrix $A_{\varrho h} = M_{\varrho h}$ is defined by (5). It is well-known that the mass matrix M_h is spectrally equivalent to $D_h = diag(M_h)$ that is nothing but the diagonal matrix with the same diagonal elements as M_h. The same is true for the scaled mass matrix $M_{\varrho h}$.

Theorem 1. *There are positive, and mesh-independent constants $\underline{c}_{MD}$ and $\overline{c}_{MD}$ such that the spectral inequalities*

$$\underline{c}_{MD} D_{\varrho h} \leq M_{\varrho h} \leq \overline{c}_{MD} D_{\varrho h} \tag{9}$$

hold, where $D_{\varrho h} = diag(d_1, \ldots, d_{n_h}) = diag(M_{\varrho h})$.

Proof. Here we skip the proof. It is easy to compute the constants on the basis of the corresponding element matrices. □

Due to the locally scaled mass matrix $M_{\varrho h}$, we cannot use the same approach for estimating the Schur complement $S_{\varrho h}$ by M_h from above as in the case $\varrho = h^4$. Fortunately, we can prove the following spectral equivalence theorem by means of another technique that allows localization.

Theorem 2. *Let us assume that $\varrho = \varrho(x)$ is chosen as in (8). Then the spectral equivalence inequalities (7) hold with $\underline{c}_{SM} = 1$ and $\overline{c}_{SM} = c + 1$, where c is a generic constant that does not depend on the mesh refinement but only on the shape regularity parameters.*

Proof. It is obviously sufficient to estimate $(K_h M_{\varrho h}^{-1} K_h v_h, v_h)$ from above. Using (9) and the fact that $D_{\varrho h}$ is spd, we get the estimate

$$(K_h M_{\varrho h}^{-1} K_h v_h, v_h) \leq \underline{c}_{MD}^{-1} (D_{\varrho h}^{-1} K_h v_h, K_h v_h) = \underline{c}_{MD}^{-1} \|D_{\varrho h}^{-1/2} K_h v_h\|^2. \tag{10}$$

Now we proceed to represent $\|D_{\varrho h}^{-1/2} K_h v_h\|$ as follows:

$$
\begin{aligned}
\|D_{\varrho h}^{-1/2} K_h v_h\| &= \sup_{w_h \in \mathbb{R}^{n_h}} \frac{(D_{\varrho h}^{-1/2} K_h v_h, w_h)}{\|w_h\|} = \sup_{w_h \in \mathbb{R}^{n_h}} \frac{(K_h v_h, D_{\varrho h}^{-1/2} w_h)}{\|w_h\|} \\
&= \sup_{w_h \in \mathbb{R}^{n_h}} \frac{(\nabla v_h, \nabla \widetilde{w}_h)_{L_2(\Omega)}}{\|w_h\|} = \sup_{w_h \in \mathbb{R}^{n_h}} \frac{\sum_{\tau \in \mathcal{T}_h} (\nabla v_h, \nabla \widetilde{w}_h)_{L_2(\tau)}}{\|w_h\|} \\
&= \sup_{w_h \in \mathbb{R}^{n_h}} \frac{\sum_{\tau \in \mathcal{T}_h} (K_\tau v_\tau, D_\tau^{-1/2} w_\tau)}{\|w_h\|} \\
&= \sup_{w_h \in \mathbb{R}^{n_h}} \frac{\sum_{\tau \in \mathcal{T}_h} (D_\tau^{-1/2} K_\tau v_\tau, w_\tau)}{\|w_h\|},
\end{aligned}
\tag{11}
$$

where the fe function $\tilde{w}_h = \sum_{i=1}^{n_h} d_i^{-1/2} w_i \varphi_{hi} \in V_h$ corresponds to the fe vector $D_{\varrho h}^{-1/2} \mathbf{w}_h \in \mathbb{R}^{n_h}$ via the fe isomorphism, $\mathbf{v}_\tau, \mathbf{w}_\tau \in \mathbb{R}^{|\tau|}$ are the corresponding local fe vectors, K_τ is the $|\tau| \times |\tau|$ element stiffness matrix, and D_τ is the $|\tau| \times |\tau|$ diagonal matrix with diagonal entries from $D_{\varrho h}$ corresponding to $\tau \in \mathcal{T}_h$. We mention that the supremum over $\mathbf{w}_h \in \mathbb{R}^{n_h}$ always means $\mathbf{w}_h \in \mathbb{R}^{n_h} \setminus \{\mathbf{0}_h\}$.

Using the Cauchy-Schwarz inequality twice and $\left(\sum_{\tau \in \mathcal{T}_h} \|\mathbf{w}_\tau\|^2\right)^{1/2} \leq c_m \|\mathbf{w}_h\|$ that follows from the shape regularity of the mesh (maximal number of simplices around the vertices) for estimating (11), we arrive at the estimate

$$\|D_{\varrho h}^{-1/2} K_h \mathbf{v}_h\| \leq \sup_{\mathbf{w}_h \in \mathbb{R}^{n_h}} \frac{\left(\sum_{\tau \in \mathcal{T}_h} \|D_\tau^{-1/2} K_\tau \mathbf{v}_\tau\|^2\right)^{1/2} \left(\sum_{\tau \in \mathcal{T}_h} \|\mathbf{w}_\tau\|^2\right)^{1/2}}{\|\mathbf{w}_h\|},$$

$$\leq c_m \left(\sum_{\tau \in \mathcal{T}_h} \|D_\tau^{-1/2} K_\tau \mathbf{v}_\tau\|^2\right)^{1/2}. \tag{12}$$

Using $\|D_\tau^{-1}\| \leq c_D h_\tau^4 h_\tau^{-d}$, $\|K_\tau\| \leq c_K h_\tau^{d-2}$, and $c_M^{-1} h_\tau^d (\mathbf{v}_\tau, \mathbf{v}_\tau) \leq (M_\tau \mathbf{v}_\tau, \mathbf{v}_\tau)$, we arrive at the estimates

$$\sum_{\tau \in \mathcal{T}_h} \|D_\tau^{-1/2} K_\tau \mathbf{v}_\tau\|^2 = \sum_{\tau \in \mathcal{T}_h} (D_\tau^{-1} K_\tau \mathbf{v}_\tau, K_\tau \mathbf{v}_\tau)$$

$$\leq \sum_{\tau \in \mathcal{T}_h} \|D_\tau^{-1}\| \|K_\tau\|^2 \|\mathbf{v}_\tau\|^2$$

$$\leq c_D c_K^2 \sum_{\tau \in \mathcal{T}_h} h_\tau^4 h_\tau^{-d} (h_\tau^{d-2})^2 \|\mathbf{v}_\tau\|^2$$

$$= c_D c_K^2 \sum_{\tau \in \mathcal{T}_h} h_\tau^d \|\mathbf{v}_\tau\|^2 \leq c_D c_K^2 c_M \sum_{\tau \in \mathcal{T}_h} (M_\tau \mathbf{v}_\tau, \mathbf{v}_\tau) \tag{13}$$

$$= c_D c_K^2 c_M (M_h \mathbf{v}_h, \mathbf{v}_h).$$

We note that the positive constants c_M, c_D, and c_K can be chosen globally due to the shape regularity assumption. Combining (10), (11), (12), and (13), we get the upper bound $\overline{c}_{SM} = c + 1$ with $c = c_{MD}^{-1} c_m^2 c_D c_K^2 c_M$. The lower bound $\underline{c}_{SM} = 1$ is trivial. This concludes the proof of the theorem.   □

Remark 1. Theorem 2 even remains valid for nonsymmetric stiffness matrices K_h arising from the fe discretization of more general elliptic equations like diffusion-convection-reaction equations. Indeed, we have to estimate $(K_h^\top M_{\varrho h}^{-1} K_h \mathbf{v}_h, \mathbf{v}_h) = (M_{\varrho h}^{-1} K_h \mathbf{v}_h, K_h \mathbf{v}_h)$ that can be done as in the proof of Theorem 2.

It is now clear from the spectral equivalence Theorems 1 and 2 that, in (7), the mass matrix M_h can be replaced by a suitable diagonal approximation D_h such as diag(M_h) or the lumped mass matrix lump(M_h).

Using these spectral equivalence results, we can now construct robust and efficient preconditioned iterative Krylov subspace solvers like the preconditioned MINRES or Bramble-Pasciak's Preconditioned Conjugate Gradient (BP-PCG) tailored to symmetric, but indefinite systems like (4). In this paper, we focus on the BP-PCG that was proposed in [4]; see also [13] for improved convergence rate estimates. Thus, we consider a properly scaled diagonal approximation

$$D_{\varrho h} = \delta \operatorname{diag}(M_{\varrho h}) \quad \text{or} \quad D_{\varrho h} = \delta \operatorname{lump}(M_{\varrho h})$$

such that (9) is valid with $1 < \underline{c}_{\text{MD}} \leq \overline{c}_{\text{MD}}$, i.e. $D_{\varrho h} < M_{\varrho h}$ as requested in BP-PCG. We note that the positive scaling parameter δ can be easily calculated on element level. Furthermore, we consider

$$D_h = \operatorname{diag}(M_h) \quad \text{or} \quad D_h = \operatorname{lump}(M_h) \tag{14}$$

that are spectrally equivalent to $S_{\varrho h}$ under the assumptions of Theorem 2. Then the BP-PCG is nothing but the PCG applied to the spd system

$$\mathcal{K}_h \begin{bmatrix} \mathbf{p}_h \\ \mathbf{y}_h \end{bmatrix} = \begin{bmatrix} \mathbf{0}_h \\ -\mathbf{y}_{dh} \end{bmatrix}, \text{ with } \mathcal{K}_h = \begin{bmatrix} M_{\varrho h} D_{\varrho h}^{-1} - I \; 0_h \\ K_h D_{\varrho h}^{-1} \quad -I \end{bmatrix} \begin{bmatrix} M_{\varrho h} \; K_h \\ K_h^\top \; -M_h \end{bmatrix},$$

which is equivalent to (4), with the preconditioner

$$\mathcal{P}_h = \begin{bmatrix} M_{\varrho h} - D_{\varrho h} \; 0_h \\ 0_h \quad\quad D_h \end{bmatrix}.$$

The BP-PCG converges with an h-independent rate in asymptotically optimal complexity $O(n_h \ln(\varepsilon^{-1}))$, where $\varepsilon \in (0, 1)$ denotes a fixed relative accuracy with respect to the preconditioned residual norm; see [4] and [13].

Alternatively, we can solve the spd Schur complement system (6), arising from (4) when the term $(\varrho^{-1} v, w)_{L_2(\Omega)}$ is discretized by the lumped mass techniques leading to $A_{\varrho h} = \operatorname{lump}(M_{\varrho h})$, by means of the PCG method preconditioned by D_h as defined in (14). This Schur complement PCG solver converges in optimal complexity; see [6] for numerical results in the case of constant $\varrho = h^4$.

3 Numerical Results

We here focus on the three-dimensional $(d = 3)$ case with the discontinuous desired state $y_d(x)$ that is equal to 1 for $x \in (0.25, 0.75)^3$ and 0 elsewhere, where the unit cube $\Omega = (0, 1)^3$ serves as computational domain. Thus, the desired state y_d is not contained in the state space $Y = H_0^1(\Omega)$, and has a low regularity of $y_d \in H^{1/2-\varepsilon}(\Omega)$ for any $\varepsilon > 0$. The same example was already numerically studied in [5] where the variable energy regularization was considered. Here we focus on the variable L_2 regularization in connection with the same simple adaptive procedure as in [5]. This adaptive procedure will considerably improve the accuracy in comparison with the uniform refinement. We expect that the preconditioners presented in Sect. 2 lead to asymptotically optimal and robust iterative solvers.

The domain $\Omega = (0, 1)^3$ is decomposed into uniformly refined tetrahedral elements. The starting mesh contains 384 tetrahedral elements and 125 vertices, leading to an initial mesh size $h = 2^{-2}$. Our numerical tests are running on 8 uniformly refined mesh levels L_i, $i = 1, ..., 8$. On the finest level, we have $135,005,697$ vertices, $270,011,394$ degrees of freedom (#Dofs), $h = 2^{-9} = 1.9531e{-}3$, and $\varrho = h^4 = 2^{-36} = 1.4552e{-}11$. For the tests performed on the adaptive meshes, we have employed the conventional red-green refinement of tetrahedral elements, and we have chosen the local regularization parameter $\varrho_\tau = h_\tau^4$ on each tetrahedral element τ. The comparison of convergence on both uniform and adaptive refinements is illustrated in Fig. 1, from which we easily see that the adaptive refinement leads to a much better convergence rate $h^{0.75}$ than the uniform one $h^{0.5}$. In all numerical tests, we run the BP-PCG iterations $(D_{\varrho h} = 0.125 \,\mathrm{lump}(M_{\varrho h}), D_h = \mathrm{lump}(M_h))$ until the preconditioned residual is reduced by a factor 10^6. A comparison of the number of iterations (Its) on both adaptive and uniform refinements is given in Table 1, cf. with [5, Table 3] for the

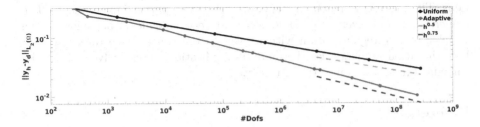

Fig. 1. Comparison of convergence using uniform and adaptive refinements.

Table 1. Comparison of the BP-PCG solver on both adaptive and uniform refinements.

Level	Adaptive			Uniform		
	#Dofs	$\|y_h - y_d\|_{L_2(\Omega)}$	Its	#Dofs	$\|y_h - y_d\|_{L_2(\Omega)}$	Its
L_1	250	3.28255e-1	12	250	3.28255e-1	12
L_2	446	2.38883e-1	118	1,458	2.30561e-1	99
L_3	2,102	1.90941e-1	171	9,826	1.63827e-1	137
L_4	9,170	1.37227e-1	204	71,874	1.15682e-1	141
L_5	21,620	1.07761e-1	202	549,250	8.16986e-2	138
L_6	65,828	8.14778e-2	201	4,293,378	5.77278e-2	132
L_7	223,162	5.93121e-2	198	33,949,186	4.08032e-2	123
L_8	330,422	5.48633e-2	201	270,011,394	2.88463e-2	114
L_9	1,084,164	4.01323e-2	199			
L_{10}	3,891,800	2.88341e-2	191			
L_{11}	4,907,338	2.77892e-2	208			
L_{12}	17,034,046	2.01781e-2	191			
L_{13}	51,731,508	1.47666e-2	186			
L_{14}	53,049,534	1.46249e-2	188			
L_{15}	234,045,680	1.01634e-2	181			

same example, but with energy regularization. It is clearly observed that our preconditioner is robust with respect to the mesh size and the local adaptivity under the choice $\varrho_\tau = h_\tau^4$.

4   Conclusions

We have shown that the mass matrix M_h and, therefore, suitable diagonal representations D_h such as diag(M_h) or lump(M_h) are robust preconditioners for the Schur complement $S_{\varrho h}$ in the case of the variable choice of the regularization parameter ϱ. This new result is very important for using adaptivity. Together with a similar, but appropriately scaled diagonal preconditioner $D_{\varrho h}$ for $M_{\varrho h}$, the corresponding BP-PCG is a robust and efficient solver for the reduced discrete optimality system (4) in the case of both constant and variable L_2 regularizations. The theoretical results are supported by the numerical experiments. The numerical results given in [5] show a similar (h-independent) behavior for the variable energy regularization with respect to the convergence rate, but with lower iteration numbers. Furthermore, in applications, these solvers should be embedded in a nested iteration strategy.

References

1. Axelsson, O., Neytcheva, M., Ström, A.: An efficient preconditioning method for state box-constrained optimal control problems. J. Numer. Math. **26**(4), 185–207 (2018)
2. Bai, Z.Z., Pan, J.Y.: Matrix Analysis and Computations. SIAM (2021)
3. Benzi, M., Golub, G., Liesen, J.: Numerical solution of saddle point problems. Acta Numer. **14**, 1–137 (2005)
4. Bramble, J.H., Pasciak, J.E.: A preconditioning technique for indefinite systems resulting from mixed approximations of elliptic problems. Math. Comp. **50**(181), 1–17 (1988)
5. Langer, U., Löscher, R., Steinbach, O., Yang, H.: An adaptive finite element method for distributed elliptic optimal control problems with variable energy regularization. Comput. Math. Appl. **160**, 1–14 (2024)
6. Langer, U., Löscher, R., Steinbach, O., Yang, H.: Robust finite element discretization and solvers for distributed elliptic optimal control problems. Comput. Methods Appl. Math. **23**(4), 989–1005 (2023). https://doi.org/10.1515/cmam-2022-0138
7. Langer, U., Steinbach, O., Yang, H.: Robust discretization and solvers for elliptic optimal control problems with energy regularization. Comput. Meth. Appl. Math. **22**(1), 97–111 (2022)
8. Lions, J.L.: Contrôle optimal de systèmes gouvernés par des équations aux dérivées partielles. Dunod Gauthier-Villars, Paris (1968)
9. Mardal, K.A., Winther, R.: Preconditioning discretizations of systems of partial differential equations. Numer. Linear Algebra Appl. **18**(1), 1–40 (2011)
10. Neumüller, M., Steinbach, O.: Regularization error estimates for distributed control problems in energy spaces. Math. Methods Appl. Sci. **44**, 4176–4191 (2021)
11. Pearson, J., Wathen, A.: A new approximation of the Schur complement in preconditioners for PDE-constrained optimization. NLAA **12**(5), 816–829 (2012)

12. Schiela, A., Ulbrich, S.: Operator preconditioning for a class of inequality constrained optimal control problems. SIAM J. Optim. **24**(1), 435–466 (2014)
13. Zulehner, W.: Analysis of iterative methods for saddle point problems: a unified approach. Math. Comp. **71**(238), 479–505 (2002)
14. Zulehner, W.: Nonstandard norms and robust estimates for saddle point problems. SIAM J. Matrix Anal. Appl. **32**(2), 536–560 (2011)

Numerical Comparison of Block Preconditioners for Poroelasticity

Tomáš Luber[(✉)][iD]

Institute of Geonics of the Academy of Sciences of the Czech Republic,
Ostrava-Poruba, Czech Republic
tomas.luber@ugn.cas.cz

Abstract. Poroelastic models describe a coupled solid-fluid interaction between a solid skeleton saturated by some fluid. We consider Biot's model that describes the deformation of the solid skeleton by linear elasticity and the fluid flow by Darcy's law. The formulation includes displacements, fluid flux and fluid pressure. After time discretization by the implicit Euler method we obtain a linear system with a matrix of 3×3 block structure. The focus of the paper is how to solve the so-arising linear system by a preconditioned iterative Krylov subspace iteration method.

We analyse a particular Schur complement-based preconditioner, constructed with respect to the pressure variable, and compare its performance with other block preconditioners that arise from various algebraic and functional settings. The numerical experiments are based on geotechnical context inspired by a particular tunnel sealing experiment and test the dependence of the performance of the considered preconditioners on mesh size and permeability.

Keywords: Numerical mathematics · poroelasticity · preconditioning · Schur complement

1 Introduction

The model that we consider here is the classical quasi-static Biot's model [4] that couples the deformation of the solid skeleton of the porous material, described by linear elasticity, with the saturated fluid flow, described by the Darcy's law. We are interested in the three-field formulation of the model, namely, in displacements u, flux v and pressure p. The classical formulation of the model in these variables on a domain $\Omega \subset \mathbb{R}^{2,3}$ reads:

$$-\operatorname{div}\sigma + \alpha_b \operatorname{grad} p = f, \tag{1}$$

$$v + \mathcal{K}\operatorname{grad} p = 0, \tag{2}$$

$$\alpha_b \frac{\partial}{\partial t}\operatorname{div} u + \operatorname{div} v + c_{pp}\frac{\partial}{\partial t}p = g, \tag{3}$$

© The Author(s), under exclusive license to Springer Nature Switzerland AG 2024
I. Lirkov and S. Margenov (Eds.): LSSC 2023, LNCS 13952, pp. 44–51, 2024.
https://doi.org/10.1007/978-3-031-56208-2_3

where σ is the effective stress tensor, given by $\sigma = \mathcal{C}\epsilon(u)$, $\mathcal{C}$ is the elastic tensor and $\epsilon(u)_{ij} = (u_{i,j} + u_{j,i})/2$ is the small strain tensor; α_b is the Biot-Willis constant, c_{pp} is the specific storage and $\mathcal{K}$ is the hydraulic conductivity. The model is completed by some suitable boundary and initial conditions.

We consider here a less general version of the model under the following assumptions: (i) all physical parameters are independent of the time variable, (ii) the elastic properties are homogeneous and isotropic meaning that $\sigma = \mathcal{C}\epsilon(u) = 2\mu\epsilon(u) + \lambda \operatorname{div} uI$, (iii) the parameter α_b is constant and $\alpha_b \in (0,1]$, (iv) the permeability is isotropic and given by a function $k(x)$ bounded from above and below, namely, $0 < k_{\min} \leq k(x) < k_{\max}$ for all x and (v) the specific storage is constant and non-negative ($c_{pp} > 0$). Assumption (v) arises from the computation of specific storage in a geomechanical setting as $c_{pp} = \frac{(\alpha_b - n)(1 - \alpha_b)}{K} + nb_v$ where K is the bulk modulus, n is porosity and b_v is water compressibility, and is critical for the construction of the proposed preconditioner.

The continuous model is discretized by suitable finite elements in space and by the implicit Euler method in time. The main focus of this paper is in constructing suitable block preconditioners for solving the algebraic system arising in each time step. We describe a block-diagonal preconditioner on the operator level, derive a bound of its relative condition number and compare this preconditioner with other preconditioners introduced in [7,12].

2   Weak Formulation

For the variational formulation we consider a domain $\Omega \subset \mathbb{R}^{2,3}$ with Lipschitz boundary Γ. We will use the following spaces for displacement, flux and pressure:

$$U = \{u \in [H^1(\Omega)]^{n_d} \mid u = 0 \text{ on } \Gamma_{u,D}\}, \tag{4}$$

$$V = \{v \in H(\operatorname{div}, \Omega) \mid v \cdot n = 0 \text{ on } \Gamma_{v,D}\}, \tag{5}$$

$$P = L^2(\Omega), \tag{6}$$

where $\Gamma_{u,D}$ and $\Gamma_{v,D}$ are parts of Γ with positive measure. We define the bilinear forms $a : U \times U \to \mathbb{R}$, $m : V \times V \to \mathbb{R}$, $d : P \times P \to \mathbb{R}$, $b_u : U \times P \to \mathbb{R}$, $b_v : V \times P \to \mathbb{R}$

$$a(u,w) = 2\mu \int_{\Omega} \epsilon(u) : \epsilon(w) \, d\Omega + \lambda \int_{\Omega} \operatorname{div} u \operatorname{div} w \, d\Omega, \tag{7}$$

$$b_u(u,p) = \int_{\Omega} p \operatorname{div} u \, d\Omega, \tag{8}$$

$$b_v(v,p) = \int_{\Omega} p \operatorname{div} v \, d\Omega, \tag{9}$$

$$m(v,z) = \int_{\Omega} k^{-1} v \cdot z \, d\Omega, \tag{10}$$

$$d_0(p,q) = \int_{\Omega} pq \, d\Omega. \tag{11}$$

After discretizing in time by the implicit Euler method the three-field variational formulation of Biot's problem in operator form reads (cf. [7,11]): find $(u, v, p) \in U \times V \times P$ such that

$$a(u, w) - \alpha_b b_u(w, p) = (\hat{f}, w) \quad \text{for all } w \in U, \tag{12}$$

$$m(v, z) - b_v(z, p) = (\hat{h}, z) \quad \text{for all } z \in V, \tag{13}$$

$$\frac{\alpha_b}{\tau} b_u(u, q) + b_v(v, q) + \frac{c_{pp}}{\tau} d_0(p, q) = (\hat{g}, q) \quad \text{for all } q \in P, \tag{14}$$

where $(\hat{f}, .) \in U^*, (\hat{h}, .) \in V^*, (\hat{g}, .) \in P^*$ denote the linear functionals that contain contributions from source terms, previous timestep and boundary conditions.

For the formulation and analysis of the preconditioner we utilize the operator form of the system (12)–(14). Let linear operators $A : U \to U^*$, $M : V \to V^*$, $D_0 : P \to P^*$, $B_u : U \to P^*$ and $B_v : V \to P^*$ represent the bilinear forms a, m, d_0, b_u and b_v, respectively, and let $B_u^* : P \to U^*$ and $B_v^* : P \to V^*$ be adjoint operators to B_u and B_v. Using these operators, and a simple symmetrization by scaling with τ, we can write the system of Eqs. (12) to (14) as

$$\mathcal{A} \begin{bmatrix} u \\ v \\ p \end{bmatrix} := \begin{bmatrix} A & & \alpha_b B_u^* \\ & \tau M & \tau B_v^* \\ \alpha_b B_u & \tau B_v & -c_{pp} D_0 \end{bmatrix} \begin{bmatrix} u \\ v \\ p \end{bmatrix} = \begin{bmatrix} \hat{f} \\ \tau \hat{h} \\ \tau \hat{g} \end{bmatrix}. \tag{15}$$

3   Preconditioners

To construct a preconditioner for (15) we work in the framework of operator preconditioning (see [10]) and generalize the main idea of the Schur complement-based approach from [3] to operator setting. More details and all proofs can be found in [9].

The approach followed here uses the Schur complement of the matrix in (15) with respect to the pressure block utilizing the assumption that $c_{pp} > 0$. We consider the two-by-two block splitting of $\mathcal{A}$ from (15) and construct block preconditioners with the full and diagonalized Schur complements

$$\mathcal{P}_U = \begin{bmatrix} S & \\ & +c_{pp} D_0 \end{bmatrix}, \quad \mathcal{P}_{UD} = \begin{bmatrix} S_D & \\ & +c_{pp} D_0 \end{bmatrix} \tag{16}$$

where S, S_D are given by

$$S = \begin{bmatrix} A + \alpha_b^2 c_{pp}^{-1} B_u^* D_0^{-1} B_u & \tau \alpha_b c_{pp}^{-1} B_u^* D_0^{-1} B_v \\ \tau \alpha_b c_{pp}^{-1} B_v^* D_0^{-1} B_u & \tau M + \tau^2 c_{pp}^{-1} B_v^* D_0^{-1} B_v \end{bmatrix}, \tag{17}$$

$$S_D = \begin{bmatrix} A + \alpha_b^2 c_{pp}^{-1} B_u^* D_0^{-1} B_u & \\ & \tau M + \tau^2 c_{pp}^{-1} B_v^* D_0^{-1} B_v \end{bmatrix}. \tag{18}$$

The following lemma is based on the observation that D_0 is in fact Riesz isomorphism. It shows that we can bypass the possible need of the inversion of the pressure mass matrix.

Lemma 1. *For any* $u \in H^1(\Omega)$, $v \in H(\mathrm{div}, \Omega)$ *holds that*

$$\langle c_{pp}^{-1} B_u^* D_0^{-1} B_u \, u, w \rangle_U = (c_{pp}^{-1} \mathrm{div} \, u, \mathrm{div} \, w) \quad \text{for all } u, w \in U, \tag{19}$$

$$\langle c_{pp}^{-1} B_u^* D_0^{-1} B_v \, v, u \rangle_U = (c_{pp}^{-1} \mathrm{div} \, v, \mathrm{div} \, u) \quad \text{for all } u \in U, \text{ for all } v \in V, \tag{20}$$

$$\langle c_{pp}^{-1} B_v^* D_0^{-1} B_u \, u, v \rangle_V = (c_{pp}^{-1} \mathrm{div} \, u, \mathrm{div} \, v) \quad \text{for all } u \in U, \text{ for all } v \in V, \tag{21}$$

$$\langle c_{pp}^{-1} B_v^* D_0^{-1} B_v \, v, z \rangle_V = (c_{pp}^{-1} \mathrm{div} \, v, \mathrm{div} \, z) \quad \text{for all } v, z \in V. \tag{22}$$

To make the estimates of the condition numbers $\kappa(\mathcal{P}_U^{-1}\mathcal{A})$, $\kappa(\mathcal{P}_{UD}^{-1}\mathcal{A})$ we first apply the main theorem from [12] and obtain an estimate of $\kappa(\mathcal{P}_U^{-1}\mathcal{A})$. Then, we derive an estimate $\kappa(\mathcal{P}_{UD}^{-1}\mathcal{P}_U)$ which, as a consequence, allows to estimate $\kappa(\mathcal{P}_{UD}^{-1}\mathcal{A})$. These results are summarized in the following theorem.

Theorem 1. *Let* $\mathcal{A}$ *and* $\mathcal{P}, \mathcal{P}_D$ *be defined as above. Let* $c_{el} = \lambda + 2\mu/n_d$. *Then*

$$\kappa\left(\mathcal{P}_U^{-1}\mathcal{A}\right) \leq \frac{\sqrt{5}+1}{\sqrt{5}-1} \tag{23}$$

and

$$(1 - \gamma)\langle S_D x, x \rangle \leq \langle S x, x \rangle \leq (1 + \gamma)\langle S_D x, x \rangle \quad \text{for all } x \in U \times V, \tag{24}$$

where $\gamma = \sqrt{\frac{1}{1+c_{el}c_{pp}/\alpha_b^2}} < 1$. *Finally*

$$\kappa\left(\mathcal{P}_{UD}^{-1}\mathcal{A}\right) \leq \frac{1+\gamma}{1-\gamma}\frac{\sqrt{5}+1}{\sqrt{5}-1}. \tag{25}$$

Remark 1. The estimate (23) can be seen as a counterpart of the standard matrix result for two-by-two saddle point systems given, e.g., in [2].

Remark 2. The estimate (24) is based on the strengthened Cauchy-Schwarz inequality and its derivation is essentially the same on operator and matrix level. The estimate shows that we can expect that $\mathcal{P}_{UD}$ will not be useful for $c_{el}c_{pp}$ close to 0.

Remark 3. The preconditioner $\mathcal{P}_U$ is of lesser practical usage as there is no obvious way how to solve the system with the block S_D. On the other hand, solving with $\mathcal{P}_{UD}$ is more viable as the first block is effectively the linear elastic operator with the shift $\lambda \mapsto \lambda + \alpha_b^2/c_{pp}$ so we can employ any robust linear elasticity solver. The second block equation is effectively a scaled $H(\mathrm{div})$ problem and we can use the approach described in, e.g., [8].

We compare the performance of the preconditioners suggested above with two other block-diagonal preconditioners, referred to as $\mathcal{P}_{BNB}$ and $\mathcal{P}_{2D}$. The preconditioner $\mathcal{P}_{BNB}$ is described in [7] where it is derived based on the theory for the existence of solutions to perturbed saddle point problems. It has the form

$$\mathcal{P}_{BNB} = \begin{bmatrix} A & \\ \tau M + \tau^2 \left(c_{pp} + \frac{\alpha_b^2}{\lambda+2\mu} + k_{min}\tau\right)^{-1} B_v^* D_0^{-1} B_v & \\ & \left(c_{pp} + \frac{\alpha_b^2}{\lambda+2\mu} + k_{min}\tau\right) D_0 \end{bmatrix}.$$

$$\tag{26}$$

The preconditioner $\mathcal{P}_{2D}$ is defined in [12] and is based on the general Schur complement construction for multiply-saddle point systems. It reads

$$
\mathcal{P}_{2D} = \begin{bmatrix} A & \\ \tau M + \tau^2 \left(c_{pp} + \frac{\alpha_b^2}{\lambda+2\mu} \right)^{-1} B_v^* D_0^{-1} B_v & \\ & \left(c_{pp} + \frac{\alpha_b^2}{\lambda+2\mu} \right) D_0 \end{bmatrix}. \tag{27}
$$

Both $\mathcal{P}_{BNB}$ and $\mathcal{P}_{2D}$ do not need the assumption $c_{pp} > 0$ and it is shown in the respective papers that their relative condition number is independent on parameters of the model.

The condition number estimate derived for $\mathcal{P}_{BNB}$ is based on the assumption that the physical parameters are constant. Since we assume that k is non-constant in numerical experiments, we replace k with its lower bound $k_{min} = \min_{x \in \Omega} k(x)$ in the flux and pressure blocks and leave k unchanged in M, similarly as it is done in [6].

4   Numerical Experiments

The numerical experiments are done in a setting motivated by the 2D cut of the so-called Tunnel Sealing Experiment (TSX), [5]. The computations are performed using FEniCS [1].

The domain Ω is a square $[0, 100] \times [0, 100]$ with an elliptical hole modeling the tunnel. The ellipse's center is in the center of the square $[50, 50]$ and its axes are parallel to x and y axis. The lengths of the semi-axes are $a_x = 4.375/2$ m, $b_y = 3.5/2$ m.

The discretization is done using conforming finite elements – Lagrange $P(2)$ for the displacement, Raviart-Thomas $RT(1)$ for the flux and discontinuous Lagrange $P^-(1)$ for the pressure.

We prescribe zero displacement $u = 0$ and initial pore pressure $p = 3 \times 10^6$ Pa. The boundary conditions on the outer (square) boundary include constant pressure $p_{do} = 3 \times 10^6$ Pa and zero normal displacements. Boundary conditions on the inner hole simulate the drilling of the tunnel for 17 days. During this time the initial stress and pore pressure linearly decrease to zero and are prescribed as the boundary conditions on the elliptical boundary.

The physical parameters used for the experiments are:

- Young modulus $E = 60$ GPa
- Poisson's ratio $\nu = 0.2$
- Biot's coefficient $\alpha_b = 0.2$
- Specific storage $c_{pp} = 7.712 \times 10^{-12}$ Pa^{-1}
- Timestep length $\tau = 6$ h

We distinguish three zones with different conductivities. One is in an ellipse again centered at $[50, 50]$ with lengths of the semi-axes $a_x = 2.9375$ m $b_y = 2.5$ m inside which the conductivity is set to 6×10^{-17} Pa^{-1}m^2/s and another

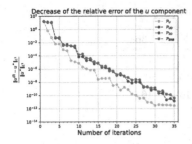 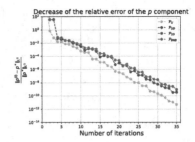

Fig. 1. Convergence history for the relative error norm of u and p during the FGMRES iterations for individual preconditioners, $\mathcal{P}_{BNB}$ and $\mathcal{P}_{2D}$ overlap

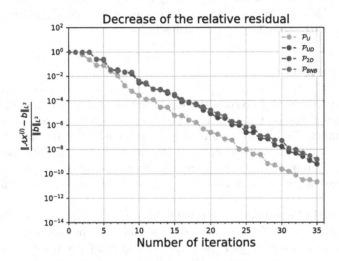

Fig. 2. Convergence history for the relative algebraic norm residual during the FGMRES iterations, $\mathcal{P}_{BNB}$ and $\mathcal{P}_{2D}$ overlap

innermost zone that is inside an ellipse with the same center and lengths of the semi-axes $a_x = 1$ m, $b_y = 2.125$ m with conductivity 6×10^{-15} Pa^{-1}m^2/s.

As an outer solver we use FGMRES preconditioned by the introduced preconditioners and a stopping criterion of the decay of the true residual by 10^{-6}.

A direct method is used to solve the inner blocks. (We consider only the first timestep of the evolution.)

In Figs. 1 and 2 we show the convergence history for the three preconditioners. The problem is computed on a mesh with 136875 vertices. We can see that for $\mathcal{P}_{BNB}$, $\mathcal{P}_{2D}$ and $\mathcal{P}_{UD}$ the decrease both in terms of errors and residual norm is almost the same. The difference between these methods and $\mathcal{P}_U$ in terms of convergence is relatively small. The here suggested preconditioner $\mathcal{P}_U$ successfully handles strongly varying parameter k, however on the cost of a somewhat higher required computational time.

Table 1. Number of iterations as a function of the level of mesh refinement

# vertices	2558	7254	36374	136875
$\mathcal{P}_{UD}$	17	15	16	16
$\mathcal{P}_{2D}$	18	18	16	16
$\mathcal{P}_{BNB}$	18	18	16	16

The number of iterations as a function of the number of the mesh vertices is reported in Table 1. We note that the mesh is refined around the jumps in permeability. The table does not show any strong dependence on the mesh refinement parameter, which confirms again the advantages of $\mathcal{P}_U$ and $\mathcal{P}_{UD}$ being efficient for problems with strongly varying permeability.

5   Conclusions

In this paper we consider preconditioning of the block matrices arising from the discretization of the Biot model, used to describe flow in porous rock massive around a tunnel. The discretization in space is done by suitable finite elements and the implicit Euler method in time. We propose two block preconditioners $\mathcal{P}_U$, $\mathcal{P}_{UD}$ and present estimates of the condition number of the corresponding preconditioned matrices, which are independent of the chosen space discretization.

The theoretical results are supported by a numerical experiments which show that the preconditioner $\mathcal{P}_{UD}$ is to be recommended for problems in the considered geomechanical setting as it performs well in comparison with other alternatives based on constructions of different types.

Acknowledgments. This work was supported by EU Horizon 2020 research and innovation programme under grant agreement number 847593 and by The Czech Radioactive Waste Repository Authority (SÚRAO) under grant agreement number SO2020-017.

References

1. Alnaes, M.S., et al.: The FEniCS project version 1.5. Arch. Numer. Softw. **3** (2015). https://doi.org/10.11588/ans.2015.100.20553
2. Axelsson, O.: Unified analysis of preconditioning methods for saddle point matrices. Numer. Linear Algebra Appl. **22**(2), 233–253 (2015). https://doi.org/10.1002/nla.1947, http://doi.wiley.com/10.1002/nla.1947
3. Axelsson, O., Blaheta, R., Luber, T.: Preconditioners for mixed FEM solution of stationary and nonstationary porous media flow problems. In: Lirkov, I., Margenov, S.D., Waśniewski, J. (eds.) LSSC 2015. LNCS, vol. 9374, pp. 3–14. Springer, Cham (2015). https://doi.org/10.1007/978-3-319-26520-9_1

4. Biot, M.A.: Theory of elasticity and consolidation for a porous anisotropic solid. J. Appl. Phys. **26**(2), 182–185 (2004). https://doi.org/10.1063/1.1721956, https://aip.scitation.org/doi/10.1063/1.1721956

5. Chandler, N.A., Cournut, A., Dixon, D.: The five year report of the tunnel sealing experiment: an international project of AECL, JNC, ANDRA and WIPP. Technical report, Atomic Energy of Canada Limited (2002). https://www.osti.gov/etdeweb/biblio/21313951, tex.ids: Chandler2002a

6. Hong, Q., Kraus, J., Lymbery, M., Philo, F.: Conservative discretizations and parameter-robust preconditioners for biot and multiple-network flux-based poroelasticity models. Numer. Linear Algebra Appl. **26**(4) (2019). https://doi.org/10.1002/nla.2242, https://onlinelibrary.wiley.com/doi/abs/10.1002/nla.2242

7. Hong, Q., Kraus, J., Lymbery, M., Philo, F.: A new practical framework for the stability analysis of perturbed saddle-point problems and applications. Math. Comput. **92**(340), 607–634 (2023). https://doi.org/10.1090/mcom/3795, https://www.ams.org/mcom/2023-92-340/S0025-5718-2022-03795-9/

8. Kraus, J., Lazarov, R., Lymbery, M., Margenov, S., Zikatanov, L.: Preconditioning heterogeneous $\boldsymbol{H}$(div) problems by additive schur complement approximation and applications. SIAM J. Sci. Comput. **38**(2), A875–A898 (2016). https://doi.org/10.1137/140974092,http://epubs.siam.org/doi/10.1137/140974092

9. Luber, T.: Efficient iterative methods and solvers for FEM analysis. phdthesis, VSB - Technical University of Ostrava (2022). http://dspace.vsb.cz/handle/10084/148531, accepted: 2022-09-01T07:49:26Z Publisher: Vysoká škola báňská - Technická univerzita Ostrava

10. Mardal, K.A., Winther, R.: Preconditioning discretizations of systems of partial differential equations. Numer. Linear Algebra Appl. **18**(1), 1–40 (2011). https://doi.org/10.1002/nla.716, https://onlinelibrary.wiley.com/doi/abs/10.1002/nla.716, _eprint: https://onlinelibrary.wiley.com/doi/pdf/10.1002/nla.716

11. Rodrigo, C., Hu, X., Ohm, P., Adler, J., Gaspar, F., Zikatanov, L.: New stabilized discretizations for poroelasticity and the stokes' equations. Comput. Methods Appl. Mech. Eng. **341**, 467–484 (2018). https://doi.org/10.1016/j.cma.2018.07.003, https://linkinghub.elsevier.com/retrieve/pii/S0045782518303347

12. Sogn, J., Zulehner, W.: Schur complement preconditioners for multiple saddle point problems of block tridiagonal form with application to optimization problems. IMA J. Numer. Anal. **39**(3), 1328–1359 (2018). https://doi.org/10.1093/imanum/dry027, https://academic.oup.com/imajna/article/39/3/1328/5000058

Two-Dimensional Semi-linear Riesz Space Fractional Diffusion Equations in Convex Domains: GLT Spectral Analysis and Multigrid Solvers

Stefano Serra-Capizzano[1,2] , Rosita L. Sormani[3(✉)] ,
and Cristina Tablino-Possio[4]

[1] Department of Science and High Technology, University of Insubria, INDAM Unit,
Via Valleggio 11, 22100 Como, Italy
stefano.serrac@uninsubria.it
[2] Department of Information Technology, Uppsala University,
Box 337, 751 05 Uppsala, Sweden
[3] Department of Theoretical and Applied Sciences, University of Insubria,
Via Dunant 3, 21100 Varese, Italy
rl.sormani@uninsubria.it
[4] Department of Mathematics and Applications, University of Milano Bicocca,
via Cozzi 53, 20125 Milan, Italy
cristina.tablinopossio@unimib.it

Abstract. The current work is devoted to the design of fast numerical methods for solving large linear systems, stemming from time-dependent Riesz space fractional diffusion equations, with a nonlinear source term in the convex (non Cartesian) domain. The problem is simpler than the distributed version in [9] and hence it is easier and more elegant to show that the sequence of coefficient matrices (as the finesse parameters decrease to zero) is a Generalized Locally Toeplitz (GLT) sequence and to compute its GLT symbol. From this study we recover important spectral information that we use for designing fast multigrid methods and for discussing the convergence speed of our multigrid solver. Numerical experiments are presented and critically discussed.

Keywords: Finite Elements · Matrix-sequences · Spectral analysis · Multigrid · Optimality

1 Problem Setting and Discretization

Let $\Omega \subset \mathbb{R}^2$ be a convex region whose left and right boundaries are $a_1(y)$ and $b_1(y)$, while the lower and upper boundaries are $a_2(x)$ and $b_2(x)$, respectively. Two-dimensional Riesz space fractional diffusion equations (RSFDEs), with the homogenous Dirichlet boundary conditions, are defined on Ω:

I. Lirkov and S. Margenov (Eds.): LSSC 2023, LNCS 13952, pp. 52–60, 2024.
https://doi.org/10.1007/978-3-031-56208-2_4

$$\frac{\partial u}{\partial t} = k_1 \frac{\partial^{\alpha_1} u}{\partial |x|^{\alpha_1}} + k_2 \frac{\partial^{\alpha_2} u}{\partial |y|^{\alpha_2}} + f(u, x, y, t), \quad (x, y, t) \in \Omega \times (0, T],$$

$$u(x, y, 0) = u_0(x, y), \quad (x, y) \in \Omega, \tag{1.1}$$

$$u(x, y, t) = 0, \quad (x, y, t) \in \Omega \times (0, T],$$

with k_1 and k_2 positive constants representing the diffusivity coefficients and $f(u, x, y, t)$ nonlinear source term supposed to verify the Lipschitz condition with respect to u and t. The Riesz fractional derivatives related to the parameters α_i $(1 < \alpha_i < 2, \ i = 1, 2)$ are given by $\frac{\partial^{\alpha_1} u}{\partial |x|^{\alpha_1}} = c_{\alpha_1} \left({}_{a_1(y)} D_x^{\alpha_1} u + {}_x D_{b_1(y)}^{\alpha_1} u \right),$ $\frac{\partial^{\alpha_2} u}{\partial |y|^{\alpha_2}} = c_{\alpha_2} \left({}_{a_2(x)} D_y^{\alpha_2} u + {}_y D_{b_2(x)}^{\alpha_2} u \right),$ where $c_{\alpha_i} = -1/(2 \cos(\alpha_i \pi / 2)) > 0$ and the above left and right Riemann-Liouville fractional derivatives defined as in [4]. Hereafter, in order to determine the numerical solution of the problem (1.1), we wrap the convex domain Ω in a rectangular domain $\bar{\Omega} = (a, b) \times (c, d) \supset \Omega$ according to the idea proposed in [4]. Hence, problem (1.1) is modified as

$$\frac{\partial u_\eta}{\partial t} = k_1 \frac{\partial^{\alpha_1} u_\eta}{\partial |x|^{\alpha_1}} + k_2 \frac{\partial^{\alpha_2} u_\eta}{\partial |y|^{\alpha_2}} + \frac{1 - 1_\Omega(x, y)}{\eta} u_\eta$$

$$+ \hat{f}(u_\eta, x, y, t), \quad (x, y, t) \in \bar{\Omega} \times (0, T],$$

$$u_\eta(x, y, 0) = \hat{u}_0(x, y), \quad (x, y) \in \bar{\Omega}, \tag{1.2}$$

$$u_\eta(x, y, t) = 0, \quad (x, y, t) \in \bar{\Omega} \times (0, T],$$

where $1_\Omega(x, y)$ is an indicator function defined as $1_\Omega(x, y) = 1$ if $(x, y) \in \Omega$, or 0 elsewhere, η is the penalty parameter, $\hat{f}(u_\eta, x, y, t)$ and $\hat{u}_0(x, y)$ are the zero extensions of the source term $f(u, x, y, t)$ and the initial condition $u_0(x, y)$ from Ω to $\bar{\Omega}$, respectively, i.e. $\hat{u}_0(x, y) = 0$, $\hat{f}(u_\eta, x, y, t) = 0$ if $(x, y) \in \bar{\Omega} \setminus \Omega$. It is quite obvious to notice that the RSFDEs in (1.2) will reduce to (1.1) if $(x, y) \in \Omega$. The role of the parameter $\eta \to 0^+$ is to force the solution to be zero outside Ω so that the solution u_η is supposed to satisfy homogeneous Dirichlet boundary conditions on $\bar{\Omega} \setminus \Omega$. Hence, following [9], the numerical solution is obtained by considering a Finite Difference discretization both with respect time and space: the time derivative is approximated by means of the backward Euler difference formula, while the shifted Grünwald-Letnikov difference scheme [12] is applied in order to approximate both Riemann-Liouville fractional derivatives, with n_1, n_2, points in x, y directions, respectively, and m points in time. The nonlinearity is mild and it is treated using a fixed point method as in [9]. By considering the standard lexicographic order of grid points from left to right, from bottom to top, the difference scheme can be expressed (see [8]) as

$$M_n \mathbf{u}^k = \mathbf{b}^{k-1}, \quad n = (n_1, n_2), \quad N(n) = n_1 n_2, \tag{1.3}$$

$$M_n = I_n - A_n + D_n \in \mathbb{R}^{N(n) \times N(n)}, \quad \mathbf{b}^{k-1} = \mathbf{u}^{k-1} + \Delta t \mathbf{f}^{k-1},$$

with $A_n = A_{x, n_1} \otimes I_{n_2} + I_{n_1} \otimes A_{y, n_2}$, where D_n represents the penalization matrix defined as $d_{i,j} = 0$ if $(x_i, y_j) \in \Omega$ and $d_{i,j} = \Delta t / \eta$ for $(x_i, y_j) \in \bar{\Omega} \setminus \Omega$, with $A_{x, n_1} = c_1 G_{n_1}^{(\alpha_1)}$, $A_{y, n_2} = c_2 G_{n_2}^{(\alpha_2)}$, and $G_m^{(\alpha)}$ being a real symmetric Toeplitz

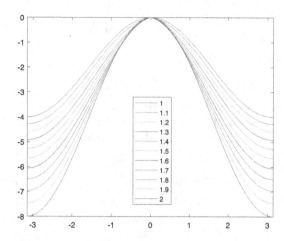

Fig. 1. Generating function of the Toeplitz matrix $G_m^{(\alpha)}$ on $[-\pi, \pi]$ for various α

matrix whose first column is $[2g_1^{(\alpha)}, g_0^{(\alpha)} + g_2^{(\alpha)}, g_3^{(\alpha)}, \ldots, g_{m-1}^{(\alpha)}, g_m^{(\alpha)}]^T$ (for the expression of the coefficients see [11]). We recall that a Toeplitz matrix is a matrix constant along the diagonal (see e.g. [6][Eq (6.1), p. 95]). Finally we recall the convergence features of the method, which are important also from numerical linear algebra viewpoint since, by Proposition 1, the optimal choice is to set $\Delta t = h_1 = h_2$.

Proposition 1. [8] *The difference scheme giving raise to* (1.3) *is uncondi-tionally stable, is convergent and satisfies* $\|e_n^k\|_\infty \leq c(\Delta t + h_1 + h_2)$, *with* $(e_n^k)_{i,j} = u_\eta(x_i, y_j, t_k) - u_{i,j}^k$, $1 \leq i \leq n_1$, $1 \leq j \leq n_2$ *and where c is a positive constant independent of the time step* Δt *and spatial steps* h_1 *and* h_2.

2   Spectral Analysis via Toeplitz and GLT Methods

By using the analysis of entries of the matrix $G_m^{(\alpha)}$ given in [11], in [10] the authors have proven that $G_m^{(\alpha)} = -T_m(f_\alpha)$, i.e., $G_m^{(\alpha)}$ is the m-th order Toeplitz matrix generated by the continuous function $-f_\alpha$, according to the concepts in [6][Sec 6.1], with f_α being a nonnegative function having a unique zero at $\theta = 0$ of order α (see Fig. 1 for a visualization of $-f_\alpha$).

As a consequence, $-A_n$ is positive definite and $M_n = I_n - A_n + D_n$ is also positive definite with $M_n = I_n - A_{x,n_1} \otimes I_{n_2} - I_{n_1} \otimes A_{y,n_2} + D_n$ i.e.

$$M_n = I_n + c_1 T_{n_1}(f_{\alpha_1}) \otimes I_{n_2} + c_2 I_{n_1} \otimes T_{n_2}(f_{\alpha_2}) + D_n.$$

The positive definite character of M_n was already observed in [8]: with our Toeplitz and GLT tools we know that the maximal eigenvalue of $G_m^{(\alpha)} = -T_m(f_\alpha)$ is uniformly bounded by the infinity norm of f_α, while the minimal eigenvalue converges to zero as m^α [6][Sec 6.6] and the latter information will be used in the design of the multigrid solvers [5]. Now we are in position to give the distributional analysis of a properly scaled matrix sequence $\{\gamma(n)M_n\}_n$ using the GLT axioms in the case of two-level structures [7]. We make reference to book [7][Chap 6] with $d = 2$, i.e. two-level structures. According to Proposition 1, as already mentioned, we choose $\Delta t = h_1 = h_2$ and without loss of generality we set $\overline{\Omega} = (0,1)^2$. Furthermore, we set $\eta = (\Delta t)^a$ with $a = \max\{\alpha_1, \alpha_2\} + 0.1$: the choice is done for giving strength to the penalization process and at the same time for making the spectral analysis easier. We observe that both $c_1 = \Delta t k_1 c_{\alpha_1}/h_1^{\alpha_1} > 0$, $c_2 = \Delta t k_2 c_{\alpha_2}/h_2^{\alpha_2} > 0$ explode to ∞ since $\alpha_1, \alpha_2 > 1$ and $\Delta t = h_1 = h_2$. Hence for studying a meaningful eigenvalue distribution, we normalize M_n by using $\gamma(n) = h_1^{a-0.1}/\Delta t = h_2^{a-0.1}/\Delta t = (\Delta t)^{\max\{\alpha_1, \alpha_2\}-1}$. The consequences are the following:

- $\gamma(n)I_n$ is infinitesimal in spectral norm and hence $\{\gamma(n)I_n\}_n$ is a GLT sequence with symbol 0 (third item of Axiom **GLT3** [7][p. 118]);
- $\{\gamma(n)D_n\}_n$ is by construction a GLT sequence with symbol 0 for $(x,y) \in \Omega$ and $+\infty$ for $(x,y) \notin \Omega$, as long as Ω is Peano-Jordan measurable so that its characteristic function is continuous almost everywhere (second item of Axiom **GLT3** [7][p. 118]);
- if $\max\{\alpha_1, \alpha_2\} = \alpha_1 > \alpha_2$ then $\gamma(n)c_1 T_{n_1}(f_{\alpha_1}) \otimes I_{n_2} = T_n(k_1 c_{\alpha_1} f_{\alpha_1}(\theta_1))$, while $\gamma(n)c_2 I_{n_1} \otimes T_{n_2}(f_{\alpha_2})$ is infinitesimal in spectral norm. As a consequence, $\{\gamma(n)c_2 I_{n_1} \otimes T_{n_2}(f_{\alpha_2})\}_n$ is a GLT sequence with symbol 0 (third item of Axiom **GLT3** [7][p. 118]) and $\{\gamma(n)c_1 T_{n_1}(f_{\alpha_1}) \otimes I_{n_2}\}_n$ is a GLT sequence with symbol $k_1 c_{\alpha_1} f_{\alpha_1}(\theta_1)$ (first item of Axiom **GLT3** [7][p. 118]);
- if $\max\{\alpha_1, \alpha_2\} = \alpha_2 > \alpha_1$ then $\gamma(n)c_1 T_{n_1}(f_{\alpha_1}) \otimes I_{n_2}$ is infinitesimal in spectral norm, while $\gamma(n)c_2 I_{n_1} \otimes T_{n_2}(f_{\alpha_2}) = T_n(k_2 c_{\alpha_2} f_{\alpha_2}(\theta_2))$. As a consequence, $\{\gamma(n)c_1 T_{n_1}(f_{\alpha_1}) \otimes I_{n_2}\}_n$ is a GLT sequence with symbol 0 (third item of Axiom **GLT3** [7][p. 118]) and $\{\gamma(n)c_2 I_{n_1} \otimes T_{n_2}(f_{\alpha_2})\}_n$ is a GLT sequence with symbol $k_2 c_{\alpha_2} f_{\alpha_2}(\theta_2)$ (first item of Axiom **GLT3** [7][p. 118]);
- if $\alpha_1 = \alpha_2 = \alpha$, then $\gamma(n)c_1 T_{n_1}(f_{\alpha_1}) \otimes I_{n_2} + \gamma(n)c_2 I_{n_1} \otimes T_{n_2}(f_{\alpha_2}) = T_n(k_1 c_\alpha f_\alpha(\theta_1) + k_2 c_\alpha f_\alpha(\theta_2)) = T_n(g_\alpha)$ and the related matrix-sequence is GLT with symbol $g_\alpha(\theta_1, \theta_2)$ (first item of Axiom **GLT3** [7][p. 118]).

As a consequence of the $*$-algebra structure of the GLT sequences (Axiom **GLT4** [7][p. 118]) and of Axiom **GLT1** [7][p. 118], we deduce that the spectral distribution is given by the function $s(\theta_1, \theta_2, x, y)$ such that

- $s(\theta_1, \theta_2, x, y) = +\infty$ if $(x,y) \notin \Omega$,
- $s(\theta_1, \theta_2, x, y) = k_1 c_{\alpha_1} f_{\alpha_1}(\theta_1)$ if $\max\{\alpha_1, \alpha_2\} = \alpha_1 > \alpha_2$ and $(x,y) \in \Omega$,
- $s(\theta_1, \theta_2, x, y) = k_2 c_{\alpha_2} f_{\alpha_2}(\theta_2)$ if $\max\{\alpha_1, \alpha_2\} = \alpha_2 > \alpha_1$ and $(x,y) \in \Omega$,
- $s(\theta_1, \theta_2, x, y) = k_1 c_\alpha f_\alpha(\theta_1) + k_2 c_\alpha f_\alpha(\theta_2)$ if $\alpha_1 = \alpha_2 = \alpha$ and $(x,y) \in \Omega$.

In this way the spectral symbol is nonnegative and has at $(\theta_1, \theta_2) = (0,0)$ a zero of order less than 2 (in fact equal to $\max\{\alpha_1, \alpha_2\}$) and therefore by the analysis in [5] we have enough information for choosing ad hoc smoothers and prolongation operators in our multigrid strategy, in order to achieve optimality and mesh independence convergence speed.

3 Multigrid Proposal and Numerical Experiments

Hereafter, we define a multigrid strategy for the solution of the linear system in (1.3) based on [5], taking into account that the symbol of the related matrix-sequence has a zero of order bounded by 2. The prolongation (restriction) matrix is set out as the product of the Toeplitz matrix generated by the polynomial $p(\vartheta_1, \vartheta_2) = 16(1 + \cos(\vartheta_1))(1 + \cos(\vartheta_2))$ and a suitable matrix (see [5] for terminology and notation) defined as $H = H_n = K_{n_1} \otimes K_{n_2}$, $n = (n_1, n_2)$, with

$$
K_m = \begin{bmatrix} 0 \ 1 \ 0 & & & \\ & 0 \ 1 \ 0 & & \\ & & \ddots \ \ddots \ \ddots & \\ & & & 0 \ 1 \ 0 \end{bmatrix} \in \mathbb{R}^{M \times (2M+1)}, \quad m = 2M + 1. \tag{3.1}
$$

Thus the stencil representing the two-level Toeplitz matrix is $S_p = \frac{1}{16}[1\ 2\ 1]^T[1\ 2\ 1]$. Such a choice of the prolongation/restriction matrices is motivated by the fact that the spectral symbol is nonnegative and with a zero in the Fourier variables of order $\max\{\alpha_1, \alpha_2\} \in (1, 2)$ so that the convergence theory in [5] can be employed. Both the presmoother and postsmoother methods are chosen as the standard pointwise Gauss-Seidel with ν_{pre} and ν_{post} set at 1 or 2 respectively. All the matrices at coarser levels $A_{s+1} = P_{m_s}^{m_s+1} A_s (P_{m_s}^{m_s+1})^H$ are computed one-time in the so called *Setup phase*. It is worth stressing that we can split A_{s+1} as $\tilde{A}_{s+1} + D_{s+1}$ in order to save the computational costs by making reference to the stencil S_p. Finally, the linear system is solved by a direct method when the dimension reaches $N(n) = 3^2$. As first example we consider the RSFDEs defined on the following elliptical domain $(x, y) \in \Omega = \{(x, y) | (x - 2)^2/4 + (y - 1)^2 \leq 1\}$, with the initial condition $u_0(x, y) = ((x - 2)^2/4 + (y - 1)/2 - 1)^2$, $(x, y) \in \Omega$, and a homogeneous Dirichlet boundary condition $u(x, y, t) = 0$, $(x, y) \in \partial\Omega$. The exact solution is assumed to be $u(x, y, t) = e^{-t}((x - 2)^2/4 + (y - 1)^2 - 1)^2$ so that the source term equals $f(u, x, y, t) = -k_1 c_{\alpha_1} \frac{e^{-t}}{16}(h(\alpha_1, x - 2 + dy, dy) + h(\alpha_1, 2 - x + dy, dy))$ $-k_2 c_{\alpha_2} e^{-t}(h(\alpha_2, y - 1 + dx, dx) + h(\alpha_2, 1 - y + dx, dx)) - u(x, y, t)$, where $h(\alpha, s, d) = \frac{24\,s^{(4-\alpha)}}{\Gamma(5-\alpha)} - \frac{24ds^{(3-\alpha)}}{\Gamma(4-\alpha)} + \frac{8d^2\,s^{(2-\alpha)}}{\Gamma(3-\alpha)}$, $dy = 2\sqrt{1 - (y - 1)^2}$, $dx = \sqrt{1 - (x - 2)^2/4}$. In this example, the domain Ω is extended to be the rectangular domain $\bar{\Omega} = (0, 4) \times (0, 2)$. Let $\alpha_1 = 1.4$, $\alpha_2 = 1.7$, $k_1 = k_2$, and $n_1 = n_2 = m - 1$. Table 1 shows the numerical results in the case $k_1 = k_2 = 1$, $\eta = 1.e - 5$ at $t = t_0 + \Delta t$. Our aim is to analize multigrid performances in the numerical solution of the involved linear system, not its efficiency in dependence of temporal steps which could be also influenced by the choice of a suitable initial guess. For such a reason, the results reported in Tables 1, 2 and 3 refer to an initial guess equal to the zero vector. The multigrid convergence is tested both with respect to the increasing dimension and to the diffusion coefficients k_1, k_2. It is worth stressing that the use of two iterations both for the presmoother and postsmoother Gauss-Seidel iteration is able to stabilize the number of required iterations for larger values of the coefficients k_1, k_2. A marginal improvement with respect to the case of just one iteration of Gauss-Seidel as presmoother and postsmoother can be obtained by considering SOR method as presmoother, with the convergence speed mildly depending on the relaxation parameter. Finally, in Table 4 we illustrate the multigrid convergence depending on the initial guess: the typical choice of an initial guess equal to the approximated solution at the previous time step is considered, i.e. $(\mathbf{u}^k)_0 = \mathbf{u}^{k-1}$. In such a case we observe that the average iteration number is practically constant, often decreasing when increasing time. Regarding complexity, we present the arithmetic cost per iteration in the V-cycle with one and two post/pre smoothing steps. In the first case the cost is $M_1(N) = \alpha N \log_2(N) + \beta N + O(1)$, while for two post/pre smoothing steps is $M_1(N) = 2\alpha N \log_2(N) + (\beta - \alpha)N + O(1)$. As a consequence the cost per iteration with two post/pre smoothing steps is asymptotically a bit less than twice the cost per iteration with one post/pre smoothing steps. However, with two post/pre smoothing steps, the number of iterations is bit more than one half with respect to the other case. As a conclusion there is no (substantial) gain in using two post/pre smoothing steps and the simplest strategy with one post/pre smoothing has to be preferred. The same considerations hold for the W-cycle. Regarding CPU timings, this is an issue to be considered in the near future, also in comparison with preconditioning strategies. But an effort is needed to program everything in FORTRAN or C++: in Matlab one Discrete Sine Transform of type I (DST-I) is implemented using two FFTs and this is far from optimal, since the theoretical cost of one DST-I is slightly greater than one half of the cost of one FFT (see [9][p. 18, lines 10–17]).

Table 1. Number of multigrid iterations for matrices of increasing dimension $N(n)$, $k_1 = k_2 = 1$, $\varepsilon = 1.e-8$ - Gauss-Seidel as presmoother and postsmoother, $\nu_{pre} = \nu_{post} = 1$ or 2, respectively.

$k_1 = k_2 = 1$							
$n_1 = n_2$	$N(n)$	$GS_{\nu_{pre}=1}$ & $GS_{\nu_{post}=1}$			$GS_{\nu_{pre}=2}$ & $GS_{\nu_{post}=2}$		
		Twogrid	Vcycle	Wcycle	Twogrid	Vcycle	Wcycle
$2^5 - 1$	961	11	11	11	6	6	6
$2^6 - 1$	3969	13	13	13	7	7	7
$2^7 - 1$	16129	14	15	14	8	8	8
$2^8 - 1$	65025	16	17	16	9	10	9

Table 2. Number of multigrid iterations for matrices of increasing dimension $N(n)$, $k_1 = k_2 = 100$, $\varepsilon = 1.e-8$ - Gauss-Seidel as presmother and postsmother, $\nu_{pre} = \nu_{post} = 1$ or 2, respectively.

$k_1 = k_2 = 100$							
$n_1 = n_2$	$N(n)$	$GS_{\nu_{pre}=1}$ & $GS_{\nu_{post}=1}$			$GS_{\nu_{pre}=2}$ & $GS_{\nu_{post}=2}$		
		Twogrid	Vcycle	Wcycle	Twogrid	Vcycle	Wcycle
$2^5 - 1$	961	18	19	17	10	11	10
$2^6 - 1$	3969	19	22	20	11	12	11
$2^7 - 1$	16129	21	26	23	11	13	11
$2^8 - 1$	65025	24	29	26	11	15	13

Table 3. Number of multigrid iterations for matrices of increasing dimension $N(n)$, $k_1 = k_2 = 0.1$, $\varepsilon = 1.e-8$ - Gauss-Seidel as presmother and postsmother, $\nu_{pre} = \nu_{post} = 1$ or 2, respectively.

$k_1 = k_2 = 0.1$							
$n_1 = n_2$	$N(n)$	$GS_{\nu_{pre}=1}$ & $GS_{\nu_{post}=1}$			$GS_{\nu_{pre}=2}$ & $GS_{\nu_{post}=2}$		
		Twogrid	Vcycle	Wcycle	Twogrid	Vcycle	Wcycle
$2^5 - 1$	961	5	5	5	3	3	3
$2^6 - 1$	3969	5	5	5	3	3	3
$2^7 - 1$	16129	6	6	6	3	3	3
$2^8 - 1$	65025	7	7	7	4	4	4

Table 4. Number and average number of multigrid iterations for matrices of increasing dimension $N(n)$, $k_1 = k_2 = 1$, $\varepsilon = 1.e-8$ - Gauss-Seidel as presmother and postsmother, $\nu_{pre} = \nu_{post} = 1$ or 2, respectively - initial guess equals the approximated solution at the previous time step. The number in the first column is the number of iterations required at the first time step, while the number in the second column is the integer rounding of the average number of iterations on all time steps to reach $T = 1$.

$k_1 = k_2 = 1$														
$n_1 = n_2$	$N(n)$	$GS_{\nu_{pre}=1}$ & $GS_{\nu_{post}=1}$						$GS_{\nu_{pre}=2}$ & $GS_{\nu_{post}=2}$						
		Twogrid		Vcycle		Wcycle		Twogrid		Vcycle		Wcycle		
$2^5 - 1$	961	9	8	9	8	9	8	5	5	5	5	5	5
$2^6 - 1$	3969	10	9	11	9	10	9	6	5	6	5	6	5
$2^7 - 1$	16129	11	8	11	9	11	8	6	5	6	5	6	5
$2^8 - 1$	65025	13	8	13	8	13	8	7	5	7	5	7	5

Conclusions. In the present work, we introduced a $O(N \log N)$ GLT multigrid algorithm for problem (1.2). As a first conclusion, we found no (substantial) gain in using more than one post/pre smoothing steps and hence the simplest strategy with one post/pre smoothing has to be preferred. The same considerations hold for the W-cycle. Regarding CPU timings, also in comparison with preconditioning strategies [9], an effort is needed to program in FORTRAN or C++. For future works we plan to explore the connections of the GLT analysis with the preconditioned Krylov methods already presented in the relevant literature [4,8,9] and the combination of the latter with multigrid techniques in the spirit of multi-iterative solvers (see [3] and references therein). In particular, we will explore the relationships among the multigrid presented in this note, the Krylov convergence theory in [1,2], and the distribution results in [7,13,14]. A further direction to investigate is the case where the measure of the domain of interest is much smaller of that of the extended one, as it may happen e.g. for star-shaped domains. In that case the GLT multigrid approach and a re-discretization strategy seem more flexible than the matrix algebra preconditioning and a careful implementation is needed for assessing which method is in practice more efficient.

Acknowledgments. The authors are supported by the Italian INdAM-GNCS agency. The work of S. Serra-Capizzano was funded from the European High-Performance Computing Joint Undertaking (JU) under grant No 955701, under INdAM — GNCS Project, CUP_E53C22001930001, and under Grant 2023 from Theory, Economics and Systems Lab, Athens U.E.B.

References

1. Axelsson, O., Lindskog, G.: On the rate of convergence of the preconditioned conjugate gradient method. Numer. Math. **48**(5), 499–523 (1986)
2. Beckermann, B., Kuijlaars, A.B.J.: Sperlinear convergence of conjugate gradients. SIAM J. Numer. Anal. **39**(1), 300–329 (2001)
3. Donatelli, M., Garoni, C., Manni, C., Serra-Capizzano, S., Speleers, H.: Robust and optimal multi-iterative techniques for IgA Galerkin linear systems. Comput. Methods Appl. Mech. Eng. **284**, 230–264 (2015)
4. Du, N., Sun, H., Wang, H.: A preconditioned fast finite difference scheme for space-fractional diffusion equations in convex domains. Comput. Appl. Math. **38**, 1–13 (2019)
5. Fiorentino, G., Serra, S.: Multigrid methods for symmetric positive definite block Toeplitz matrices with nonnegative generating functions. SIAM J. Sci. Comput. **17**(5), 1068–1081 (1996)
6. Garoni, C., Serra-Capizzano, S.: Generalized Locally Toeplitz Sequences: Theory and Applications, vol. I. Springer, Cham (2017). https://doi.org/10.1007/978-3-319-53679-8

7. Garoni, C., Serra-Capizzano, S.: Generalized Locally Toeplitz Sequences: Theory and Applications, vol. II. Springer, Cham (2018). https://doi.org/10.1007/978-3-030-02233-4

8. Huang, X., Sun, H.: A preconditioner based on sine transform for two-dimensional semi-linear Riesz space fractional diffusion equations in convex domains. Appl. Num. Math. **169**, 289–302 (2021)

9. Mazza, M., Serra-Capizzano, S., Sormani, R.L.: Algebra preconditionings for $2D$ Riesz distributed-order space-fractional diffusion equations on convex domains. Numer. Linear Algebra Appl. - Special Volume in Memory of Prof. Owe Axelsson, paper e2536 (2024). https://doi.org/10.1002/nla.2536. Accessed 23 Oct 2023

10. Mazza, M., Serra-Capizzano, S., Usman, M.: Symbol-based preconditioning for Riesz distributed-order space-fractional diffusion equations. Electron. Trans. Numer. Anal. **54**, 499–513 (2021)

11. Meerschaert, M., Tadjeran, C.: Finite difference approximations for fractional advection-dispersion flows equations. J. Comput. Appl. Math. **172**, 65–77 (2004)

12. Podlubny, I.: Fractional Differential Equations. Academic Press, San Diego, CA (1999)

13. Serra Capizzano, S.: Generalized locally Toeplitz sequences: spectral analysis and applications to discretized partial differential equations. Linear Algebra Appl. **366**, 371–402 (2003)

14. Serra-Capizzano, S.: The GLT class as a generalized Fourier analysis and applications. Linear Algebra Appl. **419**(1), 180–233 (2006)

Continuation Newton Methods with Applications to Plasticity

Stanislav Sysala$^{(\boxtimes)}$ (iD)

Institute of Geonics of the Czech Academy of Sciences, Studenska 1768,
70800 Ostrava, Czech Republic
stanislav.sysala@ugn.cas.cz

Abstract. This contribution is focused on severely nonlinear systems of equations with nonsmooth operators. A continuation Newton method with a smoothing operator is suggested and its convergence analyzed. Then the method is applied to elasto-plasticity with hardening. Finally, another continuation method convenient for an elastic-perfectly plastic problem is introduced and used for finding the so-called limit load.
Dedication: This paper is dedicated to the memory of Professor Owe Axelsson. The presented results arise from joint work with Owe.

Keywords: Continuation Newton method · Smoothing functions · Elasto-plasticity

1 Introduction

In this paper, we consider a nonlinear algebraic equation

$$\text{find } u \in \mathbb{R}^n : \qquad F(u) = b, \tag{1}$$

with severely nonlinear and nonsmooth function $F \colon \mathbb{R}^n \to \mathbb{R}^n$. For the sake of brevity, we assume that $F(0) = 0$. In such a case, it is convenient to parametrize this problem by the following way:

$$F(\hat{u}(t)) = tb, \qquad 0 \le t \le 1. \tag{2}$$

Clearly, the solution of (1) satisfies $u = \hat{u}(1)$ and can be achieved by continuation over the parameter t (the so-called increasing load method). In mechanics, $b \in \mathbb{R}^n$ usually represents the load vector and one can expect that each problem for $t < 1$ is solvable with much less computational cost than for $t = 1$.

In this paper, we follow on [1] where a continuation Newton method with a smoothing operator was suggested for solving the systems (1) or (2) and its convergence was analyzed. We revise selected assumptions from [1] to be more

Supported by European Union's Horizon 2020 research and innovation programme under grant agreement number 847593 and by The Czech Radioactive Waste Repository Authority (SÚRAO) under grant agreement number SO2020-017.

transparent and simplify the proof of the convergence result in comparison to [1], see Sect. 2. In Sect. 3, we show that these assumptions are satisfied for an elastic-plastic problem with hardening. In particular, the results from [1] are summarized and slightly corrected. Section 4 is devoted to an elastic-perfectly plastic problem which is related to the previous problem such that the hardening parameter tends to zero. For such a problem, the continuation defined by (2) is not convenient and we introduce another one, which was originally introduced in [3,9].

2   Continuation Newton Method and Its Convergence

We consider a sequence $\{K_\epsilon\}_\epsilon$, $\epsilon \in (0, \epsilon_0)$, of smooth functions $K_\epsilon \colon \mathbb{R}^n \to \mathbb{R}^n$ satisfying $K_\epsilon \to F$ as $\epsilon \to 0$. In addition, we assume that the following properties of F and K_ϵ hold.

1. There exists a constant $M > 0$ such that

$$\|K_\epsilon(u) - F(u) - (K_\epsilon(v) - F(v))\| \le M\epsilon \|u - v\| \quad \forall u, v \in \mathbb{R}^n,\ \forall \epsilon \in (0, \epsilon_0). \quad (3)$$

2. There exists a constant $L > 0$ such that

$$\|K_\epsilon(v) - K_\epsilon(u) - K_\epsilon'(v)(v - u)\| \le \frac{L}{2\epsilon}\|u - v\|^2 \quad \forall u, v \in \mathbb{R}^n,\ \forall \epsilon \in (0, \epsilon_0). \quad (4)$$

3. There exists a constant $q > 0$ such that

$$[F(u) - F(v)]^T(u - v) \ge q\|u - v\|^2 \qquad \forall u, v \in \mathbb{R}^n, \quad (5)$$

$$[K_\epsilon(u) - K_\epsilon(v)]^T(u - v) \ge q\|u - v\|^2 \qquad \forall u, v \in \mathbb{R}^n,\ \forall \epsilon \in (0, \epsilon_0). \quad (6)$$

We complete these assumptions by the following remarks.

- The inequality (3) implies that F is locally Lipschitz continuous. One can easily generalize (3) such that M depends on ϵ and $M(\epsilon) \to 0$ as $\epsilon \to 0$.
- If K_ϵ' is Lipschitz continuous with respect to the constant L/ϵ then (4) holds. The Lipschitz constant depends on ϵ due to the fact that F is nonsmooth.
- Assume that for any $v \in \mathbb{R}^n$ there exists the limit of $K_\epsilon'(v)$ and define

$$F^o \colon \mathbb{R}^n \to \mathbb{R}^{n \times n}, \quad F^o(v) = \lim_{\epsilon \to 0} K_\epsilon'(v). \quad (7)$$

Since F is locally lipschitz, it is almost everywhere differentiable and thus F^o represents a generalized derivative of F. In many applications (including the elastoplastic problems introduced below), the following bound holds for any $u \in \mathbb{R}^n$ and any $v \in \mathbb{R}^n$ belonging to a vicinity of u:

$$\|F(v) - F(u) - F^o(v)(v - u)\| = o(\|u - v\|) \quad \forall u \in \mathbb{R}^n. \quad (8)$$

This bound implies that the function F is semismooth at $u \in \mathbb{R}^n$. For more details to smoothing and semismooth functions, we refer to e.g. [6,7].

– The assumptions (5) and (6) mean that the functions F and K_ϵ are uniformly strongly monotone. In addition, (5) and (6) imply, respectively:

$$\|\hat{u}(t) - \hat{u}(\bar{t})\| \leq \frac{1}{q}\|F(\hat{u}(t)) - F(\hat{u}(\bar{t}))\| \leq \frac{\|b\|}{q}|t - \bar{t}| \quad \forall t, \bar{t} \in [0,1], \quad (9)$$

$$\|(K_\epsilon'(v))^{-1}\| \leq \frac{1}{q} \quad \forall v \in \mathbb{R}^n. \quad (10)$$

– For purposes of the convergence result below, it suffices to assume that (3)–(6) hold only along the solution path $\{\hat{u}(t)\}$, $t \in [0,1]$, similarly as in [1].

Next, consider the following partition of the interval $[0,1]$:

$$0 = t_0 < t_1 < \ldots < t_k < \ldots < t_N = 1, \quad \tau_k := t_{k+1} - t_k, \quad (11)$$

and the following sequence of Newton approximation steps:

$$K_\epsilon'(u^k)(u^{k+1} - u^k) = t_{k+1}b - F(u^k), \quad k = 0, 1, \ldots N-1, \quad u^0 = 0, \quad (12)$$

for a fixed value of ϵ. It can be interpreted as a one-step smoothing Newton method since just a single smoothing Newton step is used for each load case.

The algorithm (12) defines the sequence $\{u^k\}_{k=0}^N$. We wish to be u^k close to $\hat{u}(t_k)$ for any $k = 0, 1, \ldots N$. This can be achieved if the increments τ_k and the parameter ϵ are sufficiently small.

Theorem 1. *Let the assumptions (3)–(6) hold with respect to the positive constants M, L, and q. Let*

$$\epsilon \leq \frac{q}{4M}, \quad \tau_k \leq \frac{q^2\epsilon}{4L\|b\|} \quad \forall k = 0, 1, \ldots, N-1. \quad (13)$$

Then the following estimate holds:

$$\|\hat{u}(t_k) - u^k\| \leq \frac{\|b\|}{q} \max_{0 \leq l \leq k-1} \tau_l, \quad k = 0, 1 \ldots. \quad (14)$$

Proof. We follow the idea of the proof of Proposition 4.1 in [1]. First, we derive the following equality:

$$\begin{aligned}
u^{k+1} - \hat{u}(t_{k+1}) &= u^{k+1} - u^k - (\hat{u}(t_{k+1}) - u^k) \\
&\overset{(12)}{=} (K_\epsilon'(u^k))^{-1}\left[F(\hat{u}(t_{k+1})) - F(u^k) - K_\epsilon'(u^k)(\hat{u}(t_{k+1}) - u^k)\right] \\
&= (K_\epsilon'(u^k))^{-1}\left[F(\hat{u}(t_{k+1})) - K_\epsilon(\hat{u}(t_{k+1})) - (F(u^k) - K_\epsilon(u^k))\right] \\
&\quad -(K_\epsilon'(u^k))^{-1}\left[K_\epsilon(u^k) - K_\epsilon(\hat{u}(t_{k+1})) - K_\epsilon'(u^k)(u^k - \hat{u}(t_{k+1}))\right].
\end{aligned}$$

Hence and from (3), (4), (10), we have

$$\|\hat{u}(t_{k+1}) - u^{k+1}\| \leq \frac{1}{q}\left[M\epsilon\|\hat{u}(t_{k+1}) - u^k\| + \frac{L}{2\epsilon}\|\hat{u}(t_{k+1}) - u^k\|^2\right]. \quad (15)$$

Next, the following estimates hold:

$$\|\hat{u}(t_{k+1}) - u^k\| \le \|\hat{u}(t_{k+1}) - \hat{u}(t_k)\| + \|\hat{u}(t_k) - u^k\|,$$

$$\|\hat{u}(t_{k+1}) - u^k\|^2 \le 2\|\hat{u}(t_{k+1}) - \hat{u}(t_k)\|^2 + 2\|\hat{u}(t_k) - u^k\|^2,$$

$$\|\hat{u}(t_{k+1}) - \hat{u}(t_k)\| \overset{(19)}{\le} \frac{\|b\|}{q}\tau_k.$$

Inserting these inequalities into (15) and using the assumption (13), we derive

$$\|\hat{u}(t_{k+1}) - u^{k+1}\| \le \frac{\|b\|}{2q}\tau_k + \left(\frac{1}{4} + \frac{L}{q\epsilon}\|\hat{u}(t_k) - u^k\|\right)\|\hat{u}(t_k) - u^k\|. \qquad (16)$$

Finally, it is easy to prove the estimate (14) by mathematical induction with the usage of (16) and (13).   □

Remark 1. If (7) and (8) hold then it offers to consider the algorithm

$$F^o(u^k)(u^{k+1} - u^k) = t_{k+1}b - F(u^k), \quad k = 0, 1, \dots N - 1, \quad u^0 = 0, \qquad (17)$$

as a limit case of (12). However, Theorem 1 cannot be straightforwardly extended due to the fact that the assumption (4) depends on ϵ.

Remark 2. There exists adaptive strategies enabling to determine the load increments τ_k, see e.g. [2]. They could be combined with the algorithm (12).

3   Elastic-Plastic Problem with Hardening

To illustrate that the assumptions (3)–(6) are meaningful, we shall consider a static problem based on the deformation theory of plasticity. The problem can arise from penalization of the Hencky perfectly plastic problem, where the penalty parameter represents a linear isotropic hardening. A similar problem also appears in the incremental theory of elastoplasticity with hardening after time discretization. We refer to [1] and [8] for more details.

We introduce the space S of all symmetric, 3×3 matrices for representation of stress and strain variables in a three dimensional domain. For elements of S, we define the Frobenius scalar product $\tau : e = \tau_{ij}e_{ij}$, $\tau, e \in S$, and the corresponding norm $\|\tau\|_F = \sqrt{\tau : \tau}$. Also, we use other norms on S suitable for the stress (τ) and strain (e) variables, respectively:

$$\|\tau\|^2_{\mathbb{C}^{-1}} := \mathbb{C}^{-1}\tau : \tau, \quad \|e\|^2_{\mathbb{C}} = \mathbb{C}e : e, \quad \tau, e \in S,$$

where

$$\mathbb{C}e = \frac{1}{3}(3\lambda + 2\mu)(tr\, e)I + 2\mu e^D \quad \forall e \in S,$$

is the elastic fourth-order tensor, $\lambda, \mu > 0$ denote the Lamé coefficients, I is the identity matrix in S, $tr\, e = I : e$ and $e^D := e - \frac{1}{3}(tr\, e)I$, $e \in S$ denote the trace and the deviatoric part of e.

Using this notation, one can define the nonlinear stress-strain function

$$T(e) := \frac{1}{3}(3\lambda + 2\mu)(tr\, e)I + (1-\alpha)2\mu e^D + \alpha j(2\mu \|e^D\|_F)\hat{n}(e^D), \quad e \in S,$$

where $\alpha \in (0,1)$ is the isotropic hardening parameter, $\hat{n}(e) := \frac{e}{\|e\|_F}$ for $e \neq 0$,

$$j : \mathbb{R} \to \mathbb{R}, \quad j(z) := \begin{cases} z, & z \leq \gamma \\ \gamma, & z \geq \gamma \end{cases}$$

and $\gamma > 0$ is the material parameter representing the initial yield stress. Notice that $T = \mathbb{C}$ in the elastic region, i.e.,

$$T(e) = \mathbb{C}e = \frac{1}{3}(3\lambda + 2\mu)(tr\, e)I + 2\mu e^D \quad \forall e \in S, \; 2\mu\|e^D\|_F \leq \gamma.$$

It is important to note that the function T is nondifferentiable at $e \in S$, $2\mu\|e^D\|_F = \gamma$. Next, we define the following smooth approximations of j and T:

$$j_\epsilon : \mathbb{R} \to \mathbb{R}, \quad j_\epsilon(z) := \begin{cases} z, & z \leq \gamma - \epsilon \\ \gamma - \frac{1}{4\epsilon}(z - \gamma - \epsilon)^2, & z \in [\gamma - \epsilon, \gamma + \epsilon] \; , \\ \gamma, & z \geq \gamma + \epsilon \end{cases}$$

$$T_\epsilon(e) := \frac{1}{3}(3\lambda + 2\mu)(tr\, e)I + (1-\alpha)2\mu e^D + \alpha j_\epsilon(2\mu\|e^D\|_F)\hat{n}(e^D), \quad e \in S,$$

respectively, where $\epsilon \in (0, \epsilon_0)$, $\epsilon_0 < \gamma$. The following properties of T and T_ϵ were derived in [1]:

Lemma 1. *For any $e, \eta \in S$,*

$$(T(e+\eta) - T(e)) : \eta \geq (1-\alpha)\|\eta\|_{\mathbb{C}}^2 \quad \forall e, \eta \in S,$$

$$(1-\alpha)\|\eta\|_{\mathbb{C}}^2 \leq T_\epsilon'(e)\eta : \eta \leq \|\eta\|_{\mathbb{C}}^2 \quad \forall \eta \in S, \; \forall \epsilon \in (0, \epsilon_0).$$

Lemma 2. *There exist positive constants $\bar{M}, \bar{L}$ independent of $\epsilon \in (0, \epsilon_0)$ and $\alpha \in (0,1)$ such that*

$$\|T(e) - T(\eta) - T_\epsilon(e) + T_\epsilon(\eta)\|_{\mathbb{C}^{-1}} \leq \bar{M}\alpha\epsilon\|e - \eta\|_{\mathbb{C}} \quad \forall e, \eta \in S, \; \forall \epsilon \in (0, \epsilon_0),$$

$$\|T_\epsilon(\eta) - T_\epsilon(e) - T_\epsilon'(\eta)(\eta - e)\|_{\mathbb{C}^{-1}} \leq \frac{\bar{L}\alpha}{2\epsilon}\|e - \eta\|_{\mathbb{C}}^2 \quad \forall e, \eta \in S, \; \forall \epsilon \in (0, \epsilon_0).$$

To be in accordance with the assumptions (3)–(6), one can set $q = 1 - \alpha$, $M = \bar{M}\alpha$ and $L = \bar{L}\alpha$. It is worth-noticing that the function T is strongly semismooth in S (see [8]), however the second uniform estimate from Lemma 2 cannot be extended for $\epsilon \to 0$.

The elastic-plastic problem with hardening leads to the following nonlinear variational equation:

$$\text{find } u \in V : \quad \int_{\Omega} T\left(e(u)\right) : e(v) dx = b(v) \quad \forall v \in V. \tag{18}$$

Here, $\Omega \subseteq \mathbb{R}^3$ is a bounded domain with Lipschitz boundary $\partial\Omega$. It is assumed that $\partial\Omega = \overline{\Gamma}_D \cup \overline{\Gamma}_N$, where Γ_D and Γ_N are open and disjoint sets, Γ_D has a positive surface measure. Surface tractions of density $g \in L^2(\Gamma_N; \mathbb{R}^3)$ are applied on Γ_N and the body is subject to a volume force $f \in L^2(\Omega; \mathbb{R}^3)$. Further,

$$V := \left\{ v \in H^1(\Omega; \mathbb{R}^3) \mid v = 0 \text{ on } \Gamma_D \right\}$$

denote the space of admissible displacement fields,

$$b(v) := \int_{\Omega} f \cdot v \, dx + \int_{\Gamma_N} g \cdot v \, ds, \quad v \in V$$

is the load functional and $e(v) = \frac{1}{2}\left(\nabla v + (\nabla v)^T\right)$ is the linearized strain tensor.

The problem (18) has a unique solution which can be found using the standard finite elements. Let V_h be the corresponding finite element subspace of V, $u, v, w, b \in \mathbb{R}^n$ denote the vector representations of functions $u_h, v_h, w_h \in V_h$, and the approximated load b_h, respectively. Then one can define the following algebraic notation:

$$F(v)^T w := \int_{\Omega} T(e(v_h)) : e(w_h) \, dx \quad \forall v_h, w_h \in V_h,$$

$$K_\epsilon(v)^T w := \int_{\Omega} T_\epsilon(e(v_h)) : e(w_h) \, dx \quad \forall v_h, w_h \in V_h,$$

The algebraic formulation of the problem (18) leads to solving the system (1) of nonlinear equations. Using Lemmas (1)–(2), we obtain the following estimates:

$$q\|v\|_e^2 \leq v^T K_\epsilon'(w) v \leq \|v\|_e^2 \quad \forall v, w \in \mathbb{R}^n, \ \forall \epsilon \in (0, \epsilon_0), \tag{19}$$

$$v^T (F(w + v) - F(w)) \geq q\|v\|_e^2 \quad \forall v, w \in \mathbb{R}^n, \tag{20}$$

$$\|F(v) - F(w) - K_\epsilon(v) + K_\epsilon(w)\|_* \leq M\epsilon\|v - w\|_e \quad \forall v, w \in \mathbb{R}^n, \ \forall \epsilon \in (0, \epsilon_0), \tag{21}$$

$$\|K_\epsilon(v) - K_\epsilon(u) - K_\epsilon'(v)(v - u)\|_* \leq \frac{L}{2\epsilon}\|v - w\|_{e, L^4}^2 \quad \forall u, w \in S, \ \forall \epsilon \in (0, \epsilon_0), \tag{22}$$

where

$$\|v\|_e := \left[\int_{\Omega} \|e(v_h)\|_{\mathbb{C}}^2 \, dx\right]^{1/2}, \quad \|v\|_{e, L^4} := \left[\int_{\Omega} \|e(v_h)\|_{\mathbb{C}}^4 \, dx\right]^{1/4},$$
$$\|v\|_* := \sup_{w \in \mathbb{R}^n, \|w\|_e = 1} |v^T w|.$$

Hence, the assumptions (3)–(6) hold. It is useful to note that the estimates (19)–(22) are independent of the discretization parameter h. This fact could enable to extend Theorem 1 to infinite-dimensional spaces.

A numerical example illustrating the convergence results (14) of the algorithm (12) for the elastic-plastic problem can be found in [1].

4 Continuation Method for the Limit Load in Perfect Plasticity

In this section, we summarize selected results from [3–5,9] for the special case $\alpha = 1$ in the problem above. Then we arrive to the Hencky elastic-perfectly plastic problem. In this case, the functions T and F are not strongly monotone and the problem (1) is not solvable for any $t \geq 0$. In particular, there exists the so-called limit load parameter t_{lim} such that for $t > t_{lim}$, the problem does not have any solution. The value t_{lim} is important in engineering practice because it defines a safety parameter of a structure. Clearly, t_{lim} should be larger than the value $t = 1$ representing the prescribed load.

In [3,9], it was shown that $b^T \hat{u}(t) \rightarrow +\infty$ as $t \rightarrow t_{lim}$. This important property enables us to introduce the following (more convenient) parametrization of the problem with respect to a given value of $\omega \geq 0$:

$$\text{find } \bar{u}(\omega) \in \mathbb{R}^n,\ t(\omega) \geq 0 : \qquad F(\bar{u}(\omega)) = t(\omega)b,\ b^T\bar{u}(\omega) = \omega. \qquad (23)$$

We have added one equation to the original system (2) and t is now an unknown scalar value. This system is solvable for any $\omega \geq 0$, $\bar{u}(\omega) = \hat{u}(t(\omega))$, and $t(\omega) \rightarrow t_{lim}$ as $\omega \rightarrow +\infty$. On the other hand, the sequence $\{\bar{u}(\omega)\}_\omega$ is unbounded as $\omega \rightarrow +\infty$. To suppress this drawback, the following transformation of the problem is considered:

$$\text{find } v(\omega) \in \mathbb{R}^n,\ t(\omega) \geq 0 : \qquad F(\omega v(\omega)) = t(\omega)b,\ b^T v(\omega) = 1. \qquad (24)$$

It holds that the sequence $\{v(\omega)\}_\omega$ is bounded and $v(\omega) = \bar{u}(\omega)/\omega$. The system (24) is related to the following minimization problem:

$$J_\omega(v(\omega)) = \min_{\substack{v \in \mathbb{R}^n \\ b^T v = 1}} J_\omega(v), \qquad J'_\omega(v) = F(\omega v).$$

Defining

$$J_\infty(v) = \lim_{\omega \to +\infty} J_\omega(v) \quad \forall v \in \mathbb{R}^n,$$

we arrive at the so-called *limit analysis problem*:

$$t_{lim} = \inf_{\substack{v \in \mathbb{R}^n \\ b^T v = 1}} J_\infty(v).$$

The infimum is used rather than the minimum, since the functional J_∞ is convex, 1-positively homogeneous but not-finite valued everywhere. It means that the problem contains hidden constraints. They represent divergence free velocity fields. Let us note that displacement fields are unbounded within limit analysis.

The limit analysis problem is also meaningful on infinite dimensional spaces. The minimizer can be found on spaces with bounded deformation, the so-called BD spaces, allowing certain discontinuities in velocity fields. Such discontinuities describe failure mechanism causing the so-called plastic collapse.

The determination of t_{lim} and related failure zones is important in geotechnics where stability of slopes, foundations or tunnels are investigated using the limit load analysis or similar techniques.

References

1. Axelsson, O., Sysala, S.: Continuation Newton methods. Comput. Math. Appl. **70**(11), 2621–2637 (2015)
2. Blaheta, R., Byczanski, P.: Performance analysis of Newton type and adaptive continuation solvers for nonlinear problems in geomechanics. Report IGAS DAM 2001/3, Institute of Geonics AS CR, Ostrava, Czech Republic (2001)
3. Cermak, M., Haslinger, J., Kozubek, T., Sysala, S.: Discretization and numerical realization of contact problems for elastic-perfectly plastic bodies. PART II - numerical realization. ZAMM **95**(12), 1348–1371 (2015)
4. Haslinger, J., Repin, S., Sysala, S.: A reliable incremental method of computing the limit load in deformation plasticity based on compliance: continuous and discrete setting. J. Comput. Appl. Math. **303**, 156–170 (2016)
5. Haslinger, J., Sysala, S., Repin, S.: Inf-sup conditions on convex cones and applications to limit load analysis. Math. Mech. Solids **24**(10), 3331–3353 (2019)
6. Qi, L., Sun, J.: A nonsmooth version of Newton's method. Math. Program. **58**, 353–367 (1993)
7. Qi, L., Sun, D.: Smoothing functions and smoothing Newton method for complementarity and variational inequality problems. J. Optim. Theory Appl. **113**(1), 121–147 (2002)
8. Sysala, S.: Application of a modified semismooth Newton method to some elastoplastic problems. Math. Comp. Sim. **82**, 2004–2021 (2012)
9. Sysala, S., Haslinger, J., Hlaváek, I., Cermak, M.: Discretization and numerical realization of contact problems for elasticperfectly plastic bodies. PART I-discretization, limit analysis. ZAMM **95**(4), 333–353 (2015)

Fractures and Mixed Dimensional Modeling: Discretizations, Solvers, and Methodology

Mixed and Nitsche's Discretizations of Frictional Contact-Mechanics in Fractured Porous Media

L. Beaude[1], F. Chouly[2,3,4], M. Laaziri[5], and R. Masson[5(✉)]

[1] BRGM, Orléans, France
laurence.beaude@brgm.fr
[2] Université de Bourgogne, Institut de Mathématiques de Bourgogne,
21078 Dijon, France
franz.chouly@u-bourgogne.fr
[3] Center for Mathematical Modeling and Department of Mathematical Engineering,
University of Chile and IRL 2807 – CNRS, Santiago, Chile
[4] Departamento de Ingeniería Matemática, CI2MA, Universidad de Concepción,
Casilla 160-C, Concepción, Chile
[5] Université Côte d'Azur, Inria, CNRS, LJAD, UMR 7351 CNRS, Team Coffee,
Nice Cedex 02, 06108 Parc Valrose, France
{mohamed.laaziri,roland.masson}@univ-cotedazur.fr

Abstract. This work deals with the discretization of single-phase Darcy flows in fractured and deformable porous media, including frictional contact at the matrix-fracture interfaces. Fractures are described as a network of planar surfaces leading to so-called mixed-dimensional models. Small displacements and a linear poro-elastic behavior are considered in the matrix. One key difficulty to simulate such coupled poro-mechanical models is related to the formulation and discretization of the contact mechanical sub-problem. Our starting point is based on the mixed formulation using facewise constant Lagrange multipliers along the fractures representing normal and tangential stresses. This is a natural choice for the discretization of the contact dual cone in order to account for complex fracture networks with corners and intersections. It leads to local expressions of the contact conditions and to efficient semi-smooth nonlinear solvers. On the other hand, such a mixed formulation requires to satisfy a compatibility condition between the discrete spaces restricting the choice of the displacement space and potentially leading to sub-optimal accuracy. This motivates the investigation of two alternative formulations based either on a stabilized mixed formulation or on the Nitsche's method. These three types of formulations are first investigated theoretically in order to reveal the connections between them. Then, they are compared numerically in terms of accuracy and nonlinear convergence on a coupled poromechanical 2D model.

Keywords: Contact mechanics · Coulomb friction · Stabilized mixed method · Nitsche's method · Poromechanics · Discrete Fracture Matrix model

© The Author(s), under exclusive license to Springer Nature Switzerland AG 2024
I. Lirkov and S. Margenov (Eds.): LSSC 2023, LNCS 13952, pp. 71–80, 2024.
https://doi.org/10.1007/978-3-031-56208-2_6

1   Mixed-Dimensional Contact Mechanics

In what follows, scalar fields are represented by lightface letters, vector fields by boldface letters. We use the overline notation $\overline{v}$ to distinguish an exact (scalar or vector) field from its discrete counterpart v. We consider a network Γ of planar fractures Γ_i, $i \in I$ immersed in the surrounding matrix domain $\Omega \backslash \overline{\Gamma}$ with $\Omega \subset \mathbb{R}^d$, $d \in \{2, 3\}$ a bounded polytopal domain (see Fig. 1).

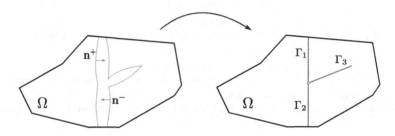

Fig. 1. Illustration of the dimension reduction in the fracture aperture for a 2D domain Ω with three intersecting fractures Γ_i, $i \in \{1, 2, 3\}$, with the equi-dimensional geometry on the left and the mixed-dimensional geometry on the right.

The two sides of a given fracture of Γ are denoted by $\pm$ in the matrix domain, with unit normal vectors $\mathbf{n}^\pm$ oriented outward from the sides $\pm$. We denote by $\gamma_\mathfrak{a}$ the trace operators on the side $\mathfrak{a} \in \{+, -\}$ of Γ for functions in $H^1(\Omega \backslash \overline{\Gamma})$. The jump operator on Γ for functions $\overline{\mathbf{u}}$ in $(H^1(\Omega \backslash \overline{\Gamma}))^d$ is defined by $[\![\overline{\mathbf{u}}]\!] = \gamma_+ \overline{\mathbf{u}} - \gamma_- \overline{\mathbf{u}}$, and we denote by

$$[\![\overline{\mathbf{u}}]\!]_n = [\![\overline{\mathbf{u}}]\!] \cdot \mathbf{n}^+ \quad \text{and} \quad [\![\overline{\mathbf{u}}]\!]_\tau = [\![\overline{\mathbf{u}}]\!] - [\![\overline{\mathbf{u}}]\!]_n \mathbf{n}^+$$

its normal and tangential components. The symmetric gradient operator ε is defined such that $\varepsilon(\overline{\mathbf{v}}) = \frac{1}{2}(\nabla \overline{\mathbf{v}} + (\nabla \overline{\mathbf{v}})^t)$ for a given vector field $\overline{\mathbf{v}} \in H^1(\Omega \backslash \overline{\Gamma})^d$.

The space for the displacement field is defined by

$$\mathbf{U}_0 = \{\overline{\mathbf{v}} \in (H^1(\Omega \backslash \overline{\Gamma}))^d \,|\, \overline{\mathbf{v}} = 0 \text{ on } \partial\Omega\},$$

endowed with the norm $\|\overline{\mathbf{v}}\|_{\mathbf{U}_0} = \|\nabla \overline{\mathbf{v}}\|_{L^2(\Omega)^d}$. We consider a static mechanical model accounting for an isotropic linear elastic behavior in the matrix domain and a Coulomb frictional contact at the matrix–fracture interface Γ:

$$\begin{cases} -\text{div}\Big(\sigma(\overline{\mathbf{u}})\Big) = \mathbf{f} & \text{on } \Omega \backslash \overline{\Gamma}, \\ \sigma(\overline{\mathbf{u}}) = \frac{E}{1+\nu}\Big(\varepsilon(\overline{\mathbf{u}}) + \frac{\nu}{1-2\nu}(\text{div } \overline{\mathbf{u}})\mathbb{I}\Big) & \text{on } \Omega \backslash \overline{\Gamma}, \\ \mathbf{T}^+(\overline{\mathbf{u}}) + \mathbf{T}^-(\overline{\mathbf{u}}) = 0 & \text{on } \Gamma, \\ T_n(\overline{\mathbf{u}}) \leqslant 0, \; [\![\overline{\mathbf{u}}]\!]_n \leqslant 0, \; [\![\overline{\mathbf{u}}]\!]_n \, T_n(\overline{\mathbf{u}}) = 0 & \text{on } \Gamma, \\ |\mathbf{T}_\tau(\overline{\mathbf{u}})| \leqslant -F \, T_n(\overline{\mathbf{u}}) & \text{on } \Gamma, \\ ([\![\overline{\mathbf{u}}]\!]_\tau) \cdot \mathbf{T}_\tau(\overline{\mathbf{u}}) - F \, T_n(\overline{\mathbf{u}})|[\![\overline{\mathbf{u}}]\!]_\tau| = 0 & \text{on } \Gamma. \end{cases} \quad (1)$$

In (1), E and ν are the Young modulus and Poisson ratio, $F \geqslant 0$ is the friction coefficient, and the contact tractions $\mathbf{T}^a(\overline{\mathbf{u}})$ are defined by

$$\begin{cases} \mathbf{T}^a(\overline{\mathbf{u}}) = \sigma(\overline{\mathbf{u}})\mathbf{n}^a, \, a \in \{+, -\}, \quad \mathbf{T}(\overline{\mathbf{u}}) = \mathbf{T}^+(\overline{\mathbf{u}}), \\ T_n(\overline{\mathbf{u}}) = \mathbf{T}(\overline{\mathbf{u}}) \cdot \mathbf{n}^+, \quad \mathbf{T}_\tau(\overline{\mathbf{u}}) = \mathbf{T}(\overline{\mathbf{u}}) - (\mathbf{T}(\overline{\mathbf{u}}) \cdot \mathbf{n}^+)\mathbf{n}^+. \end{cases}$$

2 Mixed, Stabilized Mixed and Nitsche's Discretizations of the Contact Mechanical Model

Let $\mathbf{U}_h$ denote a family of Finite Element subspaces of $\mathbf{U}_0$, indexed by h coming from a family of simplicial (to fix ideas) meshes $\mathcal{M}_h$ of the domain Ω. The mesh $\mathcal{M}_h$ is assumed conforming to the fracture network and we denote by $\mathcal{F}_\Gamma$ the subset of faces of the mesh such that $\overline{\Gamma} = \bigcup_{\sigma \in \mathcal{F}_\Gamma} \overline{\sigma}$. The family of meshes $\mathcal{M}_h$ is assumed to be shape regular in the sense that the shape regularity parameter $S_R = \max_h \max_{K \in \mathcal{M}_h} \frac{h_K}{\rho_K}$ is bounded, where h_K denotes the diameter of the cell K and ρ_K is the radius of the inscribed ball in K.

The subspace $M_h \subset L^2(\Gamma)$ denotes the set of piecewise constant functions on the partition $\mathcal{F}_\Gamma$ and we set $\mathbf{M}_h = (M_h)^d$. For $\boldsymbol{\lambda}$ in $\mathbf{M}_h$, we will use the decomposition $\boldsymbol{\lambda} = (\lambda_n, \boldsymbol{\lambda}_\tau)$ with $\lambda_n = \boldsymbol{\lambda} \cdot \mathbf{n}^+$, $\boldsymbol{\lambda}_\tau = \boldsymbol{\lambda} - \lambda_n \mathbf{n}^+$, and identify $\boldsymbol{\lambda}_\tau$ to an element of $(M_h)^{d-1}$ based on an orthonormal basis local to each planar fracture. We denote by $\boldsymbol{\lambda}_\sigma$ the constant value of $\boldsymbol{\lambda} \in \mathbf{M}_h$ on the face $\sigma \in \mathcal{F}_\Gamma$ and by $\lambda_{n,\sigma}$ and $\boldsymbol{\lambda}_{\tau,\sigma}$ its normal and tangential components. The orthogonal projection from $L^2(\Gamma)$ to M_h is denoted by $\pi_\mathcal{F}^0$. For a face $\sigma \in \mathcal{F}_\Gamma$, the face average projection will be denoted by π_σ^0 in both cases.

The following notations will be used. For all $x \in \mathbb{R}$, we set

$$[x]_{\mathbb{R}^-} = \min(0, x), \quad [x]_{\mathbb{R}^+} = \max(0, x) = -[-x]_{\mathbb{R}^-}.$$

For all $\mathbf{x} \in \mathbb{R}^{d-1}$ and $\alpha \geqslant 0$, we denote by $[\mathbf{x}]_\alpha$ the projection of $\mathbf{x}$ on the ball $\mathcal{B}(\mathbf{0}, \alpha) \subset \mathbb{R}^{d-1}$ where $\mathcal{B}(\mathbf{0}, \alpha)$ is the closed ball of origin $\mathbf{0}$ and radius α, such that

$$[\mathbf{x}]_\alpha = \begin{cases} \mathbf{x} & \text{if } |\mathbf{x}| \leqslant \alpha, \\ \alpha \frac{\mathbf{x}}{|\mathbf{x}|} & \text{else .} \end{cases}$$

Mixed Formulation: This work focuses on mixed formulations using facewise constant Lagrange multipliers, which can handle fracture intersections, corners, and tips and facilitate the expression of discrete contact conditions and semismooth Newton solvers. However, these formulations require assuming the uniform inf-sup condition between the space of displacement fields and the space of Lagrange multipliers $\mathbf{U}_h \times \mathbf{M}_h$ (see e.g. [6]). There should exists $c_\star$ independent on the mesh such that

$$\inf_{\boldsymbol{\mu} \in \mathbf{M}_h} \sup_{\mathbf{v} \in \mathbf{U}_h} \frac{\int_\Gamma \boldsymbol{\mu} \cdot [\![\mathbf{v}]\!] d\sigma}{\|\mathbf{v}\|_{\mathbf{U}_0} \|\boldsymbol{\mu}\|_{H^{-\frac{1}{2}}(\Gamma)^d}} \geqslant c_\star > 0. \tag{2}$$

Let us define the discrete dual cone of vectorial Lagrange multipliers as

$$\Lambda_h(\lambda_n) = \{\boldsymbol{\mu} = (\mu_n, \boldsymbol{\mu}_\tau) \in \mathbf{M}_h \,|\, \mu_n \geqslant 0, |\boldsymbol{\mu}_\tau| \leqslant F\lambda_n\},$$

given $\lambda_n \in \Lambda_h = \{\lambda_n \in M_h \,|\, \lambda_n \geqslant 0\}$. Note that the friction coefficient will be assumed to be facewise constant on the partition $\mathcal{F}_\Gamma$. The mixed discretization of the static contact mechanical model reads: find $(\mathbf{u}, \boldsymbol{\lambda} = (\lambda_n, \boldsymbol{\lambda}_\tau)) \in \mathbf{U}_h \times \Lambda_h(\lambda_n)$ such that

$$\begin{cases} \displaystyle\int_\Omega \sigma(\mathbf{u}) : \varepsilon(\mathbf{v})\mathrm{d}\mathbf{x} + \int_\Gamma \boldsymbol{\lambda} \cdot [\![\mathbf{v}]\!]d\sigma = \int_\Omega \mathbf{f} \cdot \mathbf{v}\,\mathrm{d}\mathbf{x}, \\[4mm] \displaystyle\int_\Gamma (\boldsymbol{\mu} - \boldsymbol{\lambda}) \cdot [\![\mathbf{u}]\!]d\sigma \leqslant 0, \end{cases} \qquad \text{for all } (\mathbf{v}, \boldsymbol{\mu}) \in \mathbf{U}_h \times \Lambda_h(\lambda_n).$$

$$(3)$$

Note that the variational inequality in (3) is equivalent to the following equations for each $\sigma \in \mathcal{F}_\Gamma$

$$\lambda_{n,\sigma} - [\lambda_{n,\sigma} + \beta_n^{\mathrm{sm}} \pi_\sigma^0 [\![\mathbf{u}]\!]_n]_{\mathbb{R}^+} = 0, \quad \boldsymbol{\lambda}_{\tau,\sigma} - [\boldsymbol{\lambda}_{\tau,\sigma} + \beta_\tau^{\mathrm{sm}} \pi_\sigma^0 [\![\mathbf{u}]\!]_\tau]_{F_\sigma[\lambda_{n,\sigma} + \beta_n^{\mathrm{sm}} \pi_\sigma^0 [\![\mathbf{u}]\!]_n]_{\mathbb{R}^+}} = 0,$$

$$(4)$$

with $\beta_n^{\mathrm{sm}} > 0$, $\beta_\tau^{\mathrm{sm}} > 0$, which are the basis of the semi-smooth Newton algorithm used in the numerical section for the solution of the mixed discretization.

Stabilized Mixed and Mean-Nitsche's Formulations: In order to circumvent the inf-sup condition (2), we consider in this subsection the following stabilized mixed formulation in the spirit of [5]. Exploiting their facewise constant approximation, the Lagrange multipliers can be eliminated leading to a Nitsche's type discretization with face averaging. It results that this stabilized mixed formulation can also be useful when $\mathbf{U}_h \times \mathbf{M}_h$ satisfies the inf-sup condition in order to eliminate the Lagrange multipliers. In that case, we can prove that the stabilized mixed solution converges to the mixed solution at the limit of large stabilization parameters.

Let us fix a parameter $\theta \in \mathbb{R}$ and two non-negative functions β_n and β_τ on Γ typically set to $\frac{\beta_n^0}{h_\sigma}$ and $\frac{\beta_\tau^0}{h_\sigma}$ on each face σ of Γ, where h_σ is the diameter of the face σ. The parameters β_n^0 and β_τ^0 play the role of stabilization parameters which must be chosen large enough to obtain the stability and convergence of the method. The parameter θ, with typically $\theta = 1, 0, -1$, provides different properties like additional symmetry for $\theta = 1$ or stability for $\theta = -1$ (see [1] for details on the choice of these parameters). By abuse of notations, in the following h will also denote the facewise constant function of $L^2(\Gamma)$ with value h_σ on each face $\sigma \in \mathcal{F}_\Gamma$. The stabilized mixed discretization of the static contact mechanical model reads: find $(\mathbf{u}, \boldsymbol{\lambda} = (\lambda_n, \boldsymbol{\lambda}_\tau)) \in \mathbf{U}_h \times \Lambda_h(\lambda_n)$ such that

$$\begin{cases} \int_\Omega \sigma(\mathbf{u}) : \varepsilon(\mathbf{v})\mathrm{dx} + \int_\Gamma \boldsymbol{\lambda} \cdot [\![\mathbf{v}]\!]d\sigma \\ \quad - \int_\Gamma \frac{\theta}{\beta_n}(\lambda_n + T_n(\mathbf{u}))\pi_{\mathcal{F}}^0 T_n(\mathbf{v})d\sigma - \int_\Gamma \frac{\theta}{\beta_\tau}(\boldsymbol{\lambda}_\tau + \mathbf{T}_\tau(\mathbf{u})) \cdot \pi_{\mathcal{F}}^0 \mathbf{T}_\tau(\mathbf{v})d\sigma = \int_\Omega \mathbf{f} \cdot \mathbf{v}\,\mathrm{dx}, \\ \int_\Gamma (\boldsymbol{\mu} - \boldsymbol{\lambda}) \cdot [\![\mathbf{u}]\!]d\sigma \\ \quad - \int_\Gamma \frac{1}{\beta_n}(\mu_n - \lambda_n)(\lambda_n + T_n(\mathbf{u}))d\sigma - \int_\Gamma \frac{1}{\beta_\tau}(\boldsymbol{\mu}_\tau - \boldsymbol{\lambda}_\tau) \cdot (\boldsymbol{\lambda}_\tau + \mathbf{T}_\tau(\mathbf{u}))d\sigma \leqslant 0, \end{cases}$$

$$(5)$$

for all $(\mathbf{v}, \boldsymbol{\mu}) \in \mathbf{U}_h \times \Lambda_h(\lambda_n)$.

Thanks to their facewise constant approximations, the Lagrange multipliers can be eliminated from the stabilized mixed formulation using that the variational inequality in (5) with $\boldsymbol{\lambda} \in \Lambda_h(\lambda_n)$ is equivalent to the following equations:

$$\begin{cases} -\lambda_n = [\pi_{\mathcal{F}}^0 \left(T_n(\mathbf{u}) - \beta_n [\![\mathbf{u}]\!]_n \right)]_{\mathbb{R}^-}, \\ -\boldsymbol{\lambda}_\tau = [\pi_{\mathcal{F}}^0 \left(\mathbf{T}_\tau(\mathbf{u}) - \beta_\tau [\![\mathbf{u}]\!]_\tau \right)]_{\left(-F \left[\pi_{\mathcal{F}}^0 \left(T_n(\mathbf{u}) - \beta_n [\![\mathbf{u}]\!]_n \right) \right]_{\mathbb{R}^-} \right)}. \end{cases}$$

$$(6)$$

From (6), we can eliminate the Lagrange multipliers leading to the following equivalent Nitsche's type formulation: find $\mathbf{u} \in \mathbf{U}_h$ such that

$$\int_\Omega \sigma(\mathbf{u}) : \varepsilon(\mathbf{v})\mathrm{dx} - \int_\Gamma \frac{\theta}{\beta_n} T_n(\mathbf{u}) \cdot \pi_{\mathcal{F}}^0 T_n(\mathbf{v})\,\mathrm{d}\sigma - \int_\Gamma \frac{\theta}{\beta_\tau} \mathbf{T}_\tau(\mathbf{u}) \cdot \pi_{\mathcal{F}}^0 \mathbf{T}_\tau(\mathbf{v})\,\mathrm{d}\sigma$$
$$+ \int_\Gamma \frac{1}{\beta_n} \left[\pi_{\mathcal{F}}^0 \left(T_n(\mathbf{u}) - \beta_n [\![\mathbf{u}]\!]_n \right) \right]_{\mathbb{R}^-} \left(\theta T_n(\mathbf{v}) - \beta_n [\![\mathbf{v}]\!]_n \right)\,\mathrm{d}\sigma$$
$$+ \int_\Gamma \frac{1}{\beta_\tau} \left[\pi_{\mathcal{F}}^0 \left(\mathbf{T}_\tau(\mathbf{u}) - \beta_\tau [\![\mathbf{u}]\!]_\tau \right) \right]_{\left(-F \left[\pi_{\mathcal{F}}^0 \left(T_n(\mathbf{u}) - \beta_n [\![\mathbf{u}]\!]_n \right) \right]_{\mathbb{R}^-} \right)} \cdot \left(\theta \mathbf{T}_\tau(\mathbf{v}) - \beta_\tau [\![\mathbf{v}]\!]_\tau \right)\,\mathrm{d}\sigma$$
$$= \int_\Omega \mathbf{f} \cdot \mathbf{v}\,\mathrm{dx},$$

$$(7)$$

for all $\mathbf{v} \in \mathbf{U}_h$. Due to the face averaging, this formulation will be termed mean-Nitsche's method in the following.

For the Coulomb frictional model we can state the following Proposition, the proof of which can be found in [1].

Proposition 1. *Let $\overline{\beta}^0 > 0$ be large enough. For $\beta^0 \geqslant \overline{\beta}^0$, there exists a solution $(\mathbf{u}_\beta, \boldsymbol{\lambda}_\beta) \in \mathbf{U}_h \times \Lambda_h(\lambda_{\beta,n})$ to (7) with $\|\mathbf{u}_\beta\|_{\mathbf{U}_0} + \|h^{1/2}\boldsymbol{\lambda}_\beta\|_{L^2(\Gamma)^d} \leqslant C$, C independent of β_n^0, β_τ^0 and h but depending on the shape regularity parameter S_R and the physical data. Furthermore, any sequence of such solutions with β^0 going to infinity admits a subsequence which converges to $(\mathbf{u}, \boldsymbol{\lambda}) \in \mathbf{U}_h \times \Lambda_h(\lambda_n)$ solution of (3).*

Nitsche's Formulation: We follow the Nitsche's formulation introduced in [4] for Coulomb frictional models. It can be obtained by dropping the face averaging operators in (7): find $\mathbf{u} \in \mathbf{U}_h$ such that for all $\mathbf{v} \in \mathbf{U}_h$, one has

$$\int_\Omega \Big(\sigma(\mathbf{u}) : \varepsilon(\mathbf{v})\Big)d\mathbf{x} - \int_\Gamma \frac{\theta}{\beta_n} T_n(\mathbf{u}) \cdot T_n(\mathbf{v})\, d\sigma - \int_\Gamma \frac{\theta}{\beta_\tau} \mathbf{T}_\tau(\mathbf{u}) \cdot \mathbf{T}_\tau(\mathbf{v})\, d\sigma$$

$$+ \int_\Gamma \frac{1}{\beta_n} \Big[T_n(\mathbf{u}) - \beta_n[\![\mathbf{u}]\!]_n\Big]_{\mathbb{R}^-} \Big(\theta T_n(\mathbf{v}) - \beta_n[\![\mathbf{v}]\!]_n\Big)\, d\sigma \tag{8}$$

$$+ \int_\Gamma \frac{1}{\beta_\tau} \Big[\mathbf{T}_\tau(\mathbf{u}) - \beta_\tau[\![\mathbf{u}]\!]_\tau\Big]\Big(-F\Big[T_n(\mathbf{u})-\beta_n[\![\mathbf{u}]\!]_n\Big]_{\mathbb{R}^-}\Big) \cdot \Big(\theta\mathbf{T}_\tau(\mathbf{v}) - \beta_\tau[\![\mathbf{v}]\!]_\tau\Big)\, d\sigma = \int_\Omega \mathbf{f} \cdot \mathbf{v}\, d\mathbf{x}.$$

3   Comparison of the Mixed and Nitsche Formulations for a Coupled Poromechanical Simulation

The objective of this section is to compare the efficiency of the Nitsche's and mixed formulations to simulate a poromechanical model coupling the quasi static contact mechanical model with a mixed-dimensional single phase incompressible flow. The flow model accounts for a Darcy flow in the porous matrix and a Poiseuilles flow along the fractures. Isotropic linear poroelastic laws are used for the matrix porosity $\overline{\phi}_m$ and the total stress tensor σ^T:

$$\partial_t \overline{\phi}_m = b\,\mathrm{div}\,\partial_t \overline{\mathbf{u}} + \frac{1}{M}\partial_t \overline{p}_m, \quad \sigma^T(\overline{\mathbf{u}}) = \sigma(\overline{\mathbf{u}}) - b\,\overline{p}_m \mathbb{I},$$

where $\overline{p}_m$ is the matrix fluid pressure, b is the Biot coefficient and M Biot's modulus. The fracture aperture and the contact surface tractions are defined as follows, assuming that the fractures are filled by the fluid with contact at fracture asperities:

$$\overline{d}_f = d_0 - [\![\overline{\mathbf{u}}]\!]_n, \quad \mathbf{T}^\mathfrak{a}(\overline{\mathbf{u}}) = (\sigma(\overline{\mathbf{u}}) - b\,\overline{p}_m\mathbb{I})\mathbf{n}^\mathfrak{a} + \overline{p}_f\mathbf{n}^\mathfrak{a}, \ \mathfrak{a} \in \{+,-\},$$

where $\overline{p}_f$ is the fracture fluid pressure and d_0 being the fracture aperture at contact state. The fracture conductivity is defined by the cubic law $C_f = \frac{(\overline{d}_f)^3}{12\eta}$ and its normal transmissivity by $\Lambda_f = \frac{\overline{d}_f}{6\eta}$, where η is the fluid dynamic viscosity. Let us refer to [1] for the full description of the model.

The fluid flow is discretized in space by a mixed-dimensional Hybrid Finite Volume (HFV) scheme [3]. A $\mathbb{P}_2$ Finite Element discretization of the displacement field is used which guarantees that the inf-sup condition (2) is satisfied for the $\mathbb{P}_2 - \mathbb{P}_0$ mixed formulation (3). It also ensures the stability of the pressure displacement coupling. At each time step, the coupled non-linear system is solved using a fixed-point method on the function

$$\mathbf{g}_p : p \underset{\substack{\text{Contact Mechanics}\\\text{Solve}}}{\longrightarrow} \mathbf{u} \underset{\substack{\text{Darcy}\\\text{Solve}}}{\longrightarrow} \widetilde{p},$$

with $p = (p_m, p_f)$, accelerated by a Newton–Krylov algorithm in order to obtain at convergence the fully coupled poromechanical solution. At each evaluation of the fixed point function $\mathbf{g}_p$, the Darcy linear problem at given fracture aperture and porosity is solved using a GMRes iterative solver preconditioned by AMG while the contact mechanical model at given pressure p is solved using a semi-smooth Newton method. In the following experiments, the Nitsche's parameters are fixed to $\beta_n^0 = 1000$ GPa, $\beta_\tau^0 = 40$ GPa and $\theta = -1$ (see [1] for more details motivating this choice).

Our test case is taken from [2, Section 4.3] considering a 2×1 m domain including a network Γ of six fractures Γ_i, $i \in \{1, \cdots, 6\}$, cf. Fig. 2. Let us refer to [1] for additional test cases with other fracture networks. We consider a family of uniformly refined meshes $m \in \{0, \cdots, 3\}$ with 2855×4^m triangular cells and 88×2^m fracture faces equally spaced for each fracture.

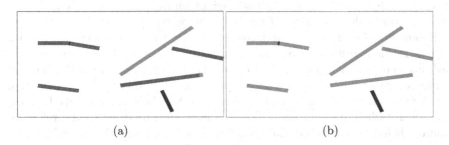

(a)   (b)

Fig. 2. Contact state along the fractures (blue: open, green: stick, red: slip) for the reference solution computed on the mesh $m = 3$ at times $t = T/4$ (a) and $t = T$ (b). (Color figure online)

We use the same values of Young's modulus and Poisson's ratio, $E = 4$ GPa and $\nu = 0.2$, and the same set of boundary conditions as in [2], that is, the two vertical sides of the domain are free, and we impose $\mathbf{u} = \mathbf{0}$ on the bottom side and

$$\mathbf{u}(t, \mathbf{x}) = \begin{cases} [0.005\,\text{m}, -0.002\,\text{m}]^\top \, 4t/T & \text{if } t \leq T/4, \\ [0.005\,\text{m}, -0.002\,\text{m}]^\top & \text{otherwise,} \end{cases} \quad \mathbf{x} \text{ on the top boundary.}$$

The friction coefficient is set to $F = 0.5$.

To fully exploit the capabilities of the HFV flow discretization, we consider the following anisotropic permeability tensor in the matrix: $\mathbb{K}_m = K_m \, \mathbf{e}_x \otimes \mathbf{e}_x + \frac{K_m}{2} \, \mathbf{e}_y \otimes \mathbf{e}_y$, $\mathbf{e}_x$ and $\mathbf{e}_y$ being the unit vectors associated with the x- and y-axes, respectively.

The permeability coefficient is set to $K_m = 10^{-15}$ m^2, the Biot coefficient to $b = 0.5$, the Biot modulus to $M = 10$ GPa, the dynamic viscosity to $\eta = 10^{-3}$ Pa$\cdot$s. The initial matrix porosity is set to $\phi_m^0 = 0.4$ and the fracture aperture corresponding to both contact state and zero displacement field is given by $d_0(\mathbf{x}) = \delta_0 \dfrac{\sqrt{\arctan(aD_i(\mathbf{x}))}}{\sqrt{\arctan(a\ell_i)}}$, $\mathbf{x} \in \Gamma_i$, $i \in \{1, \ldots, 6\}$, where $D_i(\mathbf{x})$ is the distance from $\mathbf{x}$ to the tips of fracture i, $\delta_0 = 10^{-4}$ m, $a = 25$ m^{-1} and ℓ_i is a fracture-dependent characteristic length: it is equal to $L_i/2$ (L_i being the length of fracture i) if fracture i is immersed, to L_i if one of its ends lies on the boundary, and to the distance of a corner from tips, if it includes a corner.

The initial pressure in the matrix and fracture network is $p_m^0 = p_f^0 = 10^5$ Pa. Notice that the initial fracture aperture differs from d_0, since it is computed by solving the mechanics given the initial pressures p_m^0 and p_f^0. The final time is set to $T = 2000$ s and the time integration uses an Euler implicit scheme with a uniform time stepping and 20 time steps. Concerning boundary conditions, for the flow, all sides are assumed impervious, except the left one, on which a pressure equal to the initial value 10^5 Pa is prescribed. Figure 2 exhibits the contact state along the fractures at times $t = T/4$ for which the pressures reach their maximum values and at final time $t = T$ for which the pressures are almost back to their initial value as shown in Fig. 3 for the matrix pressure. Given the Biot coefficient $b = 0.5$, this pressure decrease explains the switch from slip to stick state along the fractures at times larger than $T/4$. Note that the fracture pressure basically matches with the traces of the matrix pressure due to the high normal transmissivity of the fractures. Figure 4 plots the mean aperture as a function of time for the different meshes and the mixed and Nitsche's formulations. It shows that the Nitsche's method has a better convergence in space to the reference solution. It has been checked that the solution obtained with the mean-Nitsche's formulation is almost the same than the one of the mixed solution as expected for the inf-sup stable $\mathbb{P}_2 - \mathbb{P}_0$ approximation. Other mean quantities like the mean matrix and fracture pressures and the mean porosity as functions of time do not exhibit significant differences between both formulations. At the fracture scale, as exhibited in Fig. 5, the Nitsche's method is clearly more accurate than the Mixed and Mean-Nitsche's methods in singularity regions, typically near the corner of fracture 1. Finally, Fig. 6 plots the total number of semi-smooth Newton iterations for the contact mechanical model as a function of time. It shows that both the Nitsche's and mean-Nitsche's methods perform rather well compared to the mixed method which is the most efficient in terms of non-linear convergence. Note that the total number of fixed point function $\mathbf{g}_p$ evaluations is the same for all methods and both meshes and is equal to 190 for 20 time steps with a tight stopping criteria of 10^{-5} on the relative residual.

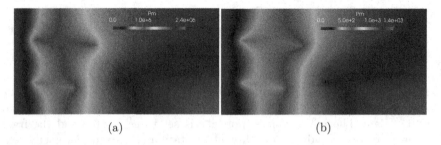

(a)   (b)

Fig. 3. Matrix overpressures (compared with the initial pressure) in Pa for the reference solution computed on the mesh $m = 3$ at times $t = T/4$ (a) and $t = T$ (b).

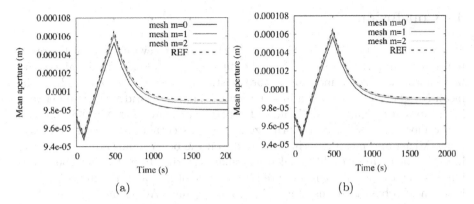

Fig. 4. Mean aperture as a function of time for the mixed (a) and Nitsche's (b) formulations. The reference solution is computed using the mesh $m = 3$.

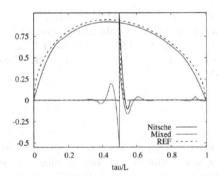

Fig. 5. Scaled normal displacement jumps at time $t = T$ for respectively fracture 1 (the upper left fracture with corner) and fracture 6 (the lower right fracture) on the coarse mesh $m = 0$ for the different discretizations (the $\mathbb{P}_2$ mean-Nitsche's basically matches with the mixed solution, hence it is not plotted).

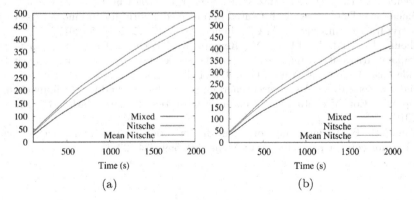

Fig. 6. Total number of semi-smooth Newton iterations for the contact mechanical model as a function of time for the mixed, Nitsche's and mean-Nitsche's formulations using the meshes $m = 0$ (a) and $m = 1$ (b).

4 Conclusion

This work compares the mixed formulation based on facewise constant Lagrange multipliers and the Nitsche's formulation of the Coulomb frictional contact mechanics in mixed-dimensional poro-elastic models. We also introduce the mean-Nitsche's method which in some sense is shown to bridge the gap between both types of formulations. Numerical experiments show that Nitsche's method is the most accurate in singularity zones such as corners or fracture intersections. Mixed and mean-Nitsche's methods, on the other hand, are less accurate in these zones due to additional oscillations resulting from contact conditions imposed on average at each fracture face. On the other hand, the mixed method is the most robust in terms of non-linear convergence. Additional work needs to be carried out to improve the robustness of the Nitsche's method in terms of nonlinear convergence.

Acknowledgements. The authors would like to thank BRGM and Andra for partially supporting this work and authorizing its publication. Franz Choulys work is partially supported by the I-Site BFC project NAANoD and the EIPHI Graduate School (contract ANR-17-EURE-0002). Franz Chouly is grateful of the Center for Mathematical Modeling grant FB20005.

References

1. Beaude, L., Chouly, F., Laaziri, M., Masson, R.: Mixed and Nitsche's discretizations of Coulomb frictional contact-mechanics for mixed-dimensional poromechanical models (2023). https://hal.science/hal-03949272. Accepted for Publication in Computer Methods in Applied Mechanics and Engineering
2. Berge, R.L., Berre, I., Keilegavlen, E., Nordbotten, J.M., Wohlmuth, B.: Finite volume discretization for poroelastic media with fractures modeled by contact mechanics. Int. J. Numer. Meth. Eng. **121**, 644–663 (2020)
3. Brenner, K., Hennicker, J., Masson, R., Samier, P.: Gradient discretization of hybrid dimensional Darcy flows in fractured porous media with discontinuous pressure at matrix fracture interfaces. IMA J. Numer. Anal. **37**, 1551–1585 (2017)
4. Chouly, F., Hild, P., Lleras, V., Renard, Y.: Nitsche method for contact with Coulomb friction: existence results for the static and dynamic finite element formulations. J. Comput. Appl. Math. **416**, 114557 (2022)
5. Hild, P., Renard, Y.: A stabilized Lagrange multiplier method for the finite element approximation of contact problems in elastostatics. Numer. Math. **115**, 101–129 (2010)
6. Wohlmuth, B.: Variationally consistent discretization schemes and numerical algorithms for contact problems. Acta Numer **20**, 569–734 (2011)

Machine Learning and Model Order Reduction for Large Scale Predictive Simulations

Towards Efficient SOT-Assisted STT-MRAM Cell Switching Using Reinforcement Learning

Johannes Ender[1,3]([✉]), Roberto Lacerda de Orio[3], Wolfgang Gös[2], and Viktor Sverdlov[1,3]

[1] Christian Doppler Laboratory for Nonvolatile Magnetoresistive Memory and Logic, Vienna, Austria
ender@iue.tuwien.ac.at
[2] Silvaco Europe Ltd., Cambridge, UK
[3] Institute for Microelectronics, Vienna University of Technology, Gußhausstraße 27-29, 1040 Vienna, Austria

Abstract. Nonvolatile memory is a promising candidate to replace CMOS devices with the two most common magnetoresistive RAM (MRAM) cell types being spin-transfer torque (STT-)MRAM and spin-orbit torque (SOT-)MRAM. Recently introduced combinations of these two mechanisms allow the amelioration of MRAM performance.

To optimize the switching, we implemented an approach based on reinforcement learning. In particular, we train an agent to switch an SOT-assisted STT-MRAM cell by applying STT or SOT current pulses independently. During this process, by means of rewards based on its actions, the agent is encouraged to either reverse the magnetization in the memory cell fast or by using little energy. After successfully training RL agents under the given constraints, the results of the two rewarding schemes applied to SOT-assisted STT-MRAM are discussed.

Keywords: Spin-orbit torque · Spin-transfer torque · Magnetoresistive RAM · Reinforcement Learning

1 Introduction

Recent years saw an increase in research interest in magnetoresistive random-access memory (MRAM). This is due to the fact that conventional semiconductor memory technologies based on CMOS suffer from increased standby-power consumption because of leakage currents that become a growing problem with the continuing down-scaling of these devices.

The financial support by the Federal Ministry of Labour and Economy, the National Foundation for Research, Technology and Development and the Christian Doppler Research Association is gratefully acknowledged.

I. Lirkov and S. Margenov (Eds.): LSSC 2023, LNCS 13952, pp. 83–90, 2024.
https://doi.org/10.1007/978-3-031-56208-2_7

Different flavors of MRAM devices exist, but a common feature they posses is that they store the information in the so-called magnetic tunnel junction (MTJ). An MTJ is a layered structure, consisting of two ferromagnetic layers sandwiching a nonmagnetic tunnel barrier. While the orientation of the magnetization in one of these layers, the reference layer, is fixed, the magnetization in the other layer is free to move. Depending on the orientation of the free layer relative to the reference layer, the electrical resistance of the MTJ is either high (anti-parallel configuration) or low (parallel configuration), allowing to store a bit of information. The various types of MRAM devices differ in the way the magnetization in the magnetic free layer is reversed and the most prominent mechanisms make use of either spin-orbit torque (SOT), spin-transfer torque (STT) or a combination of the two.

This work uses a reinforcement learning (RL) approach previously demonstrated to allow for the learning of reliable switching of an MRAM cell requiring SOT pulses originating from two orthogonally arranged current wires attached to its magnetic free layer [1,2,7]. Adjustments to the environment code as well as extensions to the underlying finite difference simulator make it possible to employ the established framework and use it for letting an agent learn efficient pulse schemes for a different type of MRAM cell which works by combining SOT and STT pulses [8].

1.1 SOT-Assisted STT-MRAM

As depicted in Fig. 1, SOT-assisted STT-MRAM cells use a combination of STT and SOT to change the stored bit of information. The STT arises from the torque generated by a spin-polarized electron current which is sent through the MTJ. The current gets polarized in the reference layer and when entering the free layer, a torque is exerted on the magnetization in there. SOT, on the other hand, stems from the so-called spin Hall effect, which describes the appearance of a transverse spin current when a charge current passes through materials which exhibit a (large) spin Hall angle. Each of these torque mechanisms, applied on its own, has drawbacks which can be mitigated by combining them [5]. For simulating the dynamics of the magnetization in the free magnetic layer, extended with expressions for the torques can be used:

$$\frac{\partial \mathbf{m}}{\partial t} = -\gamma \mu_0 \mathbf{m} \times \mathbf{H}_{\text{eff}} + \alpha \mathbf{m} \times \frac{\partial \mathbf{m}}{\partial t} + \mathbf{T}_{\text{SOT}} + \mathbf{T}_{\text{STT}} \qquad (1)$$

the Landau-Lifshitz-Gilbert equation $\mathbf{m}$ is the position-dependent, normalized magnetization, γ the gyromagnetic ratio, μ_0 the vacuum permeability, and α the Gilbert damping factor. $\mathbf{H}_{\text{eff}}$ is the effective magnetic field consisting of several contributions, among which some contain spatial derivatives of the magnetization $\mathbf{m}$. $\mathbf{T}_{\text{SOT}}$ and $\mathbf{T}_{\text{STT}}$ are the two torque terms.

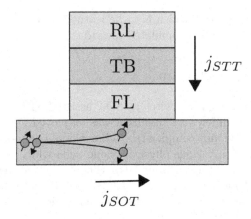

Fig. 1. SOT-assisted STT-MRAM cell.

1.2   Reinforcement Learning

Besides the growing interest in nonvolatile memory technologies like MRAM, machine learning algorithms have gained broad attention and are penetrating all possible scientific domains, often achieving staggering results. The machine learning sub-field of RL became more well-known when first applied to computer and board games [10], but has since proven helpful in many other disciplines [3]. It enables a virtual agent to learn how to optimally interact with an environment to achieve a certain goal by rewarding or punishing it for performed actions. Figure 2 shows the typical RL setup, consisting of agent and environment, which interact by exchanging a set of signals.

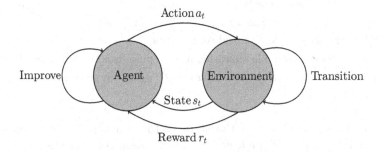

Fig. 2. Basic setup of a reinforcement learning system.

By sending an action signal to the environment, the agent triggers a transition from the current state to another one, about which information is subsequently returned from the environment to the agent, together with a reward, indicating how good or bad the performed action and that specific state was. The rewarding scheme encodes the objective which is to be learned by the agent, as the agent tries to maximize its overall accumulated reward.

2   Results and Discussion

In order to train an RL agent to switch an SOT-assisted STT-MRAM cell, an in-house developed finite differences simulator was employed [4]. This simulator is capable of solving Eq. (1) for complex three-dimensional memory cell structures, taking into consideration various physical phenomena which influence the magnetization's behavior. For the performed experiments, a time step of 0.3 ps and a space discretization length of 2 nm were used. By building a Python package from this simulator, a custom environment was created, which could subsequently be used by the RL framework *Stable-Baselines3* [9].

It was sufficient to allow the RL agent to perform four distinct actions: both pulses turned off, both pulses turned on, or each pulse turned on individually. Some restrictions, however, are imposed on the pulses. The voltage and current value for the pulses are fixed, and to prevent arbitrarily fast on and off pulses, a minimum pulse width of 100 ps is defined, which shall also reflect technical limitations concerning the frequency of generated pulses. Thus, if either a SOT or STT pulse is turned on, it cannot be turned off for 100 ps and if a pulse was turned off, it cannot be turned on for 100 ps. The SOT and STT pulses, however, are completely independent and there is no restriction on the concurrent application of the pulses.

Table 1. Input state variables for the RL agent.

	State variable	Size
$m_{x,y,z}$	Averaged magnetization components	3
$\Delta m_{x,y,z}$	Change in averaged magnetization components	3
$H^{eff}_{x,y,z}$	Averaged effective field components	3
P_{STT}	Boolean variable indicating if STT pulse can be turned on	1
P_{SOT}	Boolean variable indicating if SOT pulse can be turned on	1

Table 1 gives a summary of the state vector which is fed into the agent. Besides the information of the average magnetization components $(m_{x,y,z})$ and its change compared to the previous iteration $(\Delta m_{x,y,z})$, also the averaged values of the effective magnetic field $(H^{eff}_{x,y,z})$ are included, as well as the two variables $P_{STT/SOT}$, which indicate whether STT and SOT pulses can currently be applied. The total number of input state variables for the RL agent thus amounts to 11. Given the real-valued state vector consisting of the above-mentioned 11 variables, and the discrete action space, the deep Q-network algorithm [6] provided by the *Stable-Baselines3* framework is a suitable candidate and was used to train the agent.

The training procedure for the presented results consisted of 40 million training steps, which equals ~1500 simulations. The initial state of the magnetization in the free layer is a uniform magnetization pointing in the positive direction of the z-axis, with the target of the reversal being uniform magnetization along the negative z-axis. Maximum duration of a single simulation during training was 8 ns, but if the threshold of −0.9, at which the memory cell is considered to be switched, is reached before, the current simulation is terminated and the next one is started.

Table 2 lists all the parameters of the DQN algorithm, for which not the pre-configured values of the framework were used. Equation (2) presents a reward function which is used to encourage an agent to choose actions which reverse the magnetization:

Table 2. DQN parameters

Parameter	Value
Size of NN layers	$11 \times 150 \times 100 \times 4$
Discount factor	0.9997
Learning rate	7.5×10^{-4}
Exploration fraction	0.2
Final exploration probability	0.01
Replay buffer size	3×10^{5}
Batch size	512

$$r_{m_z} = m_{z,target} - m_{z,current} = -1 - m_{z,current} \tag{2}$$

Thus, the reward calculated from Eq. (2) is always negative, but reduces in magnitude until the target value is reached, at which it eventually becomes zero. This reward function was shown to be sufficient for training an RL agent to learn how to reverse the magnetization in a different type of MRAM cell [7]. Figure 3 shows the behavior of an exemplary representative of an RL agent which was trained using the reward function given in Eq. (2) for switching an SOT-assisted STT-MRAM cell. This reward function encourages the agent to find a pulse scheme for switching the memory cell as fast as possible. The middle panel of Fig. 3 shows the SOT and STT pulses applied by the agent. It has successfully learned that an initial SOT pulse is beneficial in accelerating the magnetization reversal. However, the applied STT pulse is never turned off. Looking at the top and bottom panel, the decision of the agent makes sense, as the z-component of the magnetization is successfully reversed and no further negative reward is accumulated, as there is no longer any deviation from the target value. The threshold of −0.9 is reached after 0.678 ns. It is clear, though, that the pulse sequence applied by the agent is sub-optimal. It achieves fast reversal, but at the cost of high energetic cost. In order to achieve fast switching while maintaining low power consumption, the reward function was extended as follows.

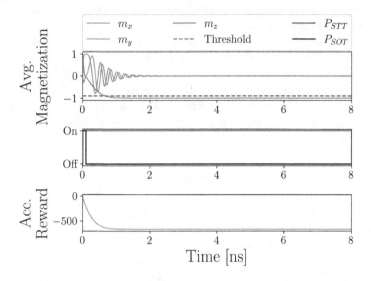

Fig. 3. Model trained without penalizing power consumption. $P_{STT/SOT}$ are boolean variables indicating whether the respective pulse is turned on or off.

$$r_{Pulse} = -(P_{STT} + P_{SOT}) \qquad (3)$$

A penalty for applying either STT or SOT current pulses can be defined as given in Eq. (3). With P_{STT}/P_{SOT} being 1 (0) when the pulse is turned on (off), this penalty is largest if both pulses are turned on and zero if both are turned off.

$$r = r_{Pulse} + r_{m_z} \qquad (4)$$

Figure 4 presents a visual representation of Eq. (4). One can see that the reward the agent receives is always negative, unless the magnetization in the memory cell is reversed completely. Furthermore, three domains with different levels of power penalty can be distinguished: when both pulses are turned on, power consumption is highest and thus the reward is the most negative, while when both pulses are turned off, the negative reward is merely a result from the deviation of the current value of the magnetization to the target value. An agent trained using the reward function with a penalty on power consumption applies pulses to reverse the magnetization as can be seen in Fig. 5. Looking at the top panel, it is apparent that magnetization reversal can still be achieved, but over a longer time period. The x- and y-component settle slower and the z-component of the magnetization crosses the threshold line after 2.76 ns, i.e. it takes three times longer than the previous agent. Again, in the beginning a short SOT pulse is applied to accelerate the magnetization reversal and the applied STT pulse is turned off after 200 ps. Examining the bottom panel of Fig. 5, one can see kinks in the accumulated reward, which indicate that the agent has moved from one of the three reward regimes presented in Fig. 4 to the next one, slowing down the rate at which negative reward is accumulated. As the reward function is a proxy for power consumption, this also shows that the agent tries to increase the energy efficiency and the write energy is reduced by three orders of magnitude as compared to the results shown in Fig. 3.

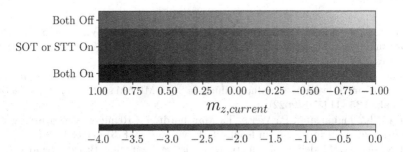

Fig. 4. Reward function with power consumption penalty. Color coding represents the reward returned by (4).

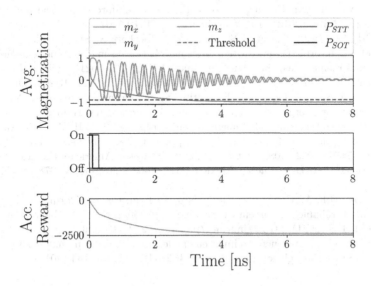

Fig. 5. Model trained with reward function penalizing power consumption. $P_{STT/SOT}$ are boolean variables indicating whether the respective pulse is turned on or off.

3 Conclusion and Outlook

In this work we presented the use of RL for the optimization of the write energy consumption of SOT-assisted STT-MRAM devices. By proper choice of the reward function an RL agent can be encouraged to learn pulse sequences for switching these memory cells in a more energy-efficient way. Thus, with the help of machine learning aided approaches like the one presented in this work, energy consumption of future devices can be greatly reduced.

References

1. Ender, J., de Orio, R.L., Fiorentini, S., Selberherr, S., Goes, W., Sverdlov, V.: Reinforcement learning approach for deterministic SOT-MRAM switching. In: Spintronics XIV, vol. 11805, pp. 56–63. SPIE (2021)
2. Ender, J., de Orio, R.L., Fiorentini, S., Selberherr, S., Goes, W., Sverdlov, V.: Reinforcement learning to reduce failures in SOT-MRAM switching. Microelectron. Reliab. **135**, 114570 (2022)
3. Fösel, T., Tighineanu, P., Weiss, T., Marquardt, F.: Reinforcement learning with neural networks for quantum feedback. Phys. Rev. X **8**(3), 031084 (2018)
4. Makarov, A.: Modeling of Emerging Resistive Switching Based Memory Cells. Ph.D. thesis, Institut für Mikroelektronik (2014). https://doi.org/10.34726/hss. 2014.23875, http://www.iue.tuwien.ac.at/phd/makarov/
5. Meo, A., Chureemart, J., Chantrell, R.W., Chureemart, P.: Magnetisation switching dynamics induced by combination of spin transfer torque and spin orbit torque. Sci. Rep. **12**(1), 3380 (2022)
6. Mnih, V., et al.: Playing atari with deep reinforcement learning. CoRR **abs/1312.5602** (2013). http://arxiv.org/abs/1312.5602
7. Orio, R., Ender, J., Fiorentini, S., Goes, W., Selberherr, S., Sverdlov, V.: Optimization of a spin-orbit torque switching scheme based on micromagnetic simulations and reinforcement learning. Micromachines **12**(4), 443 (2021). https://doi.org/10. 3390/mi12040443, https://www.iue.tuwien.ac.at/pdf/ib_2021/JB2021_Orio_2.pdf
8. Orio, R., Ender, J., Goes, W., Fiorentini, S., Selberherr, S., Sverdlov, V.: About the switching energy of a magnetic tunnel junction determined by spin-orbit torque and voltage-controlled magnetic anisotropy. In: 2022 IEEE Latin American Electron Devices Conference (LAEDC), pp. 1–4 (2022). https://doi.org/10.1109/ LAEDC54796.2022.9908222, talk: 2022 IEEE Latin American Electron Devices Conference (LAEDC), Puebla, Mexico; 2022-06-04 – 2022-06-06, 978-1-6654-9768-8
9. Raffin, A., Hill, A., Gleave, A., Kanervisto, A., Ernestus, M., Dormann, N.: Stable-baselines3: reliable reinforcement learning implementations. J. Mach. Learn. Res. **22**(268), 1–8 (2021). http://jmlr.org/papers/v22/20-1364.html
10. Silver, D., et al.: A general reinforcement learning algorithm that masters chess, shogi, and go through self-play. Science **362**(6419), 1140–1144 (2018)

Machine Learning Algorithms for Parameter Identification for Reactive Flow in Porous Media

Daria Fokina[1,2][(✉)] , Vasiliy V. Grigoriev[3] , Oleg Iliev[1] ,
and Ivan Oseledets[4,5]

[1] Fraunhofer Institute for Industrial Mathematics, Kaiserslautern, Germany
{daria.fokina,oleg.iliev}@itwm.fraunhofer.de
[2] Rheinland-Pfälzische Technische Universität Kaiserslautern-Landau,
Kaiserslautern, Germany
[3] North-Eastern Federal University, Yakutsk, Russia
v.v.grigorev@s-vfu.ru
[4] Skolkovo Institute of Science and Technology, Moscow, Russia
i.oseledets@skoltech.ru
[5] Artificial Intelligence Research Institute, Moscow, Russia

Abstract. Earlier we have explored different deterministic, stochastic and metaheuristic methods for identifying parameters of heterogeneous reactions for diffusion dominated and reaction dominated regimes [2–5]. Pore scale reactive transport was studied, breakthrough curves were the additional information used in identifying the parameters in Henry or Langmuir isotherms. All methods were time consuming, requiring multiple solution of the direct problem. Various surrogate models are used in the literature to reduce the computational burden related to parameter identification problems. In this paper we explore surrogate models based on neural network, Gaussian process, and cross approximation approaches. We also extend the number of the sought parameters. The achieved accuracy and the performance of the surrogate models were studied.

Keywords: Pore scale reactive flow · Parameter identification · Machine learning and Data Science

1 Motivation and Goals

Parameter identification for pore scale reactive flows is a task with significant scientific and practical importance. In our previous papers different deterministic, stochastic and heuristic algorithms were used to identify parameters for pore scale reactive flow. Breakthrough curves (time dependent curves of the outlet

---

DF and OI were supported by BMBF under grant 05M20AMD ML-MORE, VVG was supported by Ministry of Education and Science of the Russian Federation under grant No. FSRG-2023-0025, and IO was supported via von Humboldt Research Award.

I. Lirkov and S. Margenov (Eds.): LSSC 2023, LNCS 13952, pp. 91–98, 2024.
https://doi.org/10.1007/978-3-031-56208-2_8

concentration) were used as additional information in the identification procedure. Heterogeneous (surface) reactions were considered. In [2,3,5] the diffusion
dominated case, characterised by small Damkhöler numbers, was investigated.
In [4] parameter identification in the case of reaction dominated flow was discussed. In all the cases a large number of direct solutions was needed, leading
to large computational times. In the last decade more and more often surrogate
models based on machine learning and data science are explored to reduce the
computational burden of parameter identification, optimization, optimal control
and other computationally intensive problems. Recently we have shown [1] that
such approaches can be successfully used in predicting breakthrough curves. In
the present paper we present the usage of the approaches from [1] to build in
surrogate models for parameter identification problems. The same mathematical model as in [2,5] is used in all cases, however in addition to identifying the
parameters in Langmuir isotherm, the target here is to identify also the Peclet
number.

2   Pore Scale Reactive Transport and Numerical Algorithm

Similar to [2,4], we will consider 2D pore scale reactive flow. A part of the
domain, namely Ω_f is occupied by a fluid, while the other part is occupied by
obstacles Ω_s. The obstacle surfaces (where the reaction occurs) are denoted by
Γ_s, while symmetry lines are denoted by Γ_{sym}. It is supposed that dissolved
substance is introduced via the inlet boundary Γ_{in}, unreacted substance flows
out via Γ_{out}. Since the adsorption reaction occurs only on the surface of obstacles,
the computational domain consists only of $\Omega_f \cup \Gamma_s$.

Flow Problem. Steady state Stokes equations are governing the flow:

$$\nabla p - \mu \Delta u = 0, \quad \nabla \cdot u = 0, \quad x \in \Omega_f, \tag{1}$$

where $u(x)$ and $p(x)$ are the fluid velocity and pressure, respectively, while $\mu > 0$
is the dynamic viscosity, which we assume to be constant and equal to one.
Standard boundary conditions are formulated, see, e.g. [2] for details.

Species Transport. The unsteady solute transport is governed by convection-
diffusion equation in dimensionless form:

$$\frac{\partial c}{\partial t} + \nabla(uc) - \frac{1}{\text{Pe}} \nabla^2 c = 0, \quad x \in \Omega_f, \quad t > 0, \tag{2}$$

where the dimensionless concentration of the solute in the fluid is denoted by
$c(x,t)$ and $\text{Pe} = \frac{l\bar{u}}{D}$ is the Peclet number. The boundary conditions read:

$$c = 1, \quad x \in \Gamma_{in}, \quad \nabla c \cdot n = 0, \quad x \in \Gamma_{sym} \cup \Gamma_{out}, \quad \frac{\partial m}{\partial t} = -\nabla c \cdot n, \quad x \in \Gamma_s. \tag{3}$$

The reactions that occur at the obstacles' surface Γ_s satisfy the mass conservation law, in this particular case meaning that the change in adsorbed surface

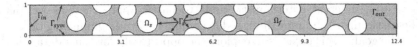

Fig. 1. Computational domain

concentration is equal to the flux from the fluid to the surface (3), where m is the surface concentration of adsorbed solute. A mixed kinetic-diffusion adsorption description, based on Langmuir isotherm, is used:

$$\frac{\partial m}{\partial t} = \text{Da}_a c \left(1 - \frac{m}{M}\right) - \text{Da}_d m, \quad x \in \Gamma_s. \tag{4}$$

where the adsorption and desorption Damköler numbers are given by $\text{Da}_a = \frac{k_a}{\bar{u}}$, $\text{Da}_d = \frac{k_d l}{\bar{u}}$. Here $k_a \geq 0$ is the rate of adsorption, measured in unit length per unit time, and $k_d \geq 0$ is the rate of desorption, measured per unit time. The dimensionless parameter M is given by $M = \frac{m_\infty}{l\bar{c}}$. Here, $M > 0$ is the maximal possible adsorbed surface concentration. Zero initial conditions are set for the concentration.

Numerical Algorithm. Finite Element Method, FEM, is used for space discretization of the above problem, together with implicit discretization in time. The algorithm used here for solving the direct problem is practically identical to the algorithm described in [2], see also the references therein. The computational domain is a rectangle with a dimensionless height of $x_2 = 1$ and dimensionless length of $x_1 = 12.4$. It is a stripe which is periodic in x_2, and it is fully resolved in x_1. Within the stripe randomly circles with different radii are placed, see Fig. 1. The radius of the non overlapping obstacles vary from 0.25 to 0.35. The porosity of the region is 58% in the active zone, which is ten dimensionless units long.

3   Parameter Identification

Consider the parameter identification problem for determining the unknown parameters in Langmuir isotherm (4), as well as the unknown Peclet number. The functional of the squared difference between computed and "measured" breakthrough curves given by

$$J(Pe, \text{Da}_a, \text{Da}_d, M) = \int_0^T (c_{out}(t) - c\widehat{(Pe, t)})^2 dt, \tag{5}$$

is to be minimized in this case. Here $\widehat{c(t)}$ is a breakthrough curve from "measurements" In fact, this curve is produced by solving the forward problem using a selected set of parameters, to be called below "exact parameters". Here $Pe = 0.6, \text{Da}_a = 0.006, \text{Da}_d = 0.06, M = 0.06$ (see Langmuir isotherm Eq.(4)) are selected as exact values. They are used to produce the "measurement breakthrough curve". Following [2,4], we consider a specific class of parameter identification problems, for which the goal is to identify admissible set of parameters for which the functional is smaller than a prescribed threshold.

In [2,4,5], different deterministic and stochastic approaches, as well as a metaheuristic algorithm, were discussed in conjunction with minimizing the functional. The goal of this paper is to explore the opportunity for using data science algorithms for deriving a cheap surrogate model which can replace the expensive direct numerical simulation. Here we consider the simplest minimization approach. A uniform grid is introduced in the parameter space, and after that a subset of parameter values for which the functional is smaller than a prescribed threshold is identified. In the case when only direct numerical simulation is used, the forward problem has to be solved in each point of the grid in the parameter space. In the case of data science algorithms target is to solve a (small) number of direct problems, and to use the produced breakthrough curves for training ML algorithms or within the cross approximation approach. Earlier in [1] the efficiency of such models was shown for reactive flow with homogeneous reactions. Here we apply the same approach for heterogeneous reaction.

3.1　Machine Learning and Data Science Algorithms

As it was mentioned above we build a surrogate model to approximate mapping $f : \mu \to c_{out}(t; \mu)$, where μ is a vector of input parameters, $c_{out}(t; \mu)$ is the obtained breakthrough curve. In our case, instead of functions $c_{out}(t; \mu)$, we consider vectors $c_{out}(\mathbf{t}; \mu)$, which are values of $c_{out}(t)$ at N_T fixed moments of time, i.e. we have a mapping $f : \Omega \to \mathbb{R}^{N_T}$. Ω is the domain of interest. There are different ways to do this in machine learning, we compare two of them: Gaussian processes, neural networks, and also a tensor-based method - cross approximation. In the previous work [1] we have shown that these methods perform well for the breakthrough curve approximation for reactive flow in porous media.

Gaussian Processes. The first technique is to consider a Gaussian process. The detailed description can be found in [6]. The main variable for this model is the kernel function. Once it is chosen, the approximation can be written as follows:

$$\hat{f}(x_*) = K(x_*, X)K(X, X)Y, \tag{6}$$

where X is a matrix of input parameter values, Y is a matrix of corresponding outputs (breakthrough curves). We consider four different kernel functions: radial basis function, rational quadratic, Matern and chi-squared. For the result provided here, a prior selection of a best kernel was performed.

Neural Networks. The next class of ML methods are artificial neural networks (NNs). In our considerations we use a classical fully-connected neural network. The approximation is as follows:

$$\hat{f}_\theta = L_d \circ g \circ \ldots \circ L_2 \circ g \circ L_1, \tag{7}$$

where $L_k(\mathbf{x}) = \mathbf{x}W_k + \mathbf{b}_k$, $W_k \in \mathbb{R}^{h_{k-1} \times h_k}$, $\mathbf{b}_k \in \mathbb{R}^{h_k}$, $\theta = \{(W_k, \mathbf{b}_k\}_{k=1}^d$, $h_0 = n$, $h_d = m$, g is a nonlinear function, which is also called an activation function. The parameters are found by solving the minimization problem:

$$l(\theta; \mathcal{D}) = \sum_{(\mathbf{x}_i, \mathbf{y}_i) \in \mathcal{D}} \|\hat{f}_\theta(\mathbf{x}_i) - \mathbf{y}_i\|^2 \to \min_\theta. \tag{8}$$

This is usually done by stochastic gradient descent or its modifications. In this work we use the Adam optimizer learning rate scheduling. We also experiment with different activation functions (ReLU, ELU, leaky ReLU). The results provided here are for the one giving the smallest error.

Cross Approximation. The third method we consider is a tensor method, a so-called cross approximation. As it follows the name suggests, it operates on tensors, which are n-dimensional matrices. The input of the model is a vector $\mu = (\mu_1, \mu_2, \mu_3, \mu_4)$, which in our case corresponds to the vector of parameters, (Pe, Da_a, Da_d, M), correspondingly. For us of interest are only μ values located on a grid. If we compute the curves c_μ for each element on the grid, we get a five-dimensional tensor $\mathsf{C} \in \mathbb{R}^{k \times k \times k \times k \times N_T}$, where k is the size of the grid, N_T is the number of simulated time steps. We assume that C can be represented in a low-rank form using Tucker decomposition:

$$\mathsf{C} = \mathsf{G} \times_1 U^1 \times_2 U^2 \times_3 U^3 \times_4 U^4, \tag{9}$$

$\mathsf{G} \in \mathbb{R}^{r_1 \times r_2 \times r_3 \times r_4 \times N_T}$, $U^1 \in \mathbb{R}^{r_1 \times k}$, $U^2 \in \mathbb{R}^{r_2 \times k}$, $U^3 \in \mathbb{R}^{r_3 \times k}$, $U^4 \in \mathbb{R}^{r_4 \times k}$. The tensor-matrix product $\times_m$ is performed along the axis m. The algorithm for choosing U^1, U^2, U^3, U^4 and G has been shown in detail in [1]. We follow the same scheme here, but for a 5-dimensional tensor instead of 3-dimensional.

3.2 Computational Experiments

The simplest approach to parameter identification, as mentioned above, is considered here. Namely, uniform grid in a proper subdomain of the parameter space is introduced. In general, this alone is a very inefficient approach, but combined with machine learning it could become attractive due to its simplicity. Taking into account the convergence studies with respect to the space and time steps (not reported here), FEM grid with of 2708 nodes and 5478 elements is used, along with time step $\tau = 0.1$

For admissible set identification we compare two sets of computational experiments. In the first set, the forward problem is solved for each point from the grid, and the functional (5) is evaluated to identify the admissible parameter set. In the second set of experiments, the forward problem is solved only in a part of the points, and the computed breakthrough curves are used to derive a cheap surrogate solution using machine learning and data science approaches described above. The latter surrogate model is used to identify the admissible parameter set, which is defined by ε: The surrogate model is used to identify the admissible parameter set, which is defined by ε:

$$J(\text{Pe}, \text{Da}_a, \text{Da}_d, M) \leq \varepsilon_J, \tag{10}$$

the threshold ε_j is selected empirically depending on the problem.

The experiments defined above are performed for a two-stage algorithm. At the first stage, the parameters in the Langmuir isotherm will be sought in the subdomain $10^{-5} \leq Da_a \leq 0.001$, $10^{-5} \leq Da_d \leq 0.01$, $10^{-5} \leq M \leq 0.01$. We seek Peclet number in the range: $10^{-5} \leq Pe \leq 1$ (diffusion dominated case). Uniform grid $11 \times 11 \times 11 \times 11$ is introduced in this subdomain. When the admissible parameter set is defined and the area where the minimum of the functional is roughly localized, we shrink the feasible parameter domain and at the second stage seek to identify parameters in this smaller domain. In our case the shrank subdomain is defined as: $0.0045 \leq Da_a \leq 0.0075$, $0.045 \leq Da_d \leq 0.08$, $0.05 \leq M \leq 0.07$, $0.5 \leq Pe \leq 0.7$, the grid is $17 \times 17 \times 17 \times 17$.

Having in mind that the visualization of a functional in four dimensional space is a challenge, bellow we show the results in (Da_a, Da_d) plane, which we find representative. In this case M, Pe are fixed at their exact values. On Fig. 2 $\varepsilon_J = 0.01$ is used for visualizing the admissible sets identified with the different approaches at the first stage, while ad the second stage $\varepsilon_J = 0.007$.

A. Minimization based on **direct solutions only**. At the first stage 14641 direct problems are solved in the initial feasible domain and 83521 problems are solved at the second stage in the smaller admissible subdomain. The solution of each problem requires 17 s. The identified admissible set is surrounded by blue line on Fig. 2. The *gray star* on the Figure denotes the exact parameters.

B.1. Minimization using **Gaussian process model**. As we mention above, one of the methods used to approximate the breakthrough curves is a Gaussian process. 1600 of curves computed on the grid solving the PDE were used for the training. The process of training took 0.409 s, prediction for the whole grid of parameters takes 0.029 s. When the curves for the parameter grid are predicted, we can estimate the functional values and identify the admissible set. The admissible set obtained with surrogate model based on Gaussian process is presented in orange color on Fig. 2.

B.2. Minimization using **neural network model**. The next used method is a fully-connected neural network with $h = 10$, $d = 5$. We start with the learning rate value 0.01, and multiply it by 0.1, when during the last 1000 iterations training error didn't decrease. This procedure is repeated 2 times. For the training set 5000 parameters and corresponding curves were randomly selected on the grid. It took 357 s to train a neural network, 0.0102 s are needed to get the breakthrough curve values for the whole grid. The green color on Fig. 2 represents the admissible set identified using the neural network.

B.3. Minimization using **cross approximation model**. The described above *cross approximation* approach is used to reconstruct the whole tensor for a given grid. Chosen ranks for the initial domain are 6 (Pe), 5 (Da$_a$), 4 (Da$_d$), 6 (Da$_a$), for the shrinked subdomain all ranks are equal to 3. 763 breakthrough

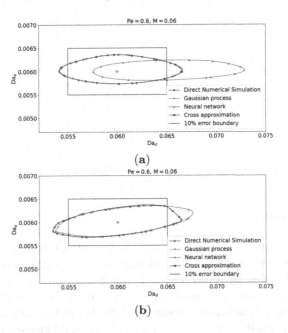

Fig. 2. Identified admissible parameters sets based on direct simulations only and using surrogate models. (a) initial domain of size $11 \times 11 \times 11 \times 11$ ($\varepsilon = 0.01$); (b) a shrinked subdomain of size $17 \times 17 \times 17 \times 17$ ($\varepsilon = 0.007$).

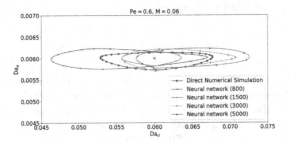

Fig. 3. NN result for the varied number of training points. The number of training points is indicated in brackets in the legend.

curves (149 for the shrinked subdomain) computed by solving the forward problems are needed, and then the breakthrough curves for the remaining parameter points on the grid can be accurately predicted, and the functional can be estimated in these points. Construction of basis vectors requires 0.0307 s, for the reconstruction of the whole grid 29 s are required. The red contour on Fig. 2 shows the set of admissible parameters.

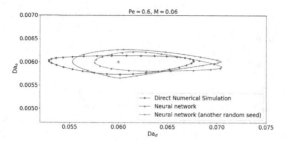

Fig. 4. NN result with different initialization. (Color figure online)

4   Conclusion

Cross approximation and Gaussian process approaches provide very good accuracy, at the same time significantly reducing the computational costs compared to using only direct simulations. Both approaches are not very sensitive to the size of the initial parameter domain. The neural network approach is, however, more time-consuming and gives a worse accuracy. We do not exclude the possibility that a better architecture could be chosen that would give a better accuracy. Based on our experience, the chosen model was good enough to give a reasonable accuracy. We would also like to note that a neural network is a non-deterministic method, and the choice of random seed before initialization and training provides different result of the training (see Figs. 4, 3).

References

1. Fokina, D., Toktaliev, P., Iliev, O., Oseledets, I.: Machine learning methods for prediction of breakthrough curves in reactive porous media (2023)
2. Grigoriev, V.V., Iliev, O., Vabishchevich, P.N.: Computational identification of adsorption and desorption parameters for pore scale transport in periodic porous media. J. Comput. Appl. Math. **370**, 112661 (2020)
3. Grigoriev, V.V., Iliev, O., Vabishchevich, P.N.: Computational identification of adsorption and desorption parameters for pore scale transport in random porous media. In: Lirkov, I., Margenov, S. (eds.) LSSC 2019. LNCS, vol. 11958, pp. 115–122. Springer, Cham (2020). https://doi.org/10.1007/978-3-030-41032-2_12
4. Grigoriev, V.V., Iliev, O., Vabishchevich, P.N.: On parameter identification for reaction-dominated pore-scale reactive transport using modified bee colony algorithm. Algorithms **15**(1), 15 (2022)
5. Grigoriev, V.V., Vabishchevich, P.N.: Bayesian estimation of adsorption and desorption parameters for pore scale transport. Mathematics **9**(16), 1974 (2021)
6. Williams, C., Rasmussen, C.: Gaussian processes for regression. In: Touretzky, D., Mozer, M., Hasselmo, M. (eds.) Advances in Neural Information Processing Systems, vol. 8. MIT Press (1995)

Randomized Symplectic Model Order Reduction for Hamiltonian Systems

R. Herkert[1][(✉)], P. Buchfink[1], B. Haasdonk[1], J. Rettberg[2], and J. Fehr[2]

[1] Institute of Applied Analysis and Numerical Simulation, University of Stuttgart,
Pfaffenwaldring 57, 70569 Stuttgart, Germany
{robin.herkert,patrick.buchfink,haasdonk}@mathematik.uni-stuttgart.de
[2] Institute of Engineering and Computational Mechanics, University of Stuttgart,
Pfaffenwaldring 9, 70569 Stuttgart, Germany
{johannes.rettberg,joerg.fehr}@itm.uni-stuttgart.de

Abstract. Simulations of large scale dynamical systems in multi-query or real-time contexts require efficient surrogate modelling techniques, as e.g. achieved via Model Order Reduction (MOR). Recently, symplectic methods like the complex singular value decomposition (cSVD) or the SVD-like decomposition have been developed for preserving Hamiltonian structure during MOR. In this contribution, we show how symplectic structure preserving basis generation can be made more efficient with randomized matrix factorizations. We present a randomized complex SVD (rcSVD) algorithm and a randomized SVD-like decomposition (rSVD-like). We demonstrate the efficiency of the approaches with numerical experiments on high dimensional systems.

Keywords: symplectic model order reduction · Hamiltonian systems · randomized algorithm

1 Introduction

Numerical simulation of large scale dynamical systems often leads to high computational costs. In a multi-query or real-time context, this requires efficient surrogate modelling techniques such as model order reduction (MOR). Additionally, models appear as Hamiltonian systems, which, for example, describe conservative dynamics and non-dissipative phenomena in classical mechanics or transport problems. The structure of a Hamiltonian system ensures conservation of energy and, under mild assumptions, stability properties. Classical MOR like the Proper Orthogonal Decomposition (POD) [12] fails to preserve this Hamiltonian structure which, in general, violates the conservation of energy and may yield unstable reduced models. Thus, one current trend in MOR is the structure-preserving MOR of Hamiltonian systems [6,9]. Recently, structure-preserving methods like the complex singular value decomposition (cSVD) [10] or the SVD-like decomposition [6,13] have been developed. Both of these methods belong to the class of data-based/snapshot-based MOR, which has the advantage that it can (i) also be used in black-box learning, (ii) be applied to closed-source nonlinear models where the underlying structure of the model is not clear.

© The Author(s), under exclusive license to Springer Nature Switzerland AG 2024
I. Lirkov and S. Margenov (Eds.): LSSC 2023, LNCS 13952, pp. 99–107, 2024.
https://doi.org/10.1007/978-3-031-56208-2_9

With projection-based MOR (e.g. POD, cSVD and SVD-like), the solution is approximated in a low-dimensional subspace. For computing a basis of such a subspace, low-rank matrix approximations, like the truncated singular value decomposition (SVD), are computed for the snapshot matrix. In many cases, a randomized approach for a low-rank matrix approximation is faster and/or more robust than its classical version [8]. How randomization can be applied to MOR is currently intensively studied: Randomized versions of classical MOR basis generation algorithms have been recently applied [1,2,7] or the reduced order model itself is approximated by its random sketch [3,4]. It has been shown that randomization can improve not only efficiency but also numerical stability. None of these approaches, in general, preserves a Hamiltonian structure of a system.

This work is focussed on how structure-preserving symplectic basis generation and efficient randomized basis generation can be combined. We present two randomized, symplectic basis generation schemes: a randomized complex SVD (rcSVD) algorithm and a randomized SVD-like decomposition (rSVD-like). For the rcSVD, we leverage randomization for complex matrices. The rSVD-like algorithm is obtained by a randomized version of the Schur decomposition. We demonstrate the efficiency of the approaches by numerical testing on high-dimensional systems such as obtained from spatial discretization of the wave equation. This work is structured as follows: In Sect. 2 essential background on MOR, Hamiltonian systems and randomized matrix factorizations is given. In Sect. 3, the new randomized, symplectic methods are presented. Section 4 is focussed on numerical experiments and comparisons with non-randomized structure-preserving methods. The work is concluded in Sect. 5.

2   Essentials

Symplectic Model Order Reduction of Hamiltonian Systems

In this section, a brief summary on MOR of Hamiltonian systems is given. For a more detailed introduction, we refer to [5] (MOR), [11] (symplectic geometry and Hamiltonian systems) and [10] (symplectic MOR of Hamiltonian systems).

Given a parametric *Hamiltonian (function)* $\mathcal{H}(\cdot; \boldsymbol{\mu}) \in \mathcal{C}^1(\mathbb{R}^{2N})$ and a parametric initial value $\boldsymbol{x}_0(\boldsymbol{\mu}) \in \mathbb{R}^{2N}$, a parametric, canonical *Hamiltonian system* reads: For a fixed (but arbitrary) parameter vector $\boldsymbol{\mu} \in \mathcal{P} \subset \mathbb{R}^{n_\mu}$ and time interval $I_t := [t_0, t_{\mathrm{end}}]^1$, find the full solution $\boldsymbol{x}(\cdot; \boldsymbol{\mu}) \in \mathcal{C}^1(I_t, \mathbb{R}^{2N})$ with

$$\frac{\mathrm{d}}{\mathrm{d}t}\boldsymbol{x}(t; \boldsymbol{\mu}) = \mathbb{J}_{2N}\nabla_{\boldsymbol{x}}\mathcal{H}(\boldsymbol{x}(t; \boldsymbol{\mu}); \boldsymbol{\mu}), \quad \boldsymbol{x}(t_0; \boldsymbol{\mu}) = \boldsymbol{x}_0(\boldsymbol{\mu}) \tag{1}$$

where

$$\mathbb{J}_{2N} := \begin{bmatrix} \mathbf{0}_N & \boldsymbol{I}_N \\ -\boldsymbol{I}_N & \mathbf{0}_N \end{bmatrix} \in \mathbb{R}^{2N \times 2N}$$

---

[1] Note that I_t may even be parameter dependent, as we will use in the experiments. For notational simplicity we keep a fixed time interval in this section.

denotes the canonical Poisson matrix and $I_N, 0_N \in \mathbb{R}^{N \times N}$ denote the identity and zero matrices. The most important property of Hamiltonian systems is that the solution conserves the Hamiltonian over time, i.e. $\frac{d}{dt}\mathcal{H}(x(t; \mu); \mu) = 0$ for all $t \in I_t$.

Given a reduced order basis (ROB) matrix $V \in \mathbb{R}^{2N \times 2n}$ and projection matrix $W \in \mathbb{R}^{2N \times 2n}$, with $W^\mathsf{T}V = I_{2n}$ and $n \ll N$, the projection-based reduction of the Hamiltonian system leads to a reduced system that reads: For a parameter vector $\mu \in \mathcal{P} \subset \mathbb{R}^{n_\mu}$, find the *reduced solution* $x_\mathrm{r}(\cdot; \mu) \in \mathcal{C}^1(I_t, \mathbb{R}^{2n})$ with

$$\frac{d}{dt}x_\mathrm{r}(t; \mu) = W^\mathsf{T}\mathbb{J}_{2N}\nabla_x\mathcal{H}(Vx_\mathrm{r}(t; \mu); \mu), \quad x_\mathrm{r}(t_0; \mu) = x_{\mathrm{r},0}(\mu) \qquad (2)$$

with the *reduced initial value* $x_{\mathrm{r},0}(\mu) := W^\mathsf{T}x_0(\mu)$. This system is not necessarily Hamiltonian and thus the conservation of the Hamiltonian over time can not be ensured. To preserve the Hamiltonian structure, symplectic MOR can be used [6,10]. Here, (i) the ROB matrix V is required to be a *symplectic matrix* which means, that for $V \in \mathbb{R}^{2N \times 2n}$ with $n \leq N$

$$V^\mathsf{T}\mathbb{J}_{2N}V = \mathbb{J}_{2n},$$

and (ii) the projection matrix W is set to be the transpose of the so-called *symplectic inverse* V^+ of the ROB matrix V, i.e.

$$W^\mathsf{T} = V^+ := \mathbb{J}_{2n}V^\mathsf{T}\mathbb{J}_{2N}^\mathsf{T}.$$

With this choice, the reduced model (2) is a low-dimensional Hamiltonian system with the *reduced Hamiltonian* $\mathcal{H}_\mathrm{r}(x_\mathrm{r}(t; \mu); \mu) := \mathcal{H}(Vx_\mathrm{r}(t; \mu); \mu)$, which is defined as the Hamiltonian of the ROB matrix times the reduced coordinates.

Randomized Matrix Factorizations

We continue with a brief summary on randomized matrix factorizations. For a more detailed presentation, we refer to [8]. The computation of a randomized factorization of a matrix $B \in \mathbb{R}^{m \times l}$ is subdivided into two stages. First, for $k \leq m$ a matrix $Q \in \mathbb{R}^{m \times k}$ with orthonormal columns is computed that approximates $B \approx QQ^\mathsf{T}B$. This task can be efficiently executed with random sampling methods. Then, a matrix decomposition (e.g. SVD, QR) of $Q^\mathsf{T}B \in \mathbb{R}^{k \times l}$ is computed and multiplied by Q. The first factor of an SVD/QR-decomposition is a matrix with orthonormal columns and this property is not changed by a multiplication with Q. For the SVD, we get $Q^\mathsf{T}B = \tilde{U}\Sigma V^\mathsf{T}$ and by setting $U := Q\tilde{U}$ we get the approximate SVD $B \approx U\Sigma V^\mathsf{T}$. Instead of the target rank k, it is known to be advantageous to introduce an oversampling parameter p_ovs and aim for $k + p_\mathrm{ovs}$ columns for Q, then for U only the first k columns are used. The matrix Q is computed via Algorithm 1. Note, that a computational advantage over a direct factorization of B will be achievable if $k \ll l$.

Algorithm 1. Random Sampling Algorithm

Input: $B \in \mathbb{R}^{m \times l}$, target rank $k \in \mathbb{N}$, oversampling parameter $p_{\text{ovs}} \in \mathbb{N}_0$
Output: Matrix with orthonormal columns $Q \in \mathbb{R}^{m \times (k + p_{\text{ovs}})}$

1: Draw a random Gaussian test matrix $\Omega \in \mathbb{R}^{l \times (k + p_{\text{ovs}})}$.
2: Compute the matrix product $Y = B\Omega$, a so-called *random sketch* of B.
3: Constr. $Q \in \mathbb{R}^{m \times (k + p_{\text{ovs}})}$ s.t. its columns are an orthon. basis for the range of Y^2.

Step 3 of this algorithm can for example be performed using a QR-decomposition or an SVD. Using a random matrix Ω, with a special factorization such as SRFT [8, Section 4.6], randomized schemes can produce an approximate SVD using only $\mathcal{O}(ml \log(k + p_{\text{ovs}}) + (m + l)(k + p_{\text{ovs}})^2)$ flops, because the multiplication $B\Omega$ can be computed in $\mathcal{O}(ml \log(k + p_{\text{ovs}}))$. In contrast, the cost of a classical approach is typically $\mathcal{O}(mlk)$ flops. For the projection error $\|B - QQ^\mathsf{T}B\|_2$, a probabalistic error bound can be proven: With probability at least $1 - 3p_{\text{ovs}}^{-p_{\text{ovs}}}$ the bound $\|B - QQ^\mathsf{T}B\|_2 \le (1 + \sqrt{9k + p_{\text{ovs}}} \min(m, l))\sigma_{k+1}$ holds under mild assumptions on p_{ovs}, with σ_{k+1} denoting the $(k+1)$th singular value of B [8, Section 10.3]. This error (bound) can be further improved by a power iteration, which means that for $q_{\text{pow}} \in \mathbb{N}_0$, the random sketch is computed as $B(B^\mathsf{T}B)^{q_{\text{pow}}}\Omega$. This is in particular useful for matrices whose singular values decay slowly. Also, randomized a posteriori error estimation is possible, even at almost no additional computational cost (see [8, Sections 4.3 and 4.4]).

3   Randomized Symplectic Model Order Reduction

In this section, two new randomized symplectic methods are presented. In the following $\mathrm{i} \in \mathbb{C}$ denotes the imaginary unit and $(\cdot)^\mathsf{H}$ the complex transpose. Moreover, we use MATLAB-style notation for matrix indexing and stacking.

Randomized Complex SVD

Consider the snapshot matrix $X_{\mathrm{s}} := (x_i^{\mathrm{s}})_{i=1}^{n_{\mathrm{s}}} \in \mathbb{R}^{2N \times n_{\mathrm{s}}}$, where X_{s} is split into $X_{\mathrm{s}} = [Q_s; P_s]$, with $Q_s, P_s \in \mathbb{R}^{N \times n_{\mathrm{s}}}$. The main idea of the cSVD algorithm is to form a complex snapshot matrix $X_{\mathrm{s}}^{\mathrm{c}} := Q_s + \mathrm{i}P_s \in \mathbb{C}^{N \times n_{\mathrm{s}}}$ and compute a truncated SVD of this complex matrix $X_{\mathrm{s}}^{\mathrm{c}} \approx U_{\mathrm{C}}\Sigma_{\mathrm{C}}V_{\mathrm{C}}^\mathsf{H}$. The matrix $U_{\mathrm{C}} \in \mathbb{C}^{N \times k}$ is then split into real and imaginary part $U_{\mathrm{C}} = V_Q + \mathrm{i}V_P$ and mapped to

$$V := \mathcal{A}(U_{\mathrm{C}}) := \begin{pmatrix} V_Q & -V_P \\ V_P & V_Q \end{pmatrix}.$$

This mapping $\mathcal{A}$ from the complex Stiefel manifold $V_k(\mathbb{C}^n)$ to $\mathbb{R}^{2N \times 2k}$ maps a complex matrix with orthonormal columns to a real symplectic matrix (see [10]). The symplectic matrix V and its symplectic inverse V^+ are then used for MOR. Instead of using a truncated SVD, we apply randomization in order to compute a rank-k approximation of $X_{\mathrm{s}}^{\mathrm{c}}$. A pseudocode is presented as Algorithm 2.

---

^{2} In the rare case of linear dependencies in Y, the number of columns in Q is reduced.

Algorithm 2. Randomized Complex SVD (rcSVD)

Input: Snapshot matrix $X_s \in \mathbb{R}^{2N \times n_s}$, target rank $2k \in \mathbb{N}$ of the ROB, oversampling parameter $p_{ovs} \in \mathbb{N}_0$, power iteration number $q_{pow} \in \mathbb{N}_0$

Output: Symplectic ROB matrix $V_{rcSVD} \in \mathbb{R}^{2N \times 2k}$

1: $X_s^c = X_s(1:N,:) + iX_s((N+1):(2N),:)$ $\qquad\triangleright$ construct complex snapshot matrix
2: $\Omega = \texttt{SRFT}(N, k + p_{ovs})$ $\qquad\triangleright$ draw a random $N \times (k + p_{ovs})$ test matrix
3: $[U_C, \Sigma_C, V_C] = \texttt{SVD}(X_s^c((X_s^c)^H X_s^c)^{q_{pow}} \Omega, k)$ $\qquad\triangleright$ basis for approximation of Y_c
4: $V_Q = \mathrm{Re}(U_C)$, $V_P = \mathrm{Im}(U_C)$ $\qquad\triangleright$ split in real and imaginary part
5: $V_{rcSVD} = [V_Q, -V_P; V_P, V_Q]$ $\qquad\triangleright$ map to symplectic matrix

---

Randomized SVD-like Decomposition

In [13] it is shown that each real $2N \times n_s$ matrix can be decomposed as $X_s = SDP^T$, with $S \in \mathbb{R}^{2N \times 2N}$ symplectic, $P \in \mathbb{R}^{n_s \times n_s}$ orthogonal,

$$
D^T = \overbrace{\begin{array}{c} p \end{array}}^{} \quad
D^T = \begin{pmatrix}
\Sigma & 0 & 0 & 0 & 0 \\
0 & I_q & 0 & 0 & 0 \\
0 & 0 & 0 & \Sigma & 0 \\
0 & 0 & 0 & 0 & 0
\end{pmatrix} \in \mathbb{R}^{n_s \times 2N}, \quad
\Sigma = \begin{pmatrix}
\sigma_1 & 0 & 0 \\
0 & \ddots & 0 \\
0 & 0 & \sigma_p
\end{pmatrix} \in \mathbb{R}^{p \times p}
$$

with column group labels: $p \quad q \quad N-p-q \quad p \quad N-p$

with $\sigma_i > 0$, $i = 1, \dots, p$. Note that due to symplecticity of S

$$
K := X_s^T J_{2N} X_s = PD^T S^T J_{2N} SDP^T = PD^T J_{2N} DP^T,
$$

where $D^T J_{2N} D \in \mathbb{R}^{n_s \times n_s}$ is a matrix with $\{\pm\sigma_i^2\}_{i=1}^p$ on the $(p + q)$th super-diagonal and subdiagonal and zeros everywhere else. Thus, the factorization $PD^T J_{2N} DP^T$ is a permutation of the real Schur decomposition $K = UTU^T$, where $\{+\sigma_i^2\}_{i=1}^p$ are on the first superdiagonal and $\{-\sigma_i^2\}_{i=1}^p$ on the first subdiagonal of T and U is orthogonal. Instead of performing a standard Schur decomposition, we randomize the Schur decomposition of K and obtain a randomized SVD-like decomposition in this way. A pseudocode is presented as Algorithm 3.

Algorithm 3. Randomized SVD-like decomposition (rSVD-like)

Input: Snapshot matrix $X_s \in \mathbb{R}^{2N \times n_s}$, target rank $2k \in \mathbb{N}$ of the ROB, oversampling parameter $p_{ovs} \in \mathbb{N}_0$, power iteration number $q_{pow} \in \mathbb{N}_0$

Output: Symplectic ROB matrix $V_{rSVD\text{-}like} \in \mathbb{R}^{2N \times 2k}$

1: $Q = \texttt{computeQ}(p_{ovs}, q_{pow}, X_s)$ $\qquad\triangleright$ see end Section 3
2: $K = X_s^T J_{2N} X_s$
3: $[U, T] = \texttt{realSchur}(Q^T KQ)$ $\qquad\triangleright$ compute real Schur decomposition
4: $p = \mathrm{rank}(T)/2$
5: $\Sigma = \mathrm{diag}(\sqrt{T_{1,2}}, \sqrt{T_{3,4}}, ..., \sqrt{T_{2p-1,2p}})$ $\qquad\triangleright$ extract real $\sigma_i, i = 1, .., p$
6: $P := [p_1, ..., p_{n_s}] := QU \cdot [I_{n_s}(:,1:2:2p-1), I_{n_s}(:,2:2:2p), I_{n_s}(:,2p+1:n_s)]$
7: $V_{rSVD\text{-}like} = X_s [P(:,1:k)\Sigma(1:k,1:k)^{-1}, P(:,p+1:p+k)\Sigma(1:k,1:k)^{-1}]$

For the function $\texttt{computeQ}(p_{\mathrm{ovs}}, q_{\mathrm{pow}}, \boldsymbol{X}_{\mathrm{s}})$ in Step 1, one of the following methods has to be inserted, where $\texttt{randn}$ generates a random Gaussian matrix:

1. $\texttt{computeQfromK}(p_{\mathrm{ovs}}, q_{\mathrm{pow}}, \boldsymbol{X}_{\mathrm{s}})$:
   $\boldsymbol{K} = \boldsymbol{X}_{\mathrm{s}}^{\mathsf{T}} \mathbb{J}_{2N} \boldsymbol{X}_{\mathrm{s}}$, $\boldsymbol{\Omega} = \texttt{randn}(n_{\mathrm{s}}, k + p_{\mathrm{ovs}})$, $\boldsymbol{Q} = \texttt{orth}(\boldsymbol{K}^{2q_{\mathrm{pow}}+1} \boldsymbol{\Omega})$
2. $\texttt{computeQfromXs}(p_{\mathrm{ovs}}, q_{\mathrm{pow}}, \boldsymbol{X}_{\mathrm{s}})$:
   $\boldsymbol{\Omega} = \texttt{randn}(2N, k + p_{\mathrm{ovs}})$, $\boldsymbol{Q} = \texttt{orth}(\boldsymbol{X}_{\mathrm{s}}^{\mathsf{T}} (\boldsymbol{X}_{\mathrm{s}} \boldsymbol{X}_{\mathrm{s}}^{\mathsf{T}})^{q_{\mathrm{pow}}} \boldsymbol{\Omega})$
3. $\texttt{computeQfromKXs}(p_{\mathrm{ovs}}, q_{\mathrm{pow}}, \boldsymbol{X}_{\mathrm{s}})$:
   $\boldsymbol{\Omega}_K = \texttt{randn}(n_{\mathrm{s}}, \lceil \frac{k+p_{\mathrm{ovs}}}{2} \rceil)$, $\boldsymbol{\Omega}_X = \texttt{randn}(2N, \lfloor \frac{k+p_{\mathrm{ovs}}}{2} \rfloor)$,
   $\boldsymbol{K} = \boldsymbol{X}_{\mathrm{s}}^{\mathsf{T}} \mathbb{J}_{2N} \boldsymbol{X}_{\mathrm{s}}$, $\boldsymbol{Q} = \texttt{orth}([\boldsymbol{K}^{2q_{\mathrm{pow}}+1} \boldsymbol{\Omega}_K, \boldsymbol{X}_{\mathrm{s}}^{\mathsf{T}} (\boldsymbol{X}_{\mathrm{s}} \boldsymbol{X}_{\mathrm{s}}^{\mathsf{T}})^{q_{\mathrm{pow}}} \boldsymbol{\Omega}_X])$.

The resulting variants of the rSVD-like algorithm will be named accordingly: rSVD-like*, where * is to be replaced by K, Xs or KXs depending on the computational variant for $\boldsymbol{Q}$.

4   Numerical Experiments

We apply the developed randomized, structure-preserving methods to a 2D linear wave equation model. The initial boundary value problem for the unknown $u(t, \boldsymbol{\xi})$ with the spatial variable $\boldsymbol{\xi} := (\xi_1, \xi_2) \in \Omega := (0, 1) \times (0, 0.2)$ and the temporal variable $t \in I_t(\boldsymbol{\mu}) := [t_0, t_{\mathrm{end}}(\boldsymbol{\mu})]$, reads

$$
\begin{aligned}
u_{tt}(t, \boldsymbol{\xi}) &= c^2 \Delta u(t, \boldsymbol{\xi}) && \text{in } I_t(\boldsymbol{\mu}) \times \Omega \\
u(0, \boldsymbol{\xi}) &= u^0(\boldsymbol{\xi}) := h(s(\boldsymbol{\xi})), \quad u_t(0, \boldsymbol{\xi}) = v^0(\boldsymbol{\xi}) := 0 && \text{in } \Omega, \\
u(t, \boldsymbol{\xi}) &= 0 && \text{in } I_t(\boldsymbol{\mu}) \times \partial\Omega,
\end{aligned}
$$

with

$$
s(\boldsymbol{\xi}) = 10 \cdot \left(\xi_2 - \frac{1}{2} \right), \qquad
h(s) = \begin{cases} 1 - \frac{3}{2}|s|^2 + \frac{3}{4}|s|^3, & 0 \le |s| \le 1 \\ \frac{1}{4}(2 - |s|)^3, & 1 < |s| \le 2 \\ 0, & |s| > 2. \end{cases}
$$

We choose $t_0 = 0$, $t_{\mathrm{end}}(\boldsymbol{\mu}) = 2/\boldsymbol{\mu}$ and as parameter (vector) $\boldsymbol{\mu} = c \in [1, 2]$. Spatial discretization via central finite differences leads to the Hamiltonian system

$$
\frac{\mathrm{d}}{\mathrm{d}t} \boldsymbol{x}(t; \boldsymbol{\mu}) = \mathbb{J}_{2N} \nabla_{\boldsymbol{x}} \mathcal{H}(\boldsymbol{x}(t; \boldsymbol{\mu}); \boldsymbol{\mu}) = \mathbb{J}_{2N} \boldsymbol{A}(\boldsymbol{\mu}) \boldsymbol{x}, \quad \boldsymbol{x}(t_0; \boldsymbol{\mu}) = \boldsymbol{x}_0(\boldsymbol{\mu}) \tag{3}
$$

with

$$
\boldsymbol{x}_0(\boldsymbol{\mu}) = [u^0(\boldsymbol{\xi}_1)); ...; u^0(\boldsymbol{\xi}_N)); \boldsymbol{0}_{N \times 1}], \quad \boldsymbol{A}(\boldsymbol{\mu}) = \begin{pmatrix} \boldsymbol{\mu}^2 (\boldsymbol{D}_{\xi_1 \xi_1} + \boldsymbol{D}_{\xi_2 \xi_2}) & \boldsymbol{0}_N \\ \boldsymbol{0}_N & \boldsymbol{I}_N \end{pmatrix},
$$

where $\{\boldsymbol{\xi}_i\}_{i=1}^{N} \subset \Omega$ are the grid points and the positive definite matrices $\boldsymbol{D}_{\xi_1, \xi_1}$, $\boldsymbol{D}_{\xi_2 \xi_2} \in \mathbb{R}^{N \times N}$ denote the three-point central difference approximations in ξ_1-direction and in ξ_2-direction. The domain is discretized equidistantly with 1000 grid points in ξ_1-direction and 20 points in ξ_2-direction which results in $N = 1000 \cdot 20 = 20000$ grid points in total. The corresponding Hamiltonian reads

$\mathcal{H}(x; \mu) = \frac{1}{2}x^{\mathsf{T}}A(\mu)x$. Temporal discretization is achieved with the implicit midpoint rule and $n_t = 1000$ equidistant time steps. This results in different time step sizes for different parameters. The experiments are performed with MATLAB R2021 on a Microsoft Surface with 8 GB RAM and a CPU with 1.1 GHz. In Figs. 1 and 2, we present basis generation times and the relative reduction error

$$e_{\text{rel}}(\mu) = \sqrt{\sum_{i=0}^{n_t} ||x_i(\mu) - Vx_{\text{r},i}(\mu)||_2^2} \Big/ \sqrt{\sum_{i=0}^{n_t} ||x_i(\mu)||_2^2}, \qquad (4)$$

with $x_i(\mu), x_{\text{r},i}(\mu), i = 0, .., n_t$ the iterates of the full model time-stepping and reduced model time-stepping, in dependence on the basis size. All results are averaged over 10 random parameters $\mu \in \mathcal{P}$. The results of the randomized methods are additionaly averaged over 5 runs for different random sketching matrices. The snapshot matrix consisting of $n_s = 2000$ snapshots is computed from the parameters $\mu_1 = 1, \mu_2 = 2$. The rcSVD is compared with three different versions of the cSVD: For 'cSVD full' a full SVD is computed and then truncated, for 'cSVD with svds' the MATLAB function svds is used that computes only the first k singular vectors and values, for 'cSVDev' the first k eigenvectors $v_1, ..., v_k$ and eigenvalues $\lambda_1, ..., \lambda_k$ of $(X_s^c)^{\mathsf{T}}X_s^c$ are computed and $U_C = X_s^c [v_1/\sqrt{\lambda_1}, ..., v_k/\sqrt{\lambda_k}]$ is set. We observe that the basis generation times are strongly reduced by randomization. With one power iteration and some oversampling (almost) the same reduction error is obtained for the rcSVD and the rSVD-likeKXs, compared to its classical version. With the other two versions of the rSVD-like either a competitive error for only small basis sizes (rSVD-likeK) or only large basis sizes (rSVD-likeXs) is obtained. Overall, the randomized techniques are able to reduce the computational time for generating a basis yielding a certain error. For example an accuracy of about 10^{-4} is achieved with about 160 basis vectors by both, the cSVD and the rcSVD with $q_{\text{pow}} = 1$ and $p_{\text{ovs}} = 10$. However, the basis generation by the rcSVD is about 2.5 times faster. One power iteration significantly improves the error, but leads to higher runtimes. Oversampling with $p_{\text{ovs}} = 10$ only slightly effects error and runtime. With about 100 basis vectors, an accuracy of 10^{-4} is reached for both, the SVD-like and the rSVD-likeKXs with one power iteration and $p_{\text{ovs}} = 30$. But the computational costs are more than 3 times less with the randomized approach. Again, one power iteration significantly improves the error, but also increases the runtimes. The code for the experiments is available under https://doi.org/10.18419/darus-3519.

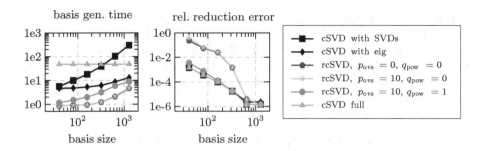

Fig. 1. Basis generation times and relative reduction errors (4), rcSVD

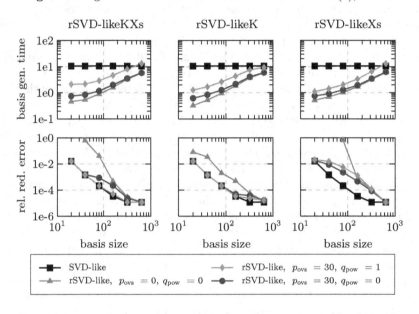

Fig. 2. Basis generation times and relative reduction errors (4), rSVD-like

5   Conclusion and Outlook

In our work, we have shown that randomized matrix factorizations can be used for the structure-preserving basis generation for symplectic MOR. The newly presented methods, the rcSVD and the rSVD-like decomposition, both provide very accurate approximations and lead to significant computational speed-ups compared to their classical version. Future work will deal with the question how a randomization of the real canonical form can be obtained for randomizing the SVD-like decomposition instead of using a randomized Schur decomposition.

Acknowledgements. Supported by Deutsche Forschungsgemeinschaft (DFG, German Research Foundation) Project No. 314733389, and under Germany's Excellence

Strategy - EXC 2075 - 390740016. We acknowledge the support by the Stuttgart Center for Simulation Science (SimTech).

References

1. Alla, A., Kutz, J.N.: Randomized model order reduction. ACOM **45**(3), 1251–1271 (2019)
2. Bach, C., Ceglia, D., Song, L., Duddeck, F.: Randomized low-rank approximation methods for projection-based model order reduction of large nonlinear dynamical problems. IJNME **118**(4), 209–241 (2019)
3. Balabanov, O.: Randomized linear algebra for model order reduction. École centrale de Nantes; Universitat politécnica de Catalunya, Theses (2019)
4. Balabanov, O., Nouy, A.: Randomized linear algebra for model reduction. Part i: Galerkin methods and error estimation. ACOM **45**(5), 2969–3019 (2019)
5. Benner, P., Ohlberger, M., Cohen, A., Willcox, K.: Model reduction and approximation. society for industrial and applied mathematics, Philadelphia, PA (2017)
6. Buchfink, P., Bhatt, A., Haasdonk, B.: Symplectic model order reduction with non-orthonormal bases. MCA **24**(2), 43 (2019)
7. Buhr, A., Smetana, K.: Randomized local model order reduction. SISC **40**(4), A2120–A2151 (2018)
8. Halko, N., Martinsson, P.G., Tropp, J.A.: Finding structure with randomness: probabilistic algorithms for constructing approximate matrix decompositions. SIAM Rev. **53**(2), 217–288 (2011)
9. Maboudi Afkham, B., Hesthaven, J.: Structure preserving model reduction of parametric Hamiltonian systems. SISC **39**(6), A2616–A2644 (2017)
10. Peng, L., Mohseni, K.: Symplectic model reduction of Hamiltonian systems. SISC **38**(1), A1–A27 (2016)
11. da Silva, A.C.: Lectures on Symplectic Geometry. Springer, Heidelberg (2008)
12. Volkwein, S.: Proper orthogonal decomposition: theory and reduced-order modelling. Lect. Notes, Univ. Konstanz **4**(4), 1–29 (2013)
13. Xu, H.: An SVD-like matrix decomposition and its applications. Linear Algebra Appl. **368**, 1–24 (2003)

Adaptive Localized Reduced Basis Methods for Large Scale PDE-Constrained Optimization

Tim Keil[(✉)], Mario Ohlberger, and Felix Schindler

Institute for Analysis and Numerics, University of Münster,
Einsteinstr. 62, 48149 Münster, Germany
{tim.keil,mario.ohlberger,felix.schindler}@uni-muenster.de
https://www.wwu.de/AMM/ohlberger

Abstract. In this contribution, we introduce and numerically evaluate a certified and adaptive localized reduced basis method as a local model in a trust-region optimization method for parameter optimization constrained by partial differential equations.

Keywords: localized reduced basis method · trust-region optimization · online enrichment

1 Introduction

We are concerned with efficient and certified approximations of multiscale or large-scale PDE-constrained parameter optimization problems. In particular, for a parameter space $\mathcal{P} \subset \mathbb{R}^P$, $P \in \mathbb{N}$, we search for a local solution of (P), i.e.

$$\min_{\mu \in \mathcal{P}} \mathcal{J}(u_\mu, \mu), \qquad \text{with } \mathcal{J} : V \times \mathcal{P} \to \mathbb{R}, \qquad \text{(P.a)}$$

where $u_\mu \in V$ solves the parameterized state equation for $\mu \in \mathcal{P}$:

$$a(u_\mu, v; \mu) = l(v; \mu) \qquad \text{for all } v \in V. \qquad \text{(P.b)}$$

Here, V is a Hilbert space and, for each admissible parameter $\mu \in \mathcal{P}$, $a(\cdot, \cdot; \mu) : V \times V \to \mathbb{R}$ denotes a continuous and coercive bilinear form and $l(\cdot; \mu) \in V'$.

Particularly, we are interested in multiscale or large-scale applications in the sense that the state equation (P.b) is a weak formulation of a PDE of the form

$$-\nabla \cdot \big(A(\mu)\nabla u_\mu\big) = f(\mu) \ \text{ in } \Omega, \qquad u_\mu = 0 \ \text{ on } \partial\Omega, \qquad (1)$$

where $\Omega \subset \mathbb{R}^d$ is a bounded domain and A denotes a diffusion tensor with a rich structure that would lead to very high dimensional approximation spaces for the state space when approximated with classical finite element type methods. On the other hand, employing model order reduction for the involved equations is problematic for high dimensional parameter spaces $\mathcal{P}$ due to large offline times.

I. Lirkov and S. Margenov (Eds.): LSSC 2023, LNCS 13952, pp. 108–116, 2024.
https://doi.org/10.1007/978-3-031-56208-2_10

Efficient adaptive model order reduction in combination with trust-region optimization methods in this context has recently been addressed in [5,10,12] employing global reduced basis models and in [6] employing a reduced basis localized orthogonal decomposition. The latter work is a first contribution based on spatial localization targeted towards problems where the computational cost of classical global methods is prohibitively large. In this work, we propose a new variant of the trust-region (TR) approach based on the localized reduced basis method (LRBM) [8,9]. We also refer to [3] for a review of localized model order reduction approaches and to [11] for recent advances in non-linear problems.

In the following, we first review the LRBM for elliptic multiscale problems. We then propose a new TR-LRBM method with certification and adaptive online enrichment for our model problem (P). Finally, in Sect. 5, we show numerical experiments demonstrating the proposed approach's full power.

2   Optimality System, Primal and Dual Equation

The primal residual of (P.b) is key for the optimization as well as for a posteriori error estimation. We define for given $u \in V$, $\mu \in \mathcal{P}$, the primal residual $r_\mu^{\mathrm{pr}}(u) \in V'$ associated with (P.b) by

$$r_\mu^{\mathrm{pr}}(u)[v] := l(v; \mu) - a(u, v; \mu) \qquad \text{for all } v \in V. \tag{2}$$

Following the approach of *first-optimize-then-discretize*, we base our discretization and model order reduction approach on the first-order necessary optimality system, i.e. (cf. [1] for details and further references)

$$r_{\bar{\mu}}^{\mathrm{pr}}(\bar{u})[v] = 0 \qquad \text{for all } v \in V, \tag{3a}$$

$$\partial_u \mathcal{J}(\bar{u}, \bar{\mu})[v] - a(v, \bar{p}; \bar{\mu}) = 0 \qquad \text{for all } v \in V, \tag{3b}$$

$$(\partial_\mu \mathcal{J}(\bar{u}, \bar{\mu}) + \nabla_\mu r_{\bar{\mu}}^{\mathrm{pr}}(\bar{u})[\bar{p}]) \cdot (\nu - \bar{\mu}) \geq 0 \qquad \text{for all } \nu \in \mathcal{P}, \tag{3c}$$

where we employ suitable differentiability assumptions. From (3b) we deduce the so-called *adjoint equation* and corresponding residual

$$r_\mu^{\mathrm{du}}(u_\mu, p_\mu)[q] := \partial_u \mathcal{J}(u_\mu, \mu)[q] - a(q, p_\mu; \mu) = 0 \qquad \text{for all } q \in V, \tag{4}$$

with solution $p_\mu \in V$ for fixed $\mu \in \mathcal{P}$ and $u_\mu \in V$ solution to (P.b).

Since (P.b) has a unique solution, we use the so-called reduced cost functional $\hat{\mathcal{J}} : \mathcal{P} \to \mathbb{R}$, $\mu \mapsto \hat{\mathcal{J}}(\mu) := \mathcal{J}(u_\mu, \mu)$, where we note that the term "reduced" is not associated with model order reduction as introduced below. For given $\mu \in \mathcal{P}$, first-order derivatives of $\hat{\mathcal{J}}$ can be computed utilizing the adjoint approach, i.e., the gradient $\nabla_\mu \hat{\mathcal{J}} : \mathcal{P} \to \mathbb{R}^P$ is given by $\nabla_\mu \hat{\mathcal{J}}(\mu) = \nabla_\mu \mathcal{J}(u_\mu, \mu) + \nabla_\mu r_\mu^{\mathrm{pr}}(u_\mu)[p_\mu]$.

3   DG-MsFEM and Localized Reduced Basis Method

Let us introduce appropriate localized discretization schemes for the primal and dual equation based on non-overlapping and non-conforming domain decomposition in the context of a Discontinuous Galerkin approach on a coarse grid. We restrict our demonstration to the primal equation as an analogous method for the dual equation is straightforward.

In order to derive a full order model (FOM) for our model reduction approach, we assume that a non-overlapping decomposition of the underlying domain Ω is given by a coarse grid $\mathcal{T}_H$ with subdomains $T_j \in \mathcal{T}_H, j = 1, \ldots N_H$. Furthermore, each cell T_j is further decomposed by a local triangulation $\tau_h(T_j)$ that resolves all fine-scale features of the multiscale problem (1). We then define the global fine-scale partition τ_h as the union of all its local contributions, i.e., $\tau_h = \bigcup_{j=1}^{N_H} \tau_h(T_j)$. For a triangulation $\tau_h(\omega)$ of some $\omega \subseteq \Omega$, let $H^p(\tau_h(\omega)) := \{v \in L^2(\omega) \mid v|_t \in H^p(t) \; \forall t \in \tau_h(\omega)\}$ denote the broken Sobolev space of order $p \in \mathbb{N}$ on $\tau_h(\omega)$. Then, $H^p(\tau_h)$ naturally inherits the decomposition $H^p(\tau_h) = \bigoplus_{j=1}^{N_H} H^p(\tau_h(T_j))$. We further define the coarse approximation space $V^c := Q^1(\mathcal{T}_H) \subset H^2(\mathcal{T}_H)$ and on each coarse element T the fine-scale correction space $V^f(T) := S^1(\tau_h(T)) \subset H^1(T)$, where $Q^1(\mathcal{T}_H)$ denotes the standard non-conforming and $S^1(\tau_h(T))$ the standard conforming piecewise linear (discontinuous) Galerkin finite element space on $\mathcal{T}_H$ and $\tau_h(T)$, respectively. Finally, we define the non-conforming solution space

$$V(\tau_h) := \bigoplus_{j=1}^{N_H} V^j \subset H^2(\tau_h) \quad \text{with} \quad V^j := V^c|_{T_j} \oplus V^f(T_j).$$

Definition 1 (Discontinuous Galerkin multiscale FEM (DG-MsFEM)). *We call $u_{h,\mu} \in V(\tau_h)$ an approximate DG-MsFEM reference solution of (1), if*

$$a_{\mathrm{DG}}(u_{h,\mu}, v; \mu) = l_{\mathrm{DG}}(v; \mu) \qquad \text{for all } v \in V(\tau_h). \tag{5}$$

Here, the DG bilinear form a_{DG} and the right hand side l_{DG} are given as

$$a_{\mathrm{DG}}(v, w; \mu) := \sum_{t \in \tau_h} \int_t A(\mu) \nabla v \cdot \nabla w + \sum_{e \in \mathcal{F}(\tau_h)} a_{\mathrm{DG}}^e(v, w; \mu), \quad l_{\mathrm{DG}}(v; \mu) := \sum_{t \in \tau_h} \int_t f(\mu) v,$$

where $\mathcal{F}(\cdot)$ denotes the set of all faces of a triangulation and the DG coupling bilinear form a_{DG}^e for a face e is given by

$$a_{\mathrm{DG}}^e(v, w; \mu) := \int_e \langle A(\mu) \nabla v \cdot n_e \rangle [w] + \langle A(\mu) \nabla w \cdot n_e \rangle [v] + \frac{\sigma_e(\mu)}{|e|^\beta} [v][w].$$

For any triangulation $\tau_h(\omega)$ of some $\omega \subseteq \Omega$, we assign to each face $e \in \mathcal{F}(\tau_h(\omega))$ a unique normal n_e pointing away from the adjacent cell t^-, where an inner face is given by $e = t^- \cap t^+$ and a boundary face is given by $e = t^- \cap \partial \omega$, for appropriate cells $t^\pm \in \tau_h(\omega)$. The average and jump of a two-valued function $v \in H^2(\tau_h(\omega))$ are given by $\langle v \rangle := \frac{1}{2}(v|_{t^-} + v|_{t^+})$ and $[v] := v|_{t^-} - v|_{t^+}$ for an inner face and by $\langle v \rangle := [v] := v$ for a boundary face, respectively. The parametric penalty function $\sigma_e(\mu)$ and the parameter β must be chosen appropriately to ensure coercivity of a_{DG} and may involve A. We restrict ourselves to the symmetric interior penalty DG scheme for simplicity. For other variants, we refer to [8] and the references therein.

Based on the definition of the DG-MsFEM above, the localized reduced basis method constructs in an iterative online enrichment procedure appropriate low dimensional local approximation spaces $V_N^j \subset V^j$ of dimensions N^j that form the global reduced solution space via

$$V_N = \bigoplus_{j=1}^{N_H} V_N^j, \qquad N := \dim(V_N) = \sum_{j=1}^{N_H} N^j. \qquad (6)$$

Once such a reduced approximation space is constructed, the LRBM approximation is defined as follows.

Definition 2 (The localized reduced basis method (LRBM)). *We call $u_{N,\mu} \in V_N$ a localized reduced basis multiscale approximation of (5) if it holds*

$$a_{\mathrm{DG}}(u_{N,\mu}, v_N; \mu) = l_{\mathrm{DG}}(v_N; \mu) \qquad \text{for all } v_N \in V_N. \qquad (7)$$

Note that $u_{N,\mu}$ solves a globally coupled reduced problem, where all arising quantities can nevertheless be locally computed w.r.t. the local reduced spaces V_N^j. For details on the construction of the local approximation spaces, e.g., with a Greedy-based procedure, we refer to [8].

The local spaces can be built in an online adaptive procedure [9], where only local patch problems need to be solved without requiring a global solve of the DG-MsFEM method. More precisely, let a reduced space V_N and the corresponding reduced approximation $u_{N,\mu} \in V_N$ of (7) for a parameter $\mu \in \mathcal{P}$ be given. For enriching the local space V_N^j associated with $T \in \mathcal{T}_H$ at μ, we consider a local oversampling domain $O_T := U(T)$, where $U(T)$ denotes a neighborhood of T, consisting of an additional layer of coarse neighbouring elements $\bar{T} \in \mathcal{T}_H$ of T. We further define $V(O_T)$ as the restriction of $V(\tau_h)$ to $O(T)$ and solve for a local correction $\varphi_T \in V(O_T)$, such that

$$a_{\mathrm{DG}}\big|_{O_T}(u_{N,\mu} + \varphi_T, v; \mu) = l_{\mathrm{DG}}\big|_{O_T}(v; \mu) \qquad \text{for all } v \in V(O_T), \qquad (8)$$

with boundary data from the preceding reduced global solution. The local reduced space V_N^j is then enriched with the restriction of φ_T to T.

In what follows, we do not precompute local approximate spaces but build them iteratively within the optimization routine to solve (P).

4   Relaxed Trust-Region Optimization with Reduced Models

A relaxed adaptive trust-region (TR) method using a localized orthogonal decomposition numerical multiscale method has recently been proposed and extensively studied in [6]. The algorithm proposed in the sequel follows this approach by replacing the FOM and the reduced-order model (ROM) from [6] by the DG-MsFEM and LRBM introduced in Sect. 3.

The overall TR-LRBM algorithm iteratively computes a first-order critical point of problem (P). In the following we denote with $\hat{\mathcal{J}}_N(\mu) := \mathcal{J}(u_{N,\mu}, \mu)$ the reduced objective functional obtained with LRBM, while $\hat{\mathcal{J}}_h(\mu) := \mathcal{J}(u_{h,\mu}, \mu)$ denotes the reduced functional obtained with DG-MsFEM. We will assume that $\hat{\mathcal{J}}_N$ admits an a posteriori error estimate of the form

$$|\hat{\mathcal{J}}_h(\mu) - \hat{\mathcal{J}}_N(\mu)| \leq \Delta_{\hat{\mathcal{J}}_N}(\mu).$$

A derivation of a suitable localizable a posteriori error estimate of this form for the LRBM is beyond the scope of this contribution and is subject to a forthcoming article.

For each outer iteration $k \geq 0$ of the relaxed TR method, we consider a model function $m^{(k)}$ as a cheap local approximation of the cost functional $\hat{\mathcal{J}}$ in the relaxed trust-region, which has radius $\delta^{(k)} + \varepsilon^{(k)}$, where $\delta^{(k)}$ can be characterized by the a posteriori error estimator and $\varepsilon^{(k)}$ denotes a relaxation parameter from an a priori chosen null sequence, where we assume the existence of $K \in \mathbb{N}$ such that $\varepsilon^{(k)} = 0$ for all $k > K$. In our approach, we choose $m^{(k)}(\cdot) := \hat{\mathcal{J}}_N^{(k)}(\mu^{(k)} + \cdot)$ for $k \geq 0$. The super-index (k) indicates that we use different localized RB spaces V_{N_k} in each iteration. Thus, we can use $\Delta_{\hat{\mathcal{J}}_N^{(k)}}(\mu)$ for characterizing the trust-region. In every outer iteration step, we solve for a local solution $\bar{s} \in \mathcal{P}$ of the following inner error-aware constrained optimization sub-problem:

$$\min_{s \in \mathcal{P}} \hat{\mathcal{J}}_N^{(k)}(\tilde{\mu}) \quad \text{s.t.} \quad \frac{\Delta_{\hat{\mathcal{J}}_N^{(k)}}(\tilde{\mu})}{\hat{\mathcal{J}}_N^{(k)}(\tilde{\mu})} \leq \delta^{(k)} + \varepsilon^{(k)}, \quad \tilde{\mu} := \mu^{(k)} + s \in \mathcal{P} \tag{9}$$
$$\text{and } r_{\tilde{\mu}}^{\mathrm{pr}}(u_{\tilde{\mu}})[v] = 0 \text{ for all } v \in V.$$

and set $\mu^{(k+1)} := \mu^{(k)} + \bar{s}$ for the next outer iterate.

In our TR algorithm, we build on the algorithm in [5]. However, we employ a conforming approach, where the reduced primal and dual spaces coincide and are only constructed from snapshots of the primal equation. In the sequel, we only summarize the main features of the algorithm and refer to [5] for more details. A deeper discussion of suitable localized a posteriori error estimates and more sophisticated construction principles are subject to future work.

As usual in the context of the LRBM, we initialize the local RB spaces V_N^j with the Lagrangian partition of unity w.r.t. $\mathcal{T}_H$, interpolated on the local grids. Typical for the adaptive TR method, we then initialize the spaces with the starting parameter $\mu^{(0)}$ by using local corrections from (8).

For every iteration point $\mu^{(k)}$, we solve (9) with the quasi-Newton projected BFGS algorithm combined with an Armijo-type condition and terminate with a standard reduced first-order critical point (FOC) criterion, modified with a projection on the parameter space $P_{\mathcal{P}}$ to account for constraints on the parameter space. Additionally, we use a second boundary termination criterion to prevent the subproblem from too many iterations on the boundary of the trust-region.

After the next iterate $\mu^{(k+1)}$ has been computed, we use the sufficient decrease condition to decide whether to accept the iterate:

$$\hat{\mathcal{J}}_N^{(k+1)}(\mu^{(k+1)}) \leq \hat{\mathcal{J}}_N^{(k)}(\mu_{\mathrm{AGC}}^{(k)}) + \varepsilon^{(k)} \qquad \text{for all } k \in \mathbb{N}, \tag{10}$$

where $\mu_{\mathrm{AGC}}^{(k)}$ denotes the approximate generalized Cauchy point computed with one gradient-descent step of (9). Condition (10) can be cheaply checked with the help of a sufficient and necessary condition. If $\mu^{(k+1)}$ is accepted, we enrich the local RB spaces $V_{N_k}^j$ by again solving (8) for every oversampling domain $O(T_j)$ (only for the primal space) and setting $V_{N_{k+1}}^j := \mathrm{span}\ \{V_{N_k}^j, \varphi_{T_j}|_{T_j}\}$. We also refer to [1], where a strategy is proposed to skip an enrichment.

Unlike in the previous works, we do not have immediate access to the FOM gradient since we do not rely on expensive global DG-MsFEM solves for enrichment. Thus, after enrichment, we evaluate the ROM-type FOC condition

$$\|\mu^{(k+1)} - \mathrm{P}_{\mathcal{P}}(\mu^{(k+1)} - \nabla_\mu \hat{\mathcal{J}}_N(\mu^{(k+1)}))\|_2 \leq \tau_{\mathrm{FOC}}. \tag{11}$$

If this condition is not fulfilled at the current outer iterate, we continue the algorithm without FOM computations. If it is fulfilled, we can not reliably terminate the algorithm. Instead, we check the overall convergence of the algorithm with the usual FOM-type FOC condition

$$\|\mu^{(k+1)} - \mathrm{P}_{\mathcal{P}}(\mu^{(k+1)} - \nabla_\mu \hat{\mathcal{J}}_h(\mu^{(k+1)}))\|_2 \leq \tau_{\mathrm{FOC}}. \tag{12}$$

If fulfilled, we terminate the algorithm. If not, we enrich the space with the primal- and dual DG-MsFEM approximations (which are available from computing $\nabla_\mu \hat{\mathcal{J}}_h(\mu^{(k+1)})$), i.e., $V_{N_{k+1}}^j := \mathrm{span}\ \{V_{N_k}^j, u_{h,\mu^{(k+1)}}|_{T_j}, p_{h,\mu^{(k+1)}}|_{T_j}\}$, for $j = 1, \ldots, N_H$. Then, we continue until (11) is fulfilled again.

Convergence of the algorithm under suitable assumptions can be proven by using Theorem 3.8 in [1]; see also [6]. Finally, we summarize the basic TR-LRBM algorithm in Algorithm 1.

Algorithm 1: Basic TR-LRBM algorithm

Data: initial data $\mu^{(0)}, \delta^{(0)}$, relaxation $(\varepsilon^{(k)})_k$, tolerances $\beta_2, \tau_{\mathrm{sub}}, \tau_{\mathrm{FOC}}$.

1 Initialize LRBM model with PoU and by solving (8) for $\mu^{(0)}$ and set $k = 0$;
2 **while** *not ROM-based criterion (11)* **do**
3   Inner iteration: Compute $\mu^{(k+1)} := \mu^{(k)} + s^{(k)}$ with $s^{(k)}$ solution of (9);
4   **if** *Sufficient decrease condition (10) is fulfilled with relaxation* $\varepsilon^{(k)}$ **then**
5   Accept $\mu^{(k+1)}$ and enrich the LRBM model by solving (8) at $\mu^{(k+1)}$;
6   Possibly enlarge the TR-radius ;
7   **else**
8   Reject $\mu^{(k+1)}$, shrink the TR radius $\delta^{(k)}$ and go to Line 3;
9   **end**
10   Set $k = k + 1$;
11 **end**
12 **if** *not FOM-based criterion (12)* **then**
13   Enrich LRBM with global solutions from $\hat{\mathcal{J}}_h$ in (12) and go back to Line 3;
14 **end**

---

5   Numerical Experiments

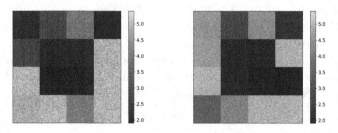

Fig. 1. Coefficient A^1 (left) and A^2 (right) for the desired state of $\mu^{\mathrm{d}} \in \mathcal{P}$.

We demonstrate the algorithm by using the multiscale benchmark problem used in [6]. The experiments have been conducted with PyMOR[1] [7] for the model reduction as well as dune-gdt[2] and the DUNE framework [2] for the DG-MsFEM discretization. We use $\Omega = [0,1]^2$ and define an L^2-misfit objective functional with a Tikhonov-regularization term as $\mathcal{J}(v,\mu) := \frac{\sigma_d}{2}\int_\Omega (v-u^{\mathrm{d}})^2 \mathrm{d}x + \frac{1}{2}\sum_{i=1}^P \sigma_i(\mu_i - \mu_i^{\mathrm{d}})^2 + 1$, where $\mu^{\mathrm{d}} \in \mathcal{P}$ denotes the desired parameter with weight $\sigma_d \in \mathbb{R}$ and $u^{\mathrm{d}} = u_h(\mu^{\mathrm{d}})$ the precomputed desired DG-MsFEM-FOM solution. Note that we do not use a coarsening as was used in [6].

For the diffusion coefficient A in (1), we consider a 4×4-thermal block problem with two different thermal block multiscale coefficients A^1 and A^2, i.e.

$$A(\mu) := A^1(\mu) + A^2(\mu) = \sum_{\xi=1}^{16} \mu_\xi A_\xi^1 + \sum_{\xi=17}^{32} \mu_\xi A_{\xi-16}^2.$$

We consider $\mathcal{P} = [1,4]^{24} \times [1,1.2]^8 \subseteq \mathbb{R}^{32}$ and the parameterized multiscale blocks are given by $A_\xi^1 = A^1\big|_{\Omega_{i,j}}$ and $A_\xi^2 = A^2\big|_{\Omega_{i,j}}$, where $\Omega_{i,j}$ denotes the (i,j)-th thermal block for $i,j = 1,\ldots,4$ enumerated by ξ. The multiscale features are uniformly distributed non-periodic values in $[0.9,1.1]$ on $N_1 \times N_1$ (for A^1) and $N_2 \times N_2$ (for A^2) quadrilateral grids; see Fig. 1 for a visualization. We set $N_1 = 150$ and $N_2 = 300$. Thus, it suffices to have $n_h \times n_h$, $n_h = 600$ fine elements for $\tau_h(\Omega)$ and $n_H \times n_H$, $n_H = 10$ coarse elements for $\mathcal{T}_H$. Both coefficients A^1 and A^2 have low-conductivity blocks in the middle of the domain, i.e., for $\Omega_{i,j}$, $i,j = 2,3$, which is enforced by a restriction on the parameter space. Further, we use a constant source $f \equiv 10$. For all other parameters, we refer to [5] and its accompanying code [4], where the same hyper-parameters for the experiment were used.

We compare two optimization algorithms: An entirely FOM-based BFGS algorithm, where only DG-MsFEM evaluations have been used, and the relaxed

---

[1] see https://pymor.org.
[2] see https://github.com/dune-community/dune-gdt.

TR method described in Sect. 4 using BFGS for the sub-problems. For termination, we use $\tau_{\text{FOC}} = 3 \cdot 10^{-6}$ in order to ensure that the FOM and ROM algorithms stop with the same optimization error in the parameter.

Table 1. Evaluations and accuracy of FOM and ROM.

	Evaluations			Iterations		μ-error
	DG-MsFEM (5)	LRBM (7)	Local (8)	outer	inner	
Cost factor	h	N_{RB}	O_T			
BFGS with DG-MsFEM	259	-	-	85	-	2.89e−3
Relaxed TR-LRBM	2	506	294	2	140	2.38e−3

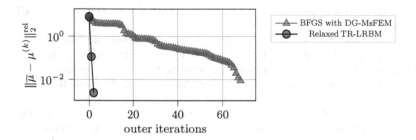

Fig. 2. Relative error decay w.r.t. the optimal parameter of selected algorithms for a single optimization run with the same random initial guess $\mu^{(0)}$ and $\tau_{\text{FOC}} = 3 \cdot 10^{-6}$.

In Fig. 2, we show the number of iterations of the algorithms. Details on the evaluations of global-, reduced- and local problems are given in Table 1. Both algorithms converged to the same point. The TR algorithm almost entirely avoids FOM evaluations, only required in the FOM-based termination (12), which is only evaluated once (corresponding to 2 FOM evaluations for primal and dual).

For large-scale problems, where FEM evaluations and even DG-MsFEM evaluations become more and more expensive, we expect our proposed TR method to be compelling. More complex numerical examples and other model problems are subject to further research.

Acknowledgements. The authors acknowledge funding by the BMBF under contracts 05M20PMA and by the Deutsche Forschungsgemeinschaft under contract OH 98/11-1 as well as under Germany's Excellence Strategy EXC 2044 390685587, Mathematics Münster: Dynamics – Geometry – Structure.

References

1. Banholzer, S., Keil, T., Ohlberger, M., Mechelli, L., Schindler, F., Volkwein, S.: An adaptive projected Newton non-conforming dual approach for trust-region reduced basis approximation of PDE-constrained parameter optimization. Pure Appl. Funct. Anal. **7**(5), 1561–1596 (2022)

2. Bastian, P., et al.: The DUNE framework: basic concepts and recent developments. Comput. Math. Appl. **81**, 75–112 (2021)

3. Buhr, A., Iapichino, L., Ohlberger, M., Rave, S., Schindler, F., Smetana, K.: Localized model reduction for parameterized problems. In: Model Order Reduction. Snapshot-Based Methods and Algorithms, vol. 2, pp. 245–305. Berlin: De Gruyter (2021)

4. Keil, T.: Software for: a relaxed localized trust-region reduced basis approach for optimization of multiscale problems (2023). https://doi.org/10.5281/zenodo.7821980 (2023)

5. Keil, T., Mechelli, L., Ohlberger, M., Schindler, F., Volkwein, S.: A non-conforming dual approach for adaptive trust-region reduced basis approximation of PDE-constrained parameter optimization. ESAIM Math. Model. Numer. Anal. **55**(3), 1239–1269 (2021)

6. Keil, T., Ohlberger, M.: A relaxed localized trust-region reduced basis approach for optimization of multiscale problems (2022). https://doi.org/10.48550/ARXIV.2203.09964

7. Milk, R., Rave, S., Schindler, F.: pyMOR–generic algorithms and interfaces for model order reduction. SIAM J. Sci. Comput. **38**(5), S194–S216 (2016)

8. Ohlberger, M., Schindler, F.: Error control for the localized reduced basis multiscale method with adaptive on-line enrichment. SIAM J. Sci. Comput. **37**(6), A2865–A2895 (2015)

9. Ohlberger, M., Schaefer, M., Schindler, F.: Localized model reduction in PDE constrained optimization. In: Schulz, V., Seck, D. (eds.) Shape Optimization, Homogenization and Optimal Control, International Series of Numerical Mathematics, vol. 169, pp. 143–163. Birkhäuser Springer, Cham (2018)

10. Qian, E., Grepl, M., Veroy, K., Willcox, K.: A certified trust region reduced basis approach to PDE-constrained optimization. SIAM J. Sci. Comput. **39**(5), S434–S460 (2017)

11. Smetana, K., Taddei, T.: Localized model reduction for nonlinear elliptic partial differential equations: localized training, partition of unity, and adaptive enrichment (2022). https://doi.org/10.48550/ARXIV.2202.09872

12. Wen, T., Zahr, M.J.: A globally convergent method to accelerate large-scale optimization using on-the-fly model hyperreduction: application to shape optimization. J. Comput. Phys. **484**, 112082 (2023)

Application of Deep Kernel Models for Certified and Adaptive RB-ML-ROM Surrogate Modeling

Tizian Wenzel[1(✉)], Bernard Haasdonk[1], Hendrik Kleikamp[2], Mario Ohlberger[2], and Felix Schindler[2]

[1] Institute for Applied Analysis and Numerical Simulation, University of Stuttgart, Stuttgart, Germany
{tizian.wenzel,bernard.haasdonk}@mathematik.uni-stuttgart.de
[2] Institute for Analysis and Numerics, Mathematics Münster, University of Münster, Münster, Germany
{hendrik.kleikamp,mario.ohlberger,felix.schindler}@uni-muenster.de

Abstract. In the framework of reduced basis methods, we recently introduced a new certified hierarchical and adaptive surrogate model, which can be used for efficient approximation of input-output maps that are governed by parametrized partial differential equations. This adaptive approach combines a full order model, a reduced order model and a machine-learning model. In this contribution, we extend the approach by leveraging novel kernel models for the machine learning part, especially structured deep kernel networks as well as two layered kernel models. We demonstrate the usability of those enhanced kernel models for the RB-ML-ROM surrogate modeling chain and highlight their benefits in numerical experiments.

Keywords: Deep Kernel Methods · Certified RB-ML-ROM Modeling · Machine Learning · Reduced Order Models · Error Estimation

1 Introduction

Model Order Reduction (MOR) is an indispensable technology in the context of multi-query, real-time or slim computing context, as full oder models (FOMs) of real physical processes may be unaffordable. Many techniques for MOR exist, cf. [1]. A large class of MOR methods are reduced basis (RB) models, which are projection-based and the projection spaces are typically constructed from samples of the FOM and allow rigorous a-posteriori error estimation by residual analysis [4]. As a drawback, the time-integration requires time-marching which may be a bottleneck in case of a fine time resolution. In recent years, also machine learning (ML) based techniques have entered the field, which enable rapid prediction without expensive time-marching [8]. Such data-based approaches, however, mostly lack any error guarantees. Recently, we presented an adaptive and

I. Lirkov and S. Margenov (Eds.): LSSC 2023, LNCS 13952, pp. 117–125, 2024.
https://doi.org/10.1007/978-3-031-56208-2_11

hierarchical RB-ML-ROM framework, which combines the benefits of those two approaches, i.e. rapid prediction of ML-models including certification from RB methods [5]. In the current presentation, we demonstrate the versatility of the RB-ML-ROM framework, by including efficient multilayer kernel models for the ML part, namely Structured Deep Kernel Networks (SDKN) and two-layered greedy kernel models (2L-VKOGA).

The paper is organized as follows: In Sect. 2 and Sect. 3 we recall some more background information on reduced order modeling and machine learning. Section 4 elaborates on novel deep kernel models and Sect. 5 reviews the recently introduced certified and adaptive RB-ML-MOR model. Subsequently, Sect. 6 provides the numerical experiments on the combination of these approaches and finally Sect. 7 summarizes and gives an outlook.

2   Reduced Order Modeling

In this paper, we consider parametrized parabolic PDEs on a bounded Lipschitz-domain $\Omega \subset \mathbb{R}^m$ over the time interval $[0, T]$ with $T > 0$. Further, let $\mathcal{P} \subset \mathbb{R}^p$ be the set of parameters. We denote by V_h a Hilbert space associated to a finite element discretization of Ω, such that $V_h \subset H^1(\Omega) \subset L^2(\Omega)$. For a parameter $\mu \in \mathcal{P}$, we consider the semi-discrete problem of finding $u_h(\mu) \in L^2(0, T; V_h)$ with $\partial_t u_h(\mu) \in L^2(0, T; V_h')$ such that

$$\langle \partial_t u_h(\mu), v_h \rangle + a\big(u_h(\mu), v_h; \mu\big) = l(v_h; \mu) \qquad \forall\, v_h \in V_h \text{ in } [0, T], \qquad (1)$$
$$u_h(0; \mu) = u_0(\mu),$$

where $u_0(\mu) \in V_h$ is the initial condition. In addition, we are particularly interested in output trajectories, i.e. we consider an output functional $f_h: \mathcal{P} \to L^2(0, T)$ defined as $f_h(t; \mu) := s(u_h(t; \mu))$ for some $s \in V_h'$. For every $\mu \in \mathcal{P}$, $l(\,\cdot\,; \mu) \in V_h'$ is assumed to be a linear and continuous functional and $a(\,\cdot\,, \cdot\,; \mu): V_h \times V_h \to \mathbb{R}$ denotes a continuous and coercive bilinear form.

Since solving the problem above for many different values of the parameter is costly, we are looking for a reduced order model that is faster to evaluate. To this end, the reduced basis (RB) method projects the problem (1) onto a well-chosen reduced space $V_{\mathrm{RB}} \subset V_h$ with $\dim V_{\mathrm{RB}} =: N_{\mathrm{RB}} \ll N_h := \dim V_h$ (for a general introduction to RB methods, see [4]). One thus searches for an approximate solution $u_{\mathrm{RB}}(t; \mu) \in V_{\mathrm{RB}}$ of $u_h(t; \mu) \in V_h$ by solving the same semi-discrete problem as in (1) with test functions from the space V_{RB}, i.e.

$$\langle \partial_t u_{\mathrm{RB}}(\mu), v_{\mathrm{RB}} \rangle + a\big(u_{\mathrm{RB}}(\mu), v_{\mathrm{RB}}; \mu\big) = l(v_{\mathrm{RB}}; \mu) \qquad \forall\, v_{\mathrm{RB}} \in V_{\mathrm{RB}} \text{ in } [0, T], \quad (2)$$
$$\langle u_{\mathrm{RB}}(0; \mu), v_{\mathrm{RB}} \rangle = \langle u_0(\mu), v_{\mathrm{RB}} \rangle \,\, \forall\, v_{\mathrm{RB}} \in V_{\mathrm{RB}}.$$

To obtain an efficient reduced order model, we pose the standard assumption that the bilinear form a and the right hand side l are parameter separable. In that case, many of the required quantities can be precomputed during the so-called *offline*-phase. In the subsequent *online*-phase, the reduced solution for a new parameter can be determined independently of the dimension N_h of V_h.

To assess the error between u_h and u_{RB}, one can consider the residual by inserting u_{RB} into Eq. (1). The norm of the residual can be computed without actually solving (1). Further, the corresponding error estimator $E_{\mathrm{RB}}(u; \mu)$ for arbitrary $u \in V_{\mathrm{RB}}$, introduced in [3], can be evaluated efficiently during the online phase if a and l fulfill the parameter separability assumption. Since the functional s is assumed to be linear and continuous, the error estimator for the state u_{RB} also results in an estimator for the error in the output functional, i.e.

$$\|f_h(\mu) - s(u_{\mathrm{RB}}(\mu))\|_{L^2(0,T)} \leq \|s(\,\cdot\,; \mu)\|_{V_h'} E_{\mathrm{RB}}(u_{\mathrm{RB}}(\mu); \mu).$$

We emphasize at this point that the error estimator computes an estimate based on the parameter μ and coefficients with respect to the reduced basis. However, these coefficients do not have to be the result of solving the problem in (2).

3   Machine Learning for Regression

In the case of regression, ML procedures typically assume that input data $X_N := \{x_1, \ldots, x_N\} \subset \mathbb{R}^d$ and corresponding output data $Y_N := \{y_1, \ldots, y_N\} \subset \mathbb{R}^b$ is given. The task is to "learn" a function which approximates the mapping $x_i \mapsto y_i$ and generalizes properly to unseen novel inputs x. One mathematically well established approach for this task is to employ kernel methods. These kind of methods can be understood as applying linear regression to a nonlinear transformation of the data, which is implicitly given by the kernel k, which is in our case a symmetric function $k \colon \mathbb{R}^d \times \mathbb{R}^d \to \mathbb{R}$. A well known representer theorem then states that f^* can be expanded as

$$s_{X_N} = \sum_{j=1}^{N} \alpha_j k(\cdot, x_j). \tag{3}$$

In order to obtain sparse models, greedy methods can be employed to select a suitable subset of centers $X_n \subset X_N$ such that the kernel expansion from Eq. (3) is smaller. An algorithm for this, which will be leveraged later on, is the vectorial kernel orthogonal greedy algorithm (VKOGA) [7].

Another lass of ML methods is deep learning, where probably neural networks are the most prominent example. Due to their deep structure, feature learning is possible which can be beneficial especially for medium to high dimensional datasets.

In order to make the best out of both approaches, the idea is to combine deep learning methods such as neural networks with the benefits of "shallow" kernel methods. In the following Sect. 4, we review two recently introduced algorithms in this direction: SDKNs [10] as well as 2L-VKOGA [9].

4   Deep Kernel Models

Both, the SDKN and the 2L-VKOGA approach, are based on a deep kernel representer theorem [2], which allows to put kernel methods into a multilayer setup in a mathematical principled way:

SDKNs [10] make use of special classes of kernels, which are used alternately in a structured way to derive a powerful setup: First, matrix-valued linear kernels are used, i.e. $k_{\text{lin}}(x, z) = \langle x, z \rangle_{\mathbb{R}^{d_0}} \cdot I_{d_1}$ with $I_{d_1} \in \mathbb{R}^{d_1 \times d_1}$ being the identity matrix. Second, single-dimensional kernels are employed that act on single components $x^{(i)}, z^{(i)}$ separately as $k_s(x, z) = \text{diag}(k(x^{(1)}, z^{(1)}), \ldots, k(x^{(d)}, z^{(d)}))$. Hereby, the one dimensional kernel $k \colon \mathbb{R} \times \mathbb{R} \to \mathbb{R}$ is e.g. the Gaussian kernel. The SDKN model is set up using the linear kernels for every odd layer and single-dimensional kernels for every even layer. Thus, the odd layers essentially allow for linear combinations of the inputs, while the even layers can be compared to activation functions of neural networks. However, due to being kernel models that can be optimized instead of fixed activation functions, they allow for more flexibility and thus enable a potentially faster optimization. As analyzed in [10], these SDKNs enjoy universal approximation properties in various limit cases.

The **2L-VKOGA** [9] combines the use of an optimized two-layered kernel with greedy kernel algorithms, where we leverage the VKOGA algorithm [7]: Radial basis function kernels are a popular class of kernels which are given as $k(x, z) = \Phi(\|x - z\|)$ for some radial basis function $\Phi \colon \mathbb{R} \to \mathbb{R}$, e.g. the Gaussian kernel $k(x, z) = \exp(-\|x - z\|^2)$. These RBF kernels usually make use of a shape parameter $\varepsilon > 0$ via $k_\varepsilon(x, z) = \Phi(\varepsilon \|x - z\|)$, which can be used to adopt the shape of the kernel to the dataset. The two-layered kernel approach generalizes this scale parameter by first allowing for a $d \times d$ matrix A instead of only a scalar valued shape parameter ε, i.e. $k_A(x, z) := \Phi(\|A(x - z)\|)$, and secondly providing an ML-based gradient descent optimization, which allows to optimize the kernel (i.e. the matrix A) efficiently. In particular, this setup can be framed within the deep kernel representer theorem [2] and thus be seen as a two-layered kernel. The optimized kernel k_A can subsequently be used with standard kernel algorithms. Here we use it in conjunction with greedy algorithms, and in particular with the VKOGA algorithm: In view of Eq. (3), the VKOGA algorithm selects a suitable subset $X_n \subset X_N$ of the dataset, such that the resulting kernel model consists only of $n \ll N$ summands and is thus sparse. For our current application, we extend the 2L-VKOGA approach that was only introduced for scalar valued target data [9] to vectorial data, i.e. $Y_N \subset \mathbb{R}^b$. Therefore, we consider for the kernel optimization the ℓ^2-norm over the errors from the single dimensions. As the VKOGA algorithm is intrinsically already vector-valued, both can be combined. The SDKN and 2L-VKOGA model are visualized in Fig. 1 respectively Fig. 2.

5   Certified and Adaptive RB-ML-ROM

We summarize the adaptive model hierarchy as specified in [5]. Instead of describing the model hierarchy in an abstract way as done in [5], we make direct use of the RB model from Sect. 2 and the deep kernel methods from Sect. 4.

The RB-ML-ROM is built as a hierarchy consisting of three models, a full-order model (FOM), a reduced basis reduced order model (RB-ROM), and a machine learning reduced order model (ML-ROM). As discussed, the error estimator of the RB-ROM (see Sect. 2) can be used to certify solutions from the

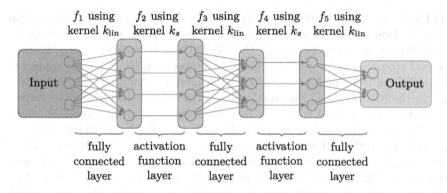

Fig. 1. Exemplary visualization of a Structured Deep Kernel Network (SDKN) with 5 layers and input dimension 3 and output dimension 2.

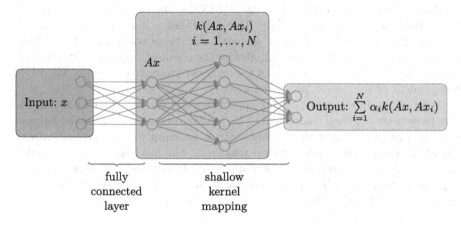

Fig. 2. Exemplary visualization of a two-layered kernel model (2L-VKOGA) with $x \in \mathbb{R}^3$, $A \in \mathbb{R}^{3 \times 3}$, $N = 7$, $x_i \in \mathbb{R}^3$ and $\alpha_i \in \mathbb{R}^2$ for $i = 1, \ldots, N$.

RB-ROM as well as from the ML-ROM. The guiding assumption is that the ML-ROM is the fastest model in the hierarchy. A query to the RB-ML-ROM now proceeds as follows: For a new parameter the ML-ROM is queried and its prediction accepted if its error estimation is sufficiently accurate, i.e. the error is below a desired tolerance $\varepsilon > 0$. Otherwise, the slightly more expensive RB-ROM is queried and its error is again validated using the error estimator. Only if the RB-ROM is also inaccurate, one has to fall back to the costly FOM, which is assumed to be accurate enough for every parameter.

Not only the prediction of the hierarchical model is adaptive, but the submodels themselves are adaptively updated as follows: When querying the adaptive RB-ML-ROM for the first time for a certain parameter $\mu \in \mathcal{P}$, the RB-ROM and the ML-ROM are initialized as empty models, i.e. they can only return zero as the solution. Consequently, most likely, the FOM is called for the parameter μ

and the FOM output trajectory $f_h(\mu)$ is returned. As intermediate quantity the FOM solution trajectory $u_h(\mu)$ is computed, which will be used to set up the RB-ROM. In our concrete setting, the RB-ROM is constructed using the hierarchical approximate proper orthogonal decomposition (HaPOD) [6] applied to data obtained from querying the FOM. Whenever solving the FOM is necessary (due to the RB-ROM and ML-ROM not being accurately enough for the current parameter), a new FOM trajectory becomes available and the RB-ROM can be improved using this new information. Finally, the ML-ROM is built upon the updated RB-ROM and is based on the same reduced basis and error estimator as the RB-ROM. More precisely, using an ML algorithm such as the deep kernel models described in Sect. 4, we learn the coefficients of the RB solution with respect to the reduced basis as a function of the parameter, see also [8]. The required training data is generated as a byproduct when the RB-ROM is queried. However, only those RB solutions that fulfill the accuracy requirements (evaluated by the error estimator of the RB-ROM) are added to the training set. Recall that the error estimator of the RB-ROM takes as input a parameter $\mu \in \mathcal{P}$ and arbitrary reduced coefficients. In particular, these coefficients do not have to stem from the RB-ROM, but can also be the result of an ML prediction. This fact enables the direct application of the error estimator to certify the ML-ROM independent of the actual ML model used.

For the ML-ROM, two different approaches are possible which differ in the way they treat the time component. The *random-access-in-time* method passes the time instance along with the parameter as an input to the machine learning model. To obtain the whole solution trajectory, the machine learning model is therefore queried for all different time instances. In contrast, the *time-vectorized* approach returns the whole time sequence for a given parameter, i.e. the output contains the reduced coefficients for all time steps at once.

6   Numerical Application

As an application for the deep kernel methods in the framework of the adaptive RB-ML-ROM model, we leverage the Monte Carlo example introduced in [5, Section 4.2] using the random-access-in-time method. Likewise, we can also use the NN used there as a baseline method for comparison. The quantity of interest is the average (in time) temperature $\overline{f} \colon \mathcal{P} \to \mathbb{R}$ within a given room of a building, represented by the domain Ω. The temperature can be computed by solving the heat equation and is depending on 28 diffusion respectively heating parameters for walls, doors and heaters. In order to determine the average and variance of this output quantity over a given probability density ρ on the 28 dimensional parameter space, we rely on a Monte Carlo estimation

$$\mathrm{E}[\overline{f}] =: \int_{\mathcal{P}} \rho(\mu)\overline{f}(\mu) \; \mathrm{d}\mu \approx \frac{1}{N_{\mathrm{mc}}} \sum_{\mu \in \mathcal{P}_{\mathrm{mc}}} \overline{f}(\mu),$$

and similarly for the variance.

For the numerical experiment, the same setup as in [5] was used, i.e. the full order model consists of 321206 DoFs and 999 timesteps and is thus quite expensive. Therefore, the use of the adaptive model introduced in Sect. 5 is desirable, which is applied with a tolerance of $\varepsilon = 5 \cdot 10^{-2}$. For the machine learning models, the SDKN followed the setup of the NN, i.e. four layers with each an inner dimension of 128 while using the Gaussian kernel for its single-dimensional kernel layers. The 2L-VKOGA used a quadratic Matérn kernel and 10 epochs with a batch size of 32 and an initial learning rate of $5 \cdot 10^{-3}$ for the Adam optimizer for the optimization of the two-layered kernel. For the greedy algorithm, the so called f-greedy selection criterion was used up to an expansion size of 500 centers. The NN and SDKN were (re)trained as soon as the data of 200 new parameter values were evaluated, the VKOGA and 2L-VKOGA models were (re)trained as soon as 40 new data samples were available. As soon as the performance of the ML-ROM models was sufficiently good (use of ML model at least in 60% of the queried parameter values), no more retrainings were performed. The results of the runs of the adaptive model are collected in Table 1 as well as Fig. 3: Table 1 summarizes the key performance numbers for the different machine learning models: In the column "Training ML-ROM", the training set size n_μ and the training duration t for the ML-ROMs are listed: It can be

Table 1. Overview of the results using the adaptive model for different kind of machine learning models: Top two boxes: Baseline NN as well as a shallow kernel VKOGA model (Sect. 3). Bottom two boxes: The deep kernel approaches SDKN and 2L-VKOGA (Sect. 4).

ML-Model	Trainings of ML-ROM		ML-ROM ratio	Num. of model evaluations		
	Param. at which training started	Training time (s)		Model	Num. evals.	Average time (s)
NN			0.0%	FOM	12	234.9
	200	536.3	36.8%	RB-ROM	876	6.7
	516	3919.1	49.1%	ML-ROM	4112	3.2
	908	6515.1	92.9%			
VKOGA			0.0%	FOM	12	235.9
	40	97.8	≤ 25%	RB-ROM	1789	8.6
	...	...	...	ML-ROM	199	4.2
	1999	3572.3	≤ 25%			
SDKN			0.0%	FOM	12	233.0
	200	3350.2	60.4%	RB-ROM	613	6.7
	703	8272.9	94.7%	ML-ROM	4375	3.3
2L-VKOGA			0.0%	FOM	12	235.9
	40	110.3	86.7%	RB-ROM	685	8.7
				ML-ROM	4303	4.1

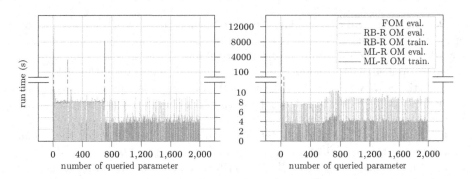

Fig. 3. Visualization of the required time (vertical axis) to build and evaluate the adaptive model M_{adapt} for each μ chosen by the random sampling (horizontal axis). Left: SDKN as ML-ROM. Right: 2L-VKOGA as ML-ROM.

seen that the VKOGA models are particularly fast, which is due to their smaller training batch sizes of 40 instead of 200, such that they are trained already earlier. Of particular interest is the column "ML-ROM ratio", which shows the ratio of ML-ROM vs. ROM evaluations: For the baseline NN, this ratio increases from 0% (before training) to up to 87% after 3 training stages. The shallow VKOGA model does not seem to be accurate at all at any time, its ML-ROM ratio never exceeds 25% – and is thus stopped after 2000 queried parameter evaluations. For the SDKN, the ML-ROM ratio is already 60.4% after the first training, thus surpassing the NN performance after two trainings. Its final ratio of 94.7% is large enough such that no further retraining is required anymore. The deep 2L-VKOGA model achieves a performance of 86.7% already after the very first training, which included only 40 evaluated parameter values. The reason for this superb performance (especially in contrast to the shallow VKOGA model) can be seen in the optimized kernel. A singular value analysis of the matrix A of the kernel k_A reveals that (quite independent of the specific run, i.e. random choice of the parameter μ) there is one large singular value of 2.5, while the second largest one is already only 0.35 and the remaining ones smaller or equal than 0.35. This furthermore highlights the amenability of the 2L-VKOGA for interpretability of the ML model. The last column "Prediction models" shows the number of FOM, ROM and ML-ROM evaluations (which is related to the ratios from the previous column) and the mean prediction times of these models. The reason for this is that the size of the ML models (in terms of the number of parameters) is approximately the same, namely approximately 80000 parameters for the NN and SDKN and approximately 100000 parameters for the VKOGA models. In terms of total runtime the NN took 48:57 h, the SDKN 48:23 h, the 2L-VKOGA 46:58 h and the VKOGA was stopped after 54:01 h (for only 2000 parameters).

The source code to carry out the experiments and generate the figures can be found online.[1]

7 Conclusion

In this paper we reviewed the certified RB-ML-ROM model and the two deep kernel models SDKN and 2L-VKOGA, whereby we extended the second one for the use with vectorial target data. These deep kernel models where then used for the machine learning part within the RB-ML-ROM model, where they outperformed the standard NN baseline as well as shallow kernel models.

Acknowledgements. Funded by BMBF under the contracts 05M20PMA and 05M20VSA. Funded by the Deutsche Forschungsgemeinschaft (DFG, German Research Foundation) under Germany's Excellence Strategy EXC 2044 –390685587, Mathematics Münster: Dynamics–Geometry–Structure, and EXC 2075 –390740016. We acknowledge the support by the Stuttgart Center for Simulation Science (SimTech).

References

1. Benner, P., Grivet-Talocia, S., Quarteroni, A., Rozza, G., Schilders, W., Silveira, L. (eds.): Model Order Reduction, vol. 1. De Gruyter, Berlin (2020)
2. Bohn, B., Rieger, C., Griebel, M.: A representer theorem for deep kernel learning. J. Mach. Learn. Res. **20**, 1–32 (2019)
3. Grepl, M., Patera, A.: A posteriori error bounds for reduced-basis approximations of parametrized parabolic partial differential equations. ESAIM Math. Model. Numer. Anal. **39**(1), 157–181 (2005)
4. Haasdonk, B.: Reduced basis methods for parametrized PDEs – a tutorial introduction for stationary and instationary problems. In: Model Reduction and Approximation: Theory and Algorithms, pp. 65–136. SIAM, Philadelphia (2017)
5. Haasdonk, B., Kleikamp, H., Ohlberger, M., Schindler, F., Wenzel, T.: A new certified hierarchical and adaptive RB-ML-ROM surrogate model for parametrized PDEs. SIAM J. Sci. Comput. **45**(3), A1039–A1065 (2023)
6. Himpe, C., Leibner, T., Rave, S.: Hierarchical approximate proper orthogonal decomposition. SIAM J. Sci. Comput. **40**(5), A3267–A3292 (2018)
7. Santin, G., Haasdonk, B.: Kernel methods for surrogate modeling. In: Model Order Reduction, vol. 2. De Gruyter (2021)
8. Wang, Q., Hesthaven, J.S., Ray, D.: Non-intrusive reduced order modeling of unsteady flows using artificial neural networks with application to a combustion problem. J. Comput. Phys. **384**, 289–307 (2019)
9. Wenzel, T., Marchetti, F., Perracchione, E.: Data-driven kernel designs for optimized greedy schemes: A machine learning perspective. SIAM J. Sci. Comput. **46**(1), C101–C126 (2024)
10. Wenzel, T., Santin, G., Haasdonk, B.: Universality and optimality of structured deep kernel networks. arXiv:2105.07228 (2021)

---

[1] https://gitlab.mathematik.uni-stuttgart.de/wenzeltn/paper-2023-deep-kernel-for-rb-ml-rom.

Fractional Differential Problems: Theoretical Aspects, Algorithms and Applications

Parametric Analysis of Space-Time Fractional Pennes Bioheat Model Using a Collocation Method Based on Radial Basis Functions and Chebyshev Polynomials

Hitesh Bansu$^{(\boxtimes)}$ and Svetozar Margenov

Institute of Information and Communication Technologies,
Bulgarian Academy of Sciences, Sofia, Bulgaria
hiteshbansu@gmail.com

Abstract. In this paper, space-time fractional Pennes bioheat model with sinusoidal heat flux condition on skin tissue is considered in Caputo sense. The model involves parameters like blood perfusion rate, specific heat, thermal conductivity, heating frequency etc. The numerical solution of the fractional bioheat model is obtained using a collocation method. The method involves a Kronecker product of two different basis functions, namely radial basis functions and Chebyshev polynomials. The heat transfer process in skin tissue is studied under the influence of space-time fractional derivatives using the presented model. The effect of various parameters, like heating frequency and thermal conductivity, on the heat transfer process is examined under fractional derivatives.

Keywords: Bioheat model · Chebyshev polynomial · Radial basis function · Kronecker product

1 Introduction

Fractional calculus and partial differential equations of fractional order have established a new research trend during last few decades because of their continuous presence in different branches of science and engineering. In modern clinical treatments like cryosurgery, cancer hyperthermia etc. behaviour of heat diffusion in living tissue is the interesting subject. Further, thermal skin damage by contact of hot substance or flash fire are very common hazards therefore heat transfer in human tissue is important topic in recent medical research [4,5]. Moreover, biothermal studies can be helpful for the design of equipments used in clinical thermal treatment, the assessment of skin burns and the formation of thermal protections for numerous intentions [4]. Because the thermal behavior of biological tissues depends on a number of processes, including blood flow

Supported by Early-stage and Postdoctoral researchers National Programme 2 - Bulgarian Academy of Sciences.

I. Lirkov and S. Margenov (Eds.): LSSC 2023, LNCS 13952, pp. 129–136, 2024.
https://doi.org/10.1007/978-3-031-56208-2_12

and metabolic heat generation. The governing mathematical models developed include the Pennes equation, which is one of the most widely used due to its efficiency [2,3].

In the present study we have considered the Pennes bioheat model in the form

$$\rho_t c_t \frac{\partial T}{\partial t} = k \frac{\partial^2 T}{\partial x^2} + W_b c_b \left(T_a - T\right) + q_{met}, \quad x \in (0, L), \ t > 0, \tag{1}$$

with initial and transient heat flux boundary conditions

$$T\left(x, 0\right) = T_a$$

$$k \left.\frac{\partial T}{\partial x}\right|_{x=0} = q_0 \cos(\omega t) \qquad k \left.\frac{\partial T}{\partial x}\right|_{x=L} = 0 \tag{2}$$

where ρ, c, T, k, t, x, T_a, W_b, q_0, q_{met} and ω are density, specific heat, temperature, thermal conductivity, time, distance, arterial temperature, blood perfusion rate, heat flux, metabolic heat and heating frequency respectively [6]. The suffix t and b are for tissue and blood respectively.

Due to it's importance in clinical research many researchers have developed and used various techniques to solve Penns bioheat equation. Shih et al. [9] focus their study on the analytic solution of Penns bioheat equation with sinusoidal heat flux. Ferras et al. [5] applied implicite finite difference scheme to solve space fractional Penns bioheat equation (FPBE) and the same method has been used by Damor et al. [3] to go through the parametric study of time FPBE with sinusoidal heat flux. Singh et al. [10] applied finite difference method and homotopy perturbation method to investigate the numerical solution of space-time fractional bioheat equation. Damor et al. [2] derived numerical solution of time FPBE with constant transient heat flux by implicit finite difference scheme. Qin and Wu [8] derived numerical solution of time fractional bioheat equation by quadratic spline collocation method.

The structure of the paper is as follows. After an introduction, Sect. 2 deals with prerequisites of fractional derivatives, radial basis functions (RBFs) and Chebyshev polynomials. Section 3 gives idea about the heat transfer model, while Sect. 4 consists of the layout of the proposed numerical scheme. Section 5 reveals results and discussion; and lastly, conclusion is reported in Sect. 6.

2 Preliminaries

Here we have discussed some preliminaries and definitions of fractional derivative, Chebyshev polynomials and Radial basis functions.

2.1 Fractional Derivatives

Definition 1. The Caputo fractional derivative of order α of function $f(x)$ for $\alpha > 0, (\alpha \in \mathbb{R})$ is [7]

$$\left(^C D_a^\alpha f\right)(x) = \frac{1}{\Gamma(n-\alpha)} \int_a^x \frac{f^{(n)}(s)\, ds}{(x-s)^{1+\alpha-n}}$$

Definition 2. Caputo fractional derivative of power function x^p, $p \geq 0$ is [7]

$$
^C D^\alpha x^p = \begin{cases} \dfrac{\Gamma(p+1)}{\Gamma(p-\alpha+1)} x^{p-\alpha}, & \text{for } p \geq \lceil \alpha \rceil \\ 0, & \text{for } p < \lceil \alpha \rceil \end{cases}
$$

2.2 Chebyshev Polynomials

The Chebyshev polynomials $T_m(x)$ have the analytic form as [1]

$$
T_m(x) = \sum_{l=0}^{\lfloor m/2 \rfloor} (-1)^l 2^{m-2l-1} \frac{m(m-l-1)!}{l!(m-2l)!} x^{m-2l}
$$

where the integer part of $m/2$ is indicated by $\lfloor m/2 \rfloor$. Now to use these polynomials on $[0,t]$ it is necessary to perform the change of variable $s = \frac{2x}{t} - 1$, $s \in [-1,1]$. Hence the shifted Chebyshev polynomials can be defined as $T_m^*(x) = T_m\left(\frac{2x}{t} - 1\right)$ and will have the analytic form as [1]

$$
T_m^*(x) = \sum_{l=0}^{m} (-1)^{m-l} 2^{2l} \frac{m(m+l-1)!}{(2l)!(m-l)!t^l} x^l.
$$

Thus, the shifted Chebyshev polynomial have the Caputo fractional derivative $D^\beta T_m^*(x)$ as

$$
D^\beta T_m^*(x) = \sum_{l=\lceil \beta \rceil}^{m} (-1)^{m-l} 2^{2l} \frac{m(m+l-1)!}{(2l)!(m-l)!t^l} \frac{\Gamma(l+1)}{\Gamma(l+1-\beta)} x^{l-\beta}, \quad m \geq \lceil \beta \rceil.
$$

2.3 Radial Basis Functions

Over the past few decades, RBFs have achieved remarkable success and have become an important tool for interpolating sparse data. In the interpolation process involving RBFs, the function $u(x)$ can be approximated as [1]

$$
u(x) = \sum_{i=1}^{N} \lambda_i \phi_i(r),
$$

where the number of data points is denoted by N. The coefficients $\{\lambda_i\}_{i=1}^{N}$ are to be investigated. $\phi(r)$ is any RBF and the Euclidean distance from point x to x_i is denoted by $r = \|x - x_i\|$.

Some commonly used RBFs are the multiquadric $\phi(r) = (r^2 + \epsilon^2)^{\alpha/2}$, $\alpha = -1, 1, 3, 5, \ldots$, the Gaussian $\phi(r) = e^{-(\epsilon r)^2}$, the conical type $\phi(r) = r^m$, $m = 1, 3, 5, \ldots$ and the polyharmonic splines $\phi(r) = r^m \log r$, $m = 2, 4, 6, \ldots$. In this study the implemented RBF is conical type with $m = 3$.

3 Heat Transfer Model

As noted in the introduction, Penns bioheat model is commonly used to study heat transfer in living tissues [6]. Here we have obtained space and time FPBE

$$\rho_t c_t \frac{\partial^\alpha T}{\partial t^\alpha} = k \frac{\partial^\beta T}{\partial x^\beta} + W_b c_b (T_a - T) + q_{met}, \quad 0 < \alpha \le 1, \quad 1 < \beta \le 2 \tag{3}$$

by replacing the space and time derivatives with Caputo fractional derivatives. For $\alpha = 1$ and $\beta = 2$ Eq. (3) coincides with classical Penns bioheat equation as in (1).

4 Numerical Solution

For the numerical solution of the biothermal model (3) with initial and boundary conditions (2), the function $T(x, t)$ is approximated in terms of RBF and Chebyshev polynomials as [1]

$$T(x, t) \approx \sum_{i=1}^{N} \sum_{j=1}^{n} \mathbb{T}_j(t) c_{ji} \Phi_i(x) = \mathbb{T}(t) \, C \, \Phi(x) \tag{4}$$

where $\Phi(x) = [\phi_1(x), \phi_2(x), \phi_3(x), \ldots, \phi_N(x)]^T$, $\mathbb{T}(t) = [T_1(t), T_2(t), T_3(t), \ldots, T_n(x)]$ are cubic RBFs and Chebyshev polynomials. N and n (both $\in N$) are discretization parameters in space and time, respectively. The unknowns c_{ji} form the matrix

$$C = [c_{ji}]_{i,j=1,1}^{N,n}.$$

We have used uniform mesh in $[c, d]$ with nodes

$$z_m = z_{m-1} + \frac{d-c}{p-1}; \quad m = 1, 2, \ldots, p, \quad z_0 = c.$$

Here, the discretizations in space and time are independent of each other. Using Eq. (4) it is easy to derive the equalities

$$_0^C D_t^\alpha T(x, t) = {}_0^C D_t^\alpha (\mathbb{T} \, C \, \Phi) = \left[{}_0^C D_t^\alpha \mathbb{T} \right] C \Phi = \mathbb{T}^\alpha \, C \, \Phi, \tag{5}$$

$$_0^C D_x^\beta T(x, t) = {}_0^C D_x^\beta (\mathbb{T} \, C \, \Phi) = \mathbb{T} \, C \left[{}_0^C D_x^\beta \Phi \right] = \mathbb{T} \, C \, \Phi^\beta. \tag{6}$$

Now, putting Eqs. (5) and (6) in (3) we obtain

$$\rho_t c_t \{\mathbb{T}^\alpha \, C \, \Phi\} = k \{\mathbb{T} \, C \, \Phi^\beta\} + W_b c_b \{T_a - \{\mathbb{T} \, C \, \Phi\}\} + q_{met}. \tag{7}$$

Collocating (7) in $N - 2$ and $n - 1$ uniform nodes will give $(N - 2)(n - 1)$ equations as

$$\rho_t c_t \{M_1 \, C \, L\} = k \{M \, C \, L_1\} + W_b c_b T_a - W_b c_b \{M \, C \, L\} + q_{met} \tag{8}$$

where

$$L = [\phi_i\,(x_j)]_{i,j=1,2}^{N,N-1}, \quad M_1 = [_0^C D_t^\alpha T_i\,(t_j)]_{i,j=1,2}^{n,n}$$

$$M = [T_i\,(t_j)]_{i,j=1,2}^{n,n}, \quad L_1 = [_0^C D_x^\beta \phi_i\,(x_j)]_{i,j=1,2}^{N,N-1}$$

Similarly, applying Eq. (4) in (2) we get

$$M_2\,C\,L_2 = f\,(x),\quad M\,C\,L_3 = g_1\,(t),\quad M\,C\,L_4 = g_2\,(t) \tag{9}$$

where

$$L_2 = [\phi_i\,(x_j)]_{i,j=1,1}^{N,N}$$

$L_3 = \left[\phi_1{}'\,(x_1)\,\phi_2{}'\,(x_1)\,\dots\,\phi_N{}'\,(x_1)\right]^{\mathrm{T}}$, $L_4 = \left[\phi_1{}'\,(x_N)\,\phi_2{}'\,(x_N)\,\dots\,\phi_N{}'\,(x_N)\right]^{\mathrm{T}}$
and $M_2 = \left[T_1\,(t_1)\,T_2\,(t_1)\,\dots\,T_n\,(t_1)\right]$.

Collocating Eq. (9) will produce $(N + 2n - 2)$ equations. Combining these equations with Eq. (8) will result in Nn equations. For the solution of Eq. (8), we convert it in a handy form using the Kronecker product

$$\left\{\rho_t c_t\,(L^{\mathrm{T}} \otimes M_1) - k\left(L_1{}^{\mathrm{T}} \otimes M\right) + W_b c_b\,(L^{\mathrm{T}} \otimes M)\right\}\vec{c} = \overrightarrow{(W_b c_b T_a + q_{met})},$$

which can be formed as

$$A_1\vec{c} = F_1 \tag{10}$$

where $(N - 2)(n - 1) \times Nn$ is the size of matrix A_1. $\vec{c}$ is acquired by putting the columns of C on top of one another and is of size $Nn \times 1$. Size of vector F_1 is $(N - 2)(n - 1) \times 1$.

An expression for initial and boundary conditions can be given as

$$(L_2^t \otimes M2)\,\vec{c} = \overrightarrow{(T_a)},\quad (L_3^t \otimes M)\,\vec{c} = \overrightarrow{(q_0)},\quad (L_4^t \otimes M)\,\vec{c} = \overrightarrow{(0)}$$

that can be rewritten as

$$A_2\vec{c} = F_2,\quad A_3\vec{c} = F_3,\quad A_4\vec{c} = F_4, \tag{11}$$

where the dimension of A_2 is $N \times Nn$, and A_3 and A_4 have equal dimension, i.e. $(n - 1) \times Nn$. Finally, the system including (10) and (11) is given by

$$A\vec{c} = F. \tag{12}$$

where $A = [A_1, A_2, A_3, A_4]^{\mathrm{T}}$ is of dimension $Nn \times Nn$. $Nn \times 1$ is the size of the vector $F = [F_1, F_2, F_3, F_4]^{\mathrm{T}}$. The coefficient $\vec{c}$ can be obtained by solving the system given in (12). Reshaping $\vec{c}$ to C and putting in Eq. (4) will produce expected approximation for $T(x,t)$ which is the solution of the problem given in (3).

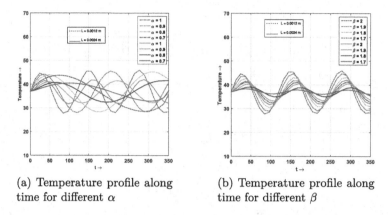

(a) Temperature profile along time for different α

(b) Temperature profile along time for different β

Fig. 1. Temperature profile for different α and β.

5   Results and Discussion

In this study, we have considered $N = 50$, $n = 31$ and the set of parameters as [2]: $L = 0.02\,\text{m}$, $\omega = 0.05$, $T_a = 37\,^\circ\text{C}$, $q_0 = 5000\,\text{W/m}^2$, $\rho_t = 1050\,\text{kg/m}^3$, $\rho_b = 1050\,\text{kg/m}^3$, $q_{met} = 368.1\,\text{W/m}^3$, $W_b = 0.5\,\text{kg/m}^3$, $c_b = 3770\,\text{JC/kg}$, $c_t = 4180\,\text{JC/kg}$ and $k = 0.5\,\text{WC/m}$.

Figure 1a represents the temperature profile for transient heat flux along time for different values of α (order of temporal derivative). We observe that the frequency decreases as α decreases. Also, the amplitude at $L = 0.0024\,\text{m}$ is higher than at $L = 0.0012\,\text{m}$. Moreover, maximum temperature in tissue decreases as distance increases.

Figure 1b shows the effect of transient heat flux on skin surface for different β (order of spatial derivative). It is observed that neither amplitude nor frequency changes as β varies. Only the highest temperature is deceasing for smaller values of β. Also, the amplitude at $L = 0.0012\,\text{m}$ is lower than at $L = 0.0024\,\text{m}$.

Figure 2a demonstrates the effect of heating frequency ω for different values of α and β at $t = 70$ s. It is observed that if the initial temperature is lower than ideal temperature $37\,^\circ\text{C}$ then there is an increment in temperature for higher and lower values of ω. Further, if the initial temperature is higher than ideal temperature then the temperature increases for lower values of ω and decreases for higher values of ω. Moreover, The temperature response after $L = 0.01\,\text{m}$ seems to be independent of the heat flux on the heating skin.

The effect of heating frequency (ω) on temperature for different α and β along time is presented in Fig. 2b. We observe that the amplitude deceases as ω increases and the amplitude increases as ω decreases. Also, the temperature is elevated for smaller values of ω and remains lower for higher values of ω. Moreover, frequency decreases for smaller values of α. The effect of thermal conductivity (k) on temperature along distance is plotted in Fig. 3a at $t = 70$ s. Here we observe that if the initial temperature is below the ideal temperature ($37\,^\circ\text{C}$) then the temperature decreases for lower values of k. Further, if the initial

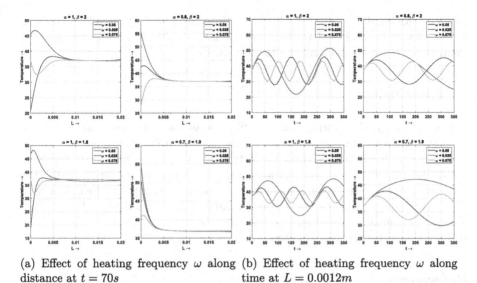

(a) Effect of heating frequency ω along distance at $t = 70s$ (b) Effect of heating frequency ω along time at $L = 0.0012m$

Fig. 2. Effect of heating frequency ω along distance and time

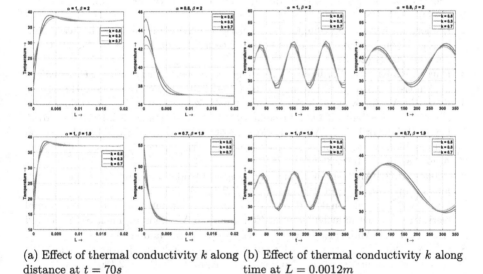

(a) Effect of thermal conductivity k along distance at $t = 70s$ (b) Effect of thermal conductivity k along time at $L = 0.0012m$

Fig. 3. Effect of thermal conductivity k along distance and time

temperature is above the ideal temperature then the temperature increases for lower values of k. But in both cases the peak temperature increases for smaller values of k and for larger values of k the temperature remains lower.

The effect of thermal conductivity (k) on temperature along time is plotted in Fig. 3b. We can conclude from the figure that there is no change in amplitude or frequency for smaller or higher values of k. But the highest temperature increases for smaller value of k and it decreases for larger value of k. The variation in frequency are due to different fractional values of α.

6 Conclusion

In present study, the temperature distribution of fractional bioheat model has been studied with transient heat flux on skin surface. The temperature profile has been obtained along distance and time with different fractional values of α, β. The numerical solution acquired in this work may be useful to predict the temperature response in fractional bioheat model with transient heat flux with different heating frequency and thermal conductivity respectively. The obtained solution may also be useful for experimental model to predict the value of α, β and applicable for thermal therapy in medical sciences.

References

1. Bansu, H., Kumar, S.: Numerical solution of space and time fractional telegraph equation: a meshless approach. Int. J. Nonlinear Sci. Numer. Simul. **20**, 325–337 (2019)
2. Damor, R., Kumar, S., Shukla, A.: Numerical solution of fractional bioheat equation with constant and sinusoidal heat flux coindition on skin tissue. Am. J. Math. Anal. **1**, 20–24 (2013)
3. Damor, R., Kumar, S., Shukla, A.: Parametric study of fractional bioheat equation in skin tissue with sinusoidal heat flux. Fract. Differ. Calc. **5**, 43–53 (2015)
4. Ezzat, M.A., AlSowayan, N.S., Al-Muhiameed, Z.I., Ezzat, S.M.: Fractional modelling of Pennes bioheat transfer equation. Heat Mass Transf. **50**, 907–914 (2014)
5. Ferrás, L.L., Ford, N.J., Morgado, M.L., Nóbrega, J.M., Rebelo, M.S.: Fractional Pennes bioheat equation: theoretical and numerical studies. Fract. Calc. Appl. Anal. **18**, 1080–1106 (2015)
6. Pennes, H.H.: Analysis of tissue and arterial blood temperatures in the resting human forearm. J. Appl. Physiol. **1**, 93–122 (1948)
7. Podlubny, I.: Fractional Differential Equations: An Introduction to Fractional Derivatives, Fractional Differential Equations, to Methods of Their Solution and Some of Their Applications, vol. 198. Elsevier, Amsterdam (1998)
8. Qin, Y., Wu, K.: Numerical solution of fractional bioheat equation by quadratic spline collocation method. J. Nonlinear Sci. Appl. **9**, 5061–5072 (2016)
9. Shih, T.C., Yuan, P., Lin, W.L., Kou, H.S.: Analytical analysis of the Pennes bioheat transfer equation with sinusoidal heat flux condition on skin surface. Med. Eng. Phys. **29**, 946–953 (2007)
10. Singh, H., Sahoo, M.R., Singh, O.P.: Numerical method based on Galerkin approximation for the fractional advection-dispersion equation. Int. J. Appl. Comput. Math. **3**, 2171–2187 (2017)

Numerical Determination of Source from Point Observation in a Time-Fractional Boundary-Value Problem on Disjoint Intervals

Miglena N. Koleva$^{(\boxtimes)}$ and Lubin G. Vulkov

University of Ruse, 8 Studentska Street, 7017 Ruse, Bulgaria
{mkoleva,lvalkov}@uni-ruse.bg

Abstract. In the present work, we solve numerically an inverse problem for identification of sources from point observations in a time-fractional diffusion-reaction problem defined on disjoint intervals. The fractional derivative is in Caputo sense with different fractional order on each of the subintervals. We propose algorithms, based on decomposition with respect to the sources on time adaptive mesh. Numerical tests illustrate the efficiency of the proposed approach.

1 Introduction

Multi-layer, multi-domain problems model processes in composite and multi-component structures. They are described by different partial differential equations in different domains with coupling interface conditions, see e.g. [3,16,20]. Interface problems on disjoint domains with integer and fractional derivatives have been subject of many studies, see e.g. [1,8,9,11,14].

In recent years fractional calculus and fractional differential equations have been often used in mechanics, physics, chemistry, engineering and finance, see e.g. [2,4,5,10,13,15,18,19].

The inverse problems for partial differential equations are very important in science and engineering. In particular, there are many models which concerns inverse problems of determining the unknown source. These problems, so-called source identification problems (SIPs) are studied from different aspects. Existence and uniqueness of solutions are investigated by many authors, see e.g. [4,6,7,10,12,17] and reference therein. The numerical solution of such problems is discussed in many papers, see e.g. [5–7,12,13,17]

In this paper we investigate inverse source problems with own fractional order of the parabolic equations on each disjoint interval.

The remaining part of the paper is organized as follows. In the next section, we formulate the direct and inverse problems. In Sect. 3, finite volume explicit-implicit approximatios of the fractional PDE problems are derived. In Sect. 4, a numerical solution decomposition method for the inverse problem is proposed. Computational examples are discussed in Sect. 5. The paper is finalized by some conclusions.

© The Author(s), under exclusive license to Springer Nature Switzerland AG 2024
I. Lirkov and S. Margenov (Eds.): LSSC 2023, LNCS 13952, pp. 137–145, 2024.
https://doi.org/10.1007/978-3-031-56208-2_13

2   Direct and Inverse Problems

Let $-\infty < a_1 < b_1 < a_2 < b_2 < +\infty$, $\Omega_j = (a_j, b_j) \times (0, T)$ and $\delta_j \in (0, 1]$, $j = 1, 2$. We consider the following time-fractional diffusion equations

$$D_t^{\delta_j} u_j - \frac{\partial}{\partial x}\left(p_j(x)\frac{\partial u_j}{\partial x}\right) + q_j u_j = f_j(x, t), \quad (x, t) \in \Omega_j, \quad j = 1, 2, \qquad (1)$$

subject to non-local interface conditions, external boundary conditions and initial conditions

$$(-1)^{j-1}p_j(c_j)\frac{\partial u_j}{\partial x}(c_j, t) + \alpha_j u_j(c_j, t) = \beta_j u_{3-j}(c_{3-j}, t) + \gamma_j(t), \qquad (2)$$

$$t \in (0, T), \quad j = 1, 2, \quad c_1 = b_1, \quad c_2 = a_2,$$

$$u_1(a_1, t) = \varphi_1(t), \quad u_2(b_2, t) = \varphi_2(t), \quad t \in (0, T), \qquad (3)$$

$$u_j(x, 0) = u_j^o(x), \quad j = 1, 2. \qquad (4)$$

Here $D_t^{\alpha} u_j$ is the Caputo fractional derivative [2, 10, 15]

$$\frac{\partial^{\delta_j} u_j}{\partial t^{\delta_j}} = \frac{1}{\Gamma(1 - \delta_j)} \int_0^t (t - s)^{-\delta_j} \frac{\partial u_j}{\partial s} ds, \quad 0 < \delta_j < 1, \quad j = 1, 2, \qquad (5)$$

where $\Gamma(\cdot)$ is Gamma function.

We assume that [8, 9, 11]

$$p_j, q_j \in L^\infty(a_j, b_j), \quad \gamma_j(t) \in L^\infty(0, T], \quad f_j(x, t) \in L^\infty(\Omega_j), \qquad (6)$$
$$p_j(x) \geq p_{j_0} > 0, \quad q_j(x) \geq 0, \quad \beta_j > 0, \quad \alpha_j > 0, \quad j = 1, 2, \quad \beta_1 \beta_2 < \alpha_1 \alpha_2.$$

The problem (1)–(6) with given initial data (initial and boundary conditions, coefficients, right-hand sides) is the *direct problem*.

The aim of the present work is to propose numerical method for solving *inverse source problem*, i.e. determination of the source term $\lambda_j(t)$, when $f_j(x, t) = \lambda_j(t)\overline{f}_j(x, t), j = 1, 2$ in (1)–(6), if the additional conditions, i.e. the measurements of the solution at points $x_j^* \in (a_j, b_j)$ are given

$$u_j(x_j^*, t) = \psi_j(t), \quad j = 1, 2. \qquad (7)$$

Through the following, we assume that there exists a solution $u(x, t) = (u_1(x, t), u_2(x, t)) \in C^{2,1}(\overline{Q}_T)$, where $C^{m,n}(\overline{Q}_T)$ is the class of functions that together with their partial derivatives of order m with respect to x and order n with respect to t are continuous on $\overline{Q}_T = \overline{\Omega}_1 \cup \overline{\Omega}_2$, $\overline{\Omega}_j = [a_j, b_j] \times (0, T]$.

3   Finite Difference Schemes

Let's denote $x_j := x \in [a_j, b_j]$. We introduce uniform spatial and nonuniform temporal meshes

$$\overline{\omega}_j = \{x_{j,i_j} : x_{j,i_j} = a_j + i_j h_j, \ i_j = 0, 1, \ldots, N_j, \ h_j = (b_j - a_j)/N_j\}, j = 1, 2,$$
$$\overline{\omega}_\tau : 0 = t_0 < t_1 < \cdots < t_M = T$$

and denote the function $y_j(x,t)$ at grid node (x_{j,i_j}, t_n) by y_{j,i_j}^n.

Following [18,19], we approximate the Caputo fractional derivative of the function u_j^{n+1} by $L1$ formula on non-uniform mesh

$$\frac{\partial^{\delta_j} u_j^{n+1}(x)}{\partial t^{\delta_j}} \approx \frac{1}{\Gamma(1-\delta_j)} \sum_{s=0}^{n} \frac{u_j^{s+1}(x) - u_j^s(x)}{\tau_{s+1}} \int\limits_{t_s}^{t_{s+1}} (t_{n+1} - \eta)^{-\delta_j} d\eta$$

$$= \sum_{s=0}^{n} (u_j^{s+1} - u_j^s)\rho_{n,s}^j,$$

$$\rho_{n,s}^j = \frac{(t_{n+1} - t_s)^{1-\delta_j} - (t_{n+1} - t_{s+1})^{1-\delta_j}}{\Gamma(2-\delta_j)\tau_{s+1}} \quad \text{and} \quad \rho_{n,n}^j = \frac{\tau_{n+1}^{-\delta_j}}{\Gamma(2-\delta_j)}.$$

Applying finite volume method with implicit-explicit approximation for the interface boundary conditions, we get the following discretization of (1)–(5)

$$\mathcal{L}^j(u_{j,i_j}^{n+1}) := \widetilde{\rho}_{n,n}^j u_{j,i_j}^{n+1} - \frac{1}{h_j}\left(p_{j,i_j+1/2}u_{j,x_{i_j}}^{n+1} - p_{j,i_j-1/2}u_{j,\overline{x}_{i_j}}^{n+1}\right) = F_{j,i_j}^{n+1} + f_{j,i_j}^{n+1},$$

$$i_j = 1, 2, \ldots, N_j - 1, \quad j = 1, 2,$$

$$\mathcal{L}^r(u_{1,N_1}^{n+1}) := \left(\widetilde{\rho}_{n,n}^1 + \frac{2\alpha_1}{h_1}\right)u_{1,N_1}^{n+1} + \frac{2}{h_1}p_{1,N_1-1/2}u_{1,\overline{x}_{N_1}}^{n+1} - \frac{2\beta_1}{h_1}u_{2,0}^n = F_{1,N_1}^{n+1} + f_{1,N_1}^{n+1}, \quad (8)$$

$$\mathcal{L}^l(u_{2,0}^{n+1}) := \left(\widetilde{\rho}_{n,n}^2 + \frac{2\alpha_2}{h_2}\right)u_{2,0}^{n+1} - \frac{2}{h_2}p_{2,1/2}u_{2,x_0}^{n+1} - \frac{2\beta_2}{h_2}u_{1,N_1}^n = F_{2,0}^{n+1} + f_{2,0}^{n+1},$$

$$u_{1,0}^{n+1} = \varphi_1^{n+1}, \quad u_{2,N_2}^{n+1} = \varphi_2^{n+1}, \quad u_{j,i_j}^0 = u_j^0(x_{j,i_j}), \quad j = 1, 2.$$

Here $\widetilde{\rho}_{n,n}^j = \rho_{n,n}^j + q_j$, $F_{j,i_j}^{n+1} = G_n^j u_{j,i_j}^n + \frac{2}{h_j}\gamma_{j,i_j}^{n+1}$, if $j=1$ and $i_j = N_1$ or $j=2$ and $i_j = 0$, $F_{j,i_j}^{n+1} = G_n^j u_{j,i_j}^n$, otherwise, and

$$G_n^j u_j^n(x) := \sum_{s=1}^{n}\left(\rho_{n,s}^j - \rho_{n,s-1}^j\right)u_j^s(x) + \rho_{n,0}^j u_j^0(x), \quad n = 0, 1, \ldots, M-1,$$

$$y_{j,x_{i_j}} = \frac{v_{j,i_j+1}^{n+1} - v_{j,i_j}^{n+1}}{h_1}, \quad y_{j,\overline{x}_{i_j}} = \frac{y_{j,i_j}^{n+1} - y_{j,i_j-1}^{n+1}}{h_1}, \quad p_{j,i_j\pm1/2} = p_j(x_{j,i_j-1/2}).$$

Thus, at each time level $n = 0, 1, \ldots, M$, we find each solution u_j^n, $j = 1, 2$ separately, solving the corresponding discretization (8) on the computational domain $\overline{\omega}_j$.

4 Numerical Solution of the Inverse Problem

We consider the problem (1)–(5), where $f_j(x,t) = \lambda_j(t)\overline{f}_j(x,t)$ and $\lambda_j(t)$ is unknown function, while $\overline{f}_j(x,t)$ is given. We examine an inverse problem for identifying u_j and $\lambda_j(t)$ for given additional condition (7).

Let us introduce the following decomposition of the solution

$$u_j^n = U_j^n + \lambda_j^n V_j^n, \quad j = 1, 2. \tag{9}$$

Substituting (9) in (8) only for the solution at new time level, in order to avoid the memory term in two of the generated problems, we derive the following discrete systems

$$\mathcal{L}^1(U_{1,i_1}^{n+1}) = F_{1,i_1}^{n+1}, \quad i_1 = 1, 2, \ldots, N_1 - 1,$$
$$\mathcal{L}^r(U_{1,N_1}^{n+1}) = F_{1,N_1}^{n+1}, \tag{10}$$
$$u_{1,0}^{n+1} = \varphi_1^{n+1}, \quad u_{1,i_1}^0 = u_1^o(x_{1,i_1}).$$

$$\mathcal{L}^2(U_{2,i_2}^{n+1}) = F_{2,i_2}^{n+1}, \quad i_2 = 1, 2, \ldots, N_2 - 1,$$
$$\mathcal{L}^l(U_{2,0}^{n+1}) = F_{2,0}^{n+1}, \tag{11}$$
$$u_{2,N_2}^{n+1} = \varphi_2^{n+1}, \quad u_{2,i_2}^0 = u_2^o(x_{2,i_2}).$$

$$\left(\tilde{\rho}_{n,n}^1 + \frac{p_{1,1/2}}{h_1^2}\right) V_{1,1}^{n+1} - \frac{p_{1,3/2}}{h_1} V_{1,x_{i_1}}^{n+1} = \overline{f}_{1,1}^{n+1},$$
$$\mathcal{L}^1(V_{1,i_1}^{n+1}) = \overline{f}_{1,i_1}^{n+1}, \quad i_1 = 2, 3, \ldots, N_1 - 1, \tag{12}$$
$$\mathcal{L}^r(V_{1,N_1}^{n+1}) = \overline{f}_{1,N_1}^{n+1},$$

$$\mathcal{L}^l(V_{2,0}^{n+1}) = \overline{f}_{2,0}^{n+1},$$
$$\mathcal{L}^2(V_{2,i_2}^{n+1}) = \overline{f}_{2,i_2}^{n+1}, \quad i_2 = 1, 2, \ldots, N_2 - 2, \tag{13}$$
$$\left(\tilde{\rho}_{n,n}^2 + \frac{p_{2,N_2-1/2}}{h_2^2} V_{2,N_2-1}^{n+1}\right) V_{2,N_2-1}^{n+1} + \frac{p_{2,N_2-3/2}}{h_2} V_{2,\overline{x}_{N_2-1}}^{n+1} = \overline{f}_{2,i_2}^{n+1}$$

Further, from (9) we express the unknown source λ_j

$$\lambda_j^n = \frac{u_{j,i_j^*}^n - U_{j,i_j^*}^n}{V_{j,i_j^*}^n}, \quad j = 1, 2, \tag{14}$$

where x_{j,i_j^*} are points of measurements and $u_{j,i_j^*}^n = \psi_j(t^n)$. Then, from (9) we determine the solution u_{j,i_j}^{n+1}, $i_1 = 1, 2, \ldots, N_1 - 1$, $i_2 = 0, 1, \ldots, N_2 - 1$, $j = 1, 2$.

Further we use the notation $\|v_j\| = \max\limits_{0 \leq i_j \leq N_j} \max\limits_{0 \leq n \leq M} |v_{j,i_j}^n|$. The following statement establishes the correctness of the approach (9)–(14).

Theorem 1. *Let the assumptions (6) hold, $\overline{f}_j(x,t) \geq 0$, $\overline{f}_j(x,t) \not\equiv 0$, $x \in \Omega_j$, $0 < t \leq T$, $j = 1, 2$, then solutions of the discretizations (12), (13) are bounded and $V_j^n > 0$, $n = 1, 2, \ldots, M$, $j = 1, 2$.*

Proof (outline). Applying maximum principle for problems (12), (13), we get $V_j^n \geq 0$ and $\|V_j\| \leq \tau^{\delta_j} \|f_j\| / \Gamma(2 - \delta_j)$, $j = 1, 2$. Then, we consider equation number $\overline{i}_j$, $0 < \overline{i}_1 \leq N_1$, $0 \leq \overline{i}_2 < N_2$ of (12), (13) for which $\overline{f}_{j,\overline{i}_j}^{n+1} \neq 0$ and deduce that $u_{j,\overline{i}_j}^{n+1} > 0$, $j = 1, 2$. We consider the remaining equations sequentially, starting from the adjacent equations up to 1-th and N_1-th for (12) and up to 0-th and $N_2 - 1$-st for (13) and conclude that $u_{j,\overline{i}_j}^{n+1} > 0$, $j = 1, 2$. $\square$

5   Numerical Results

In this section we illustrate the efficiency of the proposed numerical method (9)–(14). All computations are performed on graded temporal mesh

$$t_n = T \left(\frac{n}{M} \right)^r, \quad r > 1, \quad n = 0, 1, \ldots, M.$$

Applying similar considerations as in [18], we deduce that for weak singular solution of (1)–(5), the convergence rate of the solution $u = (u_1, u_2)$ of (8) is

$$\|u(x_i, t_n) - u_i^n\| \leq O \left(h_1^2 + h_2^2 + M^{-\min_{j=1,2}\{\min\{2-\delta_j, r\delta_j\}\}} \right). \tag{15}$$

In view of (15), in order to obtain optimal accuracy

$$\|u(x_i, t_n) - u_i^n\| \leq O \left(h_1^2 + h_2^2 + M^{-\min_{j=1,2}\{2-\delta_j\}} \right), \tag{16}$$

we set $r = (2 - \delta)/\delta$, where $\delta = \min\{\delta_1, \delta_2\}$.

We consider the following model parameters and functions

$$a_1 = 1, \quad b_1 = 2, \quad p_1(x) = 2x + 3, \quad \alpha_1 = 3, \quad \beta_1 = 2, \quad q_1 = 2x, \quad x_1^* = 1.5,$$
$$a_2 = 3, \quad b_2 = 5, \quad p_2(x) = 3x^2 + 1, \quad \alpha_2 = 1, \quad \beta_2 = 0.5, \quad q_2 = 1, \quad x_2^* = 4, \quad T = 1.$$

Errors of the numerical u_j, λ_j, $j = 1, 2$, and order of convergence in time and space are given by

$$\mathcal{E}_j = \mathcal{E}_j(M, N_j) = \|u_j(x_j, t_n) - u_j^n\|, \quad \varepsilon_j = \varepsilon_j(M, N_j) = \max_{0 \leq n \leq M} |\lambda_j(t_n) - \lambda_j^n|,$$
$$\mathcal{CR}_j^\tau = \log_2 \frac{\mathcal{E}_j(M, N_j)}{\mathcal{E}_j(2M, N_j)}, \quad \mathcal{CR}_j^h = \log_2 \frac{\mathcal{E}_j(M, N_j)}{\mathcal{E}_j(M, 2N_j)}, \quad \mathcal{CR}_j^\lambda = \log_2 \frac{\varepsilon_j(M, N_j)}{\varepsilon_j(2M, N_j)}.$$

Example 1 (Exact data). We test the accuracy and order of convergence for the exact values of the measurements $\psi_j(t) = u_j(x_j^*, t)$. We determine functions γ_j, $\overline{f}_j$, such that $\lambda_j = t^{\delta_j}$, $u_j(x, t) = (t^{2\delta_j} + t^{3\delta_j} + t^2) \cos(4\pi x/j)$.

First, we verify the temporal order of convergence. In view of (16) we chose $N_j \approx j M^{(2-\delta)/2}$, such that x_j^*, $j = 1, 2$, to be grid nodes. The expected temporal order of convergence is $\min_{j=1,2} \{2 - \delta_j\}$. In Table 1 we present the computational results for numerical solutions of the inverse problem in the case of equal and different values of fractional orders.

In order to test the spatial convergence rate we compute the solution of (9)–(14), $\delta_1 = \delta_2 = 0.5$ for $N_2 = N_1$, $M = N_2^{2/(2-\delta)}$. The expected spatial convergence rate is second. The outcomes are listed in Table 2. The results in Tables 1, 2 confirm the order of convergence of the numerical solution of the inverse problem stated in (15), (16).

Table 1. Errors and temporal order of convergence of u_j and λ_j, $j = 1, 2$, Example 1

δ_1	δ_2		$M = 40$	$M = 80$	$M = 160$	$M = 320$	$M = 640$	$M = 1280$	$M = 2560$
0.5	0.5	$\mathcal{E}_1$	7.997e-4	2.992e-4	1.083e-4	3.766e-5	1.363e-5	4.782e-6	1.6920e-6
		$\mathcal{CR}_1^\tau$	1.418	1.467	1.523	1.467	1.511	1.499	
		$\mathcal{E}_2$	5.516e-3	1.936e-3	6.970e-4	2.443e-4	8.750e-5	3.081e-5	1.089e-5
		$\mathcal{CR}_2^\tau$	1.510	1.474	1.512	1.481	1.506	1.501	
		ε_1	1.014e-3	3.806e-4	1.384e-4	4.791e-5	1.737e-5	6.077e-6	2.146e-6
		$\mathcal{CR}_1^\lambda$	1.414	1.459	1.531	1.464	1.515	1.502	
		ε_2	8.493e-5	3.040e-5	1.099e-5	3.898e-6	1.396e-6	4.945e-7	1.753e-7
		$\mathcal{CR}_2^\lambda$	1.482	1.469	1.495	1.482	1.497	1.496	
0.8	0.2	$\mathcal{E}_1$	6.614e-4	2.707e-4	1.111e-4	4.604e-5	1.931e-5	8.192e-6	3.502e-6
		$\mathcal{CR}_\infty$	1.289	1.285	1.271	1.253	1.237	1.226	
		$\mathcal{E}_2$	1.802e-3	5.224e-4	1.503e-4	4.376e-5	1.262e-5	3.663e-6	1.065e-6
		$\mathcal{CR}_2^h$	1.786	1.797	1.780	1.794	1.784	1.782	
		ε_1	8.345e-3	3.970e-3	1.817e-3	8.138e-4	3.602e-4	1.583e-4	6.929e-5
		$\mathcal{CR}_1^\lambda$	1.072	1.128	1.159	1.176	1.186	1.192	
		ε_2	2.842e-4	1.318e-4	5.239e-5	1.881e-5	6.377e-6	2.087e-6	6.667e-7
		$\mathcal{CR}_2^\lambda$	1.108	1.331	1.478	1.561	1.612	1.646	

Table 2. Errors and spatial order of convergence of u_j, $j = 1, 2$, Example 1

	$N_1 = 10$	$N_1 = 20$	$N_1 = 40$	$N_1 = 80$	$N_1 = 160$	$N_1 = 320$	$N_1 = 640$
$\mathcal{E}_1$	2.064e-3	5.319e-4	1.346e-4	3.382e-5	8.472e-6	2.120e-06	5.301e-7
$\mathcal{CR}_1^h$	1.956	1.982	1.993	1.997	1.999	2.000	
$\mathcal{E}_2$	1.420e-2	3.549e-3	8.872e-4	2.218e-4	5.544e-5	1.386e-5	3.465e-6
$\mathcal{CR}_2^h$	2.001	2.000	2.000	2.000	2.000	2.000	

Example 2 (Discrete data). We consider the same test problem as in Example 1. Now the measurements are solutions of the discrete direct problem (8) at points (x_1^*, t^n), (x_2^*, t^n), $n = 1, 2, \ldots, M$. To estimate the error of the solution u_j we use numerical solution of (8) as exact solution. In Table 3, we present the results, obtained by computations with the same mesh parameters as in Example 1 and Table 1. Numerical test shows that only roundoff error appears, namely the restored solution u_j and λ_j by numerically solving inverse problem (9)–(14) are almost the same as discrete solution of the direct problem (8) and exact source, respectively, independently of the mesh step size and values of δ_j.

Example 3 (Perturbed data). We take measurement with noise [17]

$$\psi_j^\epsilon(t^{n^*}) = u_j(x_j^*, t^{n^*}) + 2\epsilon_j(\sigma_j(t^{n^*}) - 0.5), \quad j = 1, 2,$$

where ϵ_j^n are noise levels, $\sigma_j(t)$ is a random function, uniformly distributed on the interval $[0, 1]$ and $u_j(x_j^*, t^{n^*})$ is the exact solution as in Example 1. On Fig. 1

Table 3. Errors of u_j and λ_j, $j = 1, 2$, Example 2

δ_1	δ_2		$M = 40$	$M = 80$	$M = 160$	$M = 320$	$M = 640$	$M = 1280$	$M = 2560$
0.5	0.5	$\mathcal{E}_1$	8.438e-15	1.465e-14	2.244e-14	1.064e-13	1.661e-13	5.367e-13	1.414e-12
		$\mathcal{E}_2$	3.997e-15	2.813e-15	5.343e-15	6.439e-15	8.493e-15	1.499e-14	1.556e-14
		ε_1	3.331e-15	8.771e-15	2.154e-14	5.684e-14	1.018e-13	3.660e-13	8.392e-13
		ε_2	4.441e-16	5.551e-16	8.882e-16	1.332e-15	2.442e-15	3.109e-15	3.331e-15
0.8	0.2	$\mathcal{E}_\infty$	8.882e-16	1.110e-15	2.442e-15	2.565e-13	5.773e-15	3.037e-12	1.297e-11
		$\mathcal{E}_2$	1.110e-15	2.422e-15	2.474e-15	1.021e-14	3.109e-15	2.454e-14	4.574e-14
		ε_1	6.661e-16	1.110e-15	9.992e-16	1.256e-13	4.730e-13	4.713e-12	6.523e-11
		ε_2	2.220e-16	3.331e-16	4.441e-16	1.887e-15	3.109e-15	3.886e-15	6.661e-15

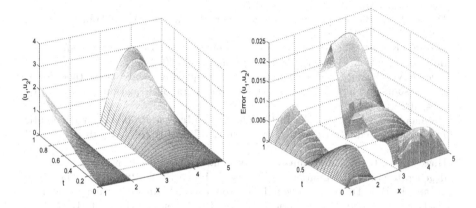

Fig. 1. Solution, computed by inverse method and absolute error of the solution

we plot the solution in the whole computational domain, obtained by inverse method (10)–(14), $M = N_1 = 40$, $N_2 = 80$, $\delta_1 = \delta_2 = 0.5$, and the corresponding absolute error with respect to the exact solution. The measurements are at 20 time levels t^{n^*} and $\epsilon_1 = 0.01$, $\epsilon_2 = 0.03$. We use polynomial curve fitting of 5-th degree in order to obtain smooth data ψ_j^ϵ at each time level.

Conclusions

In this article we are interested in inverse problem about the numerical identification of sources from point observations in a one-dimensional parabolic problem on disjoint intervals. The estimation is based on decomposition of the numerical solution with respect to the sources. Numerical tests are discussed.

There are several extensions of the present research. Firstly, the proposed scheme can be adapted to two- and three- dimensional problems. Secondly, the proposed scheme can be extended to the case of time-dependent coefficients in the equations and the boundary conditions are unknown. Thirdly, the method of fundamental solutions should be implemented to this non-standard parabolic interface problems.

Acknowledgements. This research is supported by the Bulgarian National Science Fund under Project KP-06-N 62/3 from 2022 and partly by FNSE-03.

References

1. Amosov, A.A.: Global solvability of a nonlinear nonstationary problem with a nonlocal boundary condition of radiation heat transfer type. Differ. Equ. **41**(1), 96–109 (2005)
2. Caputo, M.: Vibrations of infinite viscoelastic layer with a dissipative memory. J. Acoust. Soc. Am. **56**(3), 897–904 (1974)
3. Datta, A.K.: Biological and Bioenvironmental Heat and Mass Transfer. Marcel Dekker, New York (2002)
4. Dib, F., Kirane, M.: An inverse source problem for a two terms time-fractional diffusion equation. Bol. Soc. Paran. Mat. **40**, 1–15 (2022)
5. Duc, N.V., Thang, N.V., Thanh, N.T.: The quasi-reversability method for an inverse sourse problem for time-space fractional parabolic equations. J. Differ. Equ. **344**, 102–130 (2023)
6. Hasanoglu, A., Romanov, V.G.: Introduction to Inverse Problems for Differential Equations, 1st edn, 261 p. Springer, Cham (2017). https://doi.org/10.1007/978-3-319-62797-7
7. Hussein, M., Lesnic, D., Kamynin, V., Kostin, A.: Direct and inverse problems for degenerate parabolic equations. J. Inverse Ill-Posed Probl. **28**(3), 425–448 (2020)
8. Jovanovic, B.S., Delic, A., Vulkov, L.G.: About some boundaery value problems for fractional PDE and their numerical solution. Proc. Appl. Math. Mech. **13**, 445–446 (2013)
9. Jovanovic, B.S., Vulkov, L.G., Delic, A.: Boundary value problems for fractional PDE and their numerical approximation. In: Dimov, I., et al. (eds.) NAA 2012. LNCS, vol. 8236, pp. 38–49. Springer, Heidelberg (2013). https://doi.org/10.1007/978-3-642-41515-9_4
10. Klibas, A.A., Sriv, H.M., Trujillo, J.J.: Theory and Applications of Fractional Differential Equations. Elsevier, Amsterdam (2006)
11. Koleva, M.N., Milovanović Jeknić, Z. D., Vulkov, L.G.: Determination of external boundary conditions of a stationary nonlinear problem on disjoint intervals at point observation. Study in Computational Intelligence. Springer, Cham (accepted)
12. Lesnic, D.: Inverse Problems with Applications in Science and Engineering, p. 349. CRC Press, Abingdon (2021)
13. Liu, Y., Li, Z., Yamamoto, M.: Inverse problem of determining sources of the fractional partial differential equations. Fract. Diff. Equat. **2**, 411–430 (2019)
14. Milovanovic, Z.: Finite difference scheme for a parabolic transmission problem in disjoint domains. In: Dimov, I., Faragó, I., Vulkov, L. (eds.) NAA 2012. LNCS, vol. 8236, pp. 403–410. Springer, Heidelberg (2013). https://doi.org/10.1007/978-3-642-41515-9_45
15. Podlubny, I.: Fractional Differential Rquations, Academic Theory and Applications of Fractional Differential Equations. Elsevier, Amsterdam (1998)
16. Qatanani, N., Barham, A., Heeh, Q.: Existence and uniqueness of the solution of the coupled conduction radiation energy transfer on diffuse gray surfaces. Surv. Math. Appl. **2**, 43–58 (2007)

17. Samarskii, A.A., Vabishchevich, P.N.: Numerical Methods for Solving Inverse Problems in Mathematical Physics, 438 p. de Gruyter, Berlin (2007)
18. Stynes, M., O'Riordan, E., Gracia, J.L.: Error analysis of a finite difference method on graded meshes for a time-fractional diffusion equation. SIAM J. Numer. Anal. **55**(2), 1057–1079 (2017)
19. Zhang, Y., Sun, Z., Liao, H.: Finite difference methods for the time fractional diffusion equation on non-uniform meshes. J. Comput. Phys. **265**, 195–210 (2014)
20. Zhuo, L., Lesnic, D., Meng, S.: Reconstruction of the heat transfer coefficient at the interface of a bi-material. Inverse Probl. Sci. Eng. **28**, 374–401 (2020)

The Wright Function – Numerical Approximation and Hypergeometric Representation

Dimiter Prodanov[1,2]([✉]) [iD]

[1] Neuroelectronics Research Flanders, IMEC, Kapeldreef 75, 3001 Leuven, Belgium
dimiter.prodanov@iict.bas.bg
[2] PAML-LN, IICT, Bulgarian Academy of Sciences, Acad. G. Bonchev St., Block 25A, 1113 Sofia, Bulgaria

Abstract. The Wright function, a special function, which arises in the theory of the space-time fractional diffusion equation, is a very general mathematical object with diverse connections to other special and elementary functions. The paper presents a numerical integration technique for approximation of the Wright function $W(a, b | x)$ using the method of stationary phase. As a second contribution, the paper exhibits a symbolic algorithm for hypergeometric (HG) representation of the Wright function. For rational values of its first argument the function is reduced to a finite sum of HG functions and polynomials. The HG functions are themselves represented by known elementary or other special functions, wherever possible. Reference implementations of the algorithms are programmed in the open-source computer algebra system Maxima.

Keywords: Wright function · hypergeometric function · Bessel function

1 Introduction

Fractional calculus models of physical phenomena have rekindled the interest in special functions. The Wright function arises in the theory of the space-time fractional diffusion equation. It is a very general mathematical entity with diverse connections to other special and elementary functions. Notably, it provides a unified treatment of several classes of special functions, such as the Gaussian, Airy, Bessel, error functions, etc. The function was originally defined by the infinite series [9]:

$$W(a, b | z) := \sum_{k=0}^{\infty} \frac{z^k}{k! \, \Gamma(ak + b)}, \quad z, b \in \mathbb{C}, \quad a > -1, \tag{1}$$

where $\Gamma(\sim)$ denotes the Euler's gamma function. In view of its broad applications, methods of its computation can be of general interest. The paper presents a numerical integration technique for approximation of the Wright function using the method of stationary phase for real values of the arguments and parameters.

I. Lirkov and S. Margenov (Eds.): LSSC 2023, LNCS 13952, pp. 146–153, 2024.
https://doi.org/10.1007/978-3-031-56208-2_14

On the second place, the paper exhibits a symbolic algorithm for hypergeometric (HG) representation of the Wright function $W(a, b|x)$. For rational values of its first argument the function is reduced to a finite sum of HG functions and polynomials. The HG functions are themselves represented by known elementary or other special functions, wherever possible. Reference implementations of the algorithms are programmed in the open-source computer algebra system Maxima.

2   Numerical Approximation

Until recently, efficient numerical algorithms for Wright function's approximation were not available. The seminal works of Luchko et al. [3, 4] investigated only the case $|b| \leq 1$. Recently, the Laplace transform inversion method was used for its approximation [1]. The present work uses a different contour determined by the requirement of stationary phase, which improves the overall numerical stability of the algorithm.

Theorem 1. *The Wright function $W(a, b|z)$ for real arguments can be computed according to the table:*

b	a		
	$a < 0$	$a = 0$	$a > 0$
$b \leq 1$	$W(a,b|z) = I_r(0)$	$W(a,b|z) = \frac{e^z}{\Gamma(b)}$	$W(a,b|z) = I_u(\epsilon^a) + P(\epsilon)$
$b = 1$	$W(a,b|z) = I_r(0) + 1$		
$b > 1$	$W(a,b|z) = I_r(\epsilon) + P(\epsilon)$		

where $\epsilon = \max\left(\left| \sqrt[a+1]{|az|} \right|, 1 \right)$ and

$$I_r(\epsilon) = \frac{1}{\pi} \int_\epsilon^\infty \frac{\sin\left(\frac{\sin(\pi a)z}{r^a} + \pi b \right)}{r^b} e^{\frac{\cos(\pi a)z}{r^a} - r} dr,$$

$$I_u(\epsilon) = \frac{1}{\pi a} \int_\epsilon^\infty u^{\frac{1-b}{a}-1} \sin(\sin(\pi a)z/u + \pi b) e^{\cos(\pi a)z/u - u^{1/a}} du,$$

$$P(\epsilon) = \frac{\epsilon^{1-b}}{2\pi} \int_{-\pi}^\pi e^{\epsilon \cos\varphi + \cos(a\varphi)z/\epsilon^a} \cos(-z\sin(a\varphi)/\epsilon^a + \epsilon\sin\varphi + (1-b)\varphi)d\varphi.$$

Proof. The canonical complex line integral representation of the Wright function is given by the integral

$$W(a, b|z) = \frac{1}{2\pi i} \int_{Ha^-} \frac{e^{\xi + z\xi^{-a}}}{\xi^b} d\xi = \frac{1}{2\pi i} \int_{Ha^-} ker(\xi)d\xi, \quad z \in \mathbb{C}, \quad (2)$$

along a Hankel contour, which surrounds the negative real semi-axis. The starting point of the proof is a deformation of the Hankel contour into a part enclosing the pole at the origin and parts extending towards negative infinity The Hankel integral contour can be split in three parts - the rays BA, ED and the arc BCD. Accordingly, the value of the integral can be evaluated as:

$$W\,(a,b|\,z) = \underbrace{I_{AB} + I_{DE}}_{I_r} + \underbrace{I_{BCD}}_{P_\epsilon} = \int_{AB} ker(\xi)d\xi + \int_{BCD} ker(\xi)d\xi + \int_{DE} ker(\xi)d\xi.$$

The branch cut will be taken along the negative real axis. Then for the two rays AB and ED we obtain

$$ker_{AB} - ker_{ED} = \frac{2ie^{\cos(\delta)r + \frac{\cos(a\delta)z}{r^a}} \sin\left(\frac{\sin(a\delta)z}{r^a} - \sin(\delta)r + b\delta\right)}{r^b}.$$

Taking the limit $\delta \to \pi$ results in the radial contribution

$$I_r(\epsilon) = \frac{1}{\pi} \int_\epsilon^\infty \frac{e^{\frac{\cos(\pi a)z}{r^a} - r} \sin\left(\frac{\sin(\pi a)z}{r^a} + \pi b\right)}{r^b} dr. \tag{3}$$

For the arc part we obtain

$$ker_{BCD} = \frac{i\epsilon e^{\epsilon \cos(\varphi) + \frac{\cos(a\varphi)z}{\epsilon^a}} \cos\left(\frac{-\sin(a\varphi)z}{\epsilon^a} + \epsilon \sin(\varphi) - b\varphi + \varphi\right)}{\epsilon^b}$$
$$- \frac{\epsilon e^{\epsilon \cos(\varphi) + \frac{\cos(a\varphi)z}{\epsilon^a}} \sin\left(\frac{-\sin(a\varphi)z}{\epsilon^a} + \epsilon \sin(\varphi) - b\varphi + \varphi\right)}{\epsilon^b}.$$

Therefore, taking the limit $\delta \to \pi$ as before we obtain the circular integral

$$P(\epsilon) = \frac{\epsilon^{1-b}}{2\pi} \int_{-\pi}^{\pi} e^{\epsilon \cos \varphi + \cos(a\varphi)z/\epsilon^a} \cos\left(-(\sin(a\varphi)z/\epsilon^a + \epsilon \sin \varphi + (1 - b)\varphi\right)d\varphi.$$

What is left is to determine the optimal value of ϵ. Here the present approach departs from the treatment of [3]. The Hölder exponent β at the origin is determined by the limit of the Hölder function $h(\xi) := \xi \frac{d}{d\xi} \log ker(\xi)$ so that $\beta = \lim_{\xi \to 0} h(\xi)$. For the Wright kernel $h(\xi) = -az/\xi^a + \xi - b$. Therefore, for $a < 0$ and $\beta = h(\xi)\big|_{\xi=0} = -b$ and corresponds with the fact that $W\,(a,b|\,0) = 1/\Gamma(b)$. In particular, if $b < 1$ the singularity at the origin is integrable and $\epsilon = 0$. Furthermore, for integer a and b the contour can be closed and the Cauchy theorem can be applied. Therefore, $W\,(-n,m|\,z)$ becomes a polynomial in z for integer n and m. In contrast, if $a > 0$ an essential singularity appears at the origin as the Hölder exponent diverges. For that latter case we look for the curve where the phase of the kernel is stationary to first order to minimize oscillations. This is given by the computation

$$\frac{d}{d\xi}\left(z/\xi^a + \xi\right)\Big|_{\xi=\xi_0} = \left(-\frac{az}{\xi^{a+1}} + 1\right)\Big|_{\xi=\xi_0} = 0 \implies \xi_0 = \sqrt[a+1]{az}.$$

Therefore, we can take $\epsilon = |\xi_0| = \sqrt[a+1]{|az|}$.

The next proposition is stated without proof in view of the space limitations.

Proposition 1. *For non negative integers n, m*

$$\Gamma(mn + b) = \Gamma(mb)\, m^{mn} \prod_{j=0}^{m-1} \left(\frac{j}{m} + b\right)_n.$$

3   Hypergeometric Representation

Wherever the first parameter is rational, the Wright function can be represented by a finite sum of hypergeometric functions. The Generalized Hypergeometric (GHG) function pFq will be used under the notation

$$_pF_q\left(a_0 \ldots a_{p-1}; b_0 \ldots b_{q-1} \middle| z\right) := \sum_{r=0}^{\infty} \frac{z^r}{r!} \frac{\prod_{j=0}^{p-1}(a_j)_r}{\prod_{j=0}^{q-1}(b_j)_r}, \tag{4}$$

where $(a)_r$ and $(b)_r$ will denote rising factorials and $(a)_0 = 1$, which assumes the normalization $_pF_q\left(\sim; \sim \middle| 0\right) = 1$. By convention, equal parameters in the numerator and denominator will cancel out. For positive, rational a one could obtain the representation [2]:

Theorem 2 (First HG Representation). *Suppose that $a = n/m > 0$, where n and m are co-prime. Then*

$$W\left(\frac{n}{m}, b \middle| z\right) = \sum_{r=0}^{m-1} \frac{z^r}{r!\, \Gamma(b + ar)} \cdot {_1}F_{m+n}\left(1; b_0 \ldots b_{n-1}, c_0 \ldots c_{m-1} \middle| \frac{z^m}{n^n m^m}\right), \tag{5}$$

where $b_j = r/m + (b+j)/n$, $c_j = (r+1+j)/m$.

The proof follows [5] and is given as a staring point for the proof of the Second Representation Theorem.

Proof. Observe that

$$W\left(n/m, b \middle| z\right) = \sum_{k=0}^{\infty} \frac{z^n}{k!\, \Gamma(ak + b)} = \sum_{q=0}^{m-1} \sum_{p \geq q/m}^{\infty} \frac{z^{mp-q}}{\Gamma(mp - q + 1)\, \Gamma(a(mp - q) + b)},$$

since the integer k can be partitioned as $k = mp - q$, where $q = 0, \ldots m - 1$. After some algebra we obtain

$$W\left(n/m, b \middle| z\right) = \frac{1}{\Gamma(b)} + \sum_{r=1}^{m} z^r \sum_{p=0}^{\infty} \frac{z^{mp}}{\Gamma(ap + ra + b)\Gamma(mp + r + 1)}.$$

Observe that for $p = 0$ the inner series coefficient is $C_r = 1/\Gamma(ra + b)\Gamma(r + 1)$, which serves as a normalization factor. Therefore, the series transforms as

$$W\left(n/m, b \middle| z\right) = \sum_{r=0}^{m} \frac{z^r}{C_r} \cdot \sum_{p=0}^{\infty} \frac{C_r}{\Gamma\left(n(p + r/m + b/n)\right) \Gamma\left(m(p + (r+1)/m)\right)}. \tag{6}$$

Further, use Proposition 1 to obtain

$$\Gamma(b+aj) = \Gamma\left(a\left(j+b/a\right)\right) = \Gamma(b) \left(\frac{b}{a}\right)_j \left(\frac{b+1}{a}\right)_j \cdots \left(\frac{b+a-1}{a}\right)_j a^{aj}.$$

From Eq. 6 we read off $b_0 = \frac{r}{m} + \frac{b}{n}, c_0 = \frac{r+1}{m}$ with increments $1/n$ and $1/m$, respectively. Therefore, finally,

$$W\left(a,b|\,z\right) = \sum_{r=0}^{m-1} \frac{z^r}{r!\,\Gamma\left(b+ar\right)} \cdot {}_1F_{m+n}\left(1; b_0 \ldots b_{n-1}, c_0 \ldots c_{m-1}\Big|\frac{z^m}{n^n m^m}\right).$$

Observe that for $r = m-1$ $c_1 = 1$. Therefore, the GHG functions reduce to ${}_0F_{m+n-1}$.

Similar formula for a negative rational a can be also given as follows:

Theorem 3 (Second HG Representation). *For $b \le 1$ and $n \le m$ non-negative co-prime integers*

$$W\left(-\frac{n}{m},b\Big|\,z\right) = \sum_{r=0}^{m-1} \frac{z^r}{r!\,\Gamma\left(b+ar\right)} {}_{n+1}F_m\left(1, b_0' \ldots b_{n-1}'; c_0 \ldots c_{m-1}\Big|\frac{(-)^n z^m}{n^n m^m}\right), \quad (7)$$

where $b_j' = 1 + r/m - (b+j)/n$, $c_j = (r+1+j)/m$, $a = -n/m$.

Proof. Suppose first that $b < 1$. Let

$$W\left(-n/m, b|\,z\right) = \frac{1}{\Gamma(b)} + \sum_{r=1}^{m} \frac{z^r}{C_r} \sum_{p=0}^{\infty} \frac{C_r}{\Gamma\left(-np - rn/m + b\right)} \frac{z^{mp}}{\Gamma\left(mp+r+1\right)},$$

where C_r was defined as above. We use the Gamma reflection formula to obtain

$$\frac{1}{\Gamma\left(-np-rn/m+b\right)} = \frac{(-1)^{np}\Gamma\left(\frac{nr}{m}+np-b+1\right)}{\Gamma\left(b-\frac{nr}{m}\right)\Gamma\left(\frac{nr}{m}-b+1\right)}. \quad (8)$$

Therefore,

$$W\left(-n/m, b|\,z\right) = \frac{1}{\Gamma(b)} + \sum_{r=1}^{m} \frac{z^r}{C_r} \sum_{p=0}^{\infty} \frac{(-1)^{np} C_r \Gamma\left(\frac{nr}{m}+np-b+1\right) z^{mp}}{\Gamma\left(b-\frac{nr}{m}\right)\Gamma\left(\frac{nr}{m}-b+1\right)\Gamma\left(m(p+(r+1)/m)\right)}.$$

Finally, we read off the parameters $b_j' = 1 - (-r/m + (b+j)/n) = 1 + r/m - (b+j)/n$. Observe that for $r = m-1$ $c_1 = 1$. Therefore, as before the GHG functions reduce to ${}_nF_{m-1}$.

For $b \ge 1$ a polynomial term P is also added as follows.

Definition 1 (Wright polynomials). *Define the polynomial $P_b(-a,z)$ by the integral recursion*

$$P_b(-a,z) = \int_0^z P_{b-a}(-a,x)dx + c_{b-1}, \quad (9)$$

where $c_{b-1} = 1/(b-1)!$ if b is an integer and 0 otherwise. Furthermore, define $P_0(z,-a) := 1$ and for $b < 0$ assign $P_b(z,-a) := 0$ identically.

A   B

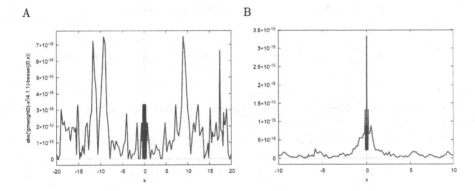

Fig. 1. Absolute error plots for $a = 1$: A $- |J_0(x) - W(1, 1| - x^2/4)|$, B $- |\sin(2x)/(\pi x) - W(1, 3/2| - x^2)|$.

Theorem 4 (Third HG representation). *Suppose that a and b are rational parameters and $b \geq 1$ and $|a| \leq 1$. Then*

$$
W(-n/m, b| z) = \sum_{r=0}^{m-1} \frac{z^r}{r!\,\Gamma(b - nr/m)} \cdot
$$
$$
{}_{n+1}F_m\left(1, b'_0 \ldots b'_{n-1}; c_0 \ldots c_{m-1}\left| \frac{(-)^n z^m}{n^n m^m}\right.\right) + P_b(n/m, z) \tag{10}
$$

where m and n are co-prime numbers.

Proof. First we prove that the arc integral results in a polynomial in z. Suppose that $b \geq 1$ is rational and $-a = n/m$ as before. Consider the arc BCD. We change variables as $\xi = \epsilon \eta^m$. We develop the integral kernel in infinite series:

$$
ker = m\,\frac{d\eta\,\epsilon^{1-b}}{\eta^{(b-1)m+1}} \sum_{j=0}^{\infty} \sum_{i=0}^{j} \frac{\epsilon^{\frac{im+(j-i)n}{m}} \eta^{im+(j-i)n} z^{j-i}}{i!\,(j-i)!} .
$$

The scale-invariant part of the series is given by the members k_j for which $\epsilon^{\frac{im+(j-i)n}{m} - b + 1} = 1$. This is given by the constraint $i = \frac{(b-1)m - jn}{m-n}$. Therefore, the monomial terms are computed by the formula

$$
c_j = \frac{z^{\frac{(j-b+1)m}{m-n}}}{\left(\frac{(b-1)m-jn}{m-n}\right)!\left(\frac{(j-b+1)m}{m-n}\right)!} = \frac{z^{\frac{j-b+1}{1-a}}}{\left(\frac{j-b+1}{1-a}\right)!\left(\frac{-aj+b-1}{1-a}\right)!}, \tag{11}
$$

where the valid indices are given by the set j : $\left(\frac{(j-b+1)m}{m-n} \in \mathbb{N}\right) \cup \left(\frac{(b-1)m-jn}{m-n} \in \mathbb{N}\right)$. Equivalently, using the a-notation it can be seen that $a < 1$ must hold for c_j not to vanish. On the other hand, $b - 1 \leq j \leq (b-1)/a \cup j \in \mathbb{N}$,

Table 1. Absolute errors at the first zeroes of $J_0(x)$

zero	w_1	w_2	w_3	w_4	w_5	w_6
error	5.9342 E-17	8.7931 E-17	2.2841 E-16	2.4453 E-17	3.204 E-17	1.2964 E-16

which is a finite set. Therefore, for a rational b the integral in Eq. 2 becomes a polynomial in z.

To derive the polynomial recursion we proceed as follows. Observe that

$$W(-a, b| z) = \int_0^z W(-a, b - a| z) + 1/\Gamma(b), \tag{12}$$

so that the equation defines a recursion relationship.

Observe that for $j = b - 1$ the constant term becomes $c_{b-1} = 1/(b-1)!$. Therefore, for non-integer b there are no constant monomials. Furthermore, consider the monomial c_j as a function of b. Differentiating Eq. 11 one obtains the recursion

$$\frac{d}{dz}c_j(b) = \frac{z^{\frac{j-b+1}{1-a}-1}}{\left(\frac{j-b+1}{1-a}-1\right)!\left(\frac{-aj+b-1}{1-a}\right)!} = \frac{z^{\frac{j-(b-a)}{1-a}}}{\left(\frac{j-(b-a)}{1-a}\right)!\left(\frac{aj-b+1}{a-1}\right)!} = c_{j-1}(b-a),$$

which is also consistent with the integral Eq. 12. Therefore, the polynomial $P_b(-a, z)$ should obey the above recursion. The second argument of the Wright function mutates and therefore it is convenient that it indexes the polynomial.

For integer values of a, the hypergeometric sum disappears from Eq. 10.

It should be noted that the Second and Third HG Representation theorems could not be traced to available literature. Based on these new results, we can classify the Wright functions as – 1st type (when $a > 0$); 2nd type (when $a \in [-1, 0]$, $b < 1$); and 3rd type when $b \geq 1$ and $a \in [-1, 0]$ or when a is a negative integer and b is a positive integer.

The utility of the above method depends on the availability of routines to calculate "ordinary" special functions, such as the Bessel, error, Gaussian etc. An illustrative example of Theorem 3 can be presented for $a = -1/3$ and $b = 1/3$:

$$W(-1/3, 1/3| - z) = -\frac{z}{3}\left(I_{2/3}\left(\frac{2z^{\frac{3}{2}}}{3^{\frac{3}{2}}}\right) - I_{-2/3}\left(\frac{2z^{\frac{3}{2}}}{3^{\frac{3}{2}}}\right)\right) =$$

$$\frac{z}{\sqrt{3\pi}}K_{2/3}\left(\frac{2z^{\frac{3}{2}}}{3^{\frac{3}{2}}}\right) = -\sqrt[3]{3}\,\mathrm{Ai}'\left(\frac{z}{\sqrt[3]{3}}\right). \tag{13}$$

4   Numerical Experiments

A reference implementation in the computer algebra system Maxima was developed and deposited in the Zenodo repository [8]. The library implements two quadrature algorithms – the function *gmwright* which is implemented using the double exponential (DE) quadrature in Lisp [6], and the function *gmwright2*, which is implemented using the default Maxima QUADPACK library translated to Lisp [7]. Both QUADPACK and DE libraries can compute with user-specified precision. In addition, the library contains the function *wrightsimp*, which implements Theorems 2, 3, and 4 and is written in the Maxima language. The library is available for download and testing at https://zenodo.org/record/7871652. Numerical experiments were performed on a 64-bit Microsoft Windows 10 Enterprise machine with configuration – Intel® Core™ i5-8350U CPU @ 1.70GHz, 1.90 GHz and 16 GB RAM. The algorithm shows good numerical stability for the computation of Bessel functions for intermediate values of the argument (Fig. 1). The plots in Fig. 1 have been produced using the QUADPACK *quad_qagi* (implementing seminifinite domain integration) and *quad_qag* (implementing Gauss-Kronrod quadrature formulae) routines in Maxima 5.46.0 running Steel Bank Common Lisp (SBCL) v. 2.2.2. The figure demonstrates that numerical quadrature with QUADPACK reaches machine precision. The absolute errors at the first zeroes of $J_0(x)$ are shown in Table 1.

Acknowledgment. The present work was funded by the European Union's Horizon Europe program under grant agreement VIBraTE, grant agreement 101086815.

References

1. Aceto, L., Durastante, F.: Efficient computation of the Wright function and its applications to fractional diffusion-wave equations. ESAIM Math. Model. Numer. Anal. **56**(6), 2181–2196 (2022). https://doi.org/10.1051/m2an/2022069
2. Apelblat, A., González-Santander, J.L.: The integral Mittag-Leffler, Whittaker and Wright functions. Mathematics **9**(24), 3255 (2021). https://doi.org/10.3390/math9243255
3. Luchko, Y.: Algorithms for evaluation of the Wright function for the real arguments' values. Fract. Calc. Appl. Anal. **11**(1), 57–75 (2008)
4. Luchko, Y., Trujillo, J., Velasco, M.: The Wright function and its numerical evaluation. Int. J. Pure Appl. Math. **64**, 567–575 (2010)
5. Miller, A.R., Moskowitz, I.S.: Reduction of a class of Fox-Wright Ψ functions for certain rational parameters. Technical report, CHACS (1995)
6. Mori, M., Sugihara, M.: The double-exponential transformation in numerical analysis. J. Comput. Appl. Math. **127**(1–2), 287–296 (2001). https://doi.org/10.1016/s0377-0427(00)00501-x
7. Piessens, R., Doncker-Kapenga, E., Überhuber, C.W., Kahaner, D.: Quadpack. Springer, Heidelberg (1983). https://doi.org/10.1007/978-3-642-61786-7
8. Prodanov, D.: Wright function approximation and representation (2023). https://doi.org/10.5281/ZENODO.7871652
9. Wright, E.M.: The asymptotic expansion of the generalized hypergeometric function. J. London Math. Soc. **s1-10**(4), 286–293 (1935). https://doi.org/10.1112/jlms/s1-10.40.286

Study of Sparsification Schemes for the FEM Stiffness Matrix of Fractional Diffusion Problems

Dimitar Slavchev[✉] and Svetozar Margenov

Institute of Information and Communication Technologies – Bulgarian Academy of Sciences, Acad. G. Bonchev St., Block 25A, 1113 Sofia, Bulgaria
{dimitargslavchev,margenov}@parallel.bas.bg

Abstract. Anomalous diffusion describes various natural and social phenomena and processes in which the Brownian motion hypothesis is violated. Such problems can be modeled with the fractional Laplace operator. We consider the finite element discretization of the integral formulation of the Fractional Laplacian. The operator is non-local and as a result the stiffness matrix $K \in \mathbb{R}^{N \times N}$ is dense. However, it can be observed that many of the off-diagonal coefficients have very small values relative to the corresponding diagonal elements. In this work, we study sparsification techniques like removing or lumping (summing into the diagonal) the coefficients that fall under a specified threshold. In this way we construct sparse approximations of the stiffness matrix K. Numerical results for a model fractional Laplacian boundary value problem are presented. Based on them, the accuracy of the approximate solutions is analysed.

Keywords: Anomalous Diffusion · Fractional Laplacian · Sparsification

1 Introduction

Anomalous diffusion is observed in many natural processes [5,6] including superconductivity, protein diffusion within cells, diffusion through porous media, image processing, machine learning, electromagnetostatics and others. It is described mathematically by the fractional Laplace operator, which is why it is often called fractional diffusion. In this work we use the Integral definition of the Fractional Laplacian defined through the Riesz potential and discretized with the finite element method. We should note that there are other definitions of the fractional Laplacian, for example in [4] the difference between the spectral and the integral definitions is studied. In this work we follow the method outlined by Acosta and Borthagaray in [2]. The Matlab implementation by the same authors in [1] is used to generate the stiffness matrix of the problem.

The fractional Laplace operator is non-local and as a result the corresponding stiffness matrix $K \in \mathbb{R}^{N \times N}$ is dense. This means that numerically solving anomalous diffusion problems is computationally expensive. In the general case solving a dense system of linear algebraic equations has computational complexity of $O(N^3)$. If K can be sparsified (approximated by a sparse matrix), then computationally cheaper specialized direct methods can be used. Such are the frontal method, nested dissection and others.

© The Author(s), under exclusive license to Springer Nature Switzerland AG 2024
I. Lirkov and S. Margenov (Eds.): LSSC 2023, LNCS 13952, pp. 154–163, 2024.
https://doi.org/10.1007/978-3-031-56208-2_15

It should be noted that a large fraction of the off-diagonal elements of K are much smaller than the diagonal elements in the corresponding row/column. This property obviously depends on the fractional power $\alpha \in (0, 1)$. As is well known, the stiffness matrix is sparse if $\alpha = 1$. It is also natural that the sufficiently small elements could be removed or lumped (added to their corresponding diagonal element) without significant loss of accuracy. In this work, we study the accuracy loss of such sparsification techniques for several values of α and for different problem sizes.

The rest of the paper is organized as follows. Some basic properties of the considered fractional diffusion problem are given in the next section. The motivation of the proposed sparsification schemes is then discussed in Sect. 3. Our experimental investigation of the relative error is presented in Sect. 4, followed by brief concluding remarks.

2 Integral Fractional Diffusion

We consider the fractional in space Laplace operator written in the following form

$$(-\Delta)^\alpha u(x) = C(d, \alpha) \, \text{P.V.} \int_{\mathbb{R}^d} \frac{u(x) - u(y)}{|x - y|^{d+2\alpha}} dy, \quad \alpha \in (0, 1), \tag{1}$$

where α is the fractional power of the Laplace operator $-\Delta$, d is the number of dimensions (in our study $d = 2$), P.V. stands for principal value and $C(d, \alpha)$ is the normalized constant (here Γ is the gamma function)

$$C(d, \alpha) = \frac{2^{2\alpha} \alpha \Gamma \left(\alpha + \frac{d}{2} \right)}{\pi^{d/2} \Gamma (1 - \alpha)}.$$

The fractional Laplace operator defined through (1) is one of the simplest pseudo-differential operators. It can be regarded as an infinitesimal generator of an α-stable Lévi operator [3, 7].

The finite element method is applied to discretize the fractional Laplacian using an admissible quasi uniform triangulation $\mathcal{T}_h$ on the problem domain Ω. Continuous piecewise linear elements form the discrete space $\mathbb{V}$, where the nodal basis $\varphi_i(x_j) = \delta^i_j$, $\{\varphi_1, \ldots, \varphi_N\} \in \mathbb{V}$ corresponds to the internal nodes $\{x_1, \ldots, x_N\}$. Thus, the stiffness matrix $K = K_{ij} \in \mathbb{R}^{N \times N}$ can be written as

$$K_{ij} = \frac{C(d, \alpha)}{2} \langle \varphi_i, \varphi_j \rangle_{H^\alpha(\mathbb{R})} = \frac{C(d, \alpha)}{2} \sum_{l=1}^{N_{\tilde{T}}} \left(\sum_{m=1}^{N_{\tilde{T}}} I_{l,m}^{i,j} + 2 J_l^{i,j} \right), \quad l, m \in [1, N_{\tilde{T}}],$$

where H is a Hilbert space , $\tilde{\mathcal{T}}_h$ is a quasi uniform admissible triangulation that extends $\mathcal{T}_h$ to an auxiliary ball domain B, that contains Ω with a suitably chosen distance between the boundary $\partial\Omega$ and the compliment B^c (an example is shown on Fig. 1), $N_{\tilde{\mathcal{T}}}$ is the number of nodes in B. The integrals $I_{l,m}^{i,j}$ and $J_l^{i,j}$ have the form

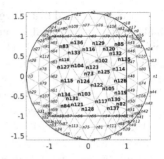

$$I_{l,m}^{i,j} = \int_{T_l}\int_{T_m} \frac{\left(\varphi_i(x)-\varphi_i(y)\right)\left(\varphi_j(x)-\varphi_j(y)\right)}{|x-y|^{d+2\alpha}}dxdy$$

$$J_l^{i,j} = \int_{T_l}\int_{B^c} \frac{\varphi_i(x)\varphi_j(x)}{|x-y|^{d+2\alpha}}dydx.$$

Fig. 1. Square domain Ω within an auxiliary ball B.

Here let us recall that the stiffness matrix is symmetric positive definite and therefore all diagonal elements are positive. In the experiments off-diagonal elements are mostly negative but a small number of positive coefficients is usually present as well. For $\alpha \leq 0.5$ the number of positive coefficients is very small: ~5 to ~50 for $N = 2131, 16184$ respectfully, while for $\alpha = 0.25 : \sim 12\,00$ to $\sim 96\,086$. In the case $\alpha = 1$ corresponds to standard (local) diffusion leading to a sparse stiffness matrix. For the fractional case $\alpha \in (0,1)$, the matrix is dense leading to higher computational complexity in solving the linear systems. We notice that many of the off-diagonal elements are very small compared to the diagonal elements in the corresponding row or column. The order of magnitude of the diagonal and off-diagonal elements can be seen in Fig. 2 for $\alpha = 0.25, 0.5, 0.75$ and 0.90. It can be clearly seen that the matrix structure appears to be closer to sparse for larger α.

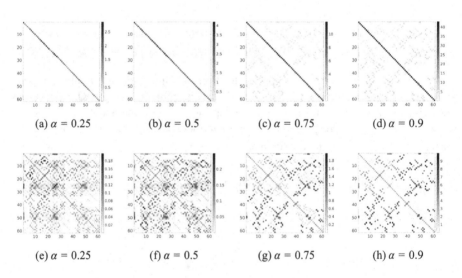

(a) $\alpha = 0.25$ (b) $\alpha = 0.5$ (c) $\alpha = 0.75$ (d) $\alpha = 0.9$

(e) $\alpha = 0.25$ (f) $\alpha = 0.5$ (g) $\alpha = 0.75$ (h) $\alpha = 0.9$

Fig. 2. Order of magnitude of the stiffness matrix elements with a small $N = 61$ problem. The second row shows the structure of the off-diagonal elements only, with the diagonal elements marked with an 'x'.

Table 1. Mean values MV^{K_D} and $MV^{K \backslash K_D}$ for various sizes N and fractional powers α.

N	$\alpha = 0.25$		$\alpha = 0.5$		$\alpha = 0.75$		$\alpha = 0.9$	
	MV^{K_D}	$MV^{K \backslash K_D}$	MV^{K_D}	$MV^{K \backslash K_D}$	MV^{K_D}	$MV^{K \backslash K_D}$	MV^{K_D}	$MV^{K \backslash K_D}$
2131	6.18e-03	3.42e-06	0.0466	1.94e-05	0.386	1.73e-04	1.44	6.59e-04
4167	3.77e-03	1.12e-06	0.0335	7.35e-06	0.326	7.58e-05	1.35	3.17e-04
8030	2.32e-03	3.75e-07	0.0243	2.82e-06	0.278	3.38e-05	1.26	1.55e-04
12805	1.64e-03	1.70e-07	0.0192	1.42e-06	0.248	1.90e-05	1.21	9.35e-05
16184	1.38e-03	1.14e-07	0.0171	1.00e-06	0.233	1.42e-05	1.18	7.22e-05

3   Sparsification Schemes

In Table 1 we present the means of the absolute values for the diagonal MV^{K_D} and off-diagonal elements $MV^{K \backslash K_D}$, calculated as follows

$$MV^{K_D} = \frac{1}{N} \sum_{i=1}^{N} K_{ii} \quad \text{and} \quad MV^{K \backslash K_D} = \frac{1}{N} \sum_{i \neq j}^{N} |K_{ij}|.$$

Here K_D is the set of diagonal coefficients of K and $K \backslash K_D$ is the set of off-diagonal coefficients. As expected MV^{K_D} and $MV^{K \backslash K_D}$ depend on the values of α and N.

Both the diagonal and off-diagonal elements decrease as the problem size increases. To account for that we use a relative sparsification threshold $\varepsilon_{\text{sparse}}$. I.e. an element $K_{ij, i \neq j}$ is considered for elimination if

$$\frac{|K_{ij}|}{K_{ii}} < \varepsilon_{\text{sparse}}, \ j \neq i \quad \text{and equivalently} \quad \frac{|K_{ij}|}{K_{jj}} < \varepsilon_{\text{sparse}}, \ i \neq j$$

are satisfied. It should also be noted that diagonal elements become larger while the off-diagonal elements become smaller as α increases.

In order to sparsify the stiffness matrix we employ and compare two schemes for removing the selected small off-diagonal coefficients.

1. *Lumping*: Eliminated off-diagonal coefficients are added to the corresponding diagonal elements and set to zero.
2. *Removing*: Eliminated off-diagonal coefficients are simply set to zero.

4   Experimental Study of the Relative Error

In the presented experimental study, we consider the linear system $K\mathbf{u} = \mathbf{f}$ obtained after finite element discretization of the fractional diffusion problem in $\Omega = (-1, 1) \times (-1, 1)$. Here $\mathbf{u} = (u_i) \in \mathbb{R}^N$ is the vector of nodal unknowns and the right hand side is $\mathbf{f} = (f_j) \in \mathbb{R}^N$, $f_j = \int_{\Omega} f \varphi_j$. The size of the discrete problem varies in the range $N \in (2\,131, 16\,184)$. The numerical experiments are run on the AVITOHOL supercomputer[1]

---

[1] http://www.hpc.acad.bg/system-1/.

Table 2. Relative error R_{lump}, percent lumped and condition numbers with different lumping thresholds for $\alpha = 0.25$.

N	$\varepsilon_{\text{sparse}}$							
	10^{-1}	10^{-2}	10^{-3}	10^{-4}	10^{-5}	10^{-6}	10^{-7}	10^{-8}
Relative error								
2 131	0.081	0.073	0.052	0.016	0.000296	0	0	0
4 167	0.085	0.080	0.063	0.031	0.002686	1.47e-07	0	0
8 030	0.087	0.084	0.072	0.045	0.009513	2.03e-05	0	0
12 805	0.088	0.086	0.077	0.054	0.017536	2.35e-04	0	0
16 184	0.089	0.087	0.079	0.059	0.022554	6.17e-04	2.27e-11	0
Percent lumped								
2 131	99.95	98.81	93.61	68.21	5.76	0	0	0
4 167	99.98	99.38	96.61	81.94	28.47	<0.01	0	0
8 030	99.99	99.68	98.19	89.91	53.21	0.42	0	0
12 805	99.99	99.80	98.85	93.41	67.06	4.66	0	0
16 184	99.99	99.84	99.09	94.71	72.78	10.33	<0.01	0

			Condition number								
N	h	$h^{-2\alpha}$	$\varepsilon_{\text{sparse}}$								dense
			10^{-1}	10^{-2}	10^{-3}	10^{-4}	10^{-5}	10^{-6}	10^{-7}	10^{-8}	
2 131	0.0433	4.80	13.80	8.93	5.54	5.24	5.25	5.25	5.25	5.25	5.25
4 167	0.0309	5.68	11.96	13.96	6.83	6.78	6.66	6.65	6.65	6.65	6.65
8 030	0.0223	6.69	15.97	17.49	7.22	7.84	7.71	7.66	7.66	7.66	7.66
12 805	0.0176	7.52	20.95	38.85	9.01	8.89	8.47	8.37	8.37	8.37	8.37
16 184	0.0157	7.97	24.09	46.67	9.52	9.48	9.39	9.21	9.21	9.21	9.21

at the Institute of Information and Communication Technologies, Bulgarian Academy of Sciences.

To measure the loss of accuracy due to the applied sparsification schemes, we use the relative error R in the form

$$R = \frac{\left\| \mathbf{u}^{\text{dense}} - \mathbf{u}^{\text{sparse}} \right\|_{l_2}}{\left\| \mathbf{u}^{\text{dense}} \right\|_{l_2}} = \frac{\sqrt{\sum_{i=1}^{N} (u_i^{\text{dense}} - u_i^{\text{sparse}})^2}}{\sqrt{\sum_{i=1}^{N} (u_i^{\text{dense}})^2}}.$$

Here, $\mathbf{u}^{\text{dense}}$ is the solution of the original system, while $\mathbf{u}^{\text{sparse}}$ is the solution of the system with the resulting sparse approximation of the stiffness matrix after *lumping* or *removal* sparsification. Direct solution methods are used in both cases.

In Tables 2, 3, 4 and 5 we show the relative error, percent of eliminated coefficients and spectral condition number when the *lumping* scheme is applied. Similar experi-

Table 3. Relative error R_{lump}, percent lumped and condition numbers with different lumping thresholds for $\alpha = 0.5$.

N	ε_{sparse}							
	10^{-1}	10^{-2}	10^{-3}	10^{-4}	10^{-5}	10^{-6}	10^{-7}	10^{-8}
				Relative error				
2 131	0.398	0.300	0.195	0.0838	0.0116	2.41e-05	0	0
4 167	0.410	0.336	0.247	0.1359	0.0366	1.02e-03	1.31e-08	0
8 030	0.416	0.362	0.290	0.1878	0.0749	8.17e-03	4.13e-06	0
12 805	0.419	0.376	0.316	0.2244	0.1081	2.05e-02	1.22e-04	8.28e-15
16 184	0.421	0.383	0.329	0.2423	0.1268	3.00e-02	4.45e-04	6.74e-10
				Percent lumped				
2 131	99.95	99.05	96.11	84.41	47.37	1.04	0.00	0.00
4 167	99.97	99.51	97.95	91.51	68.06	14.24	<0.01	<0.01
8 030	99.99	99.74	98.92	95.39	81.52	39.54	0.26	<0.01
12 805	99.99	99.84	99.31	97.04	87.71	56.04	3.56	<0.01
16 184	99.99	99.87	99.45	97.63	90.07	63.21	8.40	<0.01

			Condition number								
N	h	$h^{-2\alpha}$	ε_{sparse}								dense
			10^{-1}	10^{-2}	10^{-3}	10^{-4}	10^{-5}	10^{-6}	10^{-7}	10^{-8}	
2 131	0.0433	23.0	49.5	25.9	25.1	22.4	20.9	20.6	20.68	20.68	20.68
4 167	0.0309	32.2	51.6	39.2	39.3	35.4	31.4	30.2	30.20	30.20	30.20
8 030	0.0223	44.8	99.2	52.4	54.2	52.0	46.2	42.8	42.44	42.44	42.44
12 805	0.0176	56.5	114.3	73.1	76.2	69.6	60.0	53.6	52.41	52.41	52.41
16 184	0.0157	63.6	141.1	81.9	86.4	79.4	70.1	62.3	60.17	60.14	60.14

ments are conducted in the case of the *removing* scheme. In general, the obtained results are not better, which is why we omit them for brevity of presentation. The first conclusion is that the relative error increases for the larger problems. This can be explained by the influence of the increased condition number of the stiffness matrix. It is also not surprising that for a given threshold ε_{sparse}, the number of removed off-diagonal elements increases as α increases. Within these numerical tests, we would draw attention to the observation that the relative errors of the solution are very close for parameter sets for which similar percentages of the coefficients are removed. This is illustrated for example by the following cases:

(i) $\alpha = 0.25$, $N = 12\ 805$, $\varepsilon_{sparse} = 10^{-5}$, with 53.21% of the elements removed, $R_{lum} = 9.51e - 3$;

(ii) $\alpha = 0.5$, $N = 8\ 030$, $\varepsilon_{sparse} = 10^{-6}$, 39.54% of the elements are removed, $R_{lum} = 8.17e - 03$;

Table 4. Relative error R_{lump}, percent lumped and condition numbers with different lumping thresholds for $\alpha = 0.75$.

N	$\varepsilon_{\text{sparse}}$							
	10^{-1}	10^{-2}	10^{-3}	10^{-4}	10^{-5}	10^{-6}	10^{-7}	10^{-8}
				Relative error				
2 131	1.14	0.549	0.339	0.174	0.0605	0.00738	2.28e-05	0
4 167	1.22	0.660	0.443	0.257	0.115	0.0281	1.03e-03	9.32e-08
8 030	1.32	0.774	0.548	0.348	0.182	0.065	8.29e-03	2.29e-05
12 805	1.38	0.849	0.625	0.418	0.239	0.101	2.11e-02	4.30e-04
16 184	1.40	0.888	0.664	0.455	0.270	0.122	3.09e-02	1.25e-03
				Percent lumped				
2 131	99.78	99.27	98.05	93.55	79.29	43.39·	1.95	0
4 167	99.88	99.62	98.98	96.57	88.53	65.13	18.14	0.05
8 030	99.94	99.80	99.46	98.17	93.72	79.68	43.68	1.94
12 805	99.96	99.88	99.66	98.84	95.95	86.43	59.46	11.25
16 184	99.97	99.90	99.73	99.07	96.76	89.02	66.21	19.27

Condition number

N	h	$h^{-2\alpha}$	$\varepsilon_{\text{sparse}}$								dense
			10^{-1}	10^{-2}	10^{-3}	10^{-4}	10^{-5}	10^{-6}	10^{-7}	10^{-8}	
2 131	0.0433	110	220	162	144	125	112	105	104	104	104
4 167	0.0309	183	383	295	266	229	199	180	175	174	174
8 030	0.0223	300	645	516	480	418	360	318	297	295	295
12 805	0.0176	425	989	782	697	602	516	448	409	399	399
16 184	0.0157	507	1177	944	849	741	633	545	491	474	473

(iii) $\alpha = 0.75$, $N = 8\,030$, $\varepsilon_{\text{sparse}} = 10^{-7}$, 43.68% of the elements are removed, $R_{lum} = 8.29e - 03$;

(iv) $\alpha = 0.9$, $N = 12\,805$, $\varepsilon_{\text{sparse}} = 10^{-8}$, 56.08% of the elements are removed, $R_{lum} = 8.28e - 03$.

In the case of spectral fractional diffusion the estimate $O(h^{-2\alpha})$ holds for the spectral condition number of the stiffness matrices, where h is the mesh parameter of the FEM discretization. Tables 2, 3, 4 and 5 show that for the fractional Laplacian defined by (1), $\kappa(K)$ has a qualitatively similar behavior. In this context, we have presented $h^{-2\alpha}$ in these tables. Given that the mesh is quasi uniform, we define the mesh parameter as $h = \sqrt{S_\Omega/N} = 2/\sqrt{N}$. In our test problem $\Omega = (-1, 1) \times (-1, 1)$ and the area $S_\Omega = 4$. This table is presented as a reference for the following tables, which present κ_s for the sparsified matrices. We should note that as $\varepsilon_{\text{sparse}}$ decreases the condition number of the sparsified matrices trends towards $\kappa_s(K)$ of the dense matrix.

Table 5. Relative error R_{lump}, percent lumped and condition numbers with different lumping thresholds for $\alpha = 0.9$.

N	$\varepsilon_{\text{sparse}}$							
	10^{-1}	10^{-2}	10^{-3}	10^{-4}	10^{-5}	10^{-6}	10^{-7}	10^{-8}

	Relative error							
2 131	0.720	0.349	0.243	0.141	0.0684	0.0209	0.00192	3.58e-06
4 167	0.789	0.424	0.309	0.193	0.107	0.045	0.00961	2.60e-04
8 030	0.916	0.503	0.378	0.251	0.153	0.0768	0.0256	2.87e-03
12 805	0.987	0.56	0.429	0.294	0.189	0.104	0.0423	8.28e-03
16 184	1.01	0.59	0.456	0.318	0.209	0.119	0.0527	1.26e-02

	Percent lumped							
2 131	99.72	99.42	98.93	96.86	90.56	73.10	35.33	1.16
4 167	99.85	99.70	99.44	98.34	94.92	84.84	58.99	14.38
8 030	99.92	99.84	99.71	99.13	97.28	91.61	75.71	39.59
12 805	99.95	99.90	99.82	99.45	98.26	94.55	83.65	56.08
16 184	99.96	99.92	99.86	99.56	98.62	95.63	86.73	63.25

Condition number

N	h	$h^{-2\alpha}$	$\varepsilon_{\text{sparse}}$								dense
			10^{-1}	10^{-2}	10^{-3}	10^{-4}	10^{-5}	10^{-6}	10^{-7}	10^{-8}	
2 131	0.0433	284	555	448	417	382	355	337	330	329	329
4 167	0.0309	519	1059	862	802	730	673	629	605	598	598
8 030	0.0223	938	2039	1628	1517	1379	1264	1170	1106	1077	1074
12 805	0.0176	1428	3239	2636	2452	2224	2033	1872	1752	1685	1670
16 184	0.0157	1763	4078	3307	3079	2791	2548	2341	2181	2085	2055

The conclusion is that in the search for appropriate criteria for the selection of the sparsification threshold we must include both the size of the problem and the power of the fractional Laplacian.

In Tables 6 and 7 we present the relative errors of the solution when the mesh parameter h is used to determine the threshold $\varepsilon_{\text{sparse}}$. The percent of lumped elements and an estimation of the condition number is also presented. The results in Table 6 correspond to $\varepsilon_{\text{sparse}} = h^3$, and in Table 7 – to $\varepsilon_{\text{sparse}} = h^4$. In these tables, we also compare the solution accuracy when *lumping* and *removing* schemes are applied, denoting the corresponding relative errors by R_{lum} and R_{rem}. We see that everywhere the *lumping* technique leads to significantly smaller relative errors.

Now, we analyze from another point of view the dependence of the relative error of the solution on α. For example, with $\varepsilon_{\text{sparse}} = h^3$ (Table 6) a very large part of the off-diagonal elements is lumped, but in some of the cases the achieved accuracy is

Table 6. Relative error R_{lum} for *lumping* and R_{rem} for *removing* schemes, $\varepsilon_{\text{sparse}} = h^3$.

N	h	h^3	$\alpha = 0.25$			$\alpha = 0.5$			$\alpha = 0.75$			$\alpha = 0.9$		
			R_{lum}	R_{rem}	%	R_{lum}	R_{rem}	%	R_{lum}	R_{rem}	%	R_{lum}	R_{rem}	%
2131	0.0433	1.016e-05	0.0139	0.0875	63.82	0.0748	0.259	82.43	0.162	0.464	92.82	0.133	0.531	96.53
4167	0.0309	3.717e-06	0.0126	0.0742	60.15	0.0781	0.259	82.57	0.176	0.487	93.49	0.145	0.566	97.02
8030	0.0223	1.389e-06	0.0107	0.0616	56.00	0.0795	0.256	82.62	0.189	0.508	94.06	0.157	0.597	97.42
12805	0.0176	6.901e-07	0.0094	0.0539	52.75	0.0801	0.257	82.57	0.199	0.526	94.43	0.165	0.622	97.66
16184	0.0157	4.857e-07	0.0089	0.0500	51.13	0.0810	0.256	82.61	0.204	0.534	94.61	0.170	0.633	97.78

Table 7. Relative error R_{lum} for *lumping* and R_{rem} for *removing* schemes, $\varepsilon_{\text{sparse}} = h^4$.

N	h	h^4	$\alpha = 0.25$			$\alpha = 0.5$			$\alpha = 0.75$			$\alpha = 0.9$		
			R_{lum}	R_{rem}	%	R_{lum}	R_{rem}	%	R_{lum}	R_{rem}	%	R_{lum}	R_{rem}	%
2131	0.0433	3.52e-06	1.62e-06	9.91e-06	0.04	1.83e-03	0.00684	20.39	0.0286	0.0747	66.34	0.0434	0.135	84.64
4167	0.0309	9.21e-07	8.39e-08	5.52e-07	<0.01	8.05e-04	0.00259	12.43	0.0262	0.0627	63.84	0.0432	0.129	84.25
8030	0.0223	2.48e-07	0	0	0	2.74e-04	0.000847	6.33	0.0229	0.0521	61.16	0.0426	0.121	83.86
12805	0.0176	9.75e-08	0	0	0	1.10e-04	0.000316	3.33	0.0206	0.0461	59.03	0.0418	0.118	83.47
16184	0.0157	6.10e-08	0	0	0	6.42e-05	0.000177	2.30	0.0197	0.0428	58.01	0.0417	0.116	83.32

definitely low. Thus, for $\alpha = 0.25$ between 50% and 60% of the coefficients are lumped and the accuracy is around 0.01. For $\alpha = 0.5$ around 82% of the off-diagonal elements are lumped with $R_{lum} \in (0.07, 0.08)$. For $\alpha = 0.75$ and $\alpha = 0.9$, the rate of sparsification increases to over 90%, but the relative accuracy drops to 0.5 – 0.6. After decreasing the threshold to $\varepsilon_{\text{sparse}} = h^4$ (Table 7), for $\alpha = 0.25$ and 0.5 we obtain very few lumped off-diagonal elements. In this case the relative error R_{lum} is significantly improved but the lower level of sparsification (2% to 20%) is unlikely to yield the targeted better computational performance if a fast sparse solver is used. For $\alpha = 0.75$ the sparsification ratio is around 60% with relative error of around 0.02, while for $\alpha = 0.9$ the relative error is around 0.04 with about 84% of the coefficients being lumped. The conclusion of this analysis is that for $\alpha = 0.25$ and 0.5 it is better to use $\varepsilon_{\text{sparse}} = h^3$. For higher values of α, the smaller threshold $\varepsilon_{\text{sparse}} = h^4$ can be recommended.

5   Concluding Remarks

This paper discusses the numerical solution of fractional diffusion problems governed by the integral representation of the fractional Laplace operator. The finite element method is used to discretize the non-local boundary value problem with the main contribution being the presented relative error analysis of the proposed approach for sparse approximation of the dense stiffness matrix. Off-diagonal elements are removed if the absolute value of their ratio with the corresponding diagonal element is less than a given threshold $\varepsilon_{\text{sparse}}$.

The efficiency of the proposed *lumping* and *removing* sparsification schemes are compared. All numerical experiments performed show the better accuracy of the *lumping* scheme. The construction of the proposed sparse approximation of the dense stiffness matrix K is controlled by the threshold parameter $\varepsilon_{\text{sparse}}$. A larger value of the selected threshold means a higher percentage of off-diagonal elements removed, but lead to lower accuracy of the solution of the linear system and vice versa. We have shown in the presented analysis that the relative error depends on the number of unknowns N (or equivalently on the mesh parameter h of the finite element triangulation $\mathcal{T}_h$) as well as on the value of the fractional power $\alpha \in (0, 1)$. An important result of our experimental study is that choosing an appropriate threshold is a complex nonlinear problem. However, some practical recommendations in this regards were given in the previous section.

Furthermore, it should be noted that the finite element approximation error of the fractional diffusion problem determines the target accuracy of the approximate solution of the considered linear system. Therefore, the regularity of the solution of the boundary value problem should also be taken into account. And it also depends on α.

In conclusion, the results presented raise the following topics for future research: (i) balancing the finite element method discretization error with the error coming from the discussed sparse approximation of the dense stiffness matrix; (ii) developing an integral approach to determine the threshold $\varepsilon_{\text{sparse}}$ reflecting both the influence of the number of unknowns N (or equivalently the mesh parameter h) and the fractional power α.

Acknowledgments. The first author is partially supported by Grant No BG05M2OP001-1.001-0003, financed by the Science and Education for Smart Growth Operational Program (2014-2020) and co-financed by the European Union through the European structural and Investment funds, and by the Bulgarian NSF under the Grant КП-06-Н52/4.

We acknowledge the provided access to the e-infrastructure of the NCHDC - part of the Bulgarian National Roadmap on RIs, with the financial support by the Grant No D01-387/18.12.2020.

References

1. Acosta, G., Bersetche, F., Borthagaray, J.: A short FE implementation for a 2D homogeneous Dirichlet problem of a fractional Laplacian. Comput. Math. Appl. **74**(4), 784–816 (2017). https://doi.org/10.1016/j.camwa.2017.05.026
2. Acosta, G., Borthagaray, J.: A fractional laplace equation: regularity of solutions and finite element approximations. SIAM J. Numer. Anal. **55**(2), 472–495 (2017). https://doi.org/10.1137/15M1033952
3. Bertoin, J.: Lévi Processes. Cambridge University Press, Cambridge (1998)
4. Harizanov, S., Margenov, S., Popivanov, N.: Spectral fractional laplacian with inhomogeneous Dirichlet data: questions, problems, solutions. In: Georgiev, I., Kostadinov, H., Lilkova, E. (eds.) BGSIAM 2018. Studies in Computational Intelligence, vol. 961, pp. 123–138. Springer, Cham (2021). https://doi.org/10.1007/978-3-030-71616-5_13
5. Klafter, J., Sokolov, I.: Anomalous diffusion spreads its wings. Phys. World **18**(8), 29–32 (2005). https://doi.org/10.1088/2058-7058/18/8/33
6. Metzler, R., Klafter, J.: The restaurant at the end of the random walk: recent developments in the description of anomalous transport by fractional dynamics. J. Phys. A Math. General **37**(31), R161–R208 (2004). https://doi.org/10.1088/0305-4470/37/31/r01
7. Valdinoci, E.: From the long jump random walk to the fractional laplacian (2009)

Fractional Diffusion Problems with Reflecting Boundaries

Ercília Sousa[(⊠)]

CMUC, Department of Mathematics, University of Coimbra, Coimbra, Portugal
ecs@mat.uc.pt

Abstract. Anomalous diffusive transport, described by fractional differential equations, arises in a large variety of physical problems. We consider a fractional diffusion equation subjected to reflecting boundary conditions. The formulation of these boundaries has sparked a controversial discussion, with questions arising about the most appropriate boundary from the physical point of view. Therefore, we start to present different physical formulations regarding the boundaries. Numerical methods are then proposed to solve these diffusive models, and it is shown how the presence of boundaries changes the general structure of the problem and of the numerical method, due to the non-locality of the problem. In the end, the impact of the different boundaries on the solutions is analysed.

1 Introduction

Anomalous diffusive transport related to Lévy flights can be formulated via fractional differential equations [10]. It frequently happens that we have to apply boundary conditions when considering experimental devices and attempting to check a model for mass transport in a given medium. Due to non-locality, it is not obvious how to incorporate a boundary condition in a scenario based on Lévy flights, since the long jumps pose certain difficulties when boundary conditions are involved. In fact, the presence of certain boundaries modifies the nonlocal spatial operator since they cannot be uncoupled from the fractional partial differential equation. In literature, when discussing Lévy flights in the one dimensional half-space the boundary conditions mainly considered have been absorbing or reflecting boundaries. Absorbing boundary conditions have been imposed by assuming zero outside the problem domain. However, regarding reflecting boundary conditions several formulations have been proposed [1,3–9].

The proper formulation of physically meaningful reflecting boundary conditions for fractional diffusion equations requires careful consideration of the nonlocal operator. In this work we discuss two types of boundaries that appear respectively in [9] and [2], showing the impact of both formulations on the numerical method and on the solution. In [9] the influence of a reflective wall is modelled within the framework of space-time fractional partial differential equations. The jumps do not interact with the wall and they are as in a free space. This is modelled by a fractional differential equation that involves a non-local operator which kernel takes into account the boundary condition. In [2] the physical

I. Lirkov and S. Margenov (Eds.): LSSC 2023, LNCS 13952, pp. 164–171, 2024.
https://doi.org/10.1007/978-3-031-56208-2_16

boundary conditions are derived using a mass balance approach and the reflecting boundary condition is formulated in terms of a fractional derivative.

2   The Models

The diffusive model associated to Lévy flights is defined in the whole real line and the governing equation involves Riemann-Liouville fractional derivatives [10]. We consider the assymetric case, that is, the diffusive operator is defined only with the left Riemann-Liouville fractional derivative.

2.1   Open Domain

We start with the open domain and then describe how to evolve to the situation of having a reflecting boundary condition.

The left Riemann-Liouville fractional derivative of order α, when $1 < \alpha < 2$, for $x \in \mathbb{R}$, is given by

$$\frac{\partial^\alpha u}{\partial x^\alpha}(x,t) = \frac{1}{\Gamma(2-\alpha)} \frac{\partial^2}{\partial x^2} \int_{-\infty}^{x} u(\xi,t)(x-\xi)^{1-\alpha} d\xi. \tag{1}$$

The fractional differential equation describing the diffusive model under consideration in the open domain, for $1 < \alpha < 2$, can be stated as

$$\frac{\partial u(x,t)}{\partial t} = D \frac{\partial^\alpha u}{\partial x^\alpha}(x,t), \tag{2}$$

where D is the diffusive parameter and the parameter α is related to the tail of the solution. In this scenario of having an open domain we assume that we have an initial condition $u(x,0) = u_0(x)$, $x \in \mathbb{R}$, and that the solution goes to zero when $|x|$ goes to infinity.

In the next sections, we describe how this problem changes in the presence of a boundary at $x = 0$ and defined in the domain $x > 0$.

2.2   The Symmetric Boundary Wall

In this section we present how to formulate the diffusive problem with a left reflecting wall. The formulation of the boundary is according to [9], where a symmetric diffusive problem on a semi-infinite domain is considered. Physically, when considering a trajectory of the particle in $[0, \infty)$ with the reflecting boundary condition at $x = 0$, the jumps that end at $x < 0$ are reflected. Therefore, the model under study consists on a reflecting wall restraining the diffusing particles to a semi-infinite domain. This barrier can be viewed as a force field applied to the particles. It is assumed that the particles arriving at the boundary are bounced back as in elastic collisions, that is, if they reach the position $x = -a$ with $a > 0$, then they will end at $x = a$, describing the mirror trajectory with respect to the wall. In a porous medium such a boundary may represent a wall permeable to the fluid, but impermeable to the tracer. Mathematically, we have a problem defined in $x > 0$ by Eq. (2) and subjected to the wall condition, $u(x,t) = u(-x,t)$, for $x < 0$. Taking in consideration this wall condition, the left Riemann-Liouville fractional derivative (1) becomes a different operator, for $x > 0$, that we define as the reflecting left Riemann-Liouville fractional derivative,

$$\frac{\partial^\alpha_{ref} u}{\partial x^\alpha}(x,t) := \frac{1}{\Gamma(2-\alpha)} \frac{\partial^2}{\partial x^2} \int_0^\infty u(\xi,t)(x+\xi)^{1-\alpha} d\xi$$

$$+ \frac{1}{\Gamma(2-\alpha)} \frac{\partial^2}{\partial x^2} \int_0^x u(\xi,t)(x-\xi)^{1-\alpha} d\xi. \tag{3}$$

Formally when subjected to a reflecting wall we have the following problem,

$$\frac{\partial u}{\partial t}(x,t) = D \frac{\partial^\alpha_{ref} u}{\partial x^\alpha}(x,t), \quad x > 0, \tag{4}$$

$$u(x,t) = u(-x,t), \quad \text{for all} \quad x < 0, \tag{5}$$

with an initial condition $u(x,0) = u_0(x)$, $x \geq 0$.

2.3   A Fractional Boundary Condition

Consider the boundary condition as defined in [2], that involves a fractional derivative of order $\alpha - 1$. The physical setup is described as having mass concentration resting at the boundary instead of having mass leaving the domain. This means mass is preserved, and moved to the boundary. Unlike the traditional diffusion setup, this mass can come from far inside the domain, not just an adjacent grid point.

Let us define the fractional derivative of order $m - 1 < \alpha < m$, starting at $x = 0$, by

$$\frac{\partial^\alpha_0 u}{\partial x^\alpha}(x,t) = \frac{1}{\Gamma(m-\alpha)} \frac{\partial^m}{\partial x^m} \int_0^x u(\xi,t)(x-\xi)^{m-1-\alpha} d\xi. \tag{6}$$

Formally when subjected to the fractional Neumann condition we have the following problem, for $1 < \alpha < 2$,

$$\frac{\partial u}{\partial t}(x,t) = D \frac{\partial^\alpha_0 u}{\partial x^\alpha}(x,t), \quad x > 0, \tag{7}$$

$$\frac{\partial^{\alpha-1}_0 u}{\partial x^{\alpha-1}}(0,t) = 0, \tag{8}$$

with an initial condition $u(x,0) = u_0(x)$, $x \geq 0$.

3   The Numerical Methods

We start to present an implementation for the case when we have an open domain and then show how to adjust it to the presence of both types of boundaries.

3.1   Open Domain

For the problem defined in the whole real line, the domain discretisation is given by $x_j = x_{j-1} + \Delta x$, $j \in \mathbb{Z}$ and the time discretization $t_n = n\Delta t$, $n \geq 0$ integer.

We approximate the left Riemann-Liouville fractional derivative by the well-known Grünwald-Letnikov approximation. Define the Grünwald-Letnikov coefficients, for all $\alpha > 0$, using the following recurrence formula

$$g_0^\alpha = 1, \quad g_{k+1}^\alpha = -\frac{\alpha - k}{k+1} g_k^\alpha, \quad k \geq 0. \tag{9}$$

The Grünwald-Letnikov approximation, at (x_j, t_n), is given by [12]

$$\frac{\partial^\alpha u}{\partial x^\alpha}(x_j, t_n) \approx \frac{1}{(\Delta x)^\alpha} \sum_{k=0}^{\infty} g_k^\alpha u(x_{j-k+1}, t_n). \tag{10}$$

Let U_j^n represent the approximate solution of $u(x_j, t_n)$ in the discrete domain and define

$$\mu_\alpha = \frac{D\Delta t}{(\Delta x)^\alpha}.$$

The Euler explicit method to approximate the fractional diffusion equation will be now given by

$$U_j^{n+1} = U_j^n + \mu_\alpha \sum_{k=0}^{\infty} g_k^\alpha U_{j-k+1}^n, \quad \text{for all } j \in \mathbb{Z}. \tag{11}$$

The matricial form of the numerical method in the open domain takes in consideration that the function goes to zero as we go to infinity and we have $\mathbf{U}^{n+1} = (\mathbf{I} + \mu_\alpha \mathbf{A})\mathbf{U}^n$, with $\mathbf{U}^n = [U_{-N}^n, \ldots, U_N^n]^T$, $\mathbf{I}$ is the identity matrix and the matrix $\mathbf{A}$ is given by

$$\mathbf{A} = \begin{bmatrix} g_1^\alpha & g_0^\alpha & 0 & \ldots & 0 & 0 \\ g_2^\alpha & g_1^\alpha & g_0^\alpha & \ldots & 0 & 0 \\ g_3^\alpha & g_2^\alpha & g_1^\alpha & \ldots & 0 & 0 \\ \vdots & \vdots & \vdots & & \vdots & \vdots \\ g_{2N+1}^\alpha & g_{2N}^\alpha & g_{2N-1}^\alpha & \ldots & g_2^\alpha & g_1^\alpha \end{bmatrix}.$$

The following result indicates that in the open domain, the approximation (10) is of order one. This result was given for a function that only depends on x, but this can be easily adjusted for the case under discussion.

Theorem 1 [12]. *Let $m - 1 < \alpha < m$, $u(\cdot, t) \in C^{[\alpha]+m+1}(\mathbb{R})$, for a fixed t, such that all the derivatives, in x, up to order $[\alpha] + m + 2$ belong to $L^1(\mathbb{R})$ and where $[\alpha]$ represents the integer part of α. Then the fractional Riemann-Liouville derivative given by (1) satisfies*

$$\frac{\partial^\alpha u}{\partial x^\alpha}(x_j, t) = \frac{1}{(\Delta x)^\alpha} \sum_{k=0}^{\infty} g_k^\alpha u(x_j - (k-1)\Delta x, t) + O((\Delta x)).$$

Since we have considered an explicit numerical method, we have a conditionally stable scheme and the stability conditions can be obtained using the von Neumann analysis or Fourier analysis [11].

Theorem 2. *If the numerical method (11) is von Neumann stable then* $\mu_\alpha \leq 2^{1-\alpha}$.

When imposing a boundary the von Neumann stability condition is a necessary condition for the stability of the numerical method with boundaries.

3.2   The Reflective Boundary

The domain discretisation is given by $x_j = x_{j-1} + \Delta x$, $j \in \mathbb{Z}$. When we have a reflecting boundary condition at $x = 0$, since the left fractional derivative is modified to (3), because $U^n_{-i+1} = U^n_{-j+i-1}$, the approximation becomes

$$\frac{\delta^\alpha_{ref} u(x_j, t)}{(\Delta x)^\alpha} \approx \frac{1}{(\Delta x)^\alpha} \sum_{i=0}^{j+1} g^\alpha_i U^n_{j-i+1} + \frac{1}{(\Delta x)^\alpha} \sum_{i=j+2}^{\infty} g^\alpha_i U^n_{j-i+1}$$

$$= \frac{1}{(\Delta x)^\alpha} \sum_{i=0}^{j+1} g^\alpha_i U^n_{j-i+1} + \frac{1}{(\Delta x)^\alpha} \sum_{i=j+2}^{\infty} g^\alpha_i U^n_{i-j-1}.$$

Consider the explicit Euler scheme to approximate Eq. (4) given by

$$U^{n+1}_j = U^n_j + \mu_\alpha \delta^\alpha_{ref} U^n_j. \tag{12}$$

In this case the matricial form of the problem is $\mathbf{U}^{n+1} = (\mathbf{I} + \mu_\alpha \mathbf{A_{Sym}})\mathbf{U}^n$, with $\mathbf{U}^n = [U^n_0, \ldots, U^n_N]^T$, $\mathbf{I}$ is the identity matrix and the matrix $\mathbf{A_{Sym}}$ is

$$\mathbf{A_{Sym}} = \begin{bmatrix} g^\alpha_1 & g^\alpha_0 + g^\alpha_2 & & g^\alpha_3 & \cdots & g^\alpha_{N-2} & g^\alpha_{N-1} \\ g^\alpha_2 & g^\alpha_1 + g^\alpha_3 & g^\alpha_0 + g^\alpha_4 & \cdots & g^\alpha_{N-1} & g^\alpha_N \\ g^\alpha_3 & g^\alpha_2 + g^\alpha_4 & g^\alpha_1 + g^\alpha_5 & \cdots & g^\alpha_N & g^\alpha_{N+1} \\ \vdots & \vdots & & \vdots & & \vdots & \vdots \\ g^\alpha_{2N+1} & g^\alpha_{2N} + g^\alpha_{N+2} & g^\alpha_{2N-1} + g^\alpha_{N+3} & \cdots & g^\alpha_2 + g^\alpha_{2N} & g^\alpha_1 + g^\alpha_{2N+1} \end{bmatrix}.$$

The changes of the entries of the matrix $\mathbf{A}$ due to the presence of this reflecting boundary are displayed in gray color.

This problem is equivalent to a problem defined in the real line and therefore the stability conditions of the numerical method are similar to those obtained for the open problem.

3.3   The Fractional Boundary

If we are in a bounded domain and consider the approximation (10) directly in that domain, we can arrive at the following approximation in the interior points

$$U^{n+1}_j = U^n_j + \mu_\alpha \sum_{i=0}^{j+1} g^\alpha_i U^n_{j-i+1}, \quad j = 0, \ldots, N-1. \tag{13}$$

However in this case we can see that no mass has been moved. To enforce the boundary condition, we modify the Euler scheme (13) at the point $x = 0$. Let us see how we can impose the boundary condition (8). We give here a mathematical approach, instead of the physical interpretation given in [2].

Taking in consideration the properties of the Riemann-Liouville fractional derivative the differential Eq. (7) at $x = 0$ can be written as

$$\frac{\partial u}{\partial t}(0,t) = D\frac{\partial}{\partial x}\left(\frac{\partial_0^{\alpha-1}u}{\partial x^{\alpha-1}}\right)(0,t).$$

We can use the Euler approximation for the time derivative, that is,

$$\frac{\partial u}{\partial t}(0,t) \approx \frac{U_0^{n+1} - U_0^n}{\Delta t}.$$

A first order approximation for the first order spatial derivative allow us to write

$$\frac{\partial}{\partial x}\left(\frac{\partial_0^{\alpha-1}u}{\partial x^{\alpha-1}}\right)(0,t) \approx \frac{1}{\Delta x}\left(\frac{\partial_0^{\alpha-1}u}{\partial x^{\alpha-1}}(x_1,t) - \frac{\partial_0^{\alpha-1}u}{\partial x^{\alpha-1}}(0,t)\right). \tag{14}$$

We know the value of the second term on the right hand side of (14), since this is the boundary condition. Additionally, we can approximate the fractional derivative at x_1 using the Grünwald-Letnikov approximation, that is,

$$\frac{\partial_0^{\alpha-1}u}{\partial x^{\alpha-1}}(x_1,t_n) \approx \frac{1}{(\Delta x)^{\alpha-1}}\sum_{k=0}^{1}g_k^{\alpha-1}U_{1-k}^n.$$

Therefore, we obtain

$$U_0^{n+1} = U_0^n + \mu_\alpha(g_1^{\alpha-1}U_0^n + g_0^{\alpha-1}U_1^n).$$

Finally, the numerical method is given by

$$U_j^{n+1} = U_j^n + \mu_\alpha\sum_{i=0}^{j+1}g_{j-i+1}^{\alpha}U_i^n,\ j = 1,\ldots,N,$$

$$U_0^{n+1} = U_0^n + \mu_\alpha(g_1^{\alpha-1}U_0^n + g_0^{\alpha-1}U_1^n).$$

In this case the matricial form of the problem is $\mathbf{U}^{n+1} = (\mathbf{I} + \mu_\alpha\mathbf{A_{Neu}})\mathbf{U}^n$, with $\mathbf{U}^n = [U_0^n,\ldots,U_N^n]^T$, $\mathbf{I}$ is the identity matrix and the matrix $\mathbf{A_{Neu}}$ is given by

$$\mathbf{A_{Neu}} = \begin{bmatrix} g_1^{\alpha-1} & g_0^{\alpha-1} & 0 & \ldots & 0 & 0 \\ g_2^{\alpha} & g_1^{\alpha} & g_0^{\alpha} & \ldots & 0 & 0 \\ g_3^{\alpha} & g_2^{\alpha} & g_1^{\alpha} & \ldots & 0 & 0 \\ \vdots & \vdots & \vdots & & \vdots & \vdots \\ g_{2N+1}^{\alpha} & g_{2N}^{\alpha} & g_{2N-1}^{\alpha} & \ldots & g_2^{\alpha} & g_1^{\alpha} \end{bmatrix}.$$

The changes of the entries of the matrix $\mathbf{A}$ due to the presence of this boundary are in the first line and in gray color.

Theorem 3. *If* $\mu_\alpha \leq 1/\alpha$ *then the eigenvalues of the matrix iteration* $\mathbf{I} +$ $\mu_\alpha \mathbf{A_{Neu}}$ *are less than one.*

The previous result can be proved using the Gersghorin theorem.

The numerical experiments show that the stability region is given by the necessary stability condition $\mu_\alpha \leq 2^{1-\alpha}$ and not by this more restrictive sufficient stability condition $\mu_\alpha \leq 1/\alpha$.

4   The Influence of the Boundaries

We illustrate the effect of the boundaries for different values of α. We consider the approximation of the solution of the fractional diffusion equation with an initial condition that is an approximation of the Dirac delta function, that is,

$$u_0(x) = \frac{1}{\epsilon\sqrt{\pi}} e^{-(x-x_0)^2/\epsilon^2},$$

for a small $\epsilon > 0$. For all figures we have taken $D = 1$, $\epsilon = 0.1$, $x_0 = 1$.

In the next figures we consider two values of α, that is, $\alpha = 1.3$ in Fig. 1 and $\alpha = 1.8$ in Fig. 2. As we evolve in time, the effect of the boundaries on the solution is quite relevant. By the figures we can also see that in the open domain we have an asymmetric case since we are only considering the left Riemann Liouville fractional derivative and therefore the wave has a significant heavy tail on the right hand side, when α is closer to 1.

Near the boundary, the behaviour of the solution with the Neumann boundary can be unexpected at first. However, the steady state general solution of the fractional diffusion Eq. (2) is the combination of the functions $x^{\alpha-1}$ and $x^{\alpha-2}$. The solution that goes to zero as x goes to infinity is $x^{\alpha-2}$. This is also the function that better describes the behaviour we observe in the previous figures, near the boundary, that is, the solution $x^{\alpha-2}$ goes to infinity as x goes to zero.

We have seen the consequences of having two types of reflecting boundaries. Near the boundary the solutions behave very differently, highlighting that they

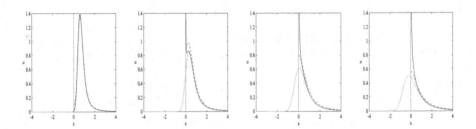

Fig. 1. Plots of $u(x,t)$ for $x_0 = 1$, $D = 1$, $\alpha = 1.3$. Open domain in green line $(- \cdot -)$; Symmetric boundary in red line $(--)$; Neumann boundary in blue line $(-)$. Evolution in time described from left to right: $t = 0.25, 0.5, 0.75, 1$.

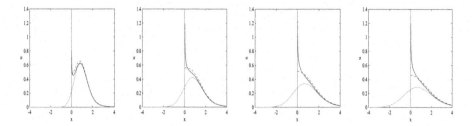

Fig. 2. Plots of $u(x,t)$ for $x_0 = 1$, $D = 1$, $\alpha = 1.8$. Open domain in green line $(- \cdot -)$; Symmetric boundary in red line $(--)$; Neumann Boundary in blue line $(-)$. Evolution in time. Left to right $t = 0.25, 0.5, 0.75, 1$.

represent completely different physical phenomena. Far away from the boundary the behaviour is similar in all three cases: open domain, symmetric reflecting boundary and the fractional Neumann boundary.

References

1. Baeumer, B., Kovács, M., Sankaranarayanan, H.: Fractional partial differential equations with boundary conditions. J. Differ. Equ. **264**, 1377–1410 (2018)
2. Baeumer, B., Kovács, M., Meerschaert, M.M., Sankaranarayanan, H.: Boundary conditions for fractional diffusion. J. Comput. Appl. Numer. Math. **336**, 408–424 (2018)
3. Cusimano, N., Burrage, K., Turner, I., Kay, D.: On reflecting boundary conditions for space-fractional equations on a finite interval: proof of the matrix transfer technique. Appl. Math. Model. **42**, 554–565 (2017)
4. Dipierro, S., Ros-Oton, X., Valdinoci, E.: Nonlocal problems with Neumann boundary conditions. Rev. Mat. Iberoam. **33**, 377–416 (2017)
5. Dybiec, B., Gudowska-Nowak, E., Hänggi, P.: Lévy-Brownian motion on finite intervals: mean first passage time analysis. Phys. Rev. E **73**, 046104 (2006)
6. Dybiec, B., Gudowska-Nowak, E., Barkai, E., Dubkov, A.A.: Lévy flights versus Lévy walks in bounded domains. Phys. Rev. E **95**, 052102 (2017)
7. Jesus, C., Sousa, E.: Superdiffusion in the presence of a reflecting boundary. Appl. Math. Lett. **112**, 106742 (2021)
8. Kelly, J.F., Sankaranarayanan, H., Meerschaert, M.M.: Boundary conditions for two-sided fractional diffusion. J. Comput. Phys. **376**, 1089–1107 (2019)
9. Krepysheva, N., Di Pietro, L., Néel, M.C.: Space-fractional advection-diffusion and reflective boundary condition. Phys. Rev. E **73**, 021104 (2006)
10. Metzler, R., Klafter, J.: The random walk's guide to anomalous diffusion: a fractional dynamics approach. Phys. Rep. **339**, 1–77 (2000)
11. Sousa, E.: Finite difference approximations for a fractional advection diffusion problem. J. Comput. Phys. **228**, 4038–4054 (2009)
12. Tuan, V.K., Gorenflo, R.: Extrapolation to the limit for numerical fractional differentiation. Z. Agnew. Math. Mech. **75**, 646–648 (1995)

Variational Analysis and Optimal Control

A Mean Field Model for Counter CBRN Threats

Rossana Capuani[1] and Antonio Marigonda[2(✉)]

[1] DEIM, University of Tuscia, Via del Paradiso 47, 01100 Viterbo, Italy
`rossana.capuani@unitus.it`
[2] Department of Computer Sciences, University of Verona, Strada Le Grazie 15,
37134 Verona, Italy
`antonio.marigonda@univr.it`

Abstract. In this paper we propose a model to optimize the allocation of resources by a decontamination team facing a CBRN emergency. Due to the possible *diffusion* of the hazardous or noxious substances in the environment, the team must implement a *dynamic* strategy which can be seen as a coupling between the countermeasures and the concentration of the threats (which can be seen as a measure of the risk). The aim is to prove the existence of an optimal strategy, and some necessary conditions.

Keywords: measurable selection · multiagent system

1 Motivation and Mathematical Description of the Model

People may be exposed to CBRN (chemical, biological, radiological and nuclear) threats as a result of unintentional disasters (eg. a chemical plant leak, nuclear power plant incidents, the spread of an infectious disease) or intentional incidents (eg. a terrorist attack). Being prepared to address the risks of such threats is a key part of the EU CBRN stockpiling strategy especially for the events that have happened in recent years. Preparing for chemical, biological, radiological and nuclear threats is an essential element in keeping the citizens safe, and a key element of a strong European Health Union. To improve the EUs preparedness and response to public health risks such as CBRN threats, the European Commission created rescEU to protect citizens from disasters and manage emerging risks. The rescEU decontamination reserve comprise expert teams, equipment, and devices to detect, decontaminate affected people or infrastructure (for more information see [7]).

We refer the reader to [1], in particular Chap. 5, for a comprehensive introduction to the concepts of measure theory used.

We suppose that the expert team has a finite amount of time $T > 0$ to detect and to neutralize as many CBRN threats as possible in the area of the operation Ω, which is supposed to be a bounded, open, connected region of the plane $\mathbb{R}^2$ with Lipschitz boundary. In order to achieve this goal, the team can

I. Lirkov and S. Margenov (Eds.): LSSC 2023, LNCS 13952, pp. 175–183, 2024.
https://doi.org/10.1007/978-3-031-56208-2_17

deploy its resources in the operation theater and use them to retrieve relevant information from the environment (using for instance sensors or equipments) and/or neutralizing the threats.

The evolution of the resources' allocation is described by a curve $\mu = \{\mu_t\}_{t\in[0,T]}$ (here and in the whole paper, the subscript t denotes the time parameter, not derivation) in the space $\mathscr{P}(\mathbb{R}^2)$ of probability measures on $\mathbb{R}^2$: given a region A and a time $t \in [0,T]$, $\mu_t(A)$ denotes the fraction of the total resources allocated in the region A at time t. At every instant of time during the operation, the team must allocate its resources in the most efficient way according to a *priority measure*, in order to secure the regions where the CBRN contamination is present. The assumption that $\mu_t \in \mathscr{P}(\mathbb{R}^2)$ amounts to say that *no resupplying is allowed* for the whole period of the operation, and that *no shortage* of decontamination resources is considered.

The priority measure $\zeta = \{\zeta_t\}_{t\in[0,T]}$ is a curve in the space of probability measures as well, that, given a region A of the theater of the operation and a time $t \in [0,T]$, denotes the estimated fraction of CBRN threats present in A at time t, requiring the decontamination procedure. In the simplest case we can assume for example that at the initial time we have some leaks of pollutant expressed by a measure ζ_0, and the evolution $t \mapsto \zeta_t$ is given by an heat equation (expressing the diffusion in the environment), with Neumann boundary condition, and initial condition ζ_0. Anyway, many other diffusion processes can be considered. Throughout this paper, ζ will be considered as a given datum satisfying supp $\zeta_t \subseteq \Omega$ for all $t \in [0,T]$.

The instantaneous strategy chosen at time $t \in [0,T]$ by the team to fullfill the requirements of the priority measure ζ_t is given by a *transport plan* $\pi_t \in \Pi(\mu_t, \zeta_t)$, i.e., a measure belonging to $\mathscr{P}(\mathbb{R}^2 \times \mathbb{R}^2)$ such that $\pi_t(A \times \mathbb{R}^2) = \mu_t(A)$ and $\pi_t(\mathbb{R}^2 \times B) = \zeta_t(B)$ for any Borel set $A, B \subseteq \mathbb{R}^2$. Roughly speaking, the measure $\pi(\mu_t, \zeta_t)(A \times B)$ expresses the fraction of the resources that are present in a region A that must be used to face the threats in the region B. The choice of a transport plan for the coupling between threats and countermeasures, from a model point of views, implies that at every time all the possible countermeasures are used against all the threats, and the choice is made about how to allocate them *properly*.

To this aim, an instantaneous cost function $\tilde{c} = \tilde{c}(x,y)$ is defined in order to express the efficiency of allocating a countermeasure resource present at the point x in order to deal with a threat located at the point y. An higher efficiency will be represented by a lower value of the cost. In the simplest cases, $\tilde{c}(x,y)$ can be assumed to be proportional to a (power of a) suitable distance function defined on Ω, for instance this case occurs when the decontamination equipment has a short range or the decontamination procedure adopted requires a close proximity with the target. In our case, we suppose that the instantaneous cost $\tilde{c}(\cdot,\cdot)$ does not depend on time (i.e., the geometry of operation theater Ω does not change during the operation, and the countermeasure protocol is independent on time). This approximation is realistic when T is sufficiently small compared with the scale of the phenomena.

The cost of the instantaneous optimal strategy at time $t \in [0, T]$ when the resources are allocated as μ_t and the threats are described by ζ_t is

$$\mathscr{W}(\mu_t, \zeta_t) := \inf_{\pi_t \in \Pi(\mu_t, \zeta_t)} \iint_{\mathbb{R}^2 \times \mathbb{R}^2} \tilde{c}(x, y) \, d\pi_t(x, y).$$

The total cost of the operation will be the *average cost* of allocating the resources according to μ during the operation

$$\mathscr{F}_\zeta(\mu) := \frac{1}{T} \int_0^T \mathscr{W}(\mu_t, \zeta_t) \, dt.$$

Considering ζ as given data of the problem, we want to optimize on the choice of the evolution of the team resource allocation $\mu = \{\mu_t\}_{t \in [0, T]}$.

Recalling that neither loss nor resupplying occurs during the operation, the macroscopic description of the resources of the team must follow (in the distributional sense) the continuity equation $\partial_t \mu_t + \operatorname{div}(v_t \mu_t) = 0$ with some initial data $\mu_{t|t=0} = \mu_0$. From a microscopic point of view, we have that $\mu_t = e_t \sharp \eta$, where $e_t(x, \gamma) = \gamma(t)$ and η is a probability measure on $\mathbb{R}^2 \times C^0([0, T])$ concentrated on pairs (x, γ) satisfying the following differential inclusion: $\dot\gamma(t) \in F(\mu_t, \zeta_t, t, \gamma(t))$ for a.e. $t \in [0, T]$, $\gamma(0) = x$. The set-valued map F can be used to represent local difficulties of the team to move itself and the equipment from the current positions, further interaction effects between the agents of the team, current priorities in the investigation. From the model point of view, for instance, the presence of many agents in the same area can *help* to remove debris, *speeding up* the movements. On the other hand, *overcrowding* of the agents in the same area could *block* the operations. We will assume that Ω is strongly invariant for F, i.e., that every trajectory starting from a point of Ω will remain in Ω for all the considered time. Notice that the definition of the microscopic dynamics depends on macroscopic effects and viceversa, therefore the resulting motion will be the outcome of a fixed point iteration (see e.g. [3, 4] for details). In this simplified setting, we will assume that F is nonempty, compact, convex valued and Lipschitz continuous. In this case, defining

$$S_I^{F,\eta}(x) := \left\{ \gamma \in AC([0, T]) : \begin{array}{l} \gamma(0) = x, \\ \gamma(t) \in \bar\Omega \text{ for all } t \in [0, T], \\ \dot\gamma(t) \in F(e_t \sharp \eta, \zeta_t, t, \gamma(t)) \text{ for a.e. } t \in [0, T] \end{array} \right\},$$

the set of admissible allocation evolutions starting from μ can be written as ($\Gamma_{[a,b]}$ denotes the set of continuous curves contained in $\bar\Omega$ defined in $[a, b]$) and

$$\mathscr{A}_{[0,T]}^F(\mu) := \{\eta \in \mathscr{P}(\bar\Omega \times \Gamma_{[0,T]}) :$$

$$e_0 \sharp \eta = \mu, \text{ for } \zeta\text{-a.e. } (x, \gamma) \in \bar\Omega \times \Gamma_{[0,T]} \text{ it holds } \gamma \in S_{[0,T]}^{F,\zeta}(x)\},$$

the set-valued map $\mu \mapsto \mathscr{A}_{[0,T]}^F(\mu)$ is Lipschitz continuous with compact convex values (see e.g. Section 3 in [6]).

2   Results

Given two continuous curves $\boldsymbol{\mu} = \{\mu_t\}_{t \in [0,T]} \subseteq \mathscr{P}_2(\mathbb{R}^d)$, $\boldsymbol{\zeta} = \{\zeta_t\}_{t \in [0,T]} \subseteq \mathscr{P}_2(\mathbb{R}^d)$ endowed with the 2-Wasserstein metric (see e.g. [1] for details on Wasserstein metric), we define the set-valued map $\Pi_{\mu,\zeta} : [0,T] \rightrightarrows \mathscr{P}(\mathbb{R}^d \times \mathbb{R}^d)$ as $\Pi_{\mu,\zeta}(t) := \Pi(\mu_t, \zeta_t)$. It can be easily proved that such a set-valued map is upper semicontinuous with compact convex images.

In the previous section we introduced the final cost functional $\mathcal{F}_\zeta(\boldsymbol{\mu})$ as the average cost paid by the team in order to allocate its resources, while minimizing the instantaneous cost $\mathcal{W}(\mu_t, \zeta_t)$ at every instant of time during the operation.

We now prove that it is possible to equivalently study the optimization problem by considering a minimization *strategy*, i.e., a time-depending dynamical allocation rather than minimizing the cost in real time (i.e., minimizing at each time the instantaneous cost).

Proposition 1. *Let $\tilde{c} : \mathbb{R}^d \times \mathbb{R}^d \to [0, +\infty]$ be l.s.c. Then there exists a measurable selection $\boldsymbol{\pi} = \{\pi_t\}_{t \in [0,T]}$ of $\Pi_{\mu,\zeta}(\cdot)$ such that*

$$\int_0^T \iint_{\mathbb{R}^d \times \mathbb{R}^d} \tilde{c}(x,y) \, d\pi_t(x,y) \, dt = \int_0^T \inf_{\pi \in \Pi_{\mu,\zeta}(t)} \iint_{\mathbb{R}^d \times \mathbb{R}^d} \tilde{c}(x,y) \, d\pi(x,y) \, dt.$$

In particular,

$$\inf \left\{ \int_0^T \iint_{\mathbb{R}^d \times \mathbb{R}^d} \tilde{c}(x,y) \, d\pi_t(x,y) \, dt : \begin{array}{l} \boldsymbol{\pi} = \{\pi_t\}_{t \in [0,T]} \subseteq \mathscr{P}(\mathbb{R}^d \times \mathbb{R}^d) \\ \pi_t \in \Pi(\mu_t, \zeta_t) \text{ for a.e. } t \in [0,T] \end{array} \right\} =$$

$$= \int_0^T \inf_{\pi \in \Pi_{\mu,\zeta}(t)} \iint_{\mathbb{R}^d \times \mathbb{R}^d} \tilde{c}(x,y) \, d\pi(x,y) \, dt.$$

Proof. To prove the result is enough to show the existence of a measurable selection $\boldsymbol{\pi} = \{\pi_t\}_{t \in [0,T]}$ of $\Pi_{\mu,\zeta}(\cdot)$ satisfying

$$\min_{\pi \in \Pi_{\mu,\zeta}(t)} \iint_{\mathbb{R}^d \times \mathbb{R}^d} \tilde{c}(x,y) \, d\pi(x,y) = \iint_{\mathbb{R}^d \times \mathbb{R}^d} \tilde{c}(x,y) \, d\pi_t(x,y),$$

since the reverse inequality is always true. Let $\tilde{g} : [0,T] \to \mathbb{R} \cup \{+\infty\}$ be measurable and satisfying for a.e. $t \in [0,T]$

$$\tilde{g}(t) \in \left\{ \iint_{\mathbb{R}^d \times \mathbb{R}^d} \tilde{c}(x,y) \, d\pi(x,y) : \pi \in \Pi_{\mu,\zeta}(t) \right\}.$$

Let $\{c_n(\cdot)\}_{n \in \mathbb{N}} \subseteq C_b^0(\mathbb{R}^d \times \mathbb{R}^d)$ be an increasing sequence, pointwise converging to $\tilde{c}$, e.g.,

$$c_n(x,y) = \min \left\{ n, \inf_{(z,w) \in \mathbb{R}^d \times \mathbb{R}^d} \{\tilde{c}(z,w) + nd((x,y),(z,w))\} \right\}.$$

Define the sequence of set-valued maps $\{F_n(\cdot)\}_{n\in\mathbb{N}}$, $F_n : [0,T] \rightrightarrows \mathscr{P}(\mathbb{R}^d \times \mathbb{R}^d)$

$$F_n(t) := \{\pi \in \mathscr{P}(\mathbb{R}^d \times \mathbb{R}^d) : \langle \pi, c_n \rangle - \tilde{g}(t) < 1/n\}$$

For every $n \in \mathbb{N}$, the function $(t,\pi) \mapsto \langle \pi, c_n \rangle - \tilde{g}(t)$ is continuous in π for every fixed $t \in [0,T]$ and measurable in t for every fixed $\pi \in \mathscr{P}(\mathbb{R}^d \times \mathbb{R}^d)$. Thus, according to Theorem 6.2 in [5], we have that $F_n(\cdot)$ is measurable, and by Proposition 2.6 in [5] we have that $\overline{F_n}(\cdot)$ is weakly measurable. Define

$$F(t) = \bigcap_{n\in\mathbb{N}} \overline{F_n(t)} \subseteq \{\pi \in \mathscr{P}(\mathbb{R}^d \times \mathbb{R}^d) : \langle \pi, c_n \rangle \le \tilde{g}(t) + 1/n \text{ for all } n \in \mathbb{N}\}$$

$$\subseteq \{\pi \in \mathscr{P}(\mathbb{R}^d \times \mathbb{R}^d) : \langle \pi, \tilde{c} \rangle \le \tilde{g}(t)\}$$

where we used monotone convergence theorem. By Theorem 4.1 in [5] we have that $t \mapsto F(t) \cap \Pi_{\mu,\varsigma}(t)$ is measurable, and by Theorem 5.7 in [5] we obtain a measurable family $\boldsymbol{\pi}^{\tilde{g}} = \{\pi_t\}_{t\in[0,T]} \subseteq \mathscr{P}(\mathbb{R}^d \times \mathbb{R}^d)$ such that $\pi_t \in F(t) \cap \Pi_{\mu,\varsigma}(t)$ for a.e. $t \in [0,T]$, i.e., $\min_{\pi\in\Pi_{\mu,\varsigma}(t)} \langle \pi, \tilde{c} \rangle \le \langle \pi_t, \tilde{c} \rangle \le \tilde{g}(t)$. By choosing $\tilde{g}(t) = \min_{\pi\in\Pi_{\mu,\varsigma}(t)} \langle \pi, \tilde{c} \rangle$, the result follows. $\qquad\square$

We notice that, in a realistic setting, the practical implementation of the optimal strategy in general is troublesome, due for instance to its irregular behaviour in time. Moreover, the effectiveness of the strategy should be compared with the consolidated protocols that the team is familiar to (and it is trained to). A possible way to perform such a comparison, is to single out, among all the possible strategies, a subset of fundamental ones, and to consider the smallest number of fundamental elements in a ε-net covering the whole set of strategies. Such concept, known as ε-entropy, introduced by Kolmogorov, has been used in several settings also to estimate the performance of numerical approximation algorithm. We refer to [2] for further details and applications.

In order to study the necessary condition for the problem, we define the *value function* by

$$u(t,\mu) := \inf \left\{ \int_t^T \mathscr{W}(\mu_\tau, \zeta_\tau) \, d\tau : \boldsymbol{\mu} = \{\mu_\tau\}_{\tau\in[t,T]} \text{ admissible}, \mu_t = \mu \right\}.$$

By standard arguments the following Dynamic Programming Principle holds.

$$u(t,\mu) = \inf \left\{ \int_t^s \mathscr{W}(\mu_\tau, \zeta_\tau) \, d\tau + u(s,\mu_s) : \begin{array}{l} \boldsymbol{\mu} = \{\mu_\tau\}_{\tau\in[t,s]} \text{ admissible} \\ s \in [t,T], \mu_\tau = \mu \end{array} \right\}.$$

Definition 1 (Viscosity subdifferential). *Let* $V : [0,T] \times \mathscr{P}_2(\mathbb{R}^d) \to \mathbb{R}$ *be a function. We say that* $(p_t, p_\mu(\cdot)) \in \mathbb{R} \times L^2_\mu$ *belongs to* $D^- V(t,\mu)$ *if and only if*

$$V(s,\hat{\mu}) - V(t,\mu) \ge p_t(s-t) + \iint_{\mathbb{R}^d \times \mathbb{R}^d} \langle p_\mu(x), x-y \rangle \, d\pi(x,y) + o\left(|t-s| + W_\pi(\mu,\hat{\mu})\right).$$

for all $\hat{\mu} \in \mathscr{P}_2(\mathbb{R}^d)$ *and* $\pi \in \Pi(\mu, \hat{\mu})$, *where* $W_\pi^2(\mu, \hat{\mu}) = \iint_{\mathbb{R}^d \times \mathbb{R}^d} |x - y|^2 \, d\pi(x, y)$. *We define also* $D^+V(t, \mu) = -D^-(-V)(t, \mu)$. *The sets* $D^-V(t, \mu)$ *and* $D^+V(t, \mu)$ *are called the viscosity subdifferential and superdifferential of* V *at* (t, μ).

Definition 2 (Viscosity solution). *Let*

$$\mathcal{T} := \{(t, \mu, p_t, p_\mu(\cdot)) : t \in [0, T], \ \mu \in \mathscr{P}_2(\mathbb{R}^d), \ p_t \in \mathbb{R}, p_\mu(\cdot) \in L_\mu^2(\mathbb{R})\}.$$

Let $\mathscr{H} : \mathcal{T} \to \mathbb{R}$ *be a function, called the Hamiltonian. We say that* $V : [0, T] \times \mathscr{P}_2(\mathbb{R}^d) \to \mathbb{R}$ *is a viscosity solution of the Hamilton-Jacobi-Bellman equation if*

$$p_t + \mathscr{H}(t, \mu, p_t, p_\mu) \leq 0, \ \text{for all } (p_t, p_\mu(\cdot)) \in D^-V(t, \mu),$$
$$p_t + \mathscr{H}(t, \mu, p_t, p_\mu) \geq 0, \ \text{for all } (p_t, p_\mu(\cdot)) \in D^+V(t, \mu).$$

Theorem 1 (Hamilton-Jacobi-Bellman equation). *The value function* $u(\cdot)$ *is a viscosity solution of the Hamilton-Jacobi-Bellman equation*

$$\partial_t u(t, \mu) + \mathscr{H}(t, \mu, \partial_t u(t, \mu), \partial_\mu u(t, \mu)) = 0,$$

where $\mathscr{H}(t, \mu, p_t, p_\mu) := \int_{\mathbb{R}^d} \sup_{v \in F(t, \mu, \zeta_t, x)} \langle p_\mu(x), v \rangle \, d\mu(x) + \mathscr{W}(\mu_t, \zeta_t)$.

Proof. According to Definition 2, for the value function $u(\cdot)$, for all $(p_t, p_\mu) \in D^+u(t, \mu)$ we have

$$p_t(s - t) + \iint_{\mathbb{R}^d \times \mathbb{R}^d} \langle p_\mu(x), x - y \rangle \, d\pi_s(x, y) + o\left(|t - s| + W_{\pi_s}(\mu, \mu_s)\right) \geq$$
$$\geq -\int_t^s \mathcal{L}(\tau, \mu_\tau) \, d\tau,$$

for all $s \in [t, T]$, all the admissible trajectories $\{\mu_\tau\}_{\tau \in [t, T]}$ with $\mu_t = \mu$, and all $\pi_s \in \Pi(\mu, \mu_s)$.

In particular, we can choose $\pi_s = (e_t, e_s)\sharp\eta$, where $\eta \in \mathscr{P}(\mathbb{R}^d \times \Gamma_{[t, T]})$ satisfies $e_t\sharp\eta = \mu$, $\mu_\tau = e_\tau\sharp\eta$ for all $\tau \in [t, T]$, for η-a.e. $(x, \gamma) \in \mathbb{R}^d \times \Gamma_{[t, T]}$ it holds $\gamma \in AC([t, T])$ with $\gamma(t) = x$, and $\gamma(\cdot)$ is an admissible microscopic trajectory, i.e., $\dot{\gamma}(\tau) \in F(\tau, \mu_\tau, \zeta_\tau, \gamma(\tau))$ for a.e. $\tau \in [t, T]$.

With this choice, we have

$$-\int_t^s \mathcal{L}(\tau, \mu_\tau) \, d\tau \leq p_t(s - t) + \iint_{\mathbb{R}^d \times \Gamma_{[t, T]}} \langle p_\mu(x), \gamma(t) - \gamma(s) \rangle \, d\eta(x, \gamma) +$$
$$+ o\left(|t - s| + \left(\iint_{\mathbb{R}^d \times \Gamma_{[t, T]}} |\gamma(s) - \gamma(t)|^2 \, d\eta\right)^{1/2}\right)$$
$$\leq p_t(s - t) + \iint_{\mathbb{R}^d \times \Gamma_{[t, T]}} \langle p_\mu(x), \gamma(t) - \gamma(s) \rangle \, d\eta(x, \gamma) + o\left(|t - s|\right),$$

recalling that, since on Ω the set-valued map is uniformly bounded, there exists $C > 0$ such that $|\gamma(s) - \gamma(t)| \leq C|s-t|$ for all $s, t \in [0, T]$ and all the trajectories of $\dot\gamma(\tau) \in F(\tau, \mu_\tau, \zeta_\tau, \gamma(\tau))$ contained in Ω. Dividing by $s - t$ yields

$$p_t - \iint_{\mathbb{R}^d \times \Gamma_{[t,T]}} \langle p_\mu(x), \frac{\gamma(s) - \gamma(t)}{s - t} \rangle \, d\eta(x, \gamma) + \frac{o(|s - t|)}{s - t} \geq -\frac{1}{s - t} \int_t^s \mathcal{L}(\tau, \mu_\tau) \, d\tau. \tag{1}$$

Given $g \in L^2_\mu(\mathbb{R}^d)$, we have

$$\langle g \circ e_t, \frac{e_s - e_t}{s - t} \rangle_{L^2_\eta} = \iint_{\mathbb{R}^d \times \Gamma_{[t,T]}} \langle g(x), \frac{\gamma(s) - \gamma(t)}{s - t} \rangle \, d\eta(x, \gamma)$$

$$= \iint_{\mathbb{R}^d \times \Gamma_{[t,T]}} \frac{1}{s - t} \int_t^s \langle g(x), \dot\gamma(\tau) \rangle \, d\tau \, d\eta(x, \gamma).$$

Recalling the assumptions on F, the set-valued map $\tau \mapsto F(\tau, \mu_\tau, \gamma(\tau))$ is continuous with compact values, and therefore

$$\tau \mapsto \inf_{v \in F(\tau, \mu_\tau, \gamma(\tau))} \langle g(x), v \rangle \, d\tau, \qquad \tau \mapsto \sup_{v \in F(\tau, \mu_\tau, \gamma(\tau))} \langle g(x), v \rangle \, d\tau$$

are continuous. By the uniform boundedness of F and the integrability of g, we can apply Dominated Convergence Theorem obtaining

$$\lim_{s \to t^+} \iint_{\mathbb{R}^d \times \Gamma_{[t,T]}} \frac{1}{s - t} \int_t^s \inf_{v \in F(\tau, \mu_\tau, \gamma(\tau))} \langle g(x), v \rangle \, d\tau \, d\eta =$$

$$= \iint_{\mathbb{R}^d \times \Gamma_{[t,T]}} \lim_{s \to t^+} \frac{1}{s - t} \int_t^s \inf_{v \in F(\tau, \mu_\tau, \gamma(\tau))} \langle g(x), v \rangle \, d\tau \, d\eta$$

$$= \iint_{\mathbb{R}^d \times \Gamma_{[t,T]}} \inf_{v \in F(t, \mu_t, \zeta_t, \gamma(t))} \langle g(x), v \rangle \, d\eta$$

and the corresponding result with inf replaced by sup.

Thus,

$$\liminf_{s \to t^+} \langle g \circ e_t, \frac{e_s - e_t}{s - t} \rangle_{L^2_\eta} \geq \int_{\mathbb{R}^d} \inf_{v \in F(t, \mu_t, \zeta_t, \gamma(t))} \langle g(x), v \rangle \, d\mu,$$

$$\limsup_{s \to t^+} \langle g \circ e_t, \frac{e_s - e_t}{s - t} \rangle_{L^2_\eta} \leq \int_{\mathbb{R}^d} \sup_{v \in F(t, \mu_t, \zeta_t, \gamma(t))} \langle g(x), v \rangle \, d\mu.$$

By taking $g(\cdot) = p_\mu(\cdot)$ we get

$$\limsup_{s \to t^+} \iint_{\mathbb{R}^d \times \Gamma_{[t,T]}} \langle p_\mu(x), \frac{\gamma(s) - \gamma(t)}{s - t} \rangle \, d\eta(x, \gamma) \leq \int_{\mathbb{R}^d} \sup_{v \in F(t, \mu_t, \zeta_t, \gamma(t))} \langle p_\mu(x), v \rangle \, d\mu.$$

Indeed, since for every measurable selection $v(\cdot)$ of $F(t, \mu_t, \zeta_t, \cdot)$ there exists an admissible trajectory μ, represented by η, such that

$$\lim_{s \to t^+} \iint_{\mathbb{R}^d \times \Gamma_{[t,T]}} \langle p_\mu(x), \frac{\gamma(s) - \gamma(t)}{s - t} \rangle \, d\eta(x, \gamma) = \int_{\mathbb{R}^d} \langle p_\mu(x), v(x) \rangle \, d\mu$$

and
$$\sup_{v(\cdot)\ \text{meas. sel. of}\ F(t,\mu_t,\zeta_t,\cdot)}\int_{\mathbb{R}^d}\langle p_\mu(x),v(x)\rangle\,d\mu = \int_{\mathbb{R}^d}\sup_{v\in F(t,\mu_t,\zeta_t,\gamma(t))}\langle p_\mu(x),v\rangle\,d\mu,$$

actually, by passing to the limit as $s \to t^+$ and taking the infimum on $v(\cdot)$ in (1), we obtain

$$p_t - \int_{\mathbb{R}^d}\sup_{v\in F(t,\mu_t,\zeta_t,\gamma(t))}\langle p_\mu(x),v\rangle\,d\mu + \mathcal{L}(t,\mu_t)\ge 0,$$

where we used also the continuity of $\tau \mapsto \mathcal{L}(\tau,\mu_\tau)$.

On the other hand, recalling the compactness of the set of admissible trajectories and the continuity of the cost function, there exists an optimal trajectory $\{\mu_\tau\}_{\tau\in[t,T]}$ such that $\mu_t = \mu$ and for all $s \in [t,T]$ it holds

$$u(s,\mu_s) - u(t,\mu) = -\int_t^s \mathcal{L}(\tau,\mu_\tau)\,d\tau.$$

In particular, for every $(p_t,p_\mu(\cdot)) \in D^- u(t,\mu)$, $s \in [t,T]$, $\boldsymbol{\pi}_s \in \Pi(\mu,\mu_s)$ it holds

$$p_t(s-t) + \iint_{\mathbb{R}^d\times\mathbb{R}^d}\langle p_\mu(x),x-y\rangle\,d\boldsymbol{\pi}_s(x,y)$$
$$+ o(|t-s| + W_{\boldsymbol{\pi}_s}(\mu,\mu_s)) \le u(s,\mu_s) - u(t,\mu) = -\int_t^s \mathcal{L}(\tau,\mu_\tau)\,d\tau.$$

By a straightforward adaption of the same arguments used before, we obtain

$$p_t - \iint_{\mathbb{R}^d\times\mathbb{R}^d}\langle p_\mu(x),\frac{\gamma(s)-\gamma(t)}{s-t}\rangle\,d\boldsymbol{\eta}(x,y) + \frac{o(|t-s|)}{s-t} \le -\frac{1}{s-t}\int_t^s \mathcal{L}(\tau,\mu_\tau^\varepsilon)\,d\tau,$$

where, as before, $\boldsymbol{\eta} \in \mathscr{P}(\mathbb{R}^d \times \Gamma_{[t,T]})$ satisfies $e_t\sharp\boldsymbol{\eta} = \mu$, $\mu_\tau = e_\tau\sharp\boldsymbol{\eta}$ for all $\tau \in [t,T]$, and for $\boldsymbol{\eta}$-a.e. $(x,\gamma) \in \mathbb{R}^d\times\Gamma_{[t,T]}$ it holds $\gamma \in AC([t,T])$ with $\gamma(t) = x$, and $\gamma(\cdot)$ is an admissible microscopic trajectory. By taking the limsup as $s \to t^+$, we have

$$p_t - \iint_{\mathbb{R}^d\times\mathbb{R}^d}\sup_{v\in F(t,\mu_t,\zeta_t,x)}\langle p_\mu(x),v\rangle\,d\mu(x) + \mathcal{L}(t,\mu_t)\le 0.$$

The thesis follows, recalling Definition 2 and the choice of $\mathscr{H}(\cdot)$. $\square$

We notice that the value function solves HJB in a suitable viscosity sense. However in order to characterize it as the *unique* solution of the equation HJB, a *comparison principle* between solutions must be proved. Some comparison principle are available in literature (e.g. [6]), but stronger regularity assumptions on the cost functions are required. Thus, in this framework, we can state the HJB equation in form of a *necessary* condition, and not as a *necessary and sufficient* condition.

Acknowledgements. This research is supported by INdAM-GNAMPA Project 2022 "Evoluzione e controllo ottimo in spazi di Wasserstein" ("Evolution and optimal control in Wasserstein spaces"), CUP_E55F22000270001. Rossana Capuani was supported by the co-financing of European Union-FSE-REACT-UE, PON "Ricerca e Innovazione 2014–2020, Azione IV.6 Contratti di ricerca su tematiche Green".

References

1. Ambrosio, L., Gigli, N., Savare, G.: Gradient Flows in Metric Spaces and in the Space of Probability Measures. Lectures in Mathematics ETH Zürich. Birkhäuser Verlag, Basel (2008)
2. Capuani, R., Dutta, P., Nguyen, K.T.: Metric entropy for functions of bounded total generalized variation. SIAM J. Math. Anal. **53**(1), 1168–1190 (2021). https://doi.org/10.1137/20M1310953
3. Capuani, R., Marigonda, A.: Constrained mean field games equilibria as fixed point of random lifting of set-valued maps. IFAC-PapersOnLine **55**(30), 180–185 (2022). https://doi.org/10.1016/j.ifacol.2022.11.049
4. Capuani, R., Marigonda, A., Mogentale, M.: Random lifting of set-valued maps. In: Lirkov, I., Margenov, S. (eds.) LSSC 2021. LNCS, vol. 13127, pp. 297–305. Springer, Cham (2022). https://doi.org/10.1007/978-3-030-97549-4_34
5. Himmelberg, C.J.: Measurable relations. Fund. Math. **87**, 53–72 (1975)
6. Jimenez, C., Marigonda, A., Quincampoix, M.: Optimal control of multiagent systems in the Wasserstein space. Calc. Var. Partial Diff. Eqn. **59**(2), 58 (2020). https://doi.org/10.1007/s00526-020-1718-6.MR4073204
7. Rimpler-Schmid, A., et al.: EU preparedness and responses to chemical, biological, radiological and nuclear (CBRN) threats (2021). ISBN: 978-92-846-8358-1

A Sufficient Condition for a Discrete-Time Optimal Control Problem

Mikhail I. Krastanov[1,2] and Boyan K. Stefanov[1,2(✉)]

[1] Institute of Mathematics and Informatics, Bulgarian Academy of Sciences,
Acad. Georgi Bonchev Str., Bl. 8, 1113 Sofia, Bulgaria
[2] Department of Mathematics and Informatics, University of Sofia,
5 James Bourchier Blvd, 1164 Sofia, Bulgaria
bojanks@fmi.uni-sofia.bg

Abstract. A discrete-time optimal control problem is considered on an infinite-time horizon. A new sufficient optimality condition is proved under suitable assumptions.

Keywords: discrete-time optimal control problem · sufficient optimality condition

1 Statement of the Problem and Preliminaries

We consider the following discrete-time control system:

$$x_{k+1} = f_k(x_k, u_k), \ u_k \in U_k, \ k \in \mathbb{N}. \tag{1}$$

Here, $\mathbb{N}$ denotes the set of all nonnegative integers, and $x_0 \in \mathbb{R}^n$ is a given starting point. We assume that each U_k, $k \in \mathbb{N}$, is a non-empty, closed, and convex subset of $\mathbb{R}^l$, and each function $f_k : \mathbb{R}^n \times \tilde{U}_k \to \mathbb{R}^n$, $k \in \mathbb{N}$, is continuously differentiable, where $\tilde{U}_k$ is an open subset in $\mathbb{R}^l$, containing the set U_k.
By

$$\mathbf{u} := (u_0, u_1, ..., u_k, ...),$$

we denote the function influencing the system, regarded as a control sequence. A control sequence is said to be **admissible** if each of its components meets the following inclusion criteria: $u_k \in U_k$, $k \in \mathbb{N}$.

This study is partly supported by the European Union-NextGenerationEU, through the National Recovery and Resilience Plan of the Republic of Bulgaria, project SUMMIT BG-RRP-2.004-0008-C01, and by the Center of Excellence in Informatics and ICT, Grant No. BG05M2OP001-1.001-0003 (financed by the Science and Education for Smart Growth Operational Program (2014–2020) and co-financed by the European Union through the European structural and investment funds), and by the Sofia University "St. K. Ohridski" under the contract No 80-10-62/25.04.2023.

I. Lirkov and S. Margenov (Eds.): LSSC 2023, LNCS 13952, pp. 184–192, 2024.
https://doi.org/10.1007/978-3-031-56208-2_18

Assumption 1. *Each admissible control sequence* **u** *belongs to the space* $\ell^2(\mathbb{N})$.

We have denoted by $\ell^2(\mathbb{N}) := \{(u_0, u_1, ..., u_k, ...) : \sum_{i=0}^{\infty} |u_i|^2 < \infty\}$ the Banach space with respect to the norm $\|\mathbf{u}\| = \left(\sum_{i=0}^{\infty} |u_i|^2\right)^{1/2}$.

For each given admissible control sequence **u**, the equality (1) generates a trajectory $\mathbf{x} = (x_0, x_1, ..., x_k, ...)$. We refer to the couple $(\mathbf{x}, \mathbf{u})$ as an **admissible process**. Given an admissible process $(\mathbf{x}, \mathbf{u})$, the state trajectory **x** can be represented (according to (1)) as

$$x_{k+1} := f_k^{(u_k)} \circ f_{k-1}^{(u_{k-1})} \circ \cdots \circ f_0^{(u_0)}(x_0), \ k \in \mathbb{N}.$$

Here, $\circ$ denotes the composition of the corresponding maps and $f_k^{(u)}(x) := f_k(x, u)$.

Over all admissible processes $(\mathbf{x}, \mathbf{u})$, we consider the following optimal control problem:

$$J(\mathbf{u}) := \sum_{k=0}^{\infty} g_k(x_k, u_k) \to \min. \tag{2}$$

Here, each function $g_k : \mathbb{R}^n \times \tilde{U}_k \to \mathbb{R}$, $k \in \mathbb{N}$, is assumed to be continuously differentiable. Infinite-horizon optimal control problems of this form are studied by various authors (cf., for example, [4,5] and the references therein).

We call an arbitrary admissible process $(\bar{\mathbf{x}}, \bar{\mathbf{u}})$ feasible if $J(\bar{\mathbf{u}})$ is a real number. Let $(\bar{\mathbf{x}}, \bar{\mathbf{u}})$ be an arbitrary feasible process. For every positive integer k and for each vector ξ, we denote by $x^{k,\xi} = (x_k, x_{k+1}, ...)$ the trajectory induced by (1) with the initial state $x_k = \xi$ at the moment of time k, i.e.,

$$x_{s+1}^{k,\xi} := f_s^{(\bar{u}_s)} \circ f_{s-1}^{(\bar{u}_{s-1})} \circ \cdots \circ f_k^{(\bar{u}_k)}(\xi), \ s = k, k+1, ...$$

Similarly to [1], we make the following assumption:

Assumption 2. *For every positive integer* k, *there exist* $\alpha_k > 0$ *and a sequence* $\{\beta_s^k\}_{s=k}^{\infty}$ *with* $\sum_{s=k}^{\infty} \beta_s^k < \infty$ *such that the following inequality holds true:*

$$\sup_{\xi \in \bar{\mathbf{B}}(\bar{x}_k; \alpha_k)} \left\|\frac{\partial}{\partial \xi} g_s(x_s^{k,\xi}, \bar{u}_s)\right\| \leq \beta_s^k, \ s = k, k+1, ... ,$$

where $\bar{\mathbf{B}}(\bar{x}_k; \alpha_k)$ *is the closed ball in* $\mathbb{R}^n$ *centered at* $\bar{x}_k$ *with radius* α_k.

Assumption 2 implies that the series

$$\sum_{s=k}^{\infty} \frac{\partial}{\partial \xi} g_s(x_s^{k,\xi}, \bar{u}_s), \ k = 1, 2, ... ,$$

is absolutely convergent, uniformly with respect to $\xi \in \bar{\mathbf{B}}(\bar{x}_k; \alpha_k)$. By using the identity

$$g_s(x_s^{k,\xi}, \bar{u}_s) = g_s\left(f_{s-1}^{(\bar{u}_{s-1})} \circ f_{s-2}^{(\bar{u}_{s-2})} \circ \cdots \circ f_k^{(\bar{u}_k)}(\xi), \bar{u}_s\right),$$

the chain rule allows us to find that

$$\frac{\partial}{\partial \xi} g_s(x_s^{k,\xi}, \bar{u}_s) = \frac{\partial}{\partial x} g_s(x_s^{k,\xi}, \bar{u}_s) \prod_{i=s-1}^{k} \frac{\partial}{\partial x} f_i(x_i^{k,\xi}, \bar{u}_i), \qquad (3)$$

where the following notation is used:

$$\prod_{i=s-1}^{k} A_i := \begin{cases} A_{s-1} A_{s-2} \cdots A_k & \text{if } s > k \\ E & \text{if } s \leq k. \end{cases}$$

Here, $A_i := \frac{\partial}{\partial x} f_i(x_i^{k,\xi}, \bar{u}_i)$, and E denotes the identity matrix of dimension $n \times n$. Also, instead of the usual symbol $\prod_{i=s}^{k}$, we use the symbol $\prod_{i=s}^{k}$ to indicate that the "increment" of the running index i is -1 (since $s \geq k$).

Following [1], we define the **adjoint sequence** $\psi := \{\psi_k\}_{k=1}^{\infty}$ as:

$$\psi_k = \sum_{s=k}^{\infty} \frac{\partial}{\partial \xi} g_s(x_s^{k,\xi}, \bar{u}_s)_{|\xi \,=\, \bar{x}_k}, \quad k = 1, 2, \dots .$$

Remark 1. Assumption 2 actually implies that $\|\psi_k\| < \infty$, $k = 1, 2, \dots$. Furthermore, taking into account that $x_s^{k,\bar{x}_k} = \bar{x}_s$, we obtain from (3) that

$$\psi_k = \sum_{s=k}^{\infty} \frac{\partial}{\partial \xi} g_s(x_s^{k,\xi}, \bar{u}_s)_{|\xi=\bar{x}_k} = \sum_{s=k}^{\infty} \frac{\partial}{\partial x} g_s(\bar{x}_s, \bar{u}_s) \prod_{i=s-1}^{k} \frac{\partial}{\partial x} f_i(\bar{x}_i, \bar{u}_i). \qquad (4)$$

Due to the second equality in (4), it turns out that the so-defined adjoint sequence solves the **adjoint equation**

$$\psi_k = \psi_{k+1} \frac{\partial}{\partial x} f_k(\bar{x}_k, \bar{u}_k) + \frac{\partial}{\partial x} g_k(\bar{x}_k, \bar{u}_k), \quad k = 1, 2, \dots .$$

Using ψ, for every $k \in \mathbb{N}$, we introduce the **Hamiltonian** function $\mathcal{H}_k :$ $\mathbb{R}^n \times \mathbb{R}^l \times \mathbb{R}^n \to \mathbb{R}$ as follows:

$$\mathcal{H}_k(x, u, \psi) := g_k(x, u) + \psi f_k(x, u).$$

In order to formulate the next assertion, we denote by $T_S(\bar{y})$ the **Bouligand tangent cone** to the closed subset S of a Banach space Y at the point $\bar{y} \in S$. Recall that $T_S(\bar{y})$ consists of all $w \in Y$ for which there exists a sequence of positive real numbers $\{t^\mu\}_{\mu=1}^{\infty} \downarrow 0$ and a sequence $\{w^\mu\}_{\mu=1}^{\infty} \subset Y$ convergent to w such that $\bar{y} + t^\mu w^\mu \in S$ for each $\mu = 1, 2, \dots$ (cf., for example, [3], Chap. 4.1).

Following [2], we introduce the next definition:

Definition 1. *Let Y be a Banach space, S be a nonempty closed subset of Y, $\bar{y}$ be an arbitrary point of S, and w be an arbitrary element of the Bouligand tangent* cone $T_S(\bar{y})$ to the set S at the point $\bar{y}$. We say that the function $h : S \to \mathbb{R}$ is differentiable in the direction w if the following limit exists:

$$\lim_{t^\mu \downarrow 0} \frac{h(\bar{y} + t^\mu w^\mu) - h(\bar{y})}{t^\mu},$$

where the sequence $\{t^\mu\}_{\mu=0}^\infty \downarrow 0$, the sequence $\{w^\mu\}_{\mu=0}^\infty \to w$ as $\mu \to +\infty$, and $\bar{y} + t^\mu w^\mu \in S$ for each $\mu = 1, 2, \dots$. We denote this limit by $dh(\bar{y}; w)$ and call it the derivative of h in the direction $w \in T_S(\bar{y})$ at the point $\bar{y} \in S$.

Remark 2. Definition 1 is equivalent to the definition of contingent derivative for the case of a single-valued mapping (cf., for example, [2]).

The following assertion is crucial for the proof of the main result:

Proposition 1. *Let Assumption 1 and Assumption 2 be satisfied, the pair $(\bar{\mathbf{x}}, \bar{\mathbf{u}})$ be a feasible process, and the corresponding adjoint sequence ψ be defined by (4). Then, for every positive integer m, for each multi-index $(k_1, k_2, \dots, k_m)$ with $0 \leq k_1 < k_2 < \cdots < k_m$, the following relation holds true:*

$$dJ(\bar{\mathbf{u}}; \mathbf{p}) = \sum_{i=1}^m \left(\frac{\partial}{\partial u} \mathcal{H}_{k_i}(\bar{x}_{k_i}, \bar{u}_{k_i}, \psi_{k_i+1}) \right) p_{k_i} \quad \text{for each } \mathbf{p} \in T_{\mathbf{U}}(\bar{\mathbf{u}}),$$

where

$$\mathbf{U} := \{(u_0, u_1, u_2, \dots, u_k, \dots) : u_s \in U_s \text{ if } s \in \mathcal{K}; u_s = \bar{u}_s \text{ if } s \notin \mathcal{K}; s \in \mathbb{N}\}$$

and $\mathcal{K} := \{k_1, k_2, \dots, k_m\}$.

Proof. We shall present here the proof only for the simplest case, $m = 1$. Let k_1 be an arbitrary nonnegative integer. We set $k_1^+ := k_1 + 1$ and define the function:

$$F^{k_1^+}(\xi) := \sum_{s=k_1^+}^\infty g_s(x_s^{k_1^+, \xi}, \bar{u}_s), \quad \xi \in \bar{\mathbf{B}}(\bar{x}_{k_1^+}; \alpha_{k_1^+}),$$

where the ball $\bar{\mathbf{B}}(\bar{x}_{k_1^+}; \alpha_{k_1^+})$ is defined by Assumption 2.
    Clearly, we have that

$$F^{k_1^+}(\bar{x}_{k_1^+}) = J(\bar{\mathbf{u}}) - \sum_{s=0}^{k_1-1} g_s(\bar{x}_s, \bar{u}_s) - g_{k_1}(\bar{x}_{k_1}, \bar{u}_{k_1}) \in (-\infty, +\infty). \tag{5}$$

If $k_1 = 0$, then (5) is reduced to

$$F^{0^+}(\bar{x}_1) = J(\bar{\mathbf{u}}) - g_0(\bar{x}_0, \bar{u}_0).$$

We set

$$\partial F^{k_1^+}(\xi) = \sum_{s=k_1^+}^\infty \frac{\partial}{\partial \xi} g_s(x_s^{k_1^+, \xi}, \bar{u}_s), \quad \xi \in \bar{\mathbf{B}}(\bar{x}_{k_1^+}; \alpha_{k_1^+}).$$

Applying Assumption 2, we obtain the following inequalities:

$$\sum_{s=k_1^+}^{\infty} \left\| \frac{\partial}{\partial \xi} g_s(x_s^{k_1^+,\xi}, \bar{u}_s) \right\| \leq \sum_{s=k_1^+}^{\infty} \beta_s^{k_1^+} < \infty.$$

Hence, $\partial F^{k_1^+}$ is a continuous function on the ball $\bar{\mathbf{B}}(\bar{x}_{k_1^+}; \alpha_{k_1^+})$ as a limit of a uniformly convergent series of continuous functions on $\bar{\mathbf{B}}(\bar{x}_{k_1^+}; \alpha_{k_1^+})$. Taking into account (5), we obtain that $F^{k_1^+}$ is also a limit of a uniformly convergent series on $\bar{\mathbf{B}}(\bar{x}_{k_1^+}; \alpha_{k_1^+})$. Moreover, it is a differentiable function such that

$$\frac{\partial}{\partial \xi} F^{k_1^+}(\xi) = \partial F^{k_1^+}(\xi) \text{ for each } \xi \in \bar{\mathbf{B}}(\bar{x}_{k_1^+}; \alpha_{k_1^+}). \tag{6}$$

Let $\mathbf{p}$ be an arbitrary element of $T_{\mathbf{U}}(\bar{u})$. Then the definition of the set $\mathbf{U}$ implies that for every $s \in \mathbb{N}$, the components of $\mathbf{p} \in T_{\mathbf{U}}(\bar{u})$ have the following representation, respectively:

$$p_s := \begin{cases} \mathbf{0} & \text{if } s \neq k_1, \\ p_{k_1} \in T_{U_{k_1}}(\bar{u}_{k_1}) & \text{if } s = k_1. \end{cases}$$

Clearly, if $\mathbf{p} = \mathbf{0}$, then Proposition 1 holds true. Let $\mathbf{p} \neq \mathbf{0}$. Then, there exists a sequence of positive real numbers $\{t^\mu\}_{\mu=1}^{\infty}$ tending to zero as $\mu \to \infty$ (without loss of generality, we may think that each t^μ is less than 1) and a sequence of points $\{p_{k_1}^\mu\}_{\mu=1}^{\infty}$ tending to p_{k_1} as $\mu \to \infty$, such that $\bar{u}_{k_1} + t^\mu p_{k_1}^\mu \in U_{k_1}$ for each $\mu = 1, 2, \dots$.

For each μ, we set $\mathbf{p}^\mu := (p_1^\mu, p_2^\mu, \dots)$ and define the control sequence $\mathbf{u}^\mu := \bar{u} + t^\mu \mathbf{p}^\mu$, i.e., for each $s \in \mathbb{N}$

$$u_s^\mu := \begin{cases} \bar{u}_s & \text{if } s \neq k_1, \\ \bar{u}_{k_1} + t^\mu p_{k_1}^\mu & \text{if } s = k_1. \end{cases}$$

It is clear that $\mathbf{u}^\mu$ is an admissible control sequence. The corresponding state trajectory $\mathbf{x}^\mu$ is given by

$$x_s^\mu := \begin{cases} \bar{x}_s, & \text{if } s \leq k_1, \\ f_{k_1}(\bar{x}_{k_1}, \bar{u}_{k_1} + t^\mu p_{k_1}^\mu), & \text{if } s = k_1 + 1, \\ f_{s-1}(x_{s-1}^\mu, \bar{u}_{s-1}), & \text{if } s > k_1 + 1. \end{cases}$$

Without loss of generality, we may think that μ_0 is so large that $x_{k_1^+}^\mu \in \bar{\mathbf{B}}(\bar{x}_{k_1^+}; \alpha_{k_1^+})$ for each $\mu \geq \mu_0$. One can directly check that for $k_1 \geq 1$,

$$\sum_{s=0}^{k_1-1} g_s(\bar{x}_s, \bar{u}_s) + g_{k_1}(\bar{x}_{k_1}, \bar{u}_{k_1} + t^\mu p_{k_1}^\mu) + F^{k_1^+}(x_{k_1+1}^\mu) = J(\mathbf{u}^\mu). \tag{7}$$

If $k_1 = 0$, then (7) takes the form

$$g_0(\bar{x}_0, \bar{u}_0 + t^\mu p_0^\mu) + F^{0^+}(x_1^\mu) = J(\mathbf{u}^\mu).$$

There exist elements θ_1^μ and θ_2^μ belonging to the interval $(0,1)$ such that

$$\lim_{t^\mu \downarrow 0} \frac{F^{k_1^+}(x_{k_1+1}^\mu) - F^{k_1^+}(\bar{x}_{k_1+1})}{t^\mu}$$

$$= \lim_{t^\mu \downarrow 0} \left(\frac{1}{t^\mu} \frac{d}{d\xi} F^{k_1^+}\left(\bar{x}_{k_1+1} + \theta_1^\mu(x_{k_1+1}^\mu - \bar{x}_{k_1+1}) \right) \left(x_{k_1+1}^\mu - \bar{x}_{k_1+1} \right) \right)$$

$$= \lim_{t^\mu \downarrow 0} \left(\frac{1}{t^\mu} \frac{d}{d\xi} F^{k_1^+}\left(\bar{x}_{k_1+1} + \theta_1^\mu(x_{k_1+1}^\mu - \bar{x}_{k_1+1}) \right) \times \right.$$

$$\left. \times \left(f_{k_1}(\bar{x}_{k_1}, \bar{u}_{k_1} + t^\mu p_{k_1}^\mu) - f_{k_1}(\bar{x}_{k_1}, \bar{u}_{k_1}) \right) \right)$$

$$= \lim_{t^\mu \downarrow 0} \left(\frac{d}{d\xi} F^{k_1^+}\left(\bar{x}_{k_1+1} + \theta_1^\mu(x_{k_1+1}^\mu - \bar{x}_{k_1+1}) \right) \frac{\partial}{\partial u} f_{k_1}(\bar{x}_{k_1}, \bar{u}_{k_1} + \theta_2^\mu t^\mu p_{k_1}^\mu) p_{k_1}^\mu \right)$$

$$= \frac{d}{d\xi} F^{k_1^+}(\bar{x}_{k_1+1}) \frac{\partial}{\partial u} f_{k_1}(\bar{x}_{k_1}, \bar{u}_{k_1}) p_{k_1} = \frac{d}{d\xi} F^{k_1^+}(\bar{x}_{k_1^+}) \frac{\partial}{\partial u} f_{k_1}(\bar{x}_{k_1}, \bar{u}_{k_1}) p_{k_1}$$

$$= \sum_{s=k_1^+}^{\infty} \frac{\partial}{\partial \xi} g_s(x_s^{k_1^+, \xi}, \bar{u}_s)|_{\xi = \bar{x}_{k_1^+}} \frac{\partial}{\partial u} f_{k_1}(\bar{x}_{k_1}, \bar{u}_{k_1}) p_{k_1}$$

$$= \psi_{k_1^+} \frac{\partial}{\partial u} f_{k_1}(\bar{x}_{k_1}, \bar{u}_{k_1}) p_{k_1} = \psi_{k_1+1} \frac{\partial}{\partial u} f_{k_1}(\bar{x}_{k_1}, \bar{u}_{k_1}) p_{k_1} \qquad (8)$$

Taking into account (5), (7), and (8), we obtain that

$$\lim_{t^\mu \downarrow 0} \frac{J(\bar{\mathbf{u}} + t^\mu \mathbf{p}^\mu) - J(\bar{\mathbf{u}})}{t^\mu}$$

$$= \lim_{t^\mu \downarrow 0} \frac{g_{k_1}(\bar{x}_{k_1}, \bar{u}_{k_1} + t^\mu p_{k_1}^\mu) - g_{k_1}(\bar{x}_{k_1}, \bar{u}_{k_1})}{t^\mu} + \lim_{t^\mu \downarrow 0} \frac{F^{k_1^+}(x_{k_1+1}^\mu) - F^{k_1^+}(\bar{x}_{k_1+1})}{t^\mu}$$

$$= \frac{\partial}{\partial u} g_{k_1}(\bar{x}_{k_1}, \bar{u}_{k_1}) p_{k_1} + \psi_{k_1+1} \frac{\partial}{\partial u} f_{k_1}(\bar{x}_{k_1}, \bar{u}_{k_1}) p_{k_1}$$

$$= \frac{\partial}{\partial u} H(\bar{x}_{k_1}, \bar{u}_{k_1}, \psi_{k_1+1}) p_{k_1}$$

This completes the proof. $\qquad\qquad \square$

Theorem 1. *Let Assumption 1 and Assumption 2 be satisfied, the pair $(\bar{\mathbf{x}}, \bar{\mathbf{u}})$ be a feasible process, the corresponding adjoint sequence $\psi = \{\psi_k\}_{k=1}^{\infty}$ be defined by (4), the criterion J be convex, and there exist constants $c, r > 0$ such that $J(\mathbf{u}) \leq c$ for all $\mathbf{u} \in \bar{\mathbf{B}}(\bar{\mathbf{u}}, r)$, and let*

$$\mathcal{H}_k(\bar{x}_k, \bar{u}_k, \psi_{k+1}) \leq \mathcal{H}_k(\bar{x}_k, u, \psi_{k+1}), \quad u \in U_k \text{ for every } k \in \mathbb{N}. \qquad (9)$$

Then, $(\bar{\mathbf{x}}, \bar{\mathbf{u}})$ ensures the minimum of the considered optimal control problem (2).

Proof. Let us fix an arbitrary $\varepsilon \in (0,1)$. Since J is bounded on $\bar{\mathbf{B}}(\bar{u}, r)$, the convexity of the criterion J implies the existence of a convex neighborhood Ω of the point $\bar{u}$ such that J is Lipschitz continuous on Ω with a Lipschitz constant $L > 0$.

Let the pair $(\mathbf{x}, \mathbf{u})$ be an arbitrary feasible process. We fix an arbitrary $\lambda_0 \in (0,1)$, which is sufficiently small so that the point $\bar{u} + \lambda(\mathbf{u} - \bar{u}) \in \Omega$ for each $\lambda \in (0, \lambda_0)$. Then the convexity of the criterion J implies that for each $\lambda \in (0, \lambda_0)$, we have that $J(\bar{u} + \lambda(\mathbf{u} - \bar{u})) \leq \lambda J(\mathbf{u}) + (1 - \lambda)J(\bar{u})$, hence

$$\frac{J(\bar{u} + \lambda(\mathbf{u} - \bar{u})) - J(\bar{u})}{\lambda} \leq J(\mathbf{u}) - J(\bar{u}). \tag{10}$$

Let us fix an arbitrary positive integer $\kappa \in \mathbb{N}$ and set

$$\mathbf{u}^\kappa := (u_0, u_1, ..., u_\kappa, 0, 0, ..., 0, ...) \text{ and } \bar{\mathbf{u}}^\kappa := (\bar{u}_0, \bar{u}_1, ..., \bar{u}_\kappa, 0, 0, ..., 0, ...).$$

Without loss of generality, we may think that κ is sufficiently large so that

$$\|\mathbf{u} - \mathbf{u}^\kappa\| < \varepsilon/2 \text{ and } \|\bar{u} - \bar{\mathbf{u}}^\kappa\| < \varepsilon/2. \tag{11}$$

Let $\mathbf{v}^\kappa := (u_0, u_1, ..., u_\kappa, \bar{u}_{\kappa+1}, \bar{u}_{\kappa+2}, ...)$. Because

$$\mathbf{v}^\kappa = \mathbf{u}^\kappa + \bar{u} - \bar{\mathbf{u}}^\kappa, \tag{12}$$

the sequence $\mathbf{v}^\kappa \in \ell^2(\mathbb{N})$, and hence $\mathbf{v}^\kappa$ is an admissible control. Moreover, by subtracting $\mathbf{u}$ from both sides of (12) and using (11), we have

$$\|\mathbf{v}^\kappa - \mathbf{u}\| \leq \|\mathbf{u} - \mathbf{u}^\kappa\| + \|\bar{u} - \bar{\mathbf{u}}^\kappa\| < \varepsilon.$$

Without loss of generality, we may think that $\bar{u} + \lambda(\mathbf{v}^\kappa - \bar{u}) \in \Omega$ for each $\lambda \in (0, \lambda_0)$. In accordance with the convexity of U_k, $k \in \{0, 1, ..., \kappa\}$, the inclusion

$$\bar{u}_k + \lambda(u_k - \bar{u}_k) \in U_k \text{ holds true for every } \lambda \in (0,1). \tag{13}$$

Moreover, the inclusion (13) implies that

$$\mathbf{p} := \mathbf{v}^\kappa - \bar{u} = (u_0 - \bar{u}_0, u_1 - \bar{u}_1, ..., u_\kappa - \bar{u}_\kappa, 0, 0, ..., 0, ...) \in T_U(\bar{u}) \tag{14}$$

with $\mathbf{U} := U_0 \times U_1 \times ... \times U_\kappa \times \{\bar{u}_{\kappa+1}\} \times \{\bar{u}_{\kappa+2}\} \times ...$. Clearly, we have that

$$J(\bar{u} + \lambda(\mathbf{u} - \bar{u})) - J(\bar{u}) = J(\bar{u} + \lambda(\mathbf{v}^\kappa - \bar{u})) - J(\bar{u})$$
$$+ J(\bar{u} + \lambda(\mathbf{u} - \bar{u})) - J(\bar{u} + \lambda(\mathbf{v}^\kappa - \bar{u}))$$

and

$$|J(\bar{u} + \lambda(\mathbf{u} - \bar{u})) - J(\bar{u} + \lambda(\mathbf{v}^\kappa - \bar{u}))| \leq L\lambda\|\mathbf{v}^\kappa - \mathbf{u}\| \leq L\lambda\varepsilon. \tag{15}$$

Taking (10) and (14) into consideration, we may conclude that

$$\frac{J(\bar{u} + \lambda\mathbf{p}) - J(\bar{u})}{\lambda} + \frac{J(\bar{u} + \lambda(\mathbf{u} - \bar{u})) - J(\bar{u} + \lambda(\mathbf{v}^\kappa - \bar{u}))}{\lambda} \leq J(\mathbf{u}) - J(\bar{u}).$$

Hence, $J(\mathbf{u}) - J(\bar{\mathbf{u}}) \geq$

$$\lim_{\lambda \downarrow 0} \frac{J(\bar{\mathbf{u}} + \lambda \mathbf{p}) - J(\bar{\mathbf{u}})}{\lambda} + \limsup_{\lambda \downarrow 0} \frac{J(\bar{\mathbf{u}} + \lambda(\mathbf{u} - \bar{\mathbf{u}})) - J(\bar{\mathbf{u}} + \lambda(\mathbf{v}^\kappa - \bar{\mathbf{u}}))}{\lambda}.$$

Using the inequality (15), we derive that

$$dJ(\bar{\mathbf{u}}; \mathbf{p}) - L\varepsilon \leq J(\mathbf{u}) - J(\bar{\mathbf{u}}). \tag{16}$$

Because the relations (9) mean that $\bar{u}_k$ provides the minimum of $\mathcal{H}_k(\bar{x}_k, \cdot, \psi_{k+1})$ on U_k, $k \in \{0, 1, ..., \kappa\}$, a first-order necessary condition for optimality gives us:

$$0 \leq \sum_{k=0}^{\kappa} \frac{\partial}{\partial u} \mathcal{H}_k(\bar{x}_k, \bar{u}_k, \psi_{k+1})(u_k - \bar{u}_k). \tag{17}$$

As a result, Proposition 1 and the inequalities (16) and (17) imply that

$$-L\varepsilon \leq \sum_{k=0}^{\kappa} \left(\frac{\partial}{\partial u} \mathcal{H}_k(\bar{x}_{k_k}, \bar{u}_k, \psi_{k+1}) \right)(u_k - \bar{u}_k) - L\varepsilon = dJ(\bar{\mathbf{u}}; \mathbf{p}) - L\varepsilon \leq J(\mathbf{u}) - J(\bar{\mathbf{u}}).$$

Finally, by letting $\varepsilon \downarrow 0$, we get $J(\mathbf{u}) \geq J(\bar{\mathbf{u}})$. This completes the proof. $\square$

Remark 3. We would like to point out that the convexity assumption in the formulation of Theorem 1 does not imply that all functions g_k, $k \in \mathbb{N}$, are convex. Indeed, let us consider the following simple illustrative example:

$$J(\mathbf{u}) = \left(\frac{1}{2}x_0^2 - \frac{1}{2}u_0^2 \right) + \left(\frac{1}{2}x_1^2 + \frac{1}{2}u_1^2 \right) \to \min,$$

subject to

$$x_{k+1} = x_k + 2u_k, \ x_0 \in \mathbb{R}, \ u_k \in [-1, 1], \ k \in \mathbb{N}.$$

Here,

$$g_0(x_0, u_0) = \frac{1}{2}x_0^2 - \frac{1}{2}u_0^2 \text{ and } g_1(x_1, u_1) = \frac{1}{2}x_1^2 + \frac{1}{2}u_1^2.$$

Clearly, g_0 is not a convex function. However, the criterion J is convex with respect to (u_0, u_1) because

$$J(\mathbf{u}) = \left(\frac{1}{2}x_0^2 - \frac{1}{2}u_0^2 \right) + \left(\frac{1}{2}(x_0 + 2u_0)^2 + \frac{1}{2}u_1^2 \right) =$$

$$= x_0^2 + 2x_0u_0 + \frac{3}{2}u_0^2 + \frac{1}{2}u_1^2.$$

References

1. Aseev, S.M., Krastanov, M.I., Veliov, V.M.: Optimality conditions for discrete-time optimal control on infinite horizon. Pure Appl. Funct. Anal. **2**, 395–409 (2017)
2. Aubin, J.P.: Lipschitz behavior of solutions to convex minimization problems. Math. Oper. Res. **9**(1), 87–111 (1984)
3. Aubin, J.P., Frankowska, H.: Set-Valued Analysis. Birkhauser, Boston (1990)
4. Carlson, D.A., Haurie, A., Leizarowitz, A.: Infinite Horizon Optimal Control. Springer, Berlin (1991). https://doi.org/10.1007/978-3-642-76755-5
5. Zaslavski, A.J.: Turnpike Phenomenon and Infinite Horizon Optimal Control. Springer Optimization and its Applications, Springer, New York (2014). https://doi.org/10.1007/978-3-319-08828-0

Stochastic Optimal Control
and Numerical Methods in Economics
and Finance

Computation of the Unknown Time-Dependent Volatility of American Options from Integral Observations

Slavi Georgiev[1,2]($\boxtimes$) (iD) and Lubin Vulkov[2]

[1] Department of Information Modeling, Institute of Mathematics and Informatics, Bulgarian Academy of Sciences, Acad. Georgi Bonchev Str., Block 8, 1113 Sofia, Bulgaria
sggeorgiev@math.bas.bg

[2] Department of Applied Mathematics and Statistics, Faculty of Natural Sciences and Education, Angel Kanchev University of Ruse, 8 Studentska Str., 7004 Ruse, Bulgaria
{sggeorgiev,lvalkov}@uni-ruse.bg

Abstract. In this paper we consider a model for American call price option with unknown time-dependent volatility. In order to determine the volatility, we impose an integral overdetermination for the option price. A front-fixing transformation is applied to the Black-Scholes equation and the overdetermination, which leads to a non-linear problem that involves homogeneous boundary conditions, independent of the free boundary. The derived explicit difference scheme is positive and monotone and its solution is easily decomposed with respect to the unknown volatility. Then, after solving the arisen subproblems, the volatility is calculated from the discretized overdetermination. Computational test examples show good performance of our method.

Keywords: Implied volatility · American option · Finite difference scheme · Inverse problem

1 Introduction

American-style options are heavily used from the investors. There is a great deal of research on the classical one-asset option, see e. g., [9,13] and references therein, for more literature. For various applications of American option modelling, different numerical methods are proposed, see e. g. [1–3,7,10–12,14,15].

Volatility identification in time-dependent Black-Scholes models, also referred to as *calibration* to market prices, has been investigated in recent years in a number of papers, see e. g., [2,4,5]. It is observed that the volatility reconstruction is *unstable* with respect to errors in data, [8,9].

The present work continues the investigation of previous results for European options with time-dependent volatility of the authors in [4,5] as well as the recovery of the time-dependent volatility in option pricing model [2].

I. Lirkov and S. Margenov (Eds.): LSSC 2023, LNCS 13952, pp. 195–202, 2024.
https://doi.org/10.1007/978-3-031-56208-2_19

In order to develop our inverse pricing model, we work with free boundary problem formulation. The front fixing transformation is often used for solution to the moving boundary value problems [6,9]. In this paper, we implement the front-fixing transformation for American put option in [1,10,12–14] to the American call.

Since there is only a limited data of market prices to recover the (time-dependent) volatility, the reconstruction problem is in general ill-posed. The reason is that small disturbances in market prices could result in large fluctuations in the volatility. This is challenging for the numerical treatment of the problem.

The paper is organized as follows. In Sect. 2 we formulate the pricing option problem and on it we perform the front-fixing transformation. Section 3 is devoted to the numerical solution of the direct problem by finite difference method (FDM). The main novelty of the study is discussed in Sect. 4, where it is solved the inverse problem. Numerical experiments are carried out to test the performance of the proposed algorithms.

2   The Pricing Option Direct and Inverse Problems

In this section we formulate the direct and inverse problems. Next, we introduce a new front-fixing transformation similar to the one used by [12] and others. This transformation translates the moving domain to a fixed one and changes the boundary conditions on the right boundary to homogeneous ones.

The American call option price model can be described [9] by the moving free boundary parabolic PDE

$$\frac{\partial C}{\partial \tau} = \frac{\sigma^2(\tau)}{2} S^2 \frac{\partial^2 C}{\partial S^2} + (r - q)S \frac{\partial C}{\partial S} - rC, \ \ 0 < S < s(t), \ \ 0 < \tau \le T, \quad (1)$$

$$C(0,\tau) = 0, \qquad C\big(s(\tau),\tau\big) = s(\tau) - K, \qquad 0 < \tau \le T, \quad (2)$$

$$\frac{\partial C}{\partial S}\big(s(\tau),\tau\big) = 1, \quad C(S,0) = \max(S - K,0), \qquad s(0) = \begin{cases} K & \text{if } r \le q, \\ \frac{r}{q}K & \text{if } r > q, \end{cases} \quad (3)$$

where $\tau = T - t$ denotes the time to maturity T, S is the asset price, $C(S,\tau)$ is the option price, $s(\tau)$ is the unknown early exercise boundary, $\sigma = \sigma(\tau)$ is the volatility of the asset, r is the risk-free interest rate, q is the continuous dividend yield and K is the strike price.

In general, there is no closed form analytical solution of problem (1)-(3), the so called *direct* problem. Thus, numerical methods must be used to determine the pair $\{C(S,\tau), s(\tau)\}$ when the volatility function $\sigma^2(\tau)$ is known.

Since the volatility parameter σ is difficult to be measured in the real markets, many authors use the theory of inverse problem in the PDEs to estimate the volatility of American option [8,9].

Here, in the free boundary formulation of the problem, in order to determine the unknown coefficient $\sigma^2(\tau)$, we impose the overdetermination condition

$$\frac{1}{s(t)} \int_0^{s(t)} C(S,t)\mathrm{d}S = \Psi(t), \quad (4)$$

$\Psi(t)$ being a given function.

The signification of the integral in (4) is the average option price. In practice, the number of traded options for underlying prices is finite, but very large: $0 = S_1 < S_2 < \ldots < S_I = S_{\max}$, $S_i - S_{i-1} = \Delta S_i$. In theoretical application, the number I could be considered tending to infinity, and thus the option prices could be averaged by the respective integral

$$\frac{1}{S_{\max}} \lim_{I \to \infty} \sum_{i=1}^{I} C(S_i, \tau) \Delta S_i = \frac{1}{S_{\max}} \int_0^{S_{\max}} C(S, \tau) dS.$$

By using the change of variable, we reduce the direct and inverse free boundary value problems into fixed boundary problems as follows. We let

$$x = \frac{S}{s(\tau)}, \quad C(S, \tau) = C(xs(\tau), \tau) \equiv c(x, \tau). \tag{5}$$

Under the transformation (5) the problem (1)–(4) can be written in the form

$$\frac{\partial c}{\partial \tau} = \frac{\sigma^2(\tau)}{2} x^2 \frac{\partial^2 c}{\partial x^2} + (r - q)x \frac{\partial c}{\partial x} + \frac{\dot{s}(\tau)}{s(\tau)} x \frac{\partial c}{\partial x} - rc, \quad 0 < x \le 1, \ 0 < \tau \le T, \tag{6}$$

with new boundary and initial conditions

$$c(0, \tau) = 0, \qquad \frac{\partial c}{\partial x}(0, \tau) = 0, \qquad 0 < \tau \le T, \tag{7}$$

$$c(1, \tau) = s(\tau) - K, \qquad \frac{\partial c}{\partial x}(1, \tau) = s(\tau), \qquad 0 < \tau \le T, \tag{8}$$

$$c(x, 0) = \max\left(xs(0) - K, 0\right), \tag{9}$$

$$\int_0^1 c(x, \tau) dx = \Psi(\tau). \tag{10}$$

3 Numerical Solution of the Direct Problem

Let us introduce the equidistant space and time meshes $1 = Ih$, $T = J\Delta\tau$, where $x_i = hi$, $i = 0, 1, \ldots, I$ and $\tau^j = j\Delta\tau$, $j = 0, 1, \ldots, J$.

The approximate value of the American option at space point x_i and time node τ^j is $c(x, \tau^j) \approx c_i^j$, and the approximate value of the free boundary at time node τ^j is $s(\tau^j) \approx s^j$. The fully discrete version of (9) with first-order in time and second-order in space discretization explicit scheme is as follows:

$$\frac{c_i^{j+1} - c_i^j}{\Delta\tau} = \frac{(\sigma^j)^2}{2} x_i^2 \frac{c_{i+1}^j - 2c_i^j + c_{i-1}^j}{h^2}$$

$$+ x_i \left(r - q + \frac{s^{j+1} - s^j}{\Delta\tau s^j}\right) \frac{c_{i+1}^j - c_{i-1}^j}{2h} - rc_i^j, \tag{11}$$

with the respective boundary conditions

$$c_0^j = 0, \qquad \frac{c_1^j - c_0^j}{h} = 0, \qquad c_I^j = s^j - K. \tag{12}$$

From (12) it directly follows that $c_1^j = 0$. Then, substituting $i = 1$ in (11), after some calculations we get that

$$s^{j+1} = d^j s^j, \qquad d^j = -\Delta\tau((\sigma^j)^2 + r - q) + 1. \tag{13}$$

In our setting, we are able to approximately calculate the free boundary s^j, $j = 0, 1, \ldots, J$ without computing the option price (11), but only knowing the parameters σ, r, q. The approximation is also independent from the space step h, in contrast to [1].

The Eq. (11) for the internal spatial nodes could be written in the form

$$c_i^{j+1} = A_i c_{i-1}^j + C_i c_i^j + B_i c_{i+1}^j,$$

$$A_i = \left(\frac{(\sigma^j)^2 x_i^2}{2h^2} - \frac{x_i}{2h} \left(r - q + \frac{s^{j+1} - s^j}{\Delta\tau s^j} \right) \right) \Delta\tau,$$

$$C_i = - \left(\frac{(\sigma^j)^2 x_i^2}{h^2} - r \right) \Delta\tau + 1,$$

$$B_i = \left(\frac{(\sigma^j)^2 x_i^2}{2h^2} + \frac{x_i}{2h} \left(r - q + \frac{s^{j+1} - s^j}{\Delta\tau s^j} \right) \right) \Delta\tau.$$

4   Numerical Solution of the Inverse Problem

One of the advantages of the explicit scheme is that the inverse problem of finding the triplet $(c(x, \tau), s(\tau), \sigma^2(\tau))$ is linear. So, we proceed directly with the decomposition as in [4, 5]

$$c^{j+1} = z^{j+1} + (\sigma^j)^2 w^{j+1}. \tag{14}$$

However, if $(\sigma^j)^2$ is unknown, so is the free boundary at the new temporal layer s^{j+1}. This makes it necessary to decompose the free boundary with (14), and substituting the decomposition in (11) yields

$$\frac{z^{j+1} + (\sigma^j)^2 w^{j+1} - c_i^j}{\Delta\tau} = \frac{(\sigma^j)^2}{2} x_i^2 \frac{c_{i+1}^j - 2c_i^j + c_{i-1}^j}{h^2} + x_i(\sigma^j)^2 \frac{c_{i+1}^j - c_{i-1}^j}{2h} - rc_i^j, \tag{15}$$

which in turn results in the explicit schemes for the auxiliary functions z^{j+1} and w^{j+1}:

$$z_i^{j+1} = (-r\Delta\tau + 1)c_i^j \tag{16}$$

and

$$w_i^{j+1} = \left(\frac{x_i^2}{2h^2} - \frac{x_i}{2h}\right)\triangle\tau c_{i-1}^j + \left(-\frac{x_i^2}{h^2}\right)\triangle\tau c_i^j + \left(\frac{x_i^2}{2h^2} + \frac{x_i}{2h}\right)\triangle\tau c_{i+1}^j. \quad (17)$$

Both left boundary conditions are zero:

$$z_0^{j+1} = w_0^{j+1} = 0, \quad (18)$$

while one needs to decompose the right boundary condition to obtain

$$z_I^{j+1} = s^j\big(-(r-q)\triangle\tau + 1\big) - K, \qquad w_I^{j+1} = -s^j\triangle\tau. \quad (19)$$

If we place the decomposition (14) into the integral measurement (10), we obtain

$$(\sigma^j)^2 = \frac{\Psi(\tau^j) - \sum\limits_{i=0}^{I} z_i^{j+1}h}{\sum\limits_{i=0}^{I} w_i^{j+1}h}. \quad (20)$$

Thus it suffices to solve both the auxiliary systems for z^{j+1} and w^{j+1}, respectively, to obtain the triplet $\big(c^{j+1}, s^{j+1}, (\sigma^2)^j\big)$.

The computations show, however, that the accuracy of reconstructing $s(\tau)$ is not high. So, although not necessary, we perform a number of iterations of the time-stepping on the new layer. We noticed that iterating the aforementioned scheme for z^{j+1} and w^{j+1} is not efficient, so after one iteration of (16)–(19), we suggest performing a couple of iterations as follows. Please note that after the first iteration, we have an approximation of the s^{j+1}, so we do not need to decompose it:

$$z_i^{j+1} = -\left(\frac{x_i}{2h}\left(r - q + \frac{s^{j+1} - s^j}{\triangle\tau s^j}\right)\right)\triangle\tau c_{i-1}^j + (-r\triangle\tau + 1)c_i^j$$

$$+ \left(\frac{x_i}{2h}\left(r - q + \frac{s^{j+1} - s^j}{\triangle\tau s^j}\right)\right)\triangle\tau c_{i+1}^j$$

and

$$w_i^{j+1} = \left(\frac{x_i^2}{2h^2}\right)\triangle\tau c_{i-1}^j - \left(\frac{x_i^2}{h^2}\right)\triangle\tau c_i^j + \left(\frac{x_i^2}{2h^2}\right)\triangle\tau c_{i+1}^j$$

with the same boundary conditions (18), (19).

5   Computational Result

Let us begin our simulations with solving the direct problem (6)–(9) with $r = 0.02$, $q = 0.03$, $T = 1$ and $K = 100$. We test with two types of volatility – a linear one and a nonlinear one:

$$\sigma(\tau) = \frac{3 + \tau}{10} \quad (21)$$

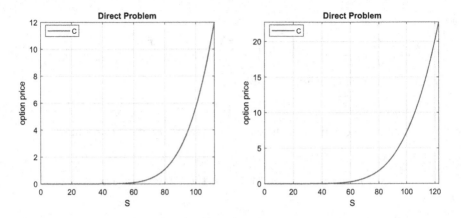

Fig. 1. The solution to the direct problem with volatility (21) (left) and (22) (right).

and

$$\sigma(\tau) = 0.5 + 0.04 \log(\tau + 0.01). \tag{22}$$

The results are given on Fig. 1.

Now we proceed to the simulation of the inverse problem (6)–(10). All the settings stay the same, except that the volatility is unknown and we are provided with measurements $\Psi(\tau)$ (10). We follow the algorithm, described in the previous section.

The tests with the linear volatility (21) are plotted on Fig. 2.

The volatility is exactly recovered and so is the reconstructed option premium. This is also the case with the nonlinear volatility (22) (Fig. 3).

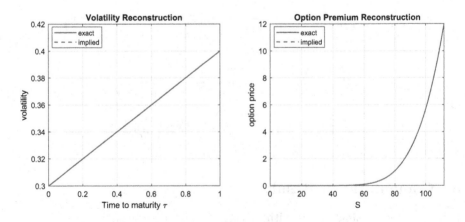

Fig. 2. The solution to the inverse problem with volatility (21).

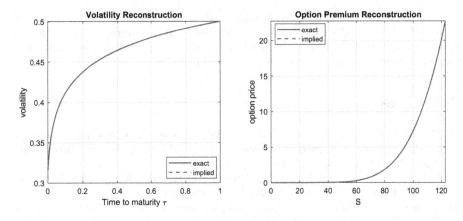

Fig. 3. The solution to the inverse problem with volatility (22).

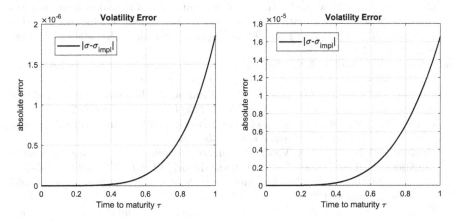

Fig. 4. The absolute errors in the recovery of the volatility (21) (left) and (22) (right).

The absolute errors in the recovered volatilities are given on Fig. 4. They are not higher that 2e–6 and 2e–5, respectively.

6   Conclusions and Future Work

In this paper we proposed an algorithm for reconstruction the time-dependent volatility from integral measurements of American call options. The volatility is identified on each temporal layer. The advantages of the method are that it is easy to implement and computationally non-intensive. The computational simulations demonstrate the accuracy of the recovery of both volatility and option premium. This study could be extended to option with multiple assets as well as with path-dependent or barrier features.

Acknowledgements. This study is supported by the Bulgarian National Science Fund under Project KP-06-N 62/3 "Numerical methods for inverse problems in evolutionary differential equations with applications to mathematical finance, heat-mass transfer, honeybee population and environmental pollution" from 2022.

References

1. Company, R., Egorova, V. N., Jódar, L.: Solving American option pricing models by the front fixing method: numerical analysis and computing. In: Abstract and Applied Analysis, Article ID 146745 (2014)
2. Deng, Z.C., Hon, Y.C., Isakov, V.: Recovery of the time-dependent volatility in option pricing model. Inv. Probl. **32**(11), 115010 (2016)
3. Ehrhardt, M., Mickens, R.: A fast, stable and accurate numerical method for the Black-Scholes equation of American options. Int. J. Theor. Appl. Finan. **11**(5), 471–501 (2008)
4. Georgiev, S.G., Vulkov, L.G.: Numerical determination of time-dependent implied volatility by a point observation. In: Georgiev, I., Kostadinov, H., Lilkova, E. (eds.) BGSIAM 2018. SCI, vol. 961, pp. 99–109. Springer, Cham (2021). https://doi.org/10.1007/978-3-030-71616-5_11
5. Georgiev, S.G., Vulkov, L.G.: Fast reconstruction of time-dependent market volatility for European options. Comp. Appl. Math. **40**, 30 (2021)
6. Ghanmi, C., Aouadi, S.M., Triki, F.: Identification of a boundary influx condition in a one-phase Stefan problem. Appl. Anal. **101**(18), 6573–6595 (2022)
7. Gyulov, T.B., Valkov, R.L.: American option pricing problem transformed on finite interval. Int. J. Comput. Math. **93**, 821–836 (2016)
8. Hasanoğlu, A.H., Romanov, V.G.: Introduction to Inverse Problems for Differential Equations. Springer, Cham (2021). https://doi.org/10.1007/978-3-319-62797-7
9. Jiang, L.: Mathematical Modeling and Methods of Option Pricing, p. 344. World Scientific, Singapore (2005)
10. Kandilarov, J.D., Valkov, R.L.: A numerical approach for the american call option pricing model. In: Dimov, I., Dimova, S., Kolkovska, N. (eds.) NMA 2010. LNCS, vol. 6046, pp. 453–460. Springer, Heidelberg (2011). https://doi.org/10.1007/978-3-642-18466-6_54
11. Kangro, R., Nicolaides, R.: Far field boundary conditions for Black-Scholes equations. SIAM J. Numer. Anal. **38**(4), 1357–1368 (2000)
12. Nielsen, B.F., Skavhaug, O., Tveito, A.: Penalty and front-fixing methods for the numerical solution of American option problems. J. Comput. Finan. **5**, 69–97 (2002)
13. Sevčovič, D.: Analysis of the free boundary for the pricing of an American call option. Eur. J. Appl. Math. **12**(1), 25–37 (2001)
14. Wu, L., Kwok, Y.K.: A front-fixing method for the valuation of American option. J. Finan. Eng. **6**(2), 83–97 (1997)
15. Zhang, K., Teo, K.L.: A penalty-based method from reconstructing smooth local volatility surface from American options. J. Ind. Manag. Optim. **11**(2), 631–644 (2015)

Tensor Methods for Big Data Analytics and Low-Rank Approximations of PDEs Solutions

Efficient Solution of Stochastic Galerkin Matrix Equations via Reduced Basis and Tensor Train Approximation

Michal Béreš[1,2][⊠] [iD]

[1] Institute of Geonics, Czech Academy of Sciences, Ostrava, Czech Republic
michal.beres@ugn.cas.cz
[2] Department of Applied Mathematics, FEECS, VŠB-TUO, Ostrava, Czech Republic

Abstract. This contribution focuses on the development of a computational method to efficiently solve matrix equations arising from stochastic Galerkin (SG) discretization of steady Darcy flow problems with uncertain and separable permeability fields. The proposed method consists of a two-step solution process. Firstly, we construct a reduced basis for the finite element portion of the discretization using the Monte Carlo (MC) method. We consider various sampling techniques for the MC method. Secondly, we use a tensor polynomial basis to handle the stochastic aspect of the problem and employ a tensor-train (TT) approximation to approximate the overall solution of the reduced SG system. To enhance the convergence of the TT approximation, we use an implicitly preconditioned system with a Kronecker-type preconditioner. Moreover, we also develop low-cost error indicators to assess the accuracy of both the reduced basis and the final solution of the reduced system.

Keywords: stochastic Galerkin method · reduced basis · tensor train approximation

1 Introduction

This paper provides a concise overview of the solution to the stationary Darcy flow problem, which involves uncertain hydraulic conductivity, using the Stochastic Galerkin (SG) method.

The SG method, initially introduced as the spectral Galerkin method in [5], offers a strategy for solving partial differential equations that incorporate additional dimensions in the form of parameterized input data. These parameters, understood as random variables, are assumed to have a specified measure. The SG method's essence lies in its ability to solve the problem in its entirety, including the additional dimensions, through the Galerkin approximation of the weak solution. The polynomial chaos expansion, an orthogonal polynomial basis on a specific measure, is often used as the basis for these additional dimensions. The SG method's strengths lie in its robust theoretical foundation, which stems from its Galerkin approximation. However, it also presents challenges due to the complexity and size of the resulting system of equations.

© The Author(s), under exclusive license to Springer Nature Switzerland AG 2024
I. Lirkov and S. Margenov (Eds.): LSSC 2023, LNCS 13952, pp. 205–214, 2024.
https://doi.org/10.1007/978-3-031-56208-2_20

The primary focus of this work is to efficiently solve the resulting system of equations in the form of matrix equations. This is done by combining the reduced basis approach [9] with the tensor-train approximation [6] of the resulting reduced system. The main contributions of this paper are the combination of those two approaches together with Monte Carlo sampling approach to the construction of the reduced basis, and the cheap error estimates computed from used Monte Carlo samples.

This paper builds upon the author's previous research, as documented in [1], and incorporates findings discussed in the author's recently submitted doctoral thesis [2].

1.1   Problem Setting

We begin by describing the problem setting. Let us consider a physical domain $\mathcal{D}$ and a random vector $\boldsymbol{Z}$, which consists of M independent standard normal random variables, defined on sample space Ω. We assume that the hydraulic conductivity field is a function of points within the domain $\mathcal{D}$ as well as the random vector $\boldsymbol{Z}$, and is expressed in the following form

$$k\left(x, \boldsymbol{Z}\right) = \sum_{m=1}^{M} \underbrace{\chi_{\mathcal{D}_m}\left(x\right)}_{k_m^D(x)} \underbrace{\exp\left(\sigma_m Z_m + \mu_m\right)}_{k_m^S(\boldsymbol{Z})} = \sum_{m=1}^{M} k_m^D\left(x\right) k_m^S\left(\boldsymbol{Z}\right). \tag{1}$$

In other words, we consider a piecewise constant function whose value is constant within each of the M subdomains $\mathcal{D}_m$, where the value is determined by the m-th element of the random vector $\boldsymbol{Z}$. The steady Darcy flow model problem can be expressed as the following partial differential equation

$$\begin{cases} -\mathrm{div}_x\left(k\left(x, \boldsymbol{Z}\right) \nabla_x u\left(x, \boldsymbol{Z}\right)\right) = f\left(x\right), & \forall x \in \mathcal{D}, \boldsymbol{Z} \in \mathbb{R}^M \\ u\left(x, \boldsymbol{Z}\right) = u_0\left(x\right), & \forall x \in \Gamma_D, \boldsymbol{Z} \in \mathbb{R}^M \\ -k\left(x, \boldsymbol{Z}\right) \frac{\partial u(x, \boldsymbol{Z})}{\partial n(x)} = g\left(x\right), & \forall x \in \Gamma_N, \boldsymbol{Z} \in \mathbb{R}^M \end{cases} \tag{2}$$

To test the use of the SG method for stated problem, we select the decomposition into subdomains using thresholding of the Gaussian random field realization, see Fig. 1.

Fig. 1. Illustration of decomposition into subdomains

2 Stochastic Galerkin Matrix Equations

In the process of creating the SG matrix equations, we begin with the weak form of the problem (2), which is expressed as

$$a\left(u_H, v\right) = b\left(v\right), \quad \forall v \in L^2\left(\Omega, H^1_{0,\Gamma_D}\left(\mathcal{D}\right)\right),$$

$$a\left(u_H, v\right) = \int_{\mathbb{R}^M} \int_{\mathcal{D}} k\left(x, \mathbf{Z}\right) \nabla_x u_H\left(x, \mathbf{Z}\right) \cdot \nabla_x v\left(x, \mathbf{Z}\right) dx \, d\mathbf{F}\mathbf{Z},$$

$$b\left(v\right) = \int_{\mathbb{R}^M} \int_{\mathcal{D}} f\left(x\right) v\left(x, \mathbf{Z}\right) dx \, d\mathbf{F}\mathbf{Z} - \int_{\mathbb{R}^M} \int_{\Gamma_N} g\left(x\right) v\left(x, \mathbf{Z}\right) dx \, d\mathbf{F}\mathbf{Z}$$

$$- \int_{\mathbb{R}^M} \int_{\mathcal{D}} k\left(x, \mathbf{Z}\right) \nabla_x u_0\left(x\right) \cdot \nabla_x v\left(x, \mathbf{Z}\right) dx \, d\mathbf{F}\mathbf{Z}.$$

The homogeneous part of the solution, denoted as u_H, belongs to $L^2(\Omega, H^1_{0,\Gamma_D}(\mathcal{D}))$, which is isometrically isomorphic to $H^1_{0,\Gamma_D}(\mathcal{D}) \otimes L^2(\Omega)$. We select the test space to have the same tensor structure, denoted as $V_{h,K} := V_h \otimes V_K$, where the discretization of $H^1_{0,\Gamma_D}(\mathcal{D})$ is finite elements, and the discretization of $L^2(\Omega)$ is in the form of polynomials

$$V_h = \{\varphi_1\left(x\right), \ldots, \varphi_{N_D}\left(x\right)\} \subset H^1_{0,\Gamma_D}\left(\mathcal{D}\right), V_K = \{\psi_1\left(\omega\right), \ldots, \psi_{N_S}\left(\omega\right)\} \subset L^2\left(\Omega\right).$$

The dimension of $V_{h,K}$ is $N_D N_S$ with the basis

$$\xi_{i,j}\left(x, \omega\right) = \varphi_i\left(x\right) \psi_j\left(\omega\right), \quad \forall i = 1, \ldots, N_D, \, j = 1, \ldots, N_S.$$

The separable format of the input data (1), in conjunction with the tensor form of $V_{h,K}$, allows the construction of the matrix in a compressed format. All terms inside the double integral are functions of either x or $\mathbf{Z}$, permitting the division of the double integral into the product of two single ones. The resultant system of equations is expressed as follows:

$$\mathbb{A}\bar{u} = \bar{b}, \quad \mathbb{A} = \sum_{m=1}^M G_m \otimes K_m, \bar{b} = \sum_{m=1}^{M_b} \bar{g}_m \otimes \bar{k}_m,$$

$$\left(K_m\right)_{il} = \int_{\mathcal{D}} k_m^D\left(x\right) \nabla \varphi_i\left(x\right) \cdot \nabla \varphi_l\left(x\right) dx,$$

$$\left(G_m\right)_{jn} = \int_{\mathbb{R}^M} k_m^S\left(\mathbf{Z}\right) \psi_j\left(\mathbf{Z}\right) \psi_n\left(\mathbf{Z}\right) d\mathbf{F}\mathbf{Z}.$$

We simplify the right-hand side as a sum over $M_b = M + 2$ terms, where M terms are for Dirichlet boundary and one each for forcing term and Neumann boundary. The vectors $\bar{g}_m$ and $\bar{k}_m$ can be assembled in a similar fashion as G_m and K_m.

The system can be viewed as a matrix equation by reshaping $\bar{u}$ into an $N_D \times N_S$ matrix $\boldsymbol{u}$

$$\sum_{m=1}^{M} K_m \boldsymbol{u} G_m^T = \sum_{m=1}^{M_b} \overline{k}_m \overline{g}_m^T. \tag{3}$$

3   Solving the Stochastic Galerkin Matrix Equations

The solution of the SG matrix equations in (3) is challenging due to the large number of degrees of freedom. To address this, we first reduce the test space using the reduced basis (RB) method, see for reference [3,9]. We then compare the performance of conjugate gradients (CG) with the complete polynomial basis and tensor-train (TT) approximation with the tensor polynomial basis. Kronecker type preconditioners, as described in [10], will be used in the comparisons.

The RB method aims to reduce the number of basis functions while maintaining the same approximation properties. In the SG method, creating a reduced basis W of V_h is sensible, as V_h is the larger part of the basis, and we have the necessary tools to form a meaningful subspace from it. The resulting SG test space takes the form of $V_{h,K} \approx W \otimes V_K$, where W is the reduced basis of V_h.

The reduced basis must satisfy all the necessary conditions for the discretized system to be well-posed, such as the discrete inf-sup condition. For our elliptic problem, we can choose any set of linearly independent basis functions for W, and the resulting system will be valid.

$$\sum_{m=1}^{M} W^T K_m W \boldsymbol{y} G_m^T = \sum_{m=1}^{M_b} W^T \overline{k}_m \overline{g}_m^T, \quad \boldsymbol{u} \approx \tilde{\boldsymbol{u}} = W \boldsymbol{y}. \tag{4}$$

The approximation error of the reduced basis W in the context of the SG system can be expressed as the residual with respect to the original system

$$R = \sum_{m=1}^{M} K_m W \boldsymbol{y} G_m^T - \sum_{m=1}^{M_b} \overline{k}_m \overline{g}_m^T. \tag{5}$$

4   Construction of the RB via Monte Carlo Sampling

The most challenging task is constructing the reduced basis itself, which we do using the Monte Carlo (MC) method. The MC approach involves iteratively refining the reduced basis. Let W_l denote the reduced basis at iteration l, where $W_0 = \emptyset$. The iterative construction can be summarized in the following steps:

1. draw N_{MC} samples $Z_1, \ldots, Z_{N_{MC}}$ of random vector $\boldsymbol{Z}$
2. for every sample Z_j assemble and solve the reduced system of deterministic counterpart

$$W_l^T A_j W_l \tilde{u}_j = W_l^T b_j$$

3. compute indicators for selecting samples based on the probability density function (pdf) of Z and the residual of reduced solutions $\widetilde{u}_j$

$$f_Z\left(Z_j\right)\left\|A_jW_l\widetilde{u}_j - b_j\right\|^2$$

4. select P (for simplicity, we use $P = 1$) highest values of indicators and compute solutions at corresponding samples Z_j

$$A_j u_j = b_j$$

5. use the collected solutions to expand the RB W_l and check if the expanded reduced basis is sufficient

Computing the reduced solutions and their residuals at samples Z_j can be quite expensive. To avoid selecting samples that do not provide enough new information, we propose modifying Step 1 of the process by using the Metropolis-Hastings algorithm to sample from a modified pdf that avoids already generated samples.

$$\widetilde{f}_l\left(Z\right) \propto f\left(Z\right)\min_{i=1,\ldots,l} w_i\left(Z\right),\ w_i\left(Z\right) = \left[1 - \exp\left(-\left\|Z - X_i\right\|^2_{\Sigma^{-1}}/2\right)\right]^\beta.$$

We choose the parameter Σ as the scaled covariance matrix of Z and the parameter $\beta > 1$.

4.1 Error Indicators for Greedy Monte Carlo Approach

We define the "true error" as the $L^2\left(\Omega, H^1\right)$ norm of the difference between the SG solution and the pathwise finite element (FE) solution. To test the accuracy of our method, we estimate the error in the stochastic space using the MC method. However, it is important to note that this approach is very costly, so we seek an efficient approximation of the "true error".

When constructing the RB using the MC approach, we rely on snapshots of deterministic counterparts. These snapshots enable us to create low-cost yet precise error indicators, which we can compare to the residual of the RB (5). We will use the snapshots (samples) to create two error indicators:

- The first error indicator, denoted as ε_1, uses all of the generated samples. It is important to note that ε_1 does not directly approximate the $L^2\left(\Omega, H^1\right)$ error since it compares the SG solution to the samples in its RB. However, it checks whether the polynomial basis used in the SG solution is rich enough to fully utilize the information in the provided RB. We will use this error indicator for adaptive refining the polynomial basis.
- The second error indicator will be computed during the enrichment phase of the RB. Specifically, we will compute this error indicator after we have computed the FE solutions of the samples selected for enrichment but before they are added to the RB. At this point, we have the FE solution of the deterministic counterpart at these samples and also the reduced solution with

respect to the old RB. The squares of the H^1 errors between the reduced solution and the FE solution of the deterministic counterpart will provide us with this error indicator, denoted as ε_2. It is important to note that ε_2 does not use the SG solution and, therefore, reflects only the quality of the RB.

By using the two error indicators together, we can solve the SG problem in two phases. Firstly, we build a sufficiently accurate reduced basis (RB) using ε_2 without the need to solve the reduced problem (4). Then, we gradually enhance the polynomial basis, solve the reduced problem, and check the precision using ε_1.

4.2   Numerical Tests

4.2.1   Construction of the RB
Figure 2 demonstrates the efficiency of the MC approach to the construction of the RB on a series of problems with an increasing number of random variables and $\mu_m = 0, \sigma_m = 0.3$. We compare two variants: $M1$, which uses crude MC sampling with $N_{MC} = 1$, and $A100$, which uses altered probability density function (pdf) and $N_{MC} = 100$. Additionally, we compare the results to the optimal case of RB constructed from the singular value decomposition of the computed full solution and point selection using Smolyak nested sparse grids (see [8]). We measure the quality of RB in terms of the "true error" $L^2\left(\Omega, H^1\right)$, which is approximated using 1000 MC samples. We use a polynomial basis that is more than sufficient so that any error is only due to the insufficiency of the RB.

4.2.2   Error Indicators
We test the MC approach in a two-phase manner, this is both computationally cheaper and allows for an adaptive selection of the maximum polynomial degree. We demonstrate the behaviour of error indicators with setting $M = 4, \sigma = 0.3, \mu = 0$, where the $L^2\left(\Omega, H^1\right)$ error is computed with complete polynomials up to degree $N = 25$ in the first phase. We use the

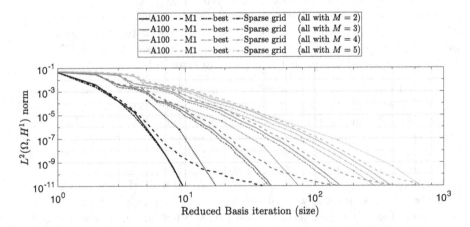

Fig. 2. Efficiency of reduced basis construction using different N_{MC}, crude sampling and sampling using altered pdf, and comparison with optimal RB and sparse grid

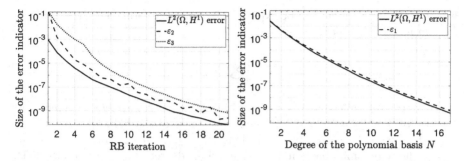

Fig. 3. Two phase solution of SGM with adaptive maximum polynomial degree selection. Left: phase 1 - construction of RB, Right: phase 2 - selection of maximum polynomial degree

MC approach with $N_{MC} = 40$ and $P = 8$. As the error indicator ε_2 tends to be noisy, we also add ε_3 as a moving geometric mean with a window length of 5.

The results of both phases can be found in Fig. 3. We can see that both error indicators ε_2 and ε_3 are rather conservative, which leads to more RB iterations than necessary but is much safer than the opposite case. The results of the second phase shows that ε_1 is very close to the "true error" while being slightly higher. When ε_1 reaches the desired precision, the problem is solved. If ε_1 stagnates, it would be necessary to go back to the first phase and expand the RB. This behavior was not observed in any numerical experiments, but it cannot be completely ruled out.

For the second phase, we use an adaptive refinement of the reduced solution via initial guess and adaptive precision. We start with a CG precision of 10^{-3} and use $\varepsilon_1 10^{-3}$ as the precision for the next CG solution. This makes the solution as cheap as the corresponding full solution of the reduced problem with the target polynomial basis.

5   Solution of Reduced System via TT Approximation

In this section, we aim to solve the problem in its multidimensional form, specifically with a spatial dimension and M dimensions of $\mathbf{Z}$. This requires the following:

- the random vector $\mathbf{Z}$ comprises of independent random variables,
- the use of tensor polynomial basis, and
- the input data must in a separable form (including dimensions of $\mathbf{Z}$), i.e.

$$k(x, \mathbf{Z}) = \sum_{m=1}^{M} k_m^D(x) \prod_{n=1}^{M} k_{m,n}^S(Z_n) = \sum_{m=1}^{M} \underbrace{\chi_{\mathcal{D}_m}(x)}_{k_m^D(x)} \underbrace{\exp(\sigma_m Z_m + \mu_m)}_{\prod_{n=1}^{M} k_{m,n}^S(Z_n)}.$$

Subsequently, the system can be expressed in canonical form. Here, the matrices G_m can now be represented as:

$$G_m = \bigotimes_{n=1}^{M} G_{m,n}.$$

Then, the solution sought is a tensor $U \in \mathbb{R}^{d_1,\dots,d_{M+1}}$ with $M+1$ dimensions. The size of dimension d_1 is determined by the size of the reduced basis, as our aim is to solve the reduced problem. The sizes $d_2,\dots,d_{M+1}$ correspond to the sizes of the polynomial bases in each dimension (degree of the polynomial + 1). We will seek for an approximation of U in a tensor-train (TT) format.

The TT decomposition of a tensor $\mathbf{A} \in \mathbb{R}^{d_1 \times d_2 \times \dots \times d_N}$ is a series of N 3-dimensional tensor cores, $C^{(n)} \in \mathbb{R}^{r_{n-1} \times d_n \times r_n}$. Each element $i_1, i_2, \dots, i_N$ of the tensor $\mathbf{A}$ is derived from a succession of matrix multiplications as follows:

$$\mathbf{A}_{i_1,i_2,\dots,i_N} = C^{(1)}(i_1)C^{(2)}(i_2)\dots C^{(N)}(i_N),$$

where $C^{(n)}(i_n) \in \mathbb{R}^{r_{n-1} \times r_n}$ is the i_n-th slice of $C^{(n)}$, and $r_0 = r_N = 1$. The rank sizes r_n dictate the compression of the original tensor.

Our objective is the TT approximation of the solution U. For a comprehensive understanding of the TT approximation of a tensor, please refer to [6]. We use the Alternating Minimal Energy method (AME) to acquire the TT approximation of the solution for linear equation systems, as proposed in [4].

In the context of the solution to the reduced problem, all segments of the system matrix's canonical form are minimal. The AME method benefits from better conditioned systems, hence we implement an implicit preconditioning approach. Specifically, we apply mean value preconditioning for the deterministic part and the sum of contributions in the canonical format at each dimension for the stochastic part.

All calculations are carried out using the TT-toolbox [7]. In particular, the `amen_solve2` function that computes TT approximation of the solution for the given system up to a specified precision, measured as residual.

5.1 Numerical Testing

We compared the TT approximation with the iterative CG solution using complete polynomials, analyzing the resulting computational time (obtained via single-threaded computation on an i7-8550U processor) and memory size of the solution.

We utilized three problem settings: $S1$: $M = 10, \sigma = 0.3, \mu = 0$; $S2$: $M = 10, \sigma = (0.1,\dots,0.3,0.3,\dots,0.1)$, $\mu = (0,\dots,-5,-5,\dots,0)$; and $S3$: $M = 10, \sigma = (0.01,\dots,0.3,0.3,\dots,0.01)$, $\mu = (0,\dots,-10,-10,\dots,0)$. Settings $S2$ and $S3$ differ in their 5th and 6th underlying normal distributions. For $S2$ and $S3$, we anticipate efficient low-rank approximation as the solution in these instances is predominantly influenced by just 2 out of the 10 input random variables. Hence, we expect that an accurate Tensor Train approximation with cores of low ranks should be available.

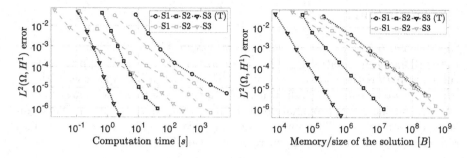

Fig. 4. Comparison of TT approximation and CG solution using complete polynomial basis. Left: computational time, Right: memory size of the solution

Table 1. Time spent on each part of the solution using RB and TT approximation

	Assembly and preparation	Construction of RB	TT approximation
$S1$	3.2 s	5121.1 s	6998.0 s
$S2$	3.2 s	1130.5 s	40.6 s
$S3$	3.2 s	150.3 s	2.3 s

Table 2. Ranks of the TT approximation of the solution

	tensor dimensions		ranks of the TT cores											
	d_1	$d_2,\ldots,d_{11}$	r_0	r_1	r_2	r_3	r_4	r_5	r_6	r_7	r_8	r_9	r_{10}	r_{11}
$S1$	2391	9	1	1104	1042	875	692	496	320	182	86	31	8	1
$S2$	531	9	1	251	236	201	167	131	76	33	21	11	4	1
$S3$	71	9	1	33	31	29	27	24	16	5	4	3	2	1

The findings are presented in Fig. 4 and Tables 2 and 1. The cited time is the cumulative duration required to arrive at the solution. The TT approximation showed its capability to leverage the specific problem structure, demanding significantly less computational and memory resources for $S3$ compared to $S2$, and, consequently, $S1$. However, the base computational expense is noticeably higher for the TT approximation due to the costly compression. Table 2 displays the dimensions and ranks of the resulting TT approximation. It's important to note that the first dimension d_1 aligns with the size of the RB and varies significantly across the problems. Table 1 indicates the time expended on each part of the solution. The time spent on the construction of the RB is largely driven by the number of deterministic system solutions required (approximately 10^6 degrees of freedom, solved using algebraic multigrid - $\approx 1.9\,$s per system). Lastly, the implicit preconditioning speeds up the TT approximation compared to the no preconditioning about $3.9\times$.

6 Conclusions

This paper presented a novel approach for solving high-dimensional stochastic partial differential equations based on the RB method and TT approximation. The main part is the construction of the RB using the MC approach with the construction of cheap error indicators. Additionally, the paper compares the CG solution and the TT approximation of the resultant reduced problem. The comparison illustrates that TT approximation significantly outperforms the standard iterative approach in instances where compression is feasible.

Acknowledgements. The work has been supported by the EURAD project, which has received funding from the European Union's Horizon 2020 research and innovation programme under grant agreement No 847593.

References

1. Béreš, M.: A comparison of approaches for the construction of reduced basis for stochastic Galerkin matrix equations. Appl. Math. **65**(2), 191–225 (2020). https://doi.org/10.21136/AM.2020.0257-19
2. Béreš, M.: Methods for the solution of differential equations with uncertainties in parameters. Ph.D. thesis, VŠB-TUO, Ostrava, Czech Republic (2023)
3. Chen, P., Quarteroni, A., Rozza, G.: Comparison between reduced basis and stochastic collocation methods for elliptic problems. J. Sci. Comput. **59**(1), 187–216 (2014). https://doi.org/10.1007/s10915-013-9764-2
4. Dolgov, S.V., Savostyanov, D.V.: Alternating minimal energy methods for linear systems in higher dimensions. SIAM J. Sci. Comput. **36**(5), A2248–A2271 (2014). https://doi.org/10.1137/140953289
5. Ghanem, R.G., Spanos, P.D.: Stochastic Finite Elements: A Spectral Approach. Springer, Heidelberg (1991). https://doi.org/10.1007/978-1-4612-3094-6
6. Oseledets, I.V.: Tensor-train decomposition. SIAM J. Sci. Comput. **33**(5), 2295–2317 (2011). https://doi.org/10.1137/090752286
7. Oseledets, I.: TT-Toolbox (2022). https://github.com/oseledets/tt-toolbox
8. Petras, K.: Smolyak cubature of given polynomial degree with few nodes for increasing dimension. Numer. Math. **93**(4), 729–753 (2003). https://doi.org/10.1007/s002110200401
9. Powell, C.E., Silvester, D., Simoncini, V.: An efficient reduced basis solver for stochastic Galerkin matrix equations. SIAM J. Sci. Comput. **39**(1), A141–A163 (2017). https://doi.org/10.1137/15M1032399
10. Ullmann, E.: A Kronecker product preconditioner for stochastic Galerkin finite element discretizations. SIAM J. Sci. Comput. **32**(2), 923–946 (2010). https://doi.org/10.1137/080742853

The Tensor-Train Mimetic Finite Difference Method for Three-Dimensional Maxwell's Wave Propagation Equations

Gianmarco Manzini[1]($\boxtimes$)(iD), Boian Alexandrov[1](iD), Phan Minh Duc Truong[2](iD), and Radoslav G. Vuchkov[3](iD)

[1] Theoretical Division, Los Alamos National Laboratory,
Los Alamos, NM 87545, USA
{gmanzini,boian}@lanl.gov
[2] Computers, Computational and Statistical Sciences Division,
Los Alamos National Laboratory, Los Alamos, NM 87545, USA
dptruong@lanl.gov
[3] Computational Mathematics Department, Sandia National Laboratories,
Albuquerque, NM 87185, USA
rgvuchk@sandia.gov

Abstract. We construct a compatible/mimetic method for solving the Maxwell wave propagation equations using the tensor train format to represent the electric and magnetic fields. We combine the space discretization with a staggered-in-time scheme to advance the discrete electric and magnetic fields in time. The resulting method is second-order accurate in time and space and compatible, so the approximation of the magnetic flux field has zero discrete divergence. Using the tensor-train format improves the solver performance by orders of magnitude in terms of CPU time and memory storage. A final set of numerical experiments confirms this expectation.

Keywords: Tensor train · Maxwell wave propagation equation · Low-rank mimetic approximation

1 Electromagnetic Wave Propagation Equations

Our work's primary goal is to reformulate the compatible/mimetic numerical method using the tensor-train format for solving the three-dimensional electromagnetic wave propagation equations. A compatible/mimetic method satisfies a discrete de Rham complex, so that important properties of the continuous calcolus, such as $\mathrm{div}\,\mathbf{curl} = 0$ and $\mathbf{curl}\,\nabla = 0$ are exactly reproduced in the discrete setting by the discrte analogs of the differential operators, i.e., ∇_h, $\mathbf{curl}_h$ and div_h, see [9]. The approximation method on regular cartesian grids minimizes the computational complexity in terms of memory storage and elapsed (or CPU) time, thus providing an extremely efficient solver.

Let $\Omega \subset \mathbb{R}^3$ be a cubic domain with boundary $\partial\Omega$. We consider the Maxwell's wave propagation equations on Ω for the electric field $\mathbf{E}(\mathbf{x},)$ and the magnetic flux field $\mathbf{B}(\mathbf{x},)$, which read as

I. Lirkov and S. Margenov (Eds.): LSSC 2023, LNCS 13952, pp. 215–222, 2024.
https://doi.org/10.1007/978-3-031-56208-2_21

$$\partial_t \mathbf{E} - \mathbf{curl}\,\mathbf{B} = \mathbf{J} \qquad \text{in } \Omega,\ \forall \in (0,T], \qquad (1a)$$

$$\varepsilon_0\mu_0\partial_t\mathbf{B} + \mathbf{curl}\,\mathbf{E} = 0 \qquad \text{in } \Omega,\ \forall \in (0,T], \qquad (1b)$$

$$\mathrm{div}\,\mathbf{E} = \mathrm{div}\,\mathbf{B} = 0 \qquad \text{in } \Omega,\ \forall \in (0,T], \qquad (1c)$$

$$\mathbf{E}(\cdot,0) = \mathbf{E}^0, \quad \mathbf{B}(\cdot,0) = \mathbf{B}^0 \qquad \text{in } \Omega, \qquad (1d)$$

$$\mathbf{E} \times \mathbf{n}_\Omega = \mathbf{g_E}, \quad \mathbf{B} \cdot \mathbf{n}_\Omega = g_\mathbf{B} \qquad \text{on } \partial\Omega, \qquad (1e)$$

where ∂_t denotes the time derivative, $\mathbf{J}$ is the electric current density, and $\mathbf{g_E}$ and $g_\mathbf{B}$ are two functions defined on $\partial\Omega$ setting the boundary conditions of the problem. The quantities ε_0 and μ_0 are the electric permittivity and the magnetic permeability, which we assume equal to 1 for simplicity.

Following [4], we design our method using discrete representations of scalar and vector fields through grid functions, which are multidimensional arrays collecting the degrees of freedom defined on the computational mesh. According to [9, Chapters 2, 6], we consider the compatible/mimetic scheme that reads as: *For any* $\in [0,T]$, *find* $(\mathbf{E}_h(t), \mathbf{B}_h(t)) \in \mathcal{E}_h \times \mathcal{F}_h$ *such that:*

$$\partial_t\mathbf{E}_h - \widetilde{\mathbf{curl}}_h\mathbf{B}_h = \mathbf{J}_h \qquad \forall \in (0,T], \qquad (2a)$$

$$\partial_t\mathbf{B}_h + \mathbf{curl}_h\mathbf{E}_h = 0 \qquad \forall \in (0,T], \qquad (2b)$$

$$\mathbf{E}_h(0) = \mathbf{E}_h^0, \quad \mathbf{B}_h(0) = \mathbf{B}_h^0, \qquad (2c)$$

where $\mathcal{E}_h$ stands for the grid function space of "edge"-type and $\mathcal{F}_h$ for the grid function space of "face"-type where we search the approximation of the electric and magnetic flux fields, respectively, and

- $\mathbf{E}_h \in \mathcal{E}_h$ and $\mathbf{B}_h \in \mathcal{F}_h$ approximate the electric and magnetic flux fields;
- $\mathbf{E}_h^0 \in \mathcal{E}_h$ and $\mathbf{B}_h^0 \in \mathcal{F}_h$ approximate the initial fields $\mathbf{E}^0$ and $\mathbf{B}^0$;
- $\mathbf{J}_h(t)$ is the grid function representing $\mathbf{J}$ in $\mathcal{E}_h$.

In this formulation, we set the boundary degrees of freedom of $\mathbf{E}_h$ and $\mathbf{B}_h$ from $\mathbf{g_E}$ and $g_\mathbf{B}$ directly according to (1d), while the divergence constraint on $\mathbf{E}_h$ and $\mathbf{B}_h$ follows from the mimetic nature of our method.

The approximation in time is given by advancing the numerical approximations $\mathbf{E}_h$ and $\mathbf{B}_h$ with a staggered time-marching scheme, cf. *leapfrog scheme* [10]. We consider the time partition with time step Δ between $t = 0$ and a final time $t = T > 0$, and denote the intermediate time instants as $^n = n\Delta$ and $^{n+\frac{1}{2}} =^n +\Delta/2 = (n+1/2)\Delta$. For $n \geq 1$, $\mathbf{E}_h^n$ and $\mathbf{J}_h^n$ are the values of $\mathbf{E}_h$ and $\mathbf{J}_h$ at the n-th time step ^{n}. Similarly, $\mathbf{B}_h^{n+\frac{1}{2}}$ is the approximate value of $\mathbf{B}_h$ at the staggered time instant $^{n+\frac{1}{2}}$. Then, the mimetic finite difference approximation in space and time for $n \geq 1$ reads as: *For $n \geq 1$, find* $(\mathbf{E}_h^n, \mathbf{B}_h^{n+\frac{1}{2}}) \in \mathcal{E}_h \times \mathcal{F}_h$ *such that:*

$$\mathbf{E}_h^n = \mathbf{E}_h^{n-1} + \Delta\big(\widetilde{\mathbf{curl}}_h\mathbf{B}_h^{n-\frac{1}{2}} + \mathbf{J}_h^{n-1}\big), \qquad (3)$$

$$\mathbf{B}_h^{n+\frac{1}{2}} = \mathbf{B}_h^{n-\frac{1}{2}} - \Delta\mathbf{curl}_h\mathbf{E}_h^n, \qquad (4)$$

assuming at the initial time $= 0$ that $\mathbf{E}_h^0$ and $\mathbf{B}_h^{\frac{1}{2}}$ are suitably accurate approximations of the initial fields $\mathbf{E}^0$ and $\mathbf{B}^0$.

2   Mimetic Calculus on Primal/dual Grids

We denote the independent variables along X, Y, and Z by x, y, and z respectively. On the open unit cube $\Omega =]0,1[\times]0,1[\times]0,1[$, we consider the equispaced univariate partitions $\{x_i\}_{i=0,...,n_x}$ with size $\Delta x = 1/n_x$ along the direction X; $\{y_j\}_{j=0,...,n_y}$ with size $\Delta y = 1/n_y$ along the direction Y; $\{z_k\}_{k=0,...,n_z}$ with size $\Delta z = 1/n_z$ along the direction Z. The tensor product of these univariate partitions produces the cartesian mesh with closed cells $C_h = \{c_{i,j,k}\}$, $h = \max(\Delta x, \Delta y, \Delta z)$, such that $\overline{\Omega} = \sum_{i,j,k} c_{i,j,k}$. Each cell is uniquely identified by the index triple (i,j,k) and defined as $c_{i,j,k} = [x_i, x_{i+1}] \times [y_j, y_{j+1}] \times [z_k, z_{k+1}]$. Then, we consider the staggered partitions $\{\widetilde{x}_i = x_i + \Delta x/2\}_{i=0,...,n_x-1}$, $\{\widetilde{y}_j = y_j + \Delta y/2\}_{j=0,...,n_y-1}$, and $\{\widetilde{z}_k = z_k + \Delta z/2\}_{k=0,...,n_z-1}$, and define the dual grid with cells $\widetilde{C}_h = \{\widetilde{c}_{i,j,k}\}$ as the cartesian mesh whose internal cells are given by $\widetilde{c}_{i,j,k} = [\widetilde{x}_i, \widetilde{x}_{i+1}] \times [\widetilde{y}_j, \widetilde{y}_{j+1}] \times [\widetilde{z}_k, \widetilde{z}_{k+1}]$. We complete the construction of the dual mesh by including the cells close to the boundary of Ω, which are defined by connecting the staggered points $(\widetilde{x}_i, \widetilde{y}_j, \widetilde{z}_k)$ with the centers of the boundary faces and the boundary edges. In such a construction, every geometric element of the dual mesh is in a bijective correspondence (duality pairing) with a geometric element of the primal mesh.

A *grid function* is a collection of degrees of freedom defined on the primal or the dual mesh. The *degrees of freedom* are real numbers associated with the vertices, edges, faces, and cells of the primal mesh Ω_h or the dual mesh $\widetilde{\Omega}_h$. A given collection of values associated with the primal edges or faces is a grid function of type edge or face, respectively. We know from [7,8] that the physical nature of a mathematical quantity is reflected by the geometric objects that support it (Fig. 1).

In particular, we characterize the electric field **E** by an edge function $\mathbf{E}_h$ and the magnetic flux field **B** by a face function $\mathbf{B}_h$. Such grid functions belong to the sets $\mathcal{E}_h$ (primal edge grid functions) and $\mathcal{F}_h$ (primal face grid functions). These two sets are linear spaces endowed with the usual operations of addition and multiplication by a scalar number. Similar definitions hold for the dual grid and we consider the linear spaces $\widetilde{\mathcal{E}}_h$ (dual edge grid functions) and $\widetilde{\mathcal{F}}_h$

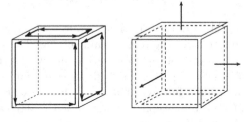

Fig. 1. Degrees of freedom of the compatible/mimetic finite difference method on a cubic cell.

(dual face grid functions). The bijective correspondence between the geometric objects of Ω_h and $\widetilde{\Omega}_h$ allows us to identify the degrees of freedom of $\widetilde{\Omega}_h$ once we set the degrees of freedom of Ω_h. For example, we identify the real number associated with a primal edge e and the corresponding dual face $\widetilde{f}_e$ as the same degree of freedom. The pairs of spaces $(\mathcal{E}_h, \widetilde{\mathcal{F}}_h)$ and $(\mathcal{F}_h, \widetilde{\mathcal{E}}_h)$ are thus in a mesh duality relation, which we exploit to define pairs of primal-dual discrete analogs of differential operators such as $\mathbf{curl}_h$ and $\widetilde{\mathbf{curl}}_h$.

The *discrete curl operator* $\mathbf{curl}_h : \mathcal{E}_h \to \mathcal{F}_h$ maps any primal edge function $\mathbf{u}_h \in \mathcal{E}_h$ with degrees of freedom $\underline{\mathbf{u}}_h = (u_e)_{e \in E_h}$ into the face function $\mathbf{curl}\,\mathbf{u}_h$. Likewise, The *discrete curl operator* $\widetilde{\mathbf{curl}}_h : \widetilde{\mathcal{E}}_h \to \widetilde{\mathcal{F}}_h$ maps any dual edge function $\widetilde{\mathbf{u}}_h \in \widetilde{\mathcal{E}}_h$ with degrees of freedom $\underline{\widetilde{\mathbf{u}}}_h = (\widetilde{u}_{\widetilde{e}})_{\widetilde{e} \in \widetilde{E}_h}$ into the dual face function $\widetilde{\mathbf{curl}}_h \widetilde{\mathbf{u}}_h$. Formally, we write:

$$\mathbf{curl}_h \mathbf{u}_{h|f} = \frac{1}{|f|} \sum_{e \in E_h^f} \sigma_{e,f} |e| u_e, \qquad \widetilde{\mathbf{curl}}_h \widetilde{\mathbf{u}}_{h|\widetilde{f}} = \frac{1}{|\widetilde{f}|} \sum_{\widetilde{e} \in \widetilde{E}_h^{\widetilde{f}}} \sigma_{\widetilde{e},\widetilde{f}} |\widetilde{e}| \widetilde{u}_{\widetilde{e}},$$

where the summation is over all the edges of the primal face $f \in F_h$ denoted as E_h^f and over all the dual edges of the dual face $\widetilde{f} \in \widetilde{F}_h$ denoted as $\widetilde{E}_h^{\widetilde{f}}$. In this definition, $\sigma_{e,f} = \mathbf{t}_e \cdot \mathbf{t}_{f,e}$ and $\sigma_{\widetilde{e},\widetilde{f}} = \mathbf{t}_{\widetilde{e}} \cdot \mathbf{t}_{\widetilde{f}\widetilde{e}}$ can be 1 or -1 depending on the mutual orientations of e and f or their dual counterparts. We express the action of the discrete curl on the edge function $\mathbf{u}_h$ by using the curl matrix $\mathbb{C}$, so that $\mathbf{curl}_h \mathbf{u}_h = \mathbb{C}\underline{\mathbf{u}}_h$, and the action of the discrete curl on the edge function $\widetilde{\mathbf{u}}_h$ by the curl matrix $\widetilde{\mathbb{C}}$, so that $\widetilde{\mathbf{curl}}_h \widetilde{\mathbf{u}}_h = \widetilde{\mathbb{C}}\underline{\widetilde{\mathbf{u}}}_h$. The matrix $\mathbb{C}$ is the connectivity matrix between edges and faces rescaled by $|e|/|f|$; the matrix $\widetilde{\mathbb{C}}$ is the connectivity matrix between dual edges and dual faces rescaled by $|\widetilde{e}|/|\widetilde{f}|$. From these definitions, it follows that $\mathbf{curl}_h \mathcal{E}_h$ is a linear subspace of $\mathcal{F}_h$ and $\widetilde{\mathbf{curl}}_h \widetilde{\mathcal{E}}_h$ is a linear subspace of $\widetilde{\mathcal{F}}_h$.

3   Tensor-Train Mimetic Method for the Maxwell Wave Propagation Equations

In this section, we present our reformulation of Maxwell's wave propagation equations in the tensor-train setting. The use of tensor formats to express data structures and operators in numerical algorithms for PDEs is relatively recent. We refer to [2,3] for a basic introduction to low-parametric tensor formats and [5,6] for a full explanation of the tensor-train format and cross-approximation of multidimensional arrays. Each scalar component of the grid functions $\mathbf{E}_h$, $\mathbf{B}_h$, and $\mathbf{J}_h$ contains three cores to represent their tensor-train decomposition. Such a reformulation requires a new design of the discrete curl operators $\mathbf{curl}_h$ and $\widetilde{\mathbf{curl}}_h$, acting respectively on the tensor-train representation of the discrete vector fields $\mathbf{E}_h$ and $\mathbf{B}_h$.

The grid function describing the electric field in $\mathcal{E}_h$ is the three-dimensional vector $\mathbf{E}_h = (E_h^x, E_h^y, E_h^z)^T$, whose components are associated with the edges parallel to the directions X, Y, and Z, respectively, and described by three-dimensional arrays indexed with i, j, k. The grid function describing the magnetic field in $\mathcal{F}_h$ is the three-dimensional vector $\mathbf{B}_h = (B_h^x, B_h^y, B_h^z)^T$, whose components are orthogonal to the directions X, Y, and Z, respectively, and described by three-dimensional arrays indexed with i, j, k. Since $\mathbf{curl}_h : \mathcal{E}_h \to \mathcal{F}_h$, for a given discrete field $\mathbf{E}_h \in \mathcal{E}_h$, we find that $\mathbf{curl}_h \mathbf{E}_h \in \mathcal{F}_h$, the discrete scalar field

$(\mathbf{curl}_h \mathbf{E}_h)_{|x}$ is associated with the primal faces orthogonal to the direction X. The X-component of $\mathbf{curl}_h \mathbf{E}_h$ reads as:

$$
\begin{aligned}
(\mathbf{curl}_h \mathbf{E}_h)_{|f^x} &= \frac{E_h^y(i,j,k+1) - E_h^y(i,j,k)}{\Delta z} - \frac{E_h^z(i,j+1,k) - E_h^z(i,j,k)}{\Delta y} \\
&= \left(\partial_z^h E_h^y\right)_{|i,j,k} - \left(\partial_y^h E_h^z\right)_{|i,j,k}
\end{aligned}
$$

This equation implies that the X-component of the discrete curl of $\mathbf{E}_h$ depends on the finite differences of E_h^y and E_h^z along the directions Z and Y, respectively. We rewrite the two finite differences above at the location indexed by (i,j,k) as the restriction of a global operation involving the *shift operators* T_1^Z and T_ℓ^Y, which in the Z and Y directions, respectively.

$$
\Delta z\left(\partial_z^h E_h^y\right)_{|i,j,k} = E_h^y(i,j,k+1) - E_h^y(i,j,k) = \left(T_1^Z E_h^y - E_h^y\right)_{i,j,k},
$$
$$
\Delta y\left(\partial_z^h E_h^z\right)_{|i,j,k} = E_h^z(i,j+1,k) - E_h^z(i,j,k) = \left(T_1^Y E_h^z - E_h^z\right)_{i,j,k}.
$$

The tensor-train representation of E_h^y and $T_1^Z E_h^y$ through the tensor-train cores $\mathcal{E}_1^Y$, $\mathcal{E}_2^Y$, and $\mathcal{E}_3^Y$ reads as

$$
(E_h^y)^{TT}(i,j,k) := \mathcal{E}_1^Y(i)\mathcal{E}_2^Y(j)\mathcal{E}_3^Y(k),
$$
$$
(T_1^Z E_h^y)^{TT}(i,j,k) := \mathcal{E}_1^Y(i)\mathcal{E}_2^Y(j)\mathcal{E}_3^Y(k+1) = \mathcal{E}_1^Y(i)\mathcal{E}_2^Y(j)\left(T_1^Z \mathcal{E}_3^Y\right)(k).
$$

Similarly, the tensor-train representation of E_h^z and $T_1^Y E_h^z$ through the tensor-train cores $\mathcal{E}_1^Z$, $\mathcal{E}_2^Z$, and $\mathcal{E}_3^Z$ reads as

$$
(E_h^z)^{TT}(i,j,k) := \mathcal{E}_1^Z(i)\mathcal{E}_2^Z(j)\mathcal{E}_3^Z(k),
$$
$$
(T_1^Y E_h^z)^{TT}(i,j,k) := \mathcal{E}_1^Z(i)\mathcal{E}_2^Z(j+1)\mathcal{E}_3^Z(k) = \mathcal{E}_1^Z(i)\left(T_1^Y \mathcal{E}_2^Z\right)(j)\mathcal{E}_3^Z(k).
$$

Using these formulas, we obtain the tensor-train representation of the two finite-difference terms in $(\mathbf{curl}_h \mathbf{E}_h)_{|f^x}$:

$$
\Delta z\left(\partial_z^h E_h^y\right)^{TT}(i,j,k) = \mathcal{E}_1^Y(i)\mathcal{E}_2^Y(j)\left(T_1^Z \mathcal{E}_3^Y - \mathcal{E}_3^Y\right)(k),
$$
$$
\Delta y\left(\partial_y^h E_h^z\right)^{TT}(i,j,k) = \mathcal{E}_1^Z(i)\left(T_1^Y \mathcal{E}_2^Z - \mathcal{E}_2^Z\right)(j)\mathcal{E}_3^Z(k).
$$

Finally, the X-component of $\mathbf{curl}_h \mathbf{E}_h$ in tensor-train format is given by

$$
(\mathbf{curl}_h \mathbf{E}_h)_{|X}^{TT} = \mathsf{rndg}\left(\left(\partial_y^h E_h^z\right)^{TT} - \left(\partial_z^h E_h^y\right)^{TT}\right).
$$

where we apply the rounding operator $\mathsf{rndg}(\cdot)$ to control the rank growth of the resulting expression. The rounding operator performs a truncated SVD on the subcores of the tensor train representation of the tensor sum on the right-hand side, thus preventing the TT rank to increase arbitrarily. We apply a similar procedure for the construction of the other components of $(\mathbf{curl}_h \mathbf{E}_h$, and, working

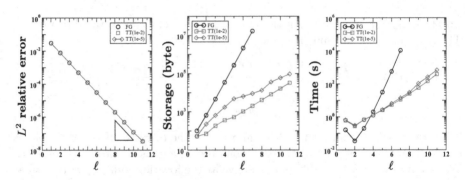

Fig. 2. Left panel: Plot of the relative error curves in the L^2 norm. Middle panel: plot of the storage in bytes. Right panel: plot of the elapsed time in seconds (s).

on the dual mesh, for the components of $\widetilde{\mathbf{curl}}_h \mathbf{B}_h$. All dependent variables forming the vectors $\mathbf{E}_h$ and $\mathbf{B}_h$ are set up from their initial values and updated during the calculation using the tensor-train format. According to Eqs. (2a) and (2b), the X-components of $\mathbf{E}_h$ and $\mathbf{B}_h$ are updated as follows:

$$\left(E_h^{x,n}\right)^{TT} = \mathsf{rndg}\left[\left(E_h^{x,n-1}\right)^{TT} + \Delta\left(\widetilde{\mathbf{curl}}_h\mathbf{B}_h^{n-\frac{1}{2}}\right)^{TT}_{|X} + \left(J_h^{x,n-1}\right)^{TT}\right],$$

$$\left(B_h^{x,n+\frac{1}{2}}\right)^{TT} = \mathsf{rndg}\left[\left(B_h^{x,n-\frac{1}{2}}\right)^{TT} - \Delta\left(\mathbf{curl}_h\mathbf{E}_h^n\right)^{TT}_{|X}\right].$$

It is worth noting that no arithmetic operation is involved here in order to determine the tensor-train representation of the shifted fields $\mathcal{T}_1^Z E_h^y$, $\mathcal{T}_1^Y E_h^z$, $\mathcal{T}_1^Z B_h^y$, and $\mathcal{T}_1^Y B_h^z$. Moreover, no change of rank takes place during the calculation of a finite difference formula through the appropriate linear combination of the shifted cores because this operation does not modify the internal dimensions of the tensor decomposition, so no rounding is needed. The rounding operation is, however, required to compute the components of $\mathbf{curl}_h\mathbf{E}_h$ and $\widetilde{\mathbf{curl}}_h\mathbf{B}_h$.

4   Numerical Experiments

We present some preliminary results on the calculation of a wave propagation problem with an exact solution that is a rank-1 elementary wave. This test case is similar too [1, Test Case 1]. We solve Eqs. (3)–(4) on the domain $\Omega = [0, 1]^3$ with $\varepsilon_0\mu_0 = 1$ by setting the initial states $\mathbf{E}^0$ and $\mathbf{B}^0$, the boundary conditions and the electric current density $\mathbf{J}$ on the right-hand side of Eq. (2a) in accordance with the exact solutions

$$\mathbf{E}(\mathbf{x},) = \left(0, 0, \sin(\pi x)\sin(\pi y)\right)^T \cos(\pi),$$

$$\mathbf{B}(\mathbf{x},) = \left(-\sin(\pi x)\cos(\pi y), \cos(\pi x)\sin(\pi y), 0\right)^T \sin(\pi).$$

We define the approximation error as

$$
\text{Error}_{L^2} := \left[\frac{||\mathbf{E} - \mathbf{E}_h||^2_{L^2(\Omega)}}{||\mathbf{E}||^2_{L^2(\Omega)}} + \frac{||\mathbf{B} - \mathbf{B}_h||^2_{L^2(\Omega)}}{||\mathbf{B}||^2_{L^2(\Omega)}} \right]^{\frac{1}{2}} .
$$

In Fig. 2, we plot the relative error curves in the L^2 norm (left panel); the storage in bytes (middle panel); the elapsed time in seconds (right panel) for the calculation using the full-grid and the tensor-train formulation of the mimetic method using a tolerance ϵ equal to 10^{-2} and 10^{-5}. All plot are versus the refinement level ℓ on the grid with $N_\ell = 2^\ell$ partitions per direction with $\ell = 1, 2, \ldots, 11$, which corresponds to a full-grid resolution with $N_\ell = 2^\ell$ partition in every space direction. The numerical convergence rate is proportional to $\mathcal{O}(h^2)$, where $h = 1/N$ with $N \in \{N_\ell, \ell \geq 1\}$ and us reflected by the slopes of the error curves. Figure 3 shows the time evolution of the L^2 norm of the divergence of the magnetic flux fields for the full-grid solver and the tensor-train solver with tolerance ϵ equal to 10^{-2}. The calculation is carried out from $t = 0.2$ to $t = 1.2$ on the grid with resolution level $\ell = 5$. The plots show that the tensor-train reformulation is still mimetic and the zero-divergence constrain is better satisfied. The three panels of Fig. 2 and Fig. 3 show that the accuracy is the same, but the tensor-train reformulation of the mimetic algorithm is much more computationally efficient. We run the full-grid solver from the refinement level $\ell = 1$, which corresponds to a grid with resolution $2 \times 2 \times 2$, to the refinement level $\ell = 6$, which corresponds to a grid with resolution $64 \times 64 \times 64$, the maximum size allowed by our computational resources.

Instead, we run our tensor-train solver from the same first refinement $\ell = 1$ to the refinement level $\ell = 11$, which corresponds to a grid with resolution $2048 \times 2048 \times 2048$. The tensor-train reformulation of the mimetic method makes it possible to solve the wave propagation problem with the same accuracy (see Fig. 2, left panel), but on much bigger meshes. The biggest calculation that the full-grid solver can run has about 2.6×10^5 degrees of freedom, while the biggest calculation that the tensor-train solver can run has about 8.6×10^9 degrees of freedom. More-

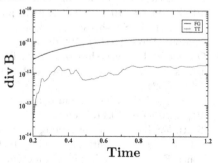

Fig. 3. Divergence-free property for the numerical solution computed with the full-grid (FG) version and the tensor-train (TT) version of the mimetic method.

over, on the grid with $\ell = 6$, the full-grid solver requires a storage of more than 10^7 bytes and an elapsed time of about 10^4 seconds. The tensor-train solver on the same grid requires a storage of about 10^4 bytes and an elapsed time of about 10 s.

Acknowledgements. This work was partially supported by the Laboratory Directed Research and Development (LDRD) program of Los Alamos National Laboratory under 20230067DR and 20210485ER. Los Alamos National Laboratory is operated by Triad National Security, LLC, for the National Nuclear Security Administration of U.S. Department of Energy (Contract No. 89233218CNA000001). This document has been approved for public release as the Los Alamos Technical Report LA-UR-23-21109.

This article has been authored by an employee of National Technology & Engineering Solutions of Sandia, LLC under Contract No. DE-NA0003525 with the U.S. Department of Energy (DOE). The employee owns all right, title and interest in and to the article and is solely responsible for its contents. The United States Government retains and the publisher, by accepting the article for publication, acknowledges that the United States Government retains a non-exclusive, paid-up, irrevocable, world-wide license to publish or reproduce the published form of this article or allows others to do so, for United States Government purposes. The DOE will provide public access to these results of federally sponsored research in accordance with the DOE Public Access Plan https://www.energy.gov/downloads/doe-public-access-plan.

References

1. D'Azevedo, E., Green, D.L., Mu, L.: Discontinuous Galerkin sparse grids methods for time domain Maxwell's equations. Comput. Phys. Commun. **256**, 107412 (2020). https://doi.org/10.1016/j.cpc.2020.107412

2. Hackbusch, W.: Tensor Spaces and Numerical Tensor Calculus. Springer Series in Computational Mathematics, vol. 42, 1st edn. Springer, Heidelberg (2012). https://doi.org/10.1007/978-3-642-28027-6

3. Hackbusch, W.: Numerical tensor calculus. Acta Numer. **23**, 651–742 (2014). https://doi.org/10.1017/S0962492914000087

4. Hyman, J.M., Shashkov, M.: Mimetic discretizations for Maxwell's equations. J. Comput. Phys. **151**(2) (1999). https://doi.org/10.1006/jcph.1999.6225

5. Oseledets, I., Tyrtyshnikov, E.: TT-cross approximation for multidimensional arrays. Linear Algebra Appl. **432**, 70–88 (2010). https://doi.org/10.1016/j.laa.2009.07.024

6. Oseledets, I.V.: Tensor-train decomposition. SIAM J. Sci. Comput. **33**(5) (2011). https://doi.org/10.1137/090752286

7. Tonti, E.: The reason for analogies between physical theories. Appl. Math. Model. **1**(1), 37–50 (1976). https://doi.org/10.1016/0307-904X(76)90023-8

8. Tonti, E.: Why starting from differential equations for computational physics? J. Comput. Phys. **257**, 1260–1290 (2014). https://doi.org/10.1016/j.jcp.2013.08.016. Physics-compatible numerical methods

9. Beirão da Veiga, L., Lipnikov, K., Manzini, G.: The Mimetic Finite Difference Method, MS&A. Modeling, Simulations and Applications, vol. 11, I edn. Springer, Cham (2014)

10. Yee, K.: Numerical solution of initial boundary value problems involving Maxwell's equations in isotropic media. IEEE Trans. Antennas Propag. **14**(3), 302–307 (1966). https://doi.org/10.1109/TAP.1966.1138693

A Functional Tensor Train Library in RUST for Numerical Integration and Resolution of Partial Differential Equations

Massimiliano Martinelli[1] and Gianmarco Manzini[2(✉)]

[1] Istituto di Matematica Applicata e Tecnologie Informatiche "E. Magenes",
Consiglio Nazionale delle Ricerche, via Ferrata 5, 27100 Pavia, Italy
martinelli@imati.cnr.it
[2] T-5 Group, Los Alamos National Laboratory, P.O. Box 1663, Los Alamos,
NM 87545, USA
gmanzini@lanl.gov

Abstract. Originally, low-rank tensor decomposition algorithms were designed to approximate high-dimensional tensors. Due to its mathematical characteristics, Tensor-Train decomposition, a type of tensor decomposition that does not necessarily suffer from the curse of dimensionality, has garnered much interest during the past decade. In recent years, Function-Train decomposition, a continuous version of Tensor-Train decomposition, was introduced. This decomposition permits the approximation of high-dimensional functions without function sampling and provides an extensible framework for function integration and differentiation. In this paper, we present a new RUST-based library designed to provide functionality for Function-Train decomposition. In addition, the library offers methods for continuous matrix factorizations and continuous multilinear algebra operations, such as addition, multiplication, integration, differentiation, etc.

Keywords: Tensor Train · Functional tensor train · Rust library

1 Introduction

Tensors as multidimensional arrays are the straightforward generalization of matrices as two-dimensional arrays. Multilinear algebra for tensor decomposition has attracted the attention of the numerical community because the strong compressive nature of tensor formats through the product of smaller objects, e.g., matrices and small tensors, the *decomposition cores*, offers a strategy to overcome the *curse of dimensionality*. The Tensor Train (TT) representation [14, 15], both exact and approximate, is among the most successful and promising techniques appeared in the last decade and available right now. Extending the tensor methodology to function approximation also paves the way to the development of new numerical methods, e.g., for the computer integration of high-dimensional functions [1, 5], and the design of more efficient solvers for Ordinary and Partial Differential equations (ODEs [4] and PDEs [3]). The Functional Tensor Train (FTT) [10] as a continuous analog of the Tensor Train combines the compressive property of tensor formats and the innovative concept of *continuous linear*

I. Lirkov and S. Margenov (Eds.): LSSC 2023, LNCS 13952, pp. 223–233, 2024.
https://doi.org/10.1007/978-3-031-56208-2_22

algebra, i.e., computing with functions as in the Chebfun library [2]. The FTT representation is an alternative to the two usual ways of compressing functions. The first one is to sample a multivariate function on a multidimensional grid to obtain a multi-way array, i.e., a tensor; then, to decompose the resulting grid function using a tensor format. The second one is to expand a multivariate function on a tensor product basis function set; then, to decompose the tensor describing the expansion coefficients using a tensor format. Currently several software libraries are available to the scientific community for working with the TT format, e.g., [12,13,18,19]. However, regarding the FTT format, the publicly available software is limited to the Compressed Continuous Computation (C3) library, see [9]. Inspired by this C/C++ library, we are developing the *Functional Tensor Library (FTL)* using the Rust programming language [11,16] and function representation based on an expansion in Chebishev polynomials as is done in the Chebfun library [2]. Rust is a modern programming languages for scientific and high-performance computing, aimed to achieve correctness, safety and performance. A preliminary version of FTL is currently available at the URL https://gitlab.com/max.martinelli/functional-tensor-library. We plan to publish FTL in the Rust community's crate registry at https://crates.io/) for our next FTT-based applications.

The paper is organized as follows. In Sect. 2, we briefly review the main concepts about the tensor-train and the functional tensor-train formats. In Sect. 3, we describe the structure and the main features of our Rust library. In Sect. 4 we present an example of the use of the library for the numerical approximation of a multi-dimensional integral. In Sect. 5, we offer our final remarks and conclusions.

2   Tensor Train and Functional Tensor Train

In this section, we review some major concepts related to the TT decomposition for tensors and FTT generalization to the continuous calculus. Hereafter, we use the following notation: capital calligraphic fonts like $\mathcal{A}$, $\mathcal{B}$, for tensors; capital roman fonts like A, B for matrices; bold fonts like $\mathbf{A}$, $\mathbf{a}$ for vectors; regular fonts A, a for scalars.

Tensor Train. Given a multidimensional array (tensor) $\mathcal{A} \in \mathbb{R}^{n_1 \times n_2 \times \dots n_d}$, we say that $\mathcal{A}$ is in the *Tensor Train (TT)* representation with TT-ranks $(r_0, r_1, \dots, r_{d-1}, r_d)$, $r_0 = r_d = 1$, if d three-dimensional tensors $\mathcal{A}_k \in \mathbb{R}^{r_{k-1} \times n_k \times r_k}$, $k = 1, 2, \dots, d$, exist such that

$$\mathcal{A}(i_1, i_2, \dots, i_d) = \sum_{\alpha_0=1}^{r_0} \sum_{\alpha_1=1}^{r_1} \cdots \sum_{\alpha_d=1}^{r_d} \mathcal{A}_1(\alpha_0, i_1, \alpha_1) \mathcal{A}_2(\alpha_1, i_2, \alpha_2) \dots \mathcal{A}_d(\alpha_{d-1}, i_d, \alpha_d),$$

for all possible choice of the d indices $i_\ell = 1, \dots, n_\ell$, $\ell = 1, \dots, d$. This representation requires a storage proportional to $\mathcal{O}(dnr^2)$, where n and r are suitable upper bounds of the space modes n_ℓ and the compressive ranks r_ℓ. The single evaluation, which we can also write in a *Matlab*-like notation as

$$\mathcal{A}[i_1, i_2, \dots, i_d] = \mathcal{A}_1[1, i_1, :] \mathcal{A}_2[:, i_2, :] \mathcal{A}_d[:, i_d, 1], \tag{1}$$

has a cost proportional to $\mathcal{O}(dr^2)$. It is also possible to construct an approximate tensor-train decomposition. To this end, we define the k-th *matrix unfolding* for $k = 1, 2, \ldots, d-1$ as the reshaped matrix

$$\mathsf{A}_k(i,j) = \mathcal{A}\Big[\underbrace{(i_1 \ldots i_k)}_{\text{index } i}, \underbrace{(i_{k+1} \ldots i_d)}_{\text{index } j}\Big] = \texttt{reshape}\Big(\mathcal{A}, \Pi_{\ell=1}^k n_\ell, \Pi_{\ell=k+1}^d n_\ell\Big),$$

where we used the Matlab function `reshape`. Tensors with a *good separability property* admit approximate *low-rank* tensor-train representations with ranks $r_k \ll n_k$ for all k. We can bound the error introduced by such approximate representation as follows. Let the matrix unfoldings be decomposable as $\mathsf{A}_k = \mathsf{G}_k + \mathsf{E}_k$ for all $k = 1, 2, \ldots, d-1$ such that $r_k = rank(\mathsf{A}_k)$ for all $k = 1, 2, \ldots, d-1$ and $\epsilon_k = \|\mathsf{E}_k\|_F$. We can obtain such approximation through a dyadic or skeleton decomposition, e.g., the truncated Singular Value Decomposition (SVD) or the CUR decomposition [8]. Then, an approximate tensor-train representation exists, e.g., $\widetilde{\mathcal{A}}^{TT} \approx \mathcal{A}$ such that TT-rank$(\widetilde{\mathcal{A}}^{TT}) = (r_1, r_2, \ldots, r_d)$ and approximation errors bounded by

$$\|\widetilde{\mathcal{A}}^{TT} - \mathcal{A}\|_F \lesssim \sum_{k=1}^{d-1} \epsilon_k^2.$$

A multi-dimensional array can be interpolated on the crossing entries of selected subsets of linearly independent rows and columns. This technique, called *cross interpolation*, expands a similar crossing algorithm that can be used to reconstruct matrices. This strategy results in a very efficient tensor train decomposition method, cf. [5, 14].

Functional Tensor Train. Likewise, we consider a real-valued multidimensional function $f : \mathcal{X} \to \mathbb{R}$. The function's domain $\mathcal{X} = \mathcal{X}_1 \times \mathcal{X}_2 \times \ldots \times \mathcal{X}_d$ is the tensor product of one-dimensional subsets $\mathcal{X}_i$ of the real line $\mathbb{R}$, which we assume to be bounded and, possibly, closed. Then, we say that f is in a *functional tensor-train* format if we expand it as follows:

$$f(x_1, x_2, \ldots, x_d) = \sum_{\alpha_0=1}^{r_0} \sum_{\alpha_1=1}^{r_1} \cdots \sum_{\alpha_d=1}^{r_d} f_1^{\alpha_0,\alpha_1}(x_1) f_2^{\alpha_1,\alpha_2}(x_2) \ldots f_d^{\alpha_{d-1},\alpha_d}(x_d), \quad (2)$$

with $r_0 = r_d = 1$, using the *core functions* $f_k^{\alpha_{k-1},\alpha_k}(x_k)$. To obtain the continuous analogue of the TT format for functions, we groups the functions associated with each input coordinates $k \in \{1, 2, \ldots, d\}$ into matrix-valued functions $\mathsf{F}_k(x_k) : \mathcal{X}_k \to \in \mathbb{R}^{r_{k-1} \times r_k}$, the so-called *FTT-cores* or *quasi-matrices*, such that

$$\mathsf{F}_k(x_k) = \begin{bmatrix} f_k^{11} & f_k^{12} & \cdots & f_k^{1r_k} \\ f_k^{21} & f_k^{22} & \cdots & f_k^{2r_k} \\ \vdots & & & \\ f_k^{r_{k-1}1} & f_k^{r_{k-1}2} & \cdots & f_k^{r_{k-1}r_k} \end{bmatrix}. \quad (3)$$

Using the core definition (3), we reformulate expansion (2) in the matrix-like compact form:

$$f(x_1, x_2, \ldots, x_N) = \mathsf{F}_1(x_1)\mathsf{F}_2(x_2)\ldots\mathsf{F}_d(x_d). \tag{4}$$

We parametrize the core functions using a linear or a non-linear parametrization. For example, a linear parametrization reads as

$$f_k^{\alpha_{k-1},\alpha_k}(x_k) = \sum_{j=1}^{n_k^{\alpha_{k-1},\alpha_k}} \theta_{k,j}^{\alpha_{k-1}\alpha_k}\phi_{k,j}^{\alpha_{k-1}\alpha_k},$$

using the set of coefficients $\{\theta_{k,j}^{\alpha_{k-1}\alpha_k}, j = 1, 2, \ldots, n_k^{\alpha_{k-1},\alpha_k}, k = 1, 2, \ldots, d\}$. FTT is equivalent to TT if all the core functions $f_k^{\alpha_{k-1},\alpha_k}(x_k)$ of a given core F_k, are linearly parameterized with the same set of basis functions $\left(\phi_{k,j}^{\alpha_{k-1}\alpha_k}(x_k)\right)_{j=1}^{n_k}$. Such a condition implies that we collect all the expansion coefficients for the k-th core in the matrix form:

$$\mathcal{A}_k[:, j, :] = \begin{bmatrix} \theta_{kj}^{11} & \theta_{kj}^{12} & \cdots & \theta_{kj}^{1r_k} \\ \theta_{kj}^{21} & \theta_{kj}^{22} & \cdots & \theta_{kj}^{2r_k} \\ \cdots & & & \\ \theta_{kj}^{r_{k-1}1} & \theta_{kj}^{r_{k-1}2} & \cdots & \theta_{kj}^{r_{k-1}r_k} \end{bmatrix} =: [\mathcal{A}_{k,j}],$$

for any $j = 1, 2, \ldots, n_k$. Similarly to TT in the discrete case, FTT provides a low-rank tensor continuous approximation of a given function f. Assume that $\widehat{\mathsf{F}}_\ell(x_\ell) \approx \mathsf{F}_\ell(x_\ell)$, $\ell = 1, 2, \ldots, d$, and

$$\widehat{f}(x_1, x_2, \ldots, x_N) = \widehat{\mathsf{F}}_1(x_1)\widehat{\mathsf{F}}_2(x_2)\ldots\widehat{\mathsf{F}}_d(x_d).$$

If $\left|f_k^{ij}(x_k)\right| \le C$ and $\left|f_k^{ij}(x_k) - \widehat{f}_k^{ij}(x_k)\right| \le C\epsilon$, then

$$\left|f(\mathbf{x}) - \widehat{f}(\mathbf{x})\right| \le ed^2 C^d r^{d-1}\epsilon.$$

We can build a low-rank FTT representation adaptively. To this end, we need to introduce the concept of *continuous tensor fibers*. We recall that a discrete fiber is obtained from a multidimensional array $\mathcal{A}$ by fixing all indices but one, so that

$$\mathcal{A}[i_1, i_2, \ldots, i_{k-1}, j, i_{k+1}\ldots i_d], \qquad j = 1, 2, \ldots, n_k.$$

Likewise, a continuous fiber is the univariate function that we can obtain from a multidimensional function f by fixing all independent unknowns but one. To construct a low-rank approximation adaptively, we work on continuous tensor fibers, and we approximate each fiber *adaptively* at a particular tolerance. We use the *continuous LU decomposition* through optimization within an approximate *maximum volume* procedure for *selecting fibers*, over a continuous rather than a discrete variable, and the *continuous QR decomposition* through an implicitly defined continuous rather than discrete inner product to have a notion of orthogonality (e.g., the SVD) and to implement the rounding algorithm. We recall that max-volume submatrices reflects a *quasi-optimality criterion*, cf. [8].

3   Rust Implementation

As mentioned in the introduction, at the moment there is only one library that performs FTT [10] and is publicly available, the *Compressed Continuous Computation (C3) library*, which is publicly available at [17]. Inspired by this library, which is implemented using the C/C++ programming language, we are currently implementing a similar library for our next FTT-based applications. To this end, we decided to use *Rust* [11,16], one of the most interesting and modern programming languages for scientific and high-performance computing. We selected Rust since we believe that this language offers several significant advantages with respect to other more common programming languages, namely C/C++, Python, Matlab. We list here some of the major edges.

- *Memory safety*: Rust ownership and borrowing system helps prevent, at compile time, common programming errors such as null or dangling pointer references, which can lead to memory corruption and security vulnerabilities.
- *Tooling*: Rust has a modern and powerful toolchain, including *Cargo*, a package manager, and built-in support for testing, documentation, and profiling. From a developer viewpoint, Cargo completely frees the developers from configuration and maintenance of external build/package-managers, such as CMake, SCons, Make, and test suites such as CTest/CDash, etc. Moreover, a Cargo plug-in application called *Clippy* flags common errors and non-idiomatic Rust expressions, thus helping writing formally more proper code. On the other end, from the user perspective, a key feature of Cargo is that Cargo downloads and installs automatically all third-party dependencies, thus making the library installation process straightforward.
- *Cross-platform support*: The Rust compiler and the standard library provide excellent support for implementig programs that can run on a wide range of computing platforms from high performance computers to embedded systems, and operating systems such as Linux, BSD, macOS, Windows.
- *Concurrency*: Rust has built-in support for concurrency and parallelism through its innovative *ownership model* and lightweight threads (called"tasks"), making it easier to write safe and efficient concurrent code.
- *Error reports*: the error messages issued by the Rust compiler are rarely criptic, and often provide suggestions for error corrections. This aspect is a substantial improvement if we compare Rust error messages and C/C++ compiler error messages, especially if these latters are related to the use of *C++ templates*.

Furthermore, the development of a scientific library intended to be used for solving multidimensional problems must also be performance-oriented. In this context Rust's performance is comparable to C++ due to its low-level control and lack of runtime overhead as Rust is a compiled language just like C++, while also providing additional safety and security guarantees.

Regarding the functionalities of the library we implemented a set of data structures and functions for the FTT decomposition of user-defined, multivariate, scalar functions.

In our library, we represent the FTT decomposition (2)-(4) by the Rust structure FunctionTrain<d, T>. This structure is parametrized by 2 arguments:

- the dimensionality d of the function domain, i.e., $\mathcal{X} = \mathcal{X}_1 \times \mathcal{X}_2 \times \ldots \times \mathcal{X}_d$;
- the type T of the core functions to be used. Technically, the type T is bounded by the *trait* FiberTrait, that is the set of functionalities (i.e. methods) that the generic type T must implement in order to be used as type for the "core function/fiber". Examples of such methods are the ones providing algebraic operations, such as addition, subtraction, multiplication, together with integration, differentiation, etc.

In the current library implementation, we consider the following types for modeling the core functions:
- global polynomials represented by the monomial (power) basis;
- global polynomials as expansion of Chebyshev polynomials of the first kind. Our implementation is similar, although at the moment with fewer capabilities, to the chebfun objects in the Chebfun library [2, 20];
- piecewise polynomials, where the single polynomial pieces can be one of the polynomial types mentioned above.

We can add other classes of univariate functions, provided that these new types implement the same common trait FiberTrait.

To build the FunctionTrain<d, T> structure, we can use an adaptive cross-approximation strategy, which requires the user to input the function f that he wants to represent in the approximate FTT format, the tolerance, and a function for creating the fibers (of type T) that approximate the univariate restriction of f in which all but one indices are fixed. The library provides a default univariate approximation through function interpolation at Chebishev points of the 2nd kind. The polynomial interpolants built using these nodes have almost ideal properties, at least for smooth functions, see, e.g., Ref. [2, Theorem 2]. Other strategies, e.g., projection techniques, will be available in the future. Furthermore, the structure FunctionTrain<d, T> implements all the principal methods defined for the continuous multilinear algebra, e.g., addition, difference, multiplication of two functions in FTT format, and also methods for the integration and differentiation of functions in FTT formats.

In the FunctionTrain<d, T> structure we store an array of d FTT-cores $F_1, \ldots, F_d$, each one of which is modeled by the structure MatrixValued Function<T>, i.e. the matrix defined in (3), whose entries are univariate core functions of type T (where, again, T is bounded by the trait FiberTrait).

For the structure MatrixValuedFunction<T> we implemented algebraic and arithmetic operations such as the standard algebraic matrix operations of addition, difference, multiplication, multiplication-by-scalar, transposition, Kronecker product, and the continuous analog of some matrix factorization algorithms such as the QR decomposition, the LU decomposition, and the SVD. Since the library implements the derivatives and integrals of core functions, we can easily compute derivatives and integrals of MatrixValuedFunction<T> structures. To understand why this is important, suppose we want to compute the multidimensional integral

$$I := \int_{a_1}^{b_1} \int_{a_2}^{b_2} \cdots \int_{a_d}^{b_d} f(x_1, x_2, \ldots, x_d) \, dx_1 dx_2 \ldots dx_d, \tag{5}$$

where f is given in FTT format. In view of (4), we can rewrite (5) as

$$I = \int_{a_1}^{b_1} \mathsf{F}_1(x_1)\,dx_1 \int_{a_2}^{b_2} \mathsf{F}_2(x_2)\,dx_2 \cdots \int_{a_d}^{b_d} \mathsf{F}_d(x_d)\,dx_d = \prod_{k=1}^{d} I_k,$$

where $I_k := \int_{a_k}^{b_k} \mathsf{F}_k(x_k)\,dx_k$ is the $r_{k-1} \times r_k$ numerical matrix whose (i,j)-th element is

$$\left[I_k\right]_{i,j} := \int_{a_k}^{b_k} \left[\mathsf{F}_k(x_k)\right]_{ij}\,dx_k = \int_{a_k}^{b_k} f_k^{ij}(x_k)\,dx_k.$$

Likewise, we can write the k-th partial derivative of f as

$$\frac{\partial f(x_1, x_2, \ldots, x_d)}{\partial x_k} = \mathsf{F}_1(x_1) \ldots \mathsf{F}_{k-1}(x_{k-1}) \frac{\partial \mathsf{F}_k(x_k)}{\partial x_k} \mathsf{F}_{k+1}(x_{k+1}) \ldots \mathsf{F}_d(x_d),$$

where $\partial \mathsf{F}_k(x_k)/\partial x_k$ is the $r_{k-1} \times r_k$ FTT-core whose (i,j)-th element is

$$\left[\frac{\partial \mathsf{F}_k(x_k)}{\partial x_k}\right]_{ij} := \frac{\partial f_k^{ij}(x_k)}{\partial^{\iota} x_k}.$$

4 Numerical Example

In order to show a use-case, suppose we want to compute a numerical approximation of the multidimensional integral $\int_{\Omega} f(x)\,dx$ where $f \colon \mathbb{R}^d \to \mathbb{R}$ is the d-dimensional Gaussian function (with zero mean and unitary variance)

$$f(x) = (2\pi)^{-\frac{d}{2}} exp\left(-\frac{\|x\|_2^2}{2}\right) \tag{6}$$

on the domain $\Omega = [-10, 10]^d$. This function has been proposed in Refs. [6,7] for testing multidimensional integration routines, and is usually referred to in the technical literature as function *Genz-2*. The exact value of $\int_{\Omega} f(x)\,dx$ can be computed as the product of d Gaussian 1D integrals over the interval $[-10, 10]$ using the error function $erf(x)$. Due to the fact that we are integrating on such wide interval for the unit Gaussian function, the numerical value of the integral is indistinguishable from the value 1.

Some preliminary results are presented in Fig. 1 where are shown the relative error of the computed integral and the required CPU time with respect to the dimension d of the domain Ω. For the tested dimensions $d \leq 27$, the relative approximation error remains below 10^{-10}, with an increasing trend with respect to the dimension d, an effect that we will study in a future work. The performance in terms of computational costs depends on the number of the evaluations of the integrand function that we need to construct for the fiber approximations. We observe that we are able to compute an integral in more than 20 dimensions in about 0.5 seconds It is also worth mentioning that the measured memory peak during the program execution is always less than 10Mb regardless of the number of dimensions considered in this test.

We report the Rust FTL-based implementation for $d = 15$ in the final appendix. For exposition sake, we omitted some preambles about the Rust modules used herein.

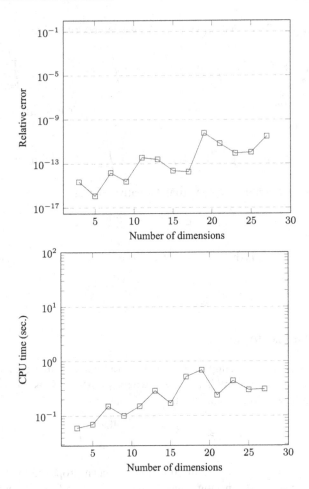

Fig. 1. Preliminary results for the computation of the integral described in Sec. 4. Relative error (on the left) and CPU time (on the right) with respect to the the problem dimension d. Regarding the CPU time, the computations were made on a single core of an Intel Xeon Gold 6242R running at 3.1 GHz.

5   Conclusions

In this work, we present a new RUST-based library that provides functionality for Function-Train decomposition. The Function Tensor-Train representation is a continuous version of the Tensor-Train representation for multidimensional arrays and combines the highly compressive nature of tensor-network formats with the continuous multilinear algebra paradigm of the Matlab Chebfun library. Our Rust library includes data structures for representing and manipulating quasi-matrices, i.e., arrays of functions, and methods for continuous matrix factorizations and continuous multilinear algebra operations such as addition, multiplication, numerical integration, differentiation. A preliminary result about the numerical approximation of a multidimensional integral is also reported.

Acknowledgements. The Rust library described in this article has only been implemented by MM, while GM contributed to the paper writing. MM was partially supported by "Programma di Ricerca CN00000013 – National Centre for HPC, Big Data and Quantum Computing – Spoke 6: Multiscale Modelling & Engineering Applications". GM was partially supported by the Laboratory Directed Research and Development (LDRD) program of Los Alamos National Laboratory under 20230067DR and 20210485ER. Los Alamos National Laboratory is operated by Triad National Security, LLC, for the National Nuclear Security Administration of U.S. Department of Energy (Contract No. 89233218CNA000001). This document has been approved for public release as the Los Alamos Technical Report LA-UR-23-22254.

Appendix: Rust Implementation of the Integration Routine

```rust
const DIM: usize = 15;

let one_div_sqrt_2_pi = 1. / (2. * std::f64::consts::PI).sqrt();

// f is the function to be integrated
let f = |x: &Point<f64, DIM>| -> f64 {
    (0..DIM).map(|i|
    {
        one_div_sqrt_2_pi * (- 0.5 * x[i].powi(2)).exp()
    }).product()
};

// here we build the DIM-dimensional domain of integration
let domain: [Interval; DIM] = std::array::from_fn(
    |_i| Interval::closed(-10.,10.));

// we fix the ranks of our FTT decomposition
let mut ranks = [1usize; DIM + 1];
for i in 1..DIM {
    ranks[i] = 3;
}

// this is a function that is responsible to compute the core functions/fibers
// (as polynomial in the Chebishev basis) using an interpolation strategy
let approximant_1D_by_interpolation =
    create_approximant_1D_by_interpolation_cheb1();

// a tolerance and an upper bound on the number of iterations
// for the cross-approximation algorithm
let tol_relative_norm_increment = RelativeError::new(10.0 * f64::EPSILON);
let max_num_iters = NumIterations::new(10);

// here we compute the FTT approximation of f,
// using the Chebishev polynomials as core functions
let ft_cross_approx_output =
    FunctionTrain::<DIM, Polynomial1DCheb1>::cross_approximation(
        &f,
        &domain,
        &ranks,
        &approximant_1D_by_interpolation,
        &tol_relative_norm_increment,
        &max_num_iters,
        &QRAlgorithm::Householder,
        &ScalarProductType::L2);

let integral_exact = 1.0;

// ft is the FTT representation of f
let ft = ft_cross_approx_output.function_train();

// here we compute the integral of the FTT approximation
```

```
// using the properties of the polynomial representation for the core functions
let integral_approx = ft.integrate_over_domain(&domain);
println!("integral (approximated) = {}", integral_approx);

// and we compute the relative error...
let rel_error = ((integral_exact - integral_approx) / integral_exact).abs();
println!("relative error: {}",rel_error);

//...that is of the order of 10^-14
// (for a 15-dimensional integral!)
assert_eq!(rel_error,2.2426505097428162e-14);
```

References

1. Alexandrov, B., Manzini, G., Skau, E.W., Truong, P.M.D., Vuchov, R.G.: Challenging the curse of dimensionality in multidimensional numerical integration by using a low-rank tensor-train format. Mathematics **11**, 534 (2023). https://doi.org/10.3390/math11030534
2. Battles, Z., Trefethen, L.N.: An extension of MATLAB to continuous functions and operators. SIAM J. Sci. Comput. **25**, 1743–1770 (2004). https://doi.org/10.1137/S1064827503430126
3. Dahmen, W., DeVore, R., Grasedyck, L., Süli, E.: Tensor-sparsity of solutions to high-dimensional elliptic partial differential equations. Found. Comput. Math. **16**, 813–874 (2015). https://doi.org/10.1007/s10208-015-9265-9
4. Dolgov, S.: A tensor decomposition algorithm for large ODEs with conservation laws. Comput. Methods Appl. Math. **19**, 23–38 (2019)
5. Dolgov, S., Savostyanov, D.: Parallel cross interpolation for high-precision calculation of high-dimensional integrals. Comput. Phys. Commun. **246**, 106869 (2019). https://doi.org/10.1016/j.cpc.2019.106869
6. Genz, A.: Testing multidimensional integration routines. In: Ford, B., Rault, J.C., (eds.) Tools, Methods, and Languages for Scientific and Engineering Computation, pp. 81–94. North-Holland (1984)
7. Genz, A.: A package for testing multiple integration subroutines. In: Keast, P., Fairweather, G. (eds.) NATO ASI Series, vol. 203, pp. 337–340. Springer, Dordrecht (1987). https://doi.org/10.1007/978-94-009-3889-2_33
8. Goreinov, S.A., Oseledets, I.V., Savostyanov, D.V., Tyrtyshnikov, E.E., Zamarashkin, N.L.: How to find a good submatrix. ICM Research Report 08-10, Institute for Computational Mathematics, Hong Kong Baptist University (2008)
9. Gorodetsky, A.: https://github.com/goroda/Compressed-Continuous-Computation
10. Gorodetsky, A., Karaman, S., Marzouk, Y.: A continuous analogue of the tensor-train decomposition. Comput. Methods Appl. Mech. Eng. **347**, 59–84 (2019). https://doi.org/10.1016/j.cma.2018.12.015
11. Matsakis, N.D., Klock II, F.S.: The rust language. In: ACM SIGAda Ada Letters, vol. 34, pp. 103–104. ACM (2014)
12. Oseledets, I.: https://github.com/oseledets/ttpy
13. Oseledets, I.: https://github.com/oseledets/TT-Toolbox
14. Oseledets, I., Tyrtyshnikov, E.: TT-cross approximation for multidimensional arrays. Linear Algebra Appl. **432**, 70–88 (2010). https://doi.org/10.1016/j.laa.2009.07.024
15. Oseledets, I.V.: Tensor-train decomposition. SIAM J. Sci. Comput. **33**(5), 2295–2317 (2011). https://doi.org/10.1137/090752286
16. Rust programming language. https://www.rust-lang.org/
17. https://github.com/goroda/Compressed-Continuous-Computation

18. Savostyanov, D.: https://github.com/savostyanov/ttcross
19. Tensor train topics. https://github.com/topics/tensor-train
20. Trefethen, L.N.: Approximation Theory and Approximation Practice. Society for Industrial and Applied Mathematics, Philadelphia, PA (2013). https://doi.org/10.1137/1.9781611975949

Applications of Metaheuristics
to Large-Scale Problems

Solving the Mountain Car Problem Using Genetic Algorithms

Amelia Bădică[1], Costin Bădică[1(✉)], Ion Buligiu[1], Liviu Ion Ciora[1],
Maria Ganzha[2,3], and Marcin Paprzycki[3]

[1] University of Craiova, Craiova, Romania
costin.badica@edu.ucv.ro
[2] Warsaw University of Technology, Warsaw, Poland
[3] Systems Research Institute, Polish Academy of Sciences, Warsaw, Poland

Abstract. In this work, suitability of using Genetic Algorithms (GA) to solve the Mountain Car problem is investigated. Two variants, involving pure policies, as well as slightly extended mixed policies are considered. Experimental results, obtained with CPU-based parallel implementation are presented. They highlight challenges of the GA approach, related to high computational cost required for accurate evaluation of the fitness function.

Keywords: Optimal Control · Genetic Algorithm · Parallel Programming

1 Introduction

The Mountain Car problem (MCP) is a standard benchmark for experimenting with the optimal control algorithms. The MCP assumes that an autonomous car is driving on a one-dimensional track that follows a mountain range. The goal is to reach the mountain top in a minimum number of steps. MCP was originally proposed in [5], while more details, including its explanation and modeling, are provided in [1]. The aim of this work is to investigate the suitability of using Genetic Algorithms (GA) for solving this problem. Note that the MCP is a challenge for GA, due to high cost of evaluation of the fitness function, which involves a complete simulation of the car movement, used to evaluate a given policy for solving the MCP.

In what follows we briefly introduce the MCP and outline the GA approach. We consider a discrete model of the problem, following approach introduced in [1]. Next, the mapping of standard components of GA to MCP is discussed. Here, the solution corresponds to the policy that defines the car actions in each system state. The fitness function represents the time required by the car to reach the goal. This value is conceptualized as the number of steps required to reach the goal.

Two variants of GA are considered: (1) with pure policies, and (2) with extended mixed policies. In the second approach the size of the search space is considerably larger, and a sequential implementation is not feasible. Hence, a parallel approach to GA [2] is evaluated. Experimental GA results are compared with those obtained with Q-learning and SARSA, as reported in [1]. This highlights challenges and limitations of application of GA to the MCP. Overall, the main contributions can be summarized as follows:

1. Formulating and implementing the GA approach for solving MCP.
2. Proposing a parallel implementation, for CPUs with multiple cores.
3. Experimental comparison of GA, Q-learning and SARSA algorithms applied to MCP, highlighting the advantages and limitations of the GA.

The paper is structured as follows. In Sect. 2, the MCP model and its finite state representation is introduced. Section 3 presents the design and implementation of the GA solving MCP, and the proposed approach to multi-core CPU-based parallel implementation. In Sect. 4, experimental results are presented and compared with results reported in [1], while in Sect. 4 conclusions are drawn and future research outlined.

2   Mountain Car Problem

2.1   Model of MCP

The MCP assumes that an autonomous car is driving on a one-dimensional track that follows a mountain range described by the equation:

$$y = \sin \omega x \quad \text{for} \quad x \in [-\frac{3\pi}{2\omega}, \frac{\pi}{2\omega}] \tag{1}$$

Equation (1) describes function that spans a full cycle of a "sinus" function that models a valley between a left and a right hill. The car starts moving at a random initial position x_0, somewhere in the valley in the vicinity of $-\frac{\pi}{2\omega}$. Its goal is to climb the rightmost hill. The goal can be described as the car reaching the position $x \geq x_g$, where the "goal position" x_g is a real value in the left vicinity of $\frac{\pi}{2\omega}$.

The car engine is controlled by a thruster that can either thrust left or right with equal force, or not thrust at all. The force of the thruster is not strong enough to defeat gravity and to accelerate up the slope to the top of the right hill "by itself". The solution assumes need of movement of the car to the left, to accumulate inertia to climb the hill on the right. Here, the car dynamics is described by the discrete system (2) using a time step $h > 0$. Note that g denotes the gravitational constant, m is the mass of the car, and k_f represents the friction coefficient. $F_t \in \{-1, 0, 1\}$ is the force of the thruster, which controls the dynamics of the model: push left, no push, and push right. Standard MCP defines $F_t = a_t - 1$, where $a_t \in \{0, 1, 2\}$ is the car action at time t.

$$\begin{aligned} \dot{x}_{t+h} &= \dot{x}_t + h \cdot (\frac{F_t}{m} - g \cdot \omega \cdot \cos \omega x_t - \frac{k_f}{m} \cdot \dot{x}_t) \\ x_{t+h} &= x_t + h \cdot \dot{x}_t \end{aligned} \tag{2}$$

As the aim of the car is to reach the top of the right hill, and each action of the car is evaluated accordingly. It is assumed that the car "does not know the goal", i.e. the only way to perceive the goal is when it was reached. So the car incurs a "small positive cost" (usually +1) for each state that does not achieve the goal.

Here, the autonomous car as an intelligent agent, acting according to its private strategy that specifies what the agent must do in each observed state. If agent takes action a in state s, then environment will transit in state s'. We assume that the environment is Markovian, i.e. its next state depends only on its current state and on agent action. If S denotes the set of environment states, and A denotes the set of agent actions, the strategy is called *policy* and it is defined as function π mapping each environment state to a probability distribution of the set of actions ($\mathfrak{P}(A)$ denotes the space of probability distributions with support A):

$$\pi : S \to \mathfrak{P}(A)$$
$$\pi(s) : A \to [0, 1] \;\; \forall s \in S \text{ is a probability distribution on } A \qquad (3)$$

Note that if $\pi(s) \in \{0, 1\}$, for all states $s \in S$, then policy π is called pure, the same action is always selected in the same environment state (i.e. it is deterministic). Otherwise the policy π is called mixed, or stochastic.

Let us assume that the cost of the agent taking action a, in state s, is denoted as $C(s, a)$ and that the initial state of the environment is s_0. An agent following a given policy π will generate the following agent run:

$$r = s_0 \xrightarrow{a_0 = \pi(s_0)} s_1 \xrightarrow{a_1 = \pi(s_1)} \ldots \xrightarrow{a_n = \pi(s_n)} s_{n+1} \ldots \qquad (4)$$

The total cost of the "agent run" accumulates all costs incurred in each visited state. Here, it is assumed that the potential length of the agent run is finite and unbounded. Agent run is terminated when a goal state is reached. The additive cost of agent runs is:

$$J^\pi(s = s_0) = J([s_0, \pi(s_0), s_1, \pi(s_1), \ldots, s_n, \pi(s_n), \ldots]) = \sum_{n \geq 0} C(s_n, \pi(s_n)) \qquad (5)$$

Note that if π is a mixed strategy, the cost of the agent run is defined by considering the expected value of the total cost J^π defined by Eq. (5). Taking this into account, the optimization problem is to determine an optimal policy (pure or mixed) that minimizes the expected value of agent run, as defined by Eq. (5).

2.2 Finite State Space Formulation

One difficulty in finding the optimal policy is that the state space of MCP is infinite. Therefore, state discretization, following the idea from [6], is applied to transform the discrete time continuous model into a finite state model. The size of the resulting model is controlled by the discretization steps of the state variables.

Let us consider a state variable with domain $[a, b)$ and a natural number $n \in \mathbb{N}$, representing the number of states of the finite state representation. The discretization step is computed as $h = (b-a)/n$. Each value $x \in [a, b)$ is mapped to a natural number $discrete(x) \in \{0, 1, \ldots, n-1\}$ according to Eq. (6).

$$discrete(x) = k \in \{0, 1, \ldots, n-1\} \iff x \in I_k = [a + k \cdot h, a + (k+1) \cdot h) \quad (6)$$

The MCP model (2) defines two state variables: position and velocity. If the finite state representation uses n_p positions and n_v velocities, then the size of the MCP state space becomes manageable, i.e. $|S| = n_p \cdot n_v$. Taking into account that the action space A has only three elements: push left, no push, and push right, the space of pure policies, defined by Eq. (3), has an exponential size $3^{n_p \cdot n_v}$.

The size of mixed policy space is still infinite after the discretization step. But, attention is restricted to only a fixed number $k \geq 3$ of mixed policies (including the pure policies) then we obtain a policy space of size $k^{n_p \cdot n_v}$.

3   System Design and Implementation

3.1   Genetic Algorithms

GAs are population-based heuristic optimization algorithms, inspired from principles of genetics and natural selection [3]. Vast majority of optimization problems, arising in engineering and science, cannot be solved directly, using standard optimization algorithms. Typically, such problems are NP-hard, and have huge search spaces. Therefore, GAs are applied as an alternative solution for such complex computational problems. The finite state formulation of the MCP, introduced in Sect. 2, is characterized by a very large state-space, suggesting possibility of employing a GA-based approach.

There are few conceptual ingredients of GA [3]. In what follows, description how they have been mapped to solve the MCP is provided:

- *Individual* (chromosome) represents a possible solution of the problem under consideration. For the MCP, an individual is represented by a pure (or mixed) car strategy and it is a sequence of genes. Each gene corresponds to a state consisting of a pair formed by the car location and velocity. The value of the gene represents the car action that is assigned to that state. An action can be pure, i.e. an element of A, or mixed, i.e. a probability distribution from $\mathfrak{P}(A)$.
- *Population* and *generation*. A population represents a set of possible solutions. A generation is the current population, at a given iteration of the GA. For the MCP, the population is a vector of policies. For each iteration, a generation represents the current value of the vector of policies under analysis.
- *Fitness function*. For the MCP, it is represented by the expected value of the total cost incurred by the car (see Eq. (5)). Our goal is to minimize the expected value of the total cost.

– *Genetic operators.* Mutation, crossover and selection are the standard genetic operators. Mutation refers to altering some of the genes of a chromosome, to produce a new chromosome. Crossover involves combination of two or more chromosomes, to generate a new chromosome. Selection refers to the definition of a new generation, based on the current generation, by using mutation and crossover and evaluating them using the fitness function. A specific number of the fittest chromosomes are maintained from one generation to another.

3.2 Design and Implementation Details

Let us start by providing the structure of the proposed GA Algorithm 1. It is described in an abstract way, avoiding technical details that are available in the implemented code.

The algorithm evolves a population of policies. Each policy is evaluated to determine the fitness function, by performing stochastic simulation (as in the most general case mixed policies are applied). The simulation is done for a maximum number of steps (car actions), or until the goal is reached (whichever occurs first). Each simulation is called a game. The simulation is repeated for a given number of games – this number is rather low to avoid high costs of computations. The fitness score of an individual (car policy) is obtained by averaging the costs determined for each evaluation game. Note that even for pure policies, the costs of each game can be different, as in the classic MCP the initial position of the car is a random value. The final generation of individuals is, however, more accurately evaluated, by repeating simulations and averaging total costs for a higher number of testing games.

As in the standard GA setting, a fixed number of fittest individuals is transferred from the given generation to the next. Moreover, this generation will be augmented with new individuals (children created using crossover and/or mutated). Parent individuals are selected for crossover using a stochastic strategy that allocates a higher selection probability to the fittest individuals of the current generation. This is known as the fitness proportional selection and it was implemented using the roulette wheel strategy [4]. Taking into account that the minimization problem is addressed, instead of using the actual fitness f, the value $\lambda \cdot f_{max} - f$ is used, where f_{max} is the maximum possible value of fitness function and $\lambda > 1$ is a scaling factor. Taking into account that fitness represents the cost incurred by the car to reach the goal in a given number of steps (car actions), f_{max} is defined as the maximum number of steps that the car is allowed to take in each game.

The population is represented as a *NumPy* array of policies. A policy is represented as a *NumPy* structured array that contains two fields:

– Field *'individual'* that represents the policy as an $n_p \times n_v \times 2$ matrix. For a given *policy*, for each $p \in 0 \ldots n_p - 1$ and $v \in 0 \ldots n_v - 1$, *policy*$[p, v]$ is a two dimensional array that represents a mixed action in state (p, v) as follows. If our mixed action (p_0, p_1, p_2) selects action i with probability p_i from $i = 0, 1, 2$ such that $p_0 + p_1 + p_2 = 1$ then in our representation the action is

---

Algorithm 1. *GASolver.* Outline of GA for solving MCP.

Input: Population size.
       Number of iterations.
       Number of fittest individuals.
       Number of evaluation games.
       Number of testing games.
       Maximum number of steps (car actions) in a game.
Output: Population.
 1: Initialize population
 2: Evaluate population using number of evaluation games
 3: Sort population in increasing order according to fitness function
 4: **for** Each iteration **do**
 5:    **if** Population is not good (goal not achieved for at least one policy) **then**
 6:       Initialize population
 7:       Evaluate population using number of evaluation games
 8:    **else**
 9:       Create new population of size equal to population size minus number of fittest
10:       **for** Each individual of new population **do**
11:          Choose two different weighted random individuals from population
12:          Compute their child using crossover
13:          Copy child to individual of the new population
14:       **end for**
15:       Transfer new population to population preserving fittest individuals in population
16:       Evaluate population using number of evaluation games except fittest
17:    **end if**
18:    Sort population in increasing order according to fitness function
19: **end for**
20: Evaluate population using number of testing games
21: **return** Population

---

captured as a two-dimensional array $[p_0, p_0 + p_1]$. For example mixed action $(1/4, 1/2, 1/4)$ is represented as $[1/4, 3/4]$. This representation facilitates the use of the roulette wheel strategy for selecting the actual car action based on the mixed action.

– Field *'weight'* that represents the fitness measure of the policy.

Crossover is a binary operator that generates a new child from two parents. There are several approaches for defining crossover operations, including a single point crossover and a uniform crossover. Single point crossover assumes choosing a random split point, splitting the parents at that point and then composing the child by taking the first part of one parent and the second half of the other parent. Uniform crossover makes a series of random choices by selecting successively genes of the child from either of the parents.

Mutation is an unary operator that alters one individual by changing some of its genes. Here, mutation is activated by a probability threshold, known as mutation rate, assessed after each crossover operation. So each child, obtained after a crossover operation, is possibly mutated when the mutation condition holds. The mutation itself involves flipping genes (i.e. mixed actions) of a policy, according to a probability threshold known as a mutation probability. It is assumed that for each state a fixed number of mixed actions is available to choose from, known as action space.

3.3 Parallel Approach

There are many possibilities for approaching the parallelization of a GA. According to [2], they can be classified, based on the hardware architecture, into CPU-based and GPU-based solutions and, based on the conceptual approach, into master-slave, island, and cellular models. In this work, a simpler, CPU-based master-slave model, that is easily mapped on a multi-core CPU was explored. According to this model, the GA maintains a single population of policies and the evaluation of their fitness measures using stochastic simulation is distributed among the available cores of the CPU. The implementation was realized using *Pool* class of Python's *multiprocessing* package that implements a pool of working processes, to which simulation jobs are submitted.

4 Experimental Results

4.1 Experimental Setup

Overall, three experiments have been completed, using the following parameter settings.

- MCP model parameters were similar with [1], see also Eq. (7).
- State discretization parameters were set to $n_p = n_v = 20$.
- One pure policy and two mixed policies with action spaces A_1, A_2, A_3 defined by Eq. (8) were explored. Note that A_1 contains only pure actions, while A_2 and A_3 contain both pure and mixed actions.
- GA parameters were set as follows: uniform crossover, population size = 20, number of fittest = 5, mutation rate = 0.05, mutation probability = 0.50, evaluation games = 20, evaluation steps = 10000, testing games = 1000, number of iterations = 100.
- Number of cores: we have used the maximum number of logical processes of our machine, i.e. 16 (value returned by *multiprocessing.cpu_count()*).

$$
\begin{aligned}
&\omega = 3.0 & m &= 1000.0 & g &= 0.0025/3.0 \\
&k_f = 0 & h &= 1.0 & x_g &= 0.5 \\
&F_t \in \{-1, 0, 1\} & x_t &\in [-1.2, 0.6] & v_t &\in [-0.07, 0.07]
\end{aligned}
\tag{7}
$$

$$
\begin{aligned}
A_1 &= \{(0,0,1), (0,1,0), (1,0,0)\} \\
A_2 &= A_1 \cup \{(0,1/2,1/2), (1/3,1/3,1/3), (1/2,0,1/2), (1/2,1/2,0)\} \\
A_3 &= A_2 \cup \{(0,1/4,3/4), (1/4,0,3/4), (1/2,1/4,1/4), (1/4,1/4,1/2), \\
&\qquad (3/4,1/4,0), (3/4,0,1/4), (1/4,1/2,1/4), (1/4,3/4,0), (0,3/4,1/4)\}
\end{aligned}
\tag{8}
$$

For each experiment the best individual in the last population, as well as the execution time (including the time required by the final more accurate evaluation of the fitness of individuals in the last population) were recorded. Each experiment was repeated 6 times and the best, and the average and standard deviation values of the total cost and total running time for each experiment were collected.

4.2   Results and Discussions

The experimental results are presented in Table 1. They show the total cost recorded for each experiment, including average, minimum, maximum and standard deviation values, as well as running times. Observe that, in terms of average costs, the best solution was obtained by experiment 3 that considers the larger space of mixed actions (marked with bold in Table 1). The costs obtained by the GA solution are slightly worse than those obtained using Q-learning and SARSA, as reported in [1].

The computational cost, in terms of running time, was rather high. On average, each GA run of an experiment took more than 10 minutes on a Windows 10 PC equipped with a 11th Gen Intel(R) Core(TM) i9-11900H ©2.50GHz CPU with 8 cores and 16 logical processes. So overall running all the experiments took around 3 h.

Although the obtained results are promising, the proposed solution has limitations.

- The slightly worse total costs obtained by GA, as compared with SARSA and Q-learning, might be caused by the rather low accuracy for evaluating the fitness of a policy. The evaluation was done by performing a number of stochastic simulations, and the number of jobs had to be maintained as low as possible, because of the high computational cost incurred.
- The parallelism of our solution is limited by the low number of available CPU cores. We plan to address this issue by using a GPU-based computing resources.

Table 1. Experimental results.

Exp.no.	Act.space (8)	Avg.cost±std.dev.	Min.&max.cost	Avg.time±std.dev. [sec]
1	A_1 pure	262.75 ± 51.07	$199.36, 346.26$	660.83 ± 87.52
2	A_2 mixed	256.49 ± 31.16	$223.86, 312.43$	704.66 ± 164.27
3	A_3 mixed	$\mathbf{234.97 \pm 32.79}$	$203.95, 283.26$	745.33 ± 180.19

5   Concluding Remarks

In this contribution, a GA-based approach for solving the MCP was explored. The main challenge was the high computational cost incurred for accurately evaluating more the fitness function, through stochastic simulation. Moreover, an initial parallel implementation was proposed and experimented with. In the future, we plan to investigate viability of use of a GPU-based computer to improve efficiency of the GA-based approach.

Acknowledgement. This work has been supported by the joint research project "Intelligent distributed systems and applications" under the agreement on scientific cooperation between the Polish Academy of Sciences and Romanian Academy for years 2023–2025.

References

1. Bădică, A., Bădică, C., Ivanović, M., Logofătu, D.: Experiments with solving mountain car problem using state discretization and q-learning. In: Nguyen, N.T., Tran, T.K., Tukayev, U., Hong, T.P., Trawiński, B., Szczerbicki, E. (eds.) ACIIDS 2022. LNCS, vol. 13757, pp. 142–155. Springer, Cham (2022). https://doi.org/10.1007/978-3-031-21743-2_12
2. Cheng, J.R., Gen, M.: Parallel genetic algorithms with GPU computing. In: Bányai, T., Petrillo, A., De Felice, F. (eds.) Industry 4.0 – Impact on Intelligent Logistics and Manufacturing, chap. 6. IntechOpen, Rijeka (2020). https://doi.org/10.5772/intechopen.89152
3. Eiben, A.E., Smith, J.E.: Introduction to Evolutionary Computing, 2nd edn. Springer, Heidelberg (2015). https://doi.org/10.1007/978-3-662-44874-8
4. Haupt, R.L., Haupt, S.E.: Practical Genetic Algorithms, 2nd edn. Wiley, Hoboken (2004)
5. Moore, A.W.: Variable resolution dynamic programming: efficiently learning action maps in multivariate real-valued state-spaces. In: Birnbaum, L.A., Collins, G.C. (eds.) Machine Learning Proceedings 1991, pp. 333–337. Morgan Kaufmann, San Francisco (1991). https://doi.org/10.1016/B978-1-55860-200-7.50069-6
6. Tabor, P.: Q learning with just numpy. Solving the mountain car. Tutorial (2019). https://www.youtube.com/watch?v=rBzOyjywtPw&t=3s. Accessed 07 Jan 2022

Ant Algorithm with Local Search Procedure for Multiple Knapsack Problem

Stefka Fidanova[1(✉)] and Krassimir Atanassov[2]

[1] Institute of Information and Communication Technologies,
Bulgarian Academy of Sciences, Sofia, Bulgaria
stefka.fidanova@gmail.com
[2] Institute of Biophysics and Biomedical Engineering,
Bulgarian Academy of Sciences, Sofia, Bulgaria

Abstract. Multiple Knapsack Problem (MKP) is a hard combinatorial optimization problem with large application. A lot of real life and industrial problems can be defined like MKP, therefore it attracts the attention of the scientists. Exact methods and traditional numerical methods are appropriate for solving small problems or problems without hard constraints. For problems, which needs non polynomial (NP) number of calculations is better to apply so called metaheuristic methods. Metaheuristics are methodology and on their basis is constructed problem dependent algorithm. Metaheuristic methods apply some stochastic rules and it helps to find faster near optimal solution even for huge problems. Ant Colony Optimization (ACO) is a nature inspired method, which follows the real ants behavior. It is between the best methods for solving combinatorial optimization problems. Sometimes the method alone is not enough to find good solutions, especially when the problem has strong constraints. In this case, one resorts to constructing an appropriate local search procedure. The aim is to find better solutions or to fasten the search process. The solutions of the problem can be represented by binary sequence. Let us consider this binary sequence as a binary number. We will calculate the average between the best solution, represented as a binary number, and any of the current solutions. The new binary number will be the new solution after local search procedure.

1 Introduction

Methods for solving mathematical problems are divided into exact and approximate. On the other hand, the approximate methods are divided into traditional numerical methods where we have guaranteed accuracy, and stochastic (metaheuristic) methods. Metaheuristic methods are applied to problems of great complexity (NP) or to problems of high dimension for which traditional numerical methods are inapplicable. A large part of metaheuristic methods are nature inspired.

The most popular metaheuristic methods are Evolutionary algorithm [14,31], which simulates the Darwinian evolutionary concept, Simulated Annealing [18]

I. Lirkov and S. Margenov (Eds.): LSSC 2023, LNCS 13952, pp. 246–252, 2024.
https://doi.org/10.1007/978-3-031-56208-2_24

and Gravitation search algorithm [26], Tabu Search [27] and Interior Search [28]. The ideas for swarm-intelligence based algorithms come from behavior of animals in the natures. The representatives of this type of algorithms are Ant Colony Optimization [3], Bee Colony Optimization [15], Bat algorithm [33], Firefly algorithm [34], Particle Swarm Optimization [17], Gray Wolf algorithm [25] and so on.

ACO is among the best methods for solving combinatorial optimization problems. The main concept of this method is imitation of real ants behavior. Dorigo first used the behavior of ants to solve combinatorial optimization problems [2–4]. Since then, this method has been repeatedly improved: elitist ants [4]; ant colony system [5]; max-min ant system [30]; rank-based ant system [4]; ant algorithm with additional reinforcement [9]. They differ in pheromone updating. For some of them is proven that they converge to the global optimum [4]. Fidanova et al. [10–12] proposed semi-random start of the ants comparing several start strategies. The method can be adapted to dynamic changes of the problem in some complex problems [7,8,30].

When the constraints are too restrictive, the algorithms hit local optima early or fail to construct a feasible solution. This situation is solved by combining several approaches. In this way, the capabilities of both algorithms are used in order to find better solutions or reduce the time to solve the problem. We apply ACO on Multiple Knapsack Problem (MKP). We have constructed a Local Search (LS) procedure to improve the performance of the algorithm.

The organization of the paper is as follow: In Sect. 2 problem is defined. Section 3 is devoted to the ACO algorithm. The description of the local search procedure is in Sect. 4. Section 5 provides concluding remarks and directions for future work.

2 Problem Definition

Let's imagine we have a knapsack and items we want to put in it. Each item has a volume and a price. The question is which items to put inside so that the knapsack has the maximum price while respecting its volume limit.

Multiple Knapsack Problem (MKP) is a generalization of single knapsack problem, where there are several knapsacks with different volume and the aim is to maximize the total price of the knapsacks. MKP is a representative of subset [24] problems and a lot of industrial problems can be described as MKP. Economical problems, which can be defined as MKP are Resource allocation in distributed systems, capital budgeting [16]. Other problem is patients scheduling [1]. Through MKP bin packing [23] and cutting stock [6] problems can be represented. Other difficult problems, which can be solved with the help of MKP are multi-processor scheduling [21] and crypto-systems and generating keys [16]. MKP models large set of binary problems with integer coefficients [19,24].

MKP is hard optimization problem (NP), which is solved with metaheuristic methods such as genetic algorithm [22], tabue search [32], swarm intelligence [20], ACO algorithm [9,13].

The most popular definition of MKP is as resource allocation problem, where m is the number of resources (the knapsacks) and n is the number of the objects. The object j has a profit p_j. Each resource has its own budget (knapsack capacity) and consumption r_{ij} of resource j by object i. The purpose is maximization of the profit within the limited budget.

The mathematical formulation of MKP can be as follows:

$$\max \sum_{j=1}^{n} p_j x_j$$

$$\text{subject to} \sum_{j=1}^{n} r_{ij} x_j \leq c_i \quad i = 1, \ldots, m \tag{1}$$

$$x_j \in \{0, 1\} \quad j = 1, \ldots, n$$

There are m constraints in this problem, so MKP is also called m-dimensional knapsack problem. Let $I = \{1, \ldots, m\}$ and $J = \{1, \ldots, n\}$, with $c_i \geq 0$ for all $i \in I$. A well-stated MKP assumes that $p_j > 0$ and $r_{ij} \leq c_i \leq \sum_{j=1}^{n} r_{ij}$ for all $i \in I$ and $j \in J$. Note that the $[r_{ij}]_{m \times n}$ matrix and $[c_i]_m$ vector are both non-negative.

The MKP partial solution is represented by $S = \{i_1, i_2, \ldots, i_j\}$ and the last element included to S, i_j is not used in the selection process for the next element. Thus the solution of MKP have not fixed length. Other representation of the solution is as binary string with length n, where 1 represents chosen items and 0 represents not chosen items.

3   Ant Colony Optimization

The ant methodology was first proposed by Marco Dorigo [3]. It mimics the foraging of ants in nature. Application is problem dependent. First the problem is represented by graph and the solutions are represented by path in a graph. The ants look for a shorter/longer path respecting problem constraints.

Every ant starts from random node and accept next node in the solution by probabilistic rule called transition probability. The transition probability $P_{i,j}$, is a product of the heuristic information $\eta_{i,j}$ and the pheromone trail level $\tau_{i,j}$ related to the selection of node j if the previous selected node is i, where $i, j = 1, \ldots, n$.

$$P_{i,j} = \frac{\tau_{i,j}^a \cdot \eta_{i,j}^b}{\sum_{k \in Unused} \tau_{i,k}^a \cdot \eta_{i,k}^b}, \tag{2}$$

where $Unused$ is the set of unused nodes.

At the beginning, the pheromone is initialized with a small constant value τ_0, $0 < \tau_0 < 1$. Every time the ants build a solution, the pheromone is bring up to date [4]. The elements of the graph with more pheromone are more tempting to the ants.

The main update rule for the pheromone is:

$$\tau_{i,j} \leftarrow \rho \cdot \tau_{i,j} + \Delta\tau_{i,j}, \tag{3}$$

where parameter ρ decreases the value of the pheromone, like evaporation in a nature decreases the quantity of old pheromone. $\Delta\tau_{i,j}$ is a new deposited pheromone, which depends on the value of the objective function, corresponding to this solution.

For MKP the items are related with the nodes of the graph and the edges fully connect the nodes. The heuristic information is an important part of transition probability function. It is an appropriate combination of problem parameters. Let $s_j = \sum_{i=1}^{m} r_{ij}$. For heuristic information we use:

$$\eta_{ij} = \begin{cases} p_j^{d_1}/s_j^{d_2} & \text{if } s_j \neq 0 \\ p_j^{d_1} & \text{if } s_j = 0 \end{cases} \tag{4}$$

where $d_1 > 0$ and $d_2 > 0$ are parameters. Hence the objects with greater profit and less average expenses will be more desirable.

4   Local Search Procedure

Sometimes using only one method is not enough to quickly find good solutions to the problem. Most often, one basic method is used and it is combined with a local search procedure. Usually, the local search procedure is applied at the end of the iteration to find neighboring solutions. If any of the neighboring solutions is better than the current one, it is accepted as a new current solution [29].

MKP solutions are represented as binary strings, where 1 represents chosen items and 0 other items. Let S^* is a global best solution and S_i is the solution, constructed by ant i during the current iteration. Let's think of the solutions as binary numbers. We generate neighbor solution as follows:

$$S_{new} = \frac{S^* + S_i}{2} \tag{5}$$

arithmetic mean between the best so far solution and the achieved by ant i solution during the iteration.

We verify if the neighbor solution S_{new} is feasible. If it is a feasible solution, it is considered as a partial solution and is supplemented by the use of ACO. We then compare the solution S_{new}, thus obtained, with the current solution S_i, and if it is better than it, it becomes the new current solution. The local search procedure is applied only once, disregarding whether the new solution S_{new} is feasible or not. Thus, the execution time of the iteration increases slightly. At the end of every iteration the quantity of the pheromone is updated according the solutions after local search procedure.

ACO algorithm for MKP, combined with proposed local search procedure is tested on 10 test instances from Operational Research Library "OR-Library" available within WWW access at http://people.brunel.ac.uk/mastjjb/jeb/info.

html. The test problems consists of 100 items and 10 constraints/knapsacks. The proposed hybrid ACO algorithm is realized as a software, coded in $C++$ programming language and is run on Pentium desktop computer at 2.8 GHz with 4 GB of memory. We fixed experimentally ACO parameters after several runs as follows:

- Number of iterations = 300
- Number of ants = 20
- $\rho = 0.5$
- $\tau_0 = 0.5$
- $a = 1$
- $b = 1$
- $d_1 = 1$
- $d_2 = 1$

Proposed local search procedure is not time consuming, because we apply it ones despite the new found solution is better or not. ACO algorithm is stochastic, therefore we perform 30 independent runs of every test problem and calculate average value of the objective function (knapsack profit) and average number of iterations to achieve this solution. We observe that our hybrid ACO algorithm achieves the same best solution as the main ACO algorithm, but it needs less number of iterations. Applying local search procedure increases the time, consumed by an individual iteration. We apply the procedure only ones and thus this added time is minimal. Application of local search decreases needed number of iterations to find best solution and thus it leads to minimization of the total calculation time. If we imagine two solutions as points in the solution space, the proposed local search procedure finds points/solutions that are the middle of the segment, connecting the best solution found so far with any of the current solutions. Thus, it finds intermediate solutions, close to the already found ones, through which the best solution of the problem can be reached more quickly.

5   Concluding Remarks

A hybrid algorithm, combination of two approaches is proposed. ACO algorithm is used as main algorithm and at the end of every iteration is applied local search procedure. Solutions are represented like binary number and the neighbor solutions are arithmetic mean between the best so far solution and the achieved solutions during iteration. The neighbor solution is accepted if it is better than current one. The proposed hybrid algorithm is tested on 10 MKP problems and shows decrease of computational time, needed to find the best solution.

The proposed local search procedure can be combined with other local search procedures to further improve the effectiveness and efficiency of algorithms for solving complex optimization problems.

Acknowledgment. The development of the proposed hybrid ACO algorithm with local search procedure has been funded by the Grant DFNI KP-06-N52/5 and Grant No BG05M2O P001-1.001-0003, financed by the Science and Education for Smart Growth Operational Program and co-financed by the European Union through the European structural and Investment funds.

References

1. Arsik, I., et al.: Effective and equitable appointment scheduling in rehabilitation centers. In: INFORMS Annual Meeting 2017 (2017)
2. Birattari, M., Stutzle, T., Paquete, L., Varrentrapp, K.: A racing algorithm for configuring metaheuristics. In: Proceedings of the Genetic and Evolutionary Computation Conference, pp. 11–18 (2002)
3. Bonabeau, E., Dorigo, M., Theraulaz, G.: Swarm Intelligence: From Natural to Artificial Systems. Oxford University Press, New York (1999)
4. Dorigo, M., Stutzle, T.: Ant Colony Optimization. MIT Press, Cambridge (2004)
5. Dorigo, M., Gambardella, L.: Ant colony system: a cooperative learning approach to the traveling salesman problem. IEEE Trans. Evol. Comput. **1**, 53–66 (1996)
6. Evtimov, G., Fidanova, S.: Heuristic algorithm for 2D cutting stock problem. In: Lirkov, I., Margenov, S. (eds.) LSSC 2017. LNCS, vol. 10665, pp. 350–357. Springer, Cham (2018). https://doi.org/10.1007/978-3-319-73441-5_37
7. Fidanova, S., Lirkov, I.: 3D protein structure prediction. In: Analele Universitatii de Vest Timisoara, vol. XLVII, pp. 33–46 (2009)
8. Fidanova, S.: An improvement of the grid-based hydrophobic-hydrophilic model. Int. J. Bioautomation **14**, 147–156 (2010)
9. Fidanova, S.: ACO algorithm with additional reinforcement. In: Dorigo, M., Di Caro, G., Sampels, M. (eds.) ANTS 2002. LNCS, vol. 2463, pp. 292–293. Springer, Heidelberg (2002). https://doi.org/10.1007/3-540-45724-0_31
10. Fidanova, S., Atanassov, K., Marinov, P.: Generalized Nets and Ant Colony Optimization. Bulg. Academy of Sciences Pub. House (2011)
11. Fidanova, S., Atanassov, K., Marinov, P.: Start strategies of ACO applied on subset problems. In: Dimov, I., Dimova, S., Kolkovska, N. (eds.) NMA 2010. LNCS, vol. 6046, pp. 248–255. Springer, Heidelberg (2011). https://doi.org/10.1007/978-3-642-18466-6_29
12. Fidanova, S., Atanassov, K., Marinov, P.: Intuitionistic fuzzy estimation of the ant colony optimization starting points. In: Lirkov, I., Margenov, S., Waśniewski, J. (eds.) LSSC 2011. LNCS, vol. 7116, pp. 222–229. Springer, Heidelberg (2012). https://doi.org/10.1007/978-3-642-29843-1_25
13. Fidanova S. Hybrid ant colony optimization algorithm for multiple knapsack problem. In: 5th IEEE International Conference on Recent Advances and Innovations in Engineering (ICRAIE). IEEE (2021). https://doi.org/10.1109/ICRAIE51050.2020.9358351,1-5
14. Goldberg, D.E., Korb, B., Deb, K.: Messy genetic algorithms: motivation analysis and first results. Complex Syst. **5**(3), 493–530 (1989)
15. Karaboga, D., Basturk, B.: Artificial bee colony (ABC) optimization algorithm for solving constrained optimization problems. In: Melin, P., Castillo, O., Aguilar, L.T., Kacprzyk, J., Pedrycz, W. (eds.) IFSA 2007. LNCS (LNAI), vol. 4529, pp. 789–798. Springer, Heidelberg (2007). https://doi.org/10.1007/978-3-540-72950-1_77

16. Kellerer, H., Pferschy, U., Pisinger, D.: Multiple knapsack problems. In: Kellerer, H., Pferschy, U., Pisinger, D. (eds.) Knapsack Problems, pp. 285–316. Springer, Berlin (2004). https://doi.org/10.1007/978-3-540-24777-7_10

17. Kennedy, J., Eberhart, R.: Particle swarm optimization. In: Proceedings of IEEE International Conference on Neural Networks, vol. IV, pp. 1942–1948 (1995)

18. Kirkpatrick, S., Gelatt, C.D., Vecchi, M.P.: Optimization by simulated annealing. Science (New York, N.Y.) **13**(220), 671–680 (1983)

19. Kochenberger, G., McCarl, G., Wymann, F.: An heuristic for general integer programming. Decis. Sci. **5**, 34–44 (1974)

20. Krause, J., Cordeiro, J., Parpinelli, R.S., Lopes, H.S.: A survey of swarm algorithms applied to discrete optimization problems. In: Swarm Intelligence and Bio-Inspired Computation, pp. 169–191. Elsevier (2013)

21. Lawler, E.L., Lenstra, J.K., Rinnooy Kan, A.H.G., Shmoys, D.B.: Sequencing and scheduling: algorithms and complexity. In: Graves, S.C., et al. (eds.) Handbooks in OR and MS, vol. 4, pp. 445–522. Elsevier Science Publishers (1993)

22. Liu, Q., Odaka, T., Kuroiwa, J., Shirai, H., Ogura, H.: A new artificial fish swarm algorithm for the multiple knapsack problem. IEICE Trans. Inf. Syst. **97**(3), 455–468 (2014)

23. Murgolo, F.D.: An efficient approximation scheme for variable-sized bin packing. SIAM J. Comput. **16**(1), 149–161 (1987)

24. Leguizamon, G., Michalevich, Z.: A new version of ant system for subset problems. In: International Conference on Evolutionary Computations, vol. 2, pp. 1459–1464 (1999)

25. Mirjalili, S., Mirjalili, S.M., Lewis, A.: Grey wolf optimizer. Adv. Eng. Softw. **69**, 46–61 (2014)

26. Mosavi, M.R., Khishe, M., Parvizi, G.R., Naseri, M.J., Ayat, M.: Training multilayer perceptron utilizing adaptive best-mass gravitational search algorithm to classify sonar dataset. Arch. Acoust. **44**(1), 137–151 (2019)

27. Osman, I.H.: Metastrategy simulated annealing and tabue search algorithms for the vehicle routing problem. Ann. Oper. Res. **41**(4), 421–451 (1993)

28. Ravakhah, S., Khishe, M., Aghababaee, M., Hashemzadeh, E.: Sonar false alarm rate suppression using classification methods based on interior search algorithm. Int. J. Comput. Sci. Netw. Secur. **17**(7), 58–65 (2017)

29. Schaffer, A.A., Yannakakis, M.: Simple local search problems that are hard to solve. Soc. Industr. Appl. Math. J. Comput. **20**, 56–87 (1991)

30. Stutzle, T., Hoos, H.: Max min ant system. Futur. Gener. Comput. Syst. **16**, 889–914 (2000)

31. Vikhar, P.A.: Evolutionary algorithms: a critical review and its future prospects. In: Proceedings of the 2016 International Conference on Global Trends in Signal Processing, Information Computing and Communication (ICGTSPICC), Jalgaon, pp. 261–265 (2016)

32. Woodcock, A.J., Wilson, J.M.: A hybrid tabue search/branch and bound approach to solving the generalized assignment problem. Eur. J. Oper. Res. **207**(2), 566–578 (2010)

33. Yang, X.S.: A new metaheuristic bat-inspired algorithm. In: González, J.R., Pelta, D.A., Cruz, C., Terrazas, G., Krasnogor, N. (eds.) Nature Inspired Cooperative Strategies for Optimization (NICSO 2010). SCI, vol. 284, pp. 65–74. Springer, Heidelberg (2010). https://doi.org/10.1007/978-3-642-12538-6_6

34. Yang, X.S.: Nature-Inspired Metaheuristic Algorithms. Luniver Press (2008)

Variable Neighborhood Search in Hamming Space

S. B. Hengeveld and A. Mucherino[✉]

IRISA, University of Rennes, Rennes, France
{simon.hengeveld,antonio.mucherino}@irisa.fr

Abstract. Variable Neighbourhood Search (VNS) is one of the most used meta-heuristics for global optimization. We focus our attention on combinatorial problems and propose a new implementation of VNS in Hamming space. Together with the basic VNS approach initially proposed by Hansen and Mladenović, we present a comparison among the most known VNS variants and we adapt them for the Hamming space. Our VNS is coded in Java programming language and implements the interfaces of the BINMETA public project, where other meta-heuristics were previously included. Our computational experiments, on a small selection of combinatorial problems, show that our new VNS implementations outperform the other meta-heuristics present in BINMETA.

1 Introduction

Meta-heuristics are widely used for the solution of hard optimization problems [16]. Variable Neighbourhood Search (VNS) is one of the most known and useful meta-heuristics for global optimization. The very first paper introducing VNS [14] appeared in 1997 and was cited, accordingly to Google Scholar, more than 5000 times in about 25 years. VNS was in fact used for attempting the solution of several problems that are otherwise intractable, and this large use encouraged the proposition in the scientific literature of several VNS variants.

This work is based on the observation that, in spite of this vast use of VNS, and in particular for the solution of combinatorial problems, there exists, to date, no VNS implementation that is specialized for this kind of problems. By specialization we mean the possibility for the meta-heuristic search to manipulate solutions expressed in binary representation, and to generate and explore neighbourhoods by using a metric different from the classical Euclidean distance. In this work, we will employ the Hamming distance.

Given a binary representation for the solutions of a given optimization problem, the smallest piece of information that we can consider is the *bit*, while a bit string of fixed length n provides the full representation. Two solutions in this representation differ therefore if at least one of their bits, when placed in the same position in the bit string, have opposite value. The Hamming distance counts the number of such different bits in two given representations.

© The Author(s), under exclusive license to Springer Nature Switzerland AG 2024
I. Lirkov and S. Margenov (Eds.): LSSC 2023, LNCS 13952, pp. 253–261, 2024.
https://doi.org/10.1007/978-3-031-56208-2_25

We point out that this is not the first contribution on meta-heuristics in Hamming space; the reader can refer for example to [1]. This is not the first contribution either where the Hamming distance is used in the conception of VNS. In [10], for example, it was already suggested to use binary strings to represent solutions when solving the simple plant location problem, and hence to define neighborhoods by using the Hamming distance. Moreover, two examples of VNS in Hamming space can be found in the context of Mixed Integer Programming (MIPs) for scheduling jobs [7] and nurse rostering problems [6]. More recently, in 2014, a VNS implementation acting on binary strings was used to perform edge-ratio network clustering in [4]. However, to the best of our knowledge, ours is the first general-purpose VNS working with binary strings and exploring neighbourhoods defined in Hamming space.

The rest of the paper is organized as follows. In Sect. 2, we will briefly recall the main ideas behind VNS, and we will focus our attention on a set of well-known VNS variants. In Sect. 3, we will present our implementation of such variants. Notice that this implementation is publicly available on the BINMETA repository, together with the Java class used for running the experiments that we will present in Sect. 4. Finally, Sect. 5 will conclude the paper.

2   Basic VNS and Variants

The basic idea behind VNS is very simple and very effective [9]. Given the current solution x to a given optimization problem with objective function f, the search attempts discovering better quality solutions x', i.e. solutions such that $f(x') < f(x)$, in a relatively small neighbourhood of x. When this is not possible (or it looks like it is not possible), larger and larger neighbourhoods can be taken into consideration. This initial phase of VNS is generally referred to as the *shaking procedure*, and it is often simply implemented by picking up a random solution in the current neighbourhood. When it fails to find better quality solutions, the neighbourhood is generally enlarged.

When a new solution x' capable of improving the objective function f is identified during the shaking phase, then the current neighbourhood may be recentered at x', because x' is then supposed to be the starting point for further explorations. If the neighbourhood had increased in size during the shaking phase, then it may also be reduced to its initial size, in order to focus the search around the nearest solutions to the current best one. We use the conditional form in this paragraph because many variants of VNS act in slightly different ways [12]; the behavior briefly mentioned here is the one of the basic VNS initially proposed in 1997 by Mladenović and Hansen [14].

Figure 1 shows, in our unique graph representation, the main steps for one iteration of the VNS variants considered in this work. The original VNS is given by the path starting with the shaking procedure, going through the local search, and then to the node performing the "sequential neighbourhood change". Subsequently, a new iteration is performed if the stopping criteria is not yet verified.

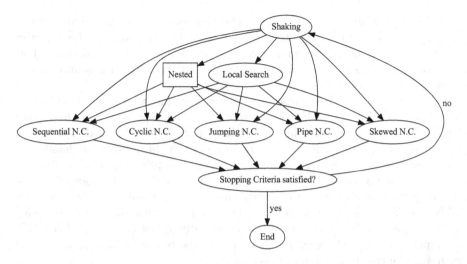

Fig. 1. A complete path on this directed graph depicts the general iteration of one VNS variant. From the shaking node, the path may go through the local search, or to point directly to one of the possible ways to update the current neighbourhood. N.C. stands for Neighbourhood Change. The graph vertex named Nested is a little special, because it represents an independent call to a VNS variant (except itself). When the stopping criteria is satisfied, VNS stops; otherwise, it cycles back to the shaking procedure for beginning with another iteration.

For lack of space, we opt in this paper for a simple description of the VNS variants providing only some main ideas behind their conception; the reader interested in a more complete and formal description of the variants can make reference to [11,12], as well as to citations given below for some specific variants.

The variations on the basic VNS basically consist in the choice on whether running or not the local search immediately after the shaking procedure, as well as in the choice of the approach for changing the current neighbourhood. In the cyclic approach [19], the size of the neighbourhoods is not reset to the initial size whenever better solutions x' are found; the neighbourhood change only consists in recentering it at x'. The adjective "cyclic" is employed here to point out the fact that the growth of the neighbourhood sizes does not depend on the encountered solutions, but it may be reset to the initial size when the maximal size has been reached and the VNS stopping criteria is not yet satisfied.

The pipe neighbourhood change only differs from the cyclic approach by the fact that the latter avoids increasing the size of the neighbourhood when a better solution x' has just been found [20]. Moreover, in the jump variant [11], the changes in the size of the neighbourhood are instead not linear: a random number of possible sizes is actually skipped, so that to produce a "jump" in the sequence of neighbourhoods.

The skewed neighbourhood change does not only take into account the values of the objective function in the previous x and the current x' solutions, but also their distance in the search domain [3]. The idea is to give the chance to solutions that are far from x to be accepted even when they do not improve the objective function, but the deterioration is somehow acceptable. In fact, this may allow the search to discover a new region of the domain where better solutions may thereafter be identified.

The last (in order of presentation) VNS variant that we consider in this work is the nested VNS [18]. The idea is to replace the local search with another VNS call, where any of the variants presented above may be used. The nested variant can be particularly interesting to use when the problem at hand does not allow to devise a local search with guarantees of optimality. It also allows us to combine, in one unique method, the potential benefits of the various variants.

We point out that we have simply chosen for this work the most popular VNS variants, but others were published in the vast scientific literature regarding VNS. One example is given by the work recently appeared in [5], where VNS is coupled with a local search particularly tailored to the problem at hand, and with a new intensified shaking procedure.

The definition of the neighbourhoods is based on the introduction of a metric in the search space. While in continuous Euclidean spaces the Euclidean norm is the evident choice, giving rise to the typical (hyper-)spherically shaped neighbourhoods, alternative metrics need instead to be considered in discrete spaces. Our choice falls on the Hamming distance, for its simplicity (and hence speed in calculation), and for its effectiveness in evaluating the dissimilarity between pairs of binary strings.

3   VNS in Hamming Space

Neighbourhoods in a Hamming space are essentially different from the sphere-like neighbourhoods we are used to define in a classical Euclidean space. First of all, the region of the space delimited by the neighbourhood is discrete, because of the discrete character of the Hamming space. Secondly, neighbourhoods containing solutions x' that are positioned farther away from the given current solution x are not necessarily larger in size. At the upper limit, in fact, the neighbourhood containing all solutions at Hamming distance n from x, where n is the number of bits necessary for representing the solutions, contains only one element. This unique element can be obtained by flipping all bits in the binary representation of x.

In some cases, therefore, the neighbourhoods defined in Hamming space are not only discrete and finite (so that an exhaustive search is potentially applicable for their exploration), but also rather small (so that the exhaustive search becomes more efficient than any other search method). This is the reason why, in our VNS implementation, the size of the neighbourhoods is systematically estimated before running the local search step; if the size is relatively small, in fact, an exhaustive search is performed, instead of invoking the local search. Notice that the idea of using an exhaustive search is not completely new, as it was already exploited in [5], for example.

In Sect. 2, we have presented some of the most used VNS variants. We point out that our list does not contain the Fleszar-Hind variant [8], because the implementation of all our VNS variants in Hamming space are, de facto, Fleszar-Hind variants. The main idea in this VNS variant is, for a selected neighbourhood, to perform the shaking procedure (and to invoke the local search) for a predetermined number of times, and then to take the best obtained improvement to define the next neighbourhood. In Hamming space, we actually have no choice but to implement this variant, because the changes in the neighbourhoods are discrete, and running one VNS iteration per neighbourhood would result in neighbourhood changes that are much too frequent.

In our VNS implementation, the neighbourhoods can vary by the following criteria. The very first neighbourhood contains all solutions from x such that the Hamming distance is either 1 or 2. We do not restrict ourselves to a Hamming distance value equal to 1 because the implemented local search (see Sect. 4) is based on the idea to look around the current solution by performing single bit flips in the binary representation. In general, therefore, a lower bound ℓ and an upper bound u on the Hamming distance are involved in the definition of the neighbourhood. When $\ell < u$, we define the next neighbourhood by approaching ℓ to u (i.e. $\ell \leftarrow \ell+1$). This allows us to intensify more and more our searches around the region of the Hamming space where the solutions admit u bit flips w.r.t. the current solution x. When $\ell = u$, instead, we create the new neighbourhood with lower bound $\ell \leftarrow 1$ and upper bound $u \leftarrow u+1$, in order to reopen the possibility to select closer solutions, and at the same time by enlarging the horizon.

We point out that necessary condition for an effective similarity (or dissimilarity) measure by Hamming distance between pairs of solutions is that there exists no solution that can be represented with two different binary strings. Other criteria may as well be taken into consideration when defining the binary representations in order to improve the search (we mention to one of these in Sect. 5).

4 Computational Experiments

We present in this work a VNS implementation that is part of the BINMETA[1] Java project, which is conceived particularly to study meta-heuristic searches for global optimization problems for which a suitable binary representation can be supplied [15]. The reason for representing solutions as bit strings stems from the fact that several combinatorial search spaces can admit such discrete representations, as for example it is the case when dealing with search trees. The BINMETA project already implements some meta-heuristic searches: the classical MultiStart heuristics [2], the Wolf Search [17], as well as a simple Random Walk. All VNS variants discussed in Sect. 2 have been implemented and added to the project while performing the work presented in this article.

The local search implemented in BINMETA, and used by most meta-heuristics, resembles to the traditional gradient descent methods in continuous spaces. In

---

[1] https://github.com/mucherino/binMeta.

Hamming space, we initially attempt flipping each bit in the binary representation, one by one, and analyze the variations on the objective function values. While bit flips that worsen the objective values are discarded, the others are kept in memory and sorted by placing the ones leading to the best improvements at the top of a stack S. When all bits have been independently considered and the stack is complete, the procedure attempts to flip all the bits in S. However, it is not guaranteed that the objective function will improve after this change (the parallel with the continuous gradient is that a full step along the gradient direction may not lead to better solutions). When there is indeed no improvement, then the bit appearing at the bottom of the stack is erased, and a new attempt with the remaining bits is performed. And so on, until there is an improvement on the objective function, or the stack is reduced to only one bit (this is the worst-case scenario). The entire procedure is repeated then and stops if the initial stack is empty (we are stuck at a local optimum).

For lack of space, we cannot describe in any detail the combinatorial problems that we use in our computational experiments. The reader is referred to [13] for the combinational problems named SUBSETSUM, NUMBERPARTITION and KNAPSACK. The instances of such problems were artificially generated by using the Java constructors recently added to the corresponding classes.

The experiments were performed on a laptop computer equipped with an processor Intel(R) Core(TM) i7-8665U CPU @1.90 GHz, running Linux Ubuntu 20.04.6 LTS. The codes were compiled and run by the free and open-source Java platform named Openjdk, version 11.0.18 2023-01-17. The version of the Java codes used in the experiments is the one related to the commit `2de34fc` on the GitHub BINMETA repository. All meta-heuristics were given 1 s of CPU time to solve each instance.

Our experiments are summarized in Table 1. In order to evaluate the quality of the found solutions, we compute the difference between the objective function value in the solution, and the known upper bound on the solutions. The latter is estimated by the procedure that is implemented by the constructors. Therefore, when the reported values are positive, this corresponds to runs where the optimal solutions were not identified; if the reported value is instead zero or even negative, this means that the expected value (or even a smaller value) was found for the objective function.

The reported experiments are rather simple to interpret. Even if there are exceptions, as one may expect when working with meta-heuristics, VNS is generally the one that is able to provide better results in the little time at disposal (recall, 1 s of CPU time is given). For the instances of larger size, we can see that the jump, skewed and nested VNS are the variants that are able to provide better results.

Table 1. A comparison among meta-heuristic searches and VNS variants.

instance objective	size	objective function value in best found solution – known upper bound									
		LocalOpt	RandomWalk	MultiStart	WolfSearch	basic VNS	cyclic VNS	pipe VNS	jump VNS	skewed VNS	nested VNS
N.Partition	600	8094.0	0.0	1328.0	0.0	0.0	0.0	0.0	0.0	4.0	0.0
N.Partition	700	434.0	0.0	0.0	8.0	0.0	0.0	0.0	0.0	4.0	0.0
N.Partition	800	4764.0	0.0	174.0	0.0	0.0	0.0	0.0	0.0	0.0	0.0
N.Partition	900	15824.0	0.0	8624.0	6.0	0.0	0.0	0.0	12.0	0.0	0.0
N.Partition	1000	15836.0	0.0	9008.0	4.0	0.0	0.0	0.0	4.0	0.0	0.0
N.Partition	1100	25774.0	0.0	20578.0	48.0	0.0	0.0	0.0	2.0	2.0	2.0
N.Partition	1200	5004.0	0.0	40.0	14.0	0.0	0.0	0.0	2.0	2.0	0.0
N.Partition	1300	1884.0	0.0	32.0	78.0	0.0	0.0	0.0	0.0	0.0	0.0
N.Partition	1400	2816.0	2.0	54.0	44.0	0.0	0.0	0.0	0.0	2.0	0.0
SubsetSum	600	25175.0	6230.0	20596.0	4093.0	0.0	0.0	0.0	2.0	3.0	0.0
SubsetSum	700	53271.0	43528.0	47219.0	51249.0	0.0	0.0	0.0	1.0	0.0	0.0
SubsetSum	800	75004.0	49977.0	70296.0	58027.0	0.0	0.0	0.0	1.0	3.0	0.0
SubsetSum	900	35987.0	7510.0	32961.0	14417.0	0.0	0.0	0.0	0.0	3.0	0.0
SubsetSum	1000	33828.0	2.0	30239.0	588.0	0.0	0.0	0.0	2.0	0.0	0.0
SubsetSum	1100	131510.0	81459.0	128598.0	99258.0	0.0	0.0	0.0	1.0	1.0	0.0
SubsetSum	1200	16670.0	0.0	12479.0	110.0	0.0	0.0	0.0	1.0	1.0	0.0
SubsetSum	1300	34867.0	7.0	31626.0	5519.0	0.0	0.0	0.0	0.0	0.0	0.0
SubsetSum	1400	197493.0	139443.0	191863.0	148870.0	1.0	0.0	0.0	2.0	1.0	0.0
Knapsack	600	69959.0	72771.2	65830.5	69959.0	−2945.5	−3948.8	−4325.0	−4729.7	−4905.2	−4976.7
Knapsack	700	43472.6	104697.0	39583.8	43472.6	−2308.5	−2688.2	−3126.8	−3432.5	−3592.9	−3592.9
Knapsack	800	55217.0	127543.9	50274.0	55217.0	−1758.5	−2462.4	−3104.9	−3346.2	−4584.4	−4584.4
Knapsack	900	177941.8	−207.0	173296.5	177941.8	15769.9	−4980.4	−6873.5	−7740.1	−9017.8	−9488.5
Knapsack	1000	219412.6	−229.6	215423.2	219412.6	67103.7	−4387.0	−6640.7	−8560.9	−10892.6	−11331.7
Knapsack	1100	135904.1	257963.6	131345.3	135904.1	−1140.5	−4212.6	−6463.7	−8253.0	−8995.3	−9564.9
Knapsack	1200	188546.0	318048.8	182676.6	188546.0	54283.7	−3602.8	−6217.9	−8645.4	−10117.0	−10272.2
Knapsack	1300	234320.5	380275.8	228962.0	234320.5	107652.4	−3604.4	−6160.9	−8346.4	−9627.6	−10497.3
Knapsack	1400	47633.4	427977.4	43082.7	47633.4	−1312.6	−1312.6	−1312.6	−2090.5	−2238.3	−2238.3

5 Conclusions

We have presented a short survey on VNS variants and adapted the most used ones to work in Hamming space. We have implemented such variants in the context of the wider BINMETA project, and our preliminary computational experiments evidently indicate that VNS can attain better performances when compared to the other meta-heuristics in the Java project.

In the near future, we intend to develop a more formal description of the meta-heuristic searches in Hamming space, for example via the use of pseudo-boolean functions. Our intention is to pay particular attention to the conception of the binary representations, in order to make sure that the induced objective function *landscapes* are smooth and hence easily inspected by the meta-heuristics. In parallel with these future works, more combinatorial problems, as well as more meta-heuristic searches, will be added to the repository.

Acknowledgments. First of all, we are thankful to all Master students that have contributed to the realization of the BINMETA project. We also wish to thank the anonymous reviewers for their fruitful comments. This work is partially supported by the MULTIBIOSTRUCT project (ANR-19-CE45-0019).

References

1. Banka, H., Dara, S.: A hamming distance based binary particle swarm optimization (HDBPSO) algorithm for high dimensional feature selection, classification and validation. Pattern Recogn. Lett. **52**, 94–100 (2015)
2. Boese, K.D., Kahng, A.B., Muddu, S.: A new adaptive multi-start technique for combinatorial global optimizations. Oper. Res. Lett. **16**(2), 101–113 (1994)
3. Brimberg, J., Mladenović, N., Urosević, D.: Solving the maximally diverse grouping problem by skewed general variable neighborhood search. Inf. Sci. **295**, 650–675 (2015)
4. Cafieri, S., Hansen, P., Mladenović, N.: Edge-ratio network clustering by variable neighborhood search. Eur. Phys. J. B **87**(5), 1–7 (2014)
5. Casado, A., Mladenović, N., Sánchez-Oro, J., Duarte, A.: Variable neighborhood search approach with intensified shake for monitor placement. Networks **81**, 1–15 (2022)
6. Della Croce, F., Salassa, F.: A variable neighborhood search based matheuristic for nurse rostering problems. Ann. Oper. Res. **218**, 185–199 (2014)
7. Driessel, R., Mönch, L.: Variable neighborhood search approaches for scheduling jobs on parallel machines with sequence-dependent setup times, precedence constraints, and ready times. Comput. Industr. Eng. **61**(2), 336–345 (2011)
8. Fleszar, K., Hindi, K.S.: Solving the resource-constrained project scheduling problem by a variable neighbourhood search. Eur. J. Oper. Res. **155**(2), 402–413 (2004)
9. Hansen, P., Mladenović, N.: Variable neighborhood search: principles and applications. Eur. J. Oper. Res. **130**(3), 449–467 (2001)
10. Hansen, P., Mladenović, N.: A tutorial on variable neighborhood search. Les Cahiers du GERAD, Montréal QC H3T 2A7, Canada, pp. 1–26 (2003)
11. Hansen, P., Mladenović, N., Moreno Pérez, J.A.: Variable neighbourhood search: methods and applications. Ann. Oper. Res. **175**, 367–407 (2010)

12. Hansen, P., Mladenović, N., Todosijević, R., Hanafi, S.: Variable neighborhood search: basics and variants. Eur. J. Comput. Optim. **5**, 423–454 (2017)
13. Korte, B.: Combinatorial Optimization, vol. 1, pp. 1–595. Springer, Heidelberg (2011)
14. Mladenović, N., Hansen, P.: Variable neighborhood search. Comput. Oper. Res. **24**(11), 1097–1100 (1997)
15. Mucherino, Antonio: BINMETA: a new java package for meta-heuristic searches. In: Lirkov, Ivan, Margenov, Svetozar (eds.) LSSC 2021. LNCS, vol. 13127, pp. 242–249. Springer, Cham (2022). https://doi.org/10.1007/978-3-030-97549-4_28
16. Mucherino, A., Seref, O.: Modeling and solving real life global optimization problems with meta-heuristic methods. In: Papajorgji, P.J., Pardalos, P.M. (eds.) Advances in Modeling Agricultural Systems. Springer Optimization and Its Applications, vol. 25, pp. 403–420. Springer, Boston (2008). https://doi.org/10.1007/978-0-387-75181-8_19
17. Tang, R., Fong, S., Yang, X.S., Deb, S.: Wolf search algorithm with ephemeral memory. In: IEEE Proceedings, 7th International Conference on Digital Information Management (ICDIM 2012), Macau, pp. 165–172 (2012)
18. Todosijević, R., Benmansour, R., Hanafi, S., Mladenović, N., Artiba, A.: Nested general variable neighborhood search for the periodic maintenance problem. Eur. J. Oper. Res. **252**(2), 385–396 (2016)
19. Todosijević, R., Mjirda, A., Mladenović, M., Hanafi, S., Gendron, B.: A general variable neighborhood search variants for the travelling salesman problem with draft limits. Optim. Lett. **11**, 1047–1056 (2017)
20. Todosijević, R., Mladenović, M., Hanafi, S., Mladenović, N., Crévits, I.: Adaptive general variable neighborhood search heuristics for solving the unit commitment problem. Int. J. Electr. Power Energy Syst. **78**, 873–883 (2016)

An Improved Algorithm for Fredholm Integral Equations

Venelin Todorov[1,2](✉)ⓘ, Slavi Georgiev[1,3]ⓘ, Stoyan Apostolov[4],
and Ivan Dimov[2]

[1] Department of Information Modeling, Institute of Mathematics and Informatics,
Bulgarian Academy of Sciences, Acad. Georgi Bonchev Str.,
Block 8, 1113 Sofia, Bulgaria
{vtodorov,sggeorgiev}@math.bas.bg
[2] Department of Parallel Algorithms, Institute of Information and Communication
Technologies, Bulgarian Academy of Sciences, Acad. Georgi Bonchev Str.,
Block 25A, 1113 Sofia, Bulgaria
venelin@parallel.bas.bg
[3] Department of Applied Mathematics and Statistics, Faculty of Natural Sciences
and Education, Angel Kanchev University of Ruse, 8 Studentska Str.,
7004 Ruse, Bulgaria
sggeorgiev@uni-ruse.bg
[4] Faculty of Mathematics and Informatics, Sofia University, Sofia, Bulgaria

Abstract. Integral equations are of high applicability in different areas
of applied mathematics, physics, engineering, geophysics, electricity and
magnetism, kinetic theory of gases, quantum mechanics, mathematical
economics, and queuing theory. That is why it is reasonable to develop
and study efficient and reliable approaches to solve integral equations.
For multidimensional problems the existing biased stochastic algorithms
based on evaluation of finite number of integrals will suffer more from
the effect of high dimensionality, because they are based on quadrature
points. So we need advanced unbiased algorithms for solving the multidimensional problem which will be developed in this paper.

Keywords: Multidimensional Fedholm integral equation · Unbiased
approach · Monte Carlo method

1 Introduction

By definition [1–3,6] the biased approach is when one is looking for a random
variable which mathematical expectation is equal to the approximation of the
solution problem by a truncated Liouville-Neumann series for a sufficiently large
number of steps. An unbiased approach assumes that the formulated random
variable is such that its mean value approaches the true solution of the problem.

Consider a Fredholm integral equation of the second kind:

$$u(\mathrm{x}) = \int_{\Omega} k(\mathrm{x}, \mathrm{x}')u(\mathrm{x}')\mathrm{dx}' + f(\mathrm{x}) \text{ or } u = \mathcal{K}u + f \quad (\mathcal{K} \text{ is an integral operator}),$$

I. Lirkov and S. Margenov (Eds.): LSSC 2023, LNCS 13952, pp. 262–270, 2024.
https://doi.org/10.1007/978-3-031-56208-2_26

where $k(x, x') \in L_2(\Omega \times \Omega)$, $f(x) \in L_2(\Omega)$ are given functions and $u(x) \in L_2(\Omega)$ is an unknown function, $x, x' \in \Omega \subset R^n$ (Ω is a bounded domain).

The key problem is to evaluate linear functionals of the solution $u(x)$ of the following type: $J(u) = \int \varphi(x) u(x) dx = (\varphi, u)$, where $\varphi(x) \in L_2(\Omega)$ is a given function.

First, we construct a random trajectory (Markov chain) T_i of length i starting from state x_0 in the domain Ω:

$$T_i : x_0 \longrightarrow x_1 \longrightarrow \ldots \longrightarrow x_i$$

according to the initial $\pi(x)$ and transition $p(x, x')$ probabilities. The functions $\pi(x)$ and $p(x, x')$ satisfy the requirements for non-negativeness, to be permissible [2] to function $\varphi(x)$ and the kernel $k(x, x')$ respectively and $\int_\Omega \pi(x) dx = 1$, $\int_\Omega p(x, x') dx' = 1$ for any $x \in \Omega \subset R^n$. The Markov chain transition probability $p(x, x')$ is chosen to be proportional to $|k(x, x')|$ following [2]. It means that $p(x, x') = c|k(x, x')|$, and the constant c is computed such that

$$c = \left(\int_\Omega |k(x, x')| dx' \right)^{-1} = 1 \quad \text{for any} \ \ x \in \Omega.$$

Form the above assumptions:

$$\mathbf{E}\theta_i[\varphi] = (\varphi, u^{(i)}), \quad \text{where} \quad \theta_i[\varphi] = \frac{\varphi(x_0)}{\pi(x_0)} \sum_{j=0}^{i} W_j f(x_j),$$

$$\text{and} \quad W_0 = 1, \quad W_j = W_{j-1} \frac{k(x_{j-1}, x_j)}{p(x_{j-1}, x_j)}, \quad j = 1, \ldots, i$$

it follows that the biased Monte Carlo estimation of $(\varphi, u^{(i)})$ is:

$$(\varphi, u^{(i)}) \approx \frac{1}{N} \sum_{l=1}^{N} \theta_i[\varphi]_l.$$

In the case of the unbiased approach the random trajectory (Markov chain) T_i defined above is infinite. The discrete Markov chain used in [4] is replaced by a continuous one with transition probabilities depending on the kernel $k(x, x')$. We will discuss both cases, namely when $0 \le k(x, x') \le 1$, where $1 - k(x, x')$ is an absorption rate and the more general kernels.

2   Unbiased Stochastic Approach

First we describe the old Unbiased Stochastic Approach (USA) which is a generalization to integral equations [3] of the algorithm developed in [4] for solving systems of linear algebraic equations.

2.1   Unbiased Old Approach for a Simplified Problem

Here the case $0 \le k(x, x') \le 1$ is described, where $1 - k(x, x')$ is an absorption rate.

Algorithm 1 :

Unbiased stochastic algorithm for $0 \le k(x, y) \le 1$, $x, y \in [0, 1]^n$

1. *Initialization* **Input** *initial data: the kernel* $k(x, y)$, *the function* $f(x)$, *and the number of random trajectories* N.
2. *Calculations:*
   2.1. **Set** *score=0*
   2.2. **Set** $x_0 = x, test = 1$
        Do $j = 1, N$
          Do While $(test \ne 0)$
            $score = score + f(x_0)$
            $U = rand[0, 1]^n$, $V = rand[0, 1]$
            If $k(x_0, U) < V$ **then**
               $test = 0$ **else**
               $x_0 = U$
            Endif
          Endwhile
        Enddo
3. *Compute the solution:*
   $u(x) = \frac{score}{N}$

2.2   Unbiased Old Approach for the General Problem

The previous USA algorithm for more general kernels is given in details in [3].

Algorithm 2 :

Unbiased stochastic algorithm for the general case:

1. *Initialization* **Input** *initial data: the kernel* $k(x, y)$, *the function* $f(x)$, *and the number of random trajectories* N.
2. *Calculations:*
   2.1. **Set** *score=0*
   2.2. **Do** $j = 1, N$
   2.3. **Set** $x_0 = x, test = 1$, *prod=1*
        Do While $(test \ne 0)$
        $score = score + f(x_0) * prod$
        $U = rand(0, 1)^n$
          If $|k(x_0, U)| > 1$ **then**
             $prod = prod * k(x_0, U)$

  else $x_0 = U$
  $V = rand[0, 1]$
  If $|k(x_0, U)| < V$ **then**
   $test = 0$ **else**
   $prod = prod * sign\,(k(x_0, U)),$ $x_0 = U$
  Endif
  Endif
 Endwhile
 Enddo

3. *Compute the solution:*
 $u(x) = \frac{score}{N}$

2.3 Description of the New Version of the Unbiased Algorithms NUSA for the Simplified Problem

Now we will describe the new unbiased stochastic approach NUSA. We start with a simple case assuming that $0 \leq k(x, x') \leq 1$ for any $x, x' \in \Omega \equiv [0, 1]^n$, $x = (x^1, \ldots, x^n)$, $x' = (x^{(1)'}, \ldots, x^{(n)'})$.

Assume that $Y \equiv U[0, 1]^n$ is a uniformly distributed random variable in $[0, 1]^n$. Then we use the following stochastic presentation:

$$u(x) = E\{k(x, Y)u(Y) + (1 - k(x, Y))0\} + f(x),$$

which suggests a representation for $u(x)$ following [4] for solving linear systems of equations. The probability for absorption is $1 - k(x, Y)$, the score is the function $f(x)$, and if the trajectory is not absorbed then it continues at point Y. We consider a random trajectory (Markov chain) T_i of length i starting from state $X = x = x_0$ in the domain Ω where we add a cemetery state $\partial \in \mathbb{R}^n \backslash \Omega$ corresponding to absorption that is

$$T_i : x_0 \longrightarrow x_1 \longrightarrow \ldots \longrightarrow x_i,$$

where

$$P(x_{i+1} \in \Omega | x_i \neq \partial) = k(x, U_i)P(U_i \in \Omega),$$
$$P(x_{i+1} = \partial | x_i \neq \partial) = 1 - k(x, U_i)$$

and

$$P(x_{i+1} = \partial | x_i = \partial) = 1, \quad \text{and} \quad f(\partial) = 0.$$

Then we have

$$u(x) = E\left(\sum_{i=0}^{\infty} f(x_i) | x_0 = x\right) = E\left(\sum_{i=0}^{\tau} f(x_i) | x_0 = x\right), \quad \text{where} \quad \tau = \inf_{i \geq 0}(x_i) = \partial.$$

Obviously, one can write

$$u(\mathbf{x}) = E\left(\sum_{i=0}^{\infty} f(\mathbf{x}_{i+1})|\mathbf{x}_0 = \mathbf{x}\right) + f(\mathbf{x}) = E\left(E\left(\sum_{i=0}^{\infty} f(\mathbf{x}_{i+1})|\mathbf{x}_1\right)\right) + f(\mathbf{x})$$
$$= \int_{\Omega} k(\mathbf{x}, \mathbf{y})u(\mathbf{y})dy + f(\mathbf{x})$$

$$(1)$$

The algorithm in this simplified case may be described as follows.

Algorithm 3 :

Unbiased stochastic algorithm NUSA for the simplified case:

1. Initialization **Input** initial data: the kernel $k(\mathbf{x}, \mathbf{y})$, the function $f(\mathbf{x})$, the number of random trajectories N and point of evaluation x_0.
2. Calculations:
3. **Set** $S = 0$
4. **for** $k = 1$ to N **do** (MC loop)
   4.1 weight $= 1$; $x = x_0$; $P_d = 0.5$.
   4.2. test $= 0$.
   4.5. **while** test $\neq 1$ **do**:
      4.5.1. $U = rand(0,1)$.
      4.5.2. **If** $U < P_d$ **Then** test=1;
      4.5.3. **Else** choose V according to the probability density function $|K(x, \cdot)|$. **If** $K(x, V) < 0$ weight $=$ weight $* \frac{(-1)}{(1-P_d)}$; $x = V$;
      4.5.4. **endif** .
   4.6 **endwhile** weight $=$ weight $* \frac{f(x)}{P_d}$; $S = S + weight$
5. **enddo**
6. Estimated value $= \frac{S}{N}$.

2.4   The New Unbiased Stochastic Algorithm NUSA for the General Problem

Now we may describe the method for the case when the kernel $k(\mathbf{x}, \mathbf{y})$ may be negative and also we allow that $|k(\mathbf{x}, \mathbf{y})| \geq 1$, but still $k(\mathbf{x}, \mathbf{y}) \in L_2(\Omega \times \Omega)$ and the Neumann series converges.

The key observation here is the greater absorbtion probability at each step p_{abs}, the shorter length of trajectories.

Algorithm 4 :

Unbiased stochastic algorithm NUSA for the general case:

1. Initialization **Input** initial data: the kernel $k(\mathbf{x}, \mathbf{y})$, the function $f(\mathbf{x})$, the number of random trajectories N and point of evaluation x_0.

2. Calculations:

3. **Set** $S = 0$

4. **for** $k = 1$ **to** N **do** *(MC loop)*

    4.1 weight = 1; $x = x_0$; $P_d = 0.5$.

    4.2. test = 0.

    4.5. **while** *test $\neq$ 1* **do***:*

        4.5.1.U=rand(0,1).

        4.5.2. **If** $U < P_d$ **Then** *test=1;*

        4.5.3. **Else** *choose V according to the probability density function*
        $M(x,V)$. weight $= weight * \frac{K(x,V)}{M(x,V)(1-P_d)}$; $x = V$;

        4.5.4. **endif** *.*

    4.6 **endwhile** *weight $= weight * \frac{f(x)}{P_d}$; $S = S + weight$*

*5.*enddo

6. Estimated value $= \frac{S}{N}$.

3 Numerical Results

3.1 Stochastic Approach for One Dimensional Problem

The numerical algorithms are tested on the following example with dimension $n = 1$ [3,5]:

$$u(x) = \int_\Omega k(x,x')u(x')\mathrm{d}x' + f(x), \qquad \Omega \equiv [0,1] \tag{2}$$

where

$$k(x,x') = x^2 e^{x'(x-1)}, \tag{3}$$

$$f(x) = x + (1-x)e^x, \tag{4}$$

$$\varphi(x) = \delta(x - x_0). \tag{5}$$

The solution of this test problem is $u(x) = e^x$. We are interested in an approximate calculation of (φ, u), where $\varphi(x) = \delta(\mathrm{x} - \mathrm{x}_0)$, $x_0 = 0.5$ (Table 1).

Table 1. Comparison of the relative errors for 1000 samples for computing different $u(\mathrm{x}_0)$ for (2)–(5).

x_0	0.1	0.2	0.3	0.4	0.5	0.6	0.7	0.8	0.9
NUSA	0.0021	0.0139	0.0164	0.0545	0.0072	0.0118	0.0186	0.0220	0.0086
USA	0.4333	0.3681	0.2874	0.1963	0.0978	0.0047	0.1099	0.2216	0.3338

For all points except the point $x_0 = 0.6$ the new unbiased algorithm for Fredholm integral equation NUSA outperforms the old algorithm USA. The difference is more pronounced for the points close to 1. For the point $x_0 = 0.9$ the difference is even more pronounced and the new algorithm NUSA outperforms USA with at least 2 orders of magnitude.

Table 2. Solution, Relative errors and Computational time for computing $u(x_0)$ at $(1.0, 1.0, \ldots, 1.0)$ for different values of norm of the kernel controlled by the parameter β.

β	Algorithm	Parameters	Problem dimension		
			$n = 3$	$n = 5$	$n = 10$
$\frac{0.1}{\|K\|_{L_2}}$		Exact solution	20.0855	148.4132	22 026.4
	USA algorithm	Relative error	**6.5e−5**	**4.1e−4**	4.1e−3
		Computational time	12.1	21.1	42.7
	NUSA algorithm	Relative error	3.7e−3	1.1e−3	**5.7e−5**
		Computational time	15.1	23.9	46.7
$\frac{0.5}{\|K\|_{L_2}}$		Exact solution	20.0855	148.4132	22 026.4
	USA algorithm	Relative error	**2.2e−3**	1.0e−2	5.9e−3
		Computational time	15.1	23.1	44.4
	NUSA algorithm	Relative error	5.7e−3	**7.5e−3**	**2.7e−4**
		Computational time	15.3	24.4	45.8
$\frac{0.95}{\|K\|_{L_2}}$		Exact solution	20.0855	148.4132	22 026.4
	USA algorithm	Relative error	**1.6e−2**	7.5e−3	7.1e−2
		Computational time	15.8	24.5	45.3
	NUSA algorithm	Relative error	6.9e−2	**5.1e−3**	**2.9e−4**
		Computational time	15.5	23.9	45.6

3.2 Stochastic Approach for Multi Dimensional Problem

Solving the multidimensional Fredholm integral equation of second kind is a great challenge to numerical algorithms. We are interested in the following multidimensional problem with dimension $n > 1$ [3]:

$$u_n(\mathbf{x}) = \int_0^1 \cdots \int_0^1 \prod_{j=1}^n \{k(x^{(j)}, x^{(j)'})\} u(\mathbf{x}') \prod_{j=1}^n dx^{(j)'} + f_n(\mathbf{x}), \qquad (6)$$

$$\mathbf{x} \equiv (x^{(1)}, \ldots, x^{(n)}), \ \mathbf{x}' \equiv (x^{(1)'}, \ldots, x^{(n)'}), \ \prod_{j=1}^n dx^{(j)'} = d\mathbf{x}',$$

where

$$k(x^{(j)}, x^{(j)'}) = \beta_j \left(2x^{(j)'} - 1\right) \{x^{(j)}\}^2 \exp\{x^{(j)'}(x^{(j)} - 1)\}, \ \beta = \prod_{j=1}^n \beta_j, \qquad (7)$$

$$f_n(x) = e^{\sum_{j=1}^n x^{(j)}} + (-1)^{(j-1)} \beta_n (2e^{x^{(j)}} - 2 - x^{(j)} e^{x^{(j)}} - x^{(j)}), \qquad (8)$$

$$\varphi(\mathbf{x}) = \delta(\mathbf{x} - \mathbf{x}_0), \qquad (9)$$

where $x_0 = (1.0, 1.0, \ldots 1.0)$ in all the numerical experiments. The solution of the above equation is a product of exponents $u(x) = \prod_{j=1}^{n} \exp\{x^{(j)}\} = \exp\{\sum_{j=1}^{n} x^{(j)}\}$. We use special parameter β to be able to deal with various difficulties, namely, with negative kernels and values of the kernel greater than one in absolute value. If the values of the parameter β increases, then we approach instability barrier when the variance is getting infinite. One can observe that for $\beta = 0.95$ in Table 2 for $n = 10$ the new unbiased approach NUSA outperforms the previous USA with at lest 2 order - the new result is $2.9e - 4$ vs the old results which is $7.1e - 2$. So to summarize that for higher values of β which is the most complicated problem from computational point of view, the new unbiased algorithm NUSA has much higher accuracy with increasing the dimensionality of the integral equation.

4 Conclusion

A new unbiased stochastic method for solving multidimensional Fredholm integral equations of second kind was proposed and analysed. We have compared the newly proposed unbiased algorithm with the old unbiased stochastic algorithm for the one dimensional problem and multidimensional problem. One can observe that for higher dimensions the new algorithm outperforms the old algorithm. Another very important observation is that the computational complexity increases only linearly with the dimension. For future work we will test the new algorithm even greater dimensions when one is looking for practical computational problems with very high dimensions.

Acknowledgements. Slavi Georgiev is supported by the Bulgarian National Science Fund (BNSF) under Project KP-06-M62/1 "Numerical deterministic, stochastic, machine and deep learning methods with applications in computational, quantitative, algorithmic finance, biomathematics, ecology and algebra" from 2022. Venelin Todorov is supported by BNSF under Project KP-06-N52/5 "Efficient methods for modeling, optimization and decision making" and BNSF under Project KP-06-N62/6 "Machine learning through physics-informed neural networks". The work is also supported by BNSF under Bilateral Project KP-06-Russia/17 "New Highly Efficient Stochastic Simulation Methods and Applications".

References

1. Curtiss, J.H.: Monte Carlo methods for the iteration of linear operators. J. Math. Phys. **32**, 209–232 (1954)
2. Dimov, I.T.: Monte Carlo Methods for Applied Scientists, p. 291. World Scientific, London (2008). ISBN: 10 981-02-2329-3
3. Dimov, I.T., Maire, S.: A new unbiased stochastic algorithm for solving linear Fredholm equations of the second kind. Adv. Comput. Math. **45**, 1499–1519 (2019)
4. Dimov, I.T., Maire, S., Sellier, J.M.: A new Walk on Equations Monte Carlo method for solving systems of linear algebraic equations. Appl. Math. Model. **39**(15), 4494–4510 (2014). https://doi.org/10.1016/j.apm.2014.12.018

5. Farnoosh, R., Ebrahimi, M.: Monte Carlo method for solving Fredholm integral equations of the second kind. Appl. Math. Comput. **195**, 309–315 (2008)
6. Kalos, M.H., Whitlock, P.A.: Monte Carlo Methods. Wiley-VCH (2008). ISBN: 978-3-527-40760-6

Optimization of the Standard Lattice Sequence for Multidimensional Integrals Regarding Large-Scale Finance Problems

Venelin Todorov[1,2,4]([envelope])[iD], Slavi Georgiev[1,3,4][iD], Borislav Chakarov[3,4],
and Svetoslav Hadzhiivanov[1,4]

[1] Department of Information Modeling, Institute of Mathematics and Informatics,
Bulgarian Academy of Sciences, Acad. Georgi Bonchev Str., Block 8, 1113 Sofia,
Bulgaria
{vtodorov,sggeorgiev}@math.bas.bg

[2] Department of Parallel Algorithms, Institute of Information and Communication
Technologies, Bulgarian Academy of Sciences, Acad. Georgi Bonchev Str., Block 25A,
1113 Sofia, Bulgaria
venelin@parallel.bas.bg

[3] Department of Applied Mathematics and Statistics, Faculty of Natural Sciences
and Education, Angel Kanchev University of Ruse, 8 Studentska Str., 7004 Ruse,
Bulgaria
sggeorgiev@uni-ruse.bg

[4] Technical University, bul. Kliment Ohridski 8, 1756 Sofia, Bulgaria

Abstract. Lots of challenges in the multidimensional option pricing exist since this is one of the fundamental discipline in large-scale finance problems today. In this paper, for the first time we develop some new highly accurate lattice sequences, based on component-by-component construction methods: construction of rank-1 lattice rules with prime number of points and with product weights; construction of rank-1 lattice sequences with prime number of points and with product weights; construction of polynomial rank-1 lattice sequences in base 2 and with product weights. Our methods show significantly optimization compared to the results produced by the standard Monte Carlo algorithms and the most widely used lattice sequence. There is optimization in the relative error as well as the computational complexity and number of operation necessary to compute the arisen multidimensional integrals. The obtained results will play an extremely principal multi-sided role.

Keywords: Multiple integrals · Optimization of lattice · Finance

1 Introduction

Recently new approaches have been designed to outperform classical Monte Carlo ones for large-scale finance problems [5,6,8]. By definition, a European call option provides its holder with the right, but not the obligation, to buy

I. Lirkov and S. Margenov (Eds.): LSSC 2023, LNCS 13952, pp. 271–278, 2024.
https://doi.org/10.1007/978-3-031-56208-2_27

some quantity of a prescribed asset (underlying) S at a prescribed price (strike or exercise price) E at a prescribed time (maturity or expiry date) T. A European put option has the same features as its call counterpart, except that holder could sell the underlying rather than buying it.

The key task [8] in option pricing is obtaining a "fair" value of the option contract $V(S,t)$, if the following parameters are given: the exercise price E, the asset price $S(t)$, the expiry time T, the risk-free interest rate r and the assumption on the dynamics model of S:

$$dS = \mu S dt + \sigma S dX, \tag{1}$$

where dX is the increment of a Wiener process, μ is the drift rate and σ is the volatility of the asset price, measuring the average growth and level of fluctuations, respectively.

The important Black-Scholes pricing formula for a European call option can be written [2], using the following parabolic partial differential equation:

$$\frac{\partial V}{\partial t} + \frac{1}{2}\sigma^2 S^2 \frac{\partial^2 V}{\partial S^2} + rS\frac{\partial V}{\partial S} - rV = 0, \tag{2}$$

with final condition
$$V(S,T) = \max(S - E, 0)$$

and Dirichlet boundary conditions

$$V(0,t) = 0, V(S,t) \sim S - Ee^{-r(T-t)}, S \to \infty.$$

In the context of option pricing, the basic idea is to represent the option premium as mathematical expectation of the random variable (European option risk-neutral valuation formula [3]):

$$V(S,t) = \mathbb{E}_{\mathbb{Q}}\left(e^{-r(T-t)}h\big(S(T)\big) \mid S(t) = S, \mu = r\right),$$

where $\mathbb{E}_{\mathbb{Q}}(\cdot)$ is the expectation operator, $h(S)$ is the payoff function, in particular $h(S) = \max(S-E, 0)$ for a call and $h(S) = \max(E-S, 0)$ for a put. We follow the idea of Lai and Spanier in [8]. Consider a European option whose payoff depends on $k > 1$ assets with prices $S_i, i = 1, ..., k$. Each asset follows the random walk $dS_i = \mu_i S_i dt + \sigma_i S_i dX_i$, where σ_i is the annualized standard deviation for the i-th asset and dX_i is Brownian motion. Suppose at expiry time T that the payoff is given by $h(S_1', \ldots, S_k')$ (where S_i' denotes the value of the i-th asset at expiry). Then the current value, V, of the option (assuming risk neutrality) will be

$$V = e^{-r(T-t)}(2\pi(T-t))^{-k/2}(\det \Sigma)^{-1/2}(\sigma_1, \ldots, \sigma_k)^{-1} \tag{3}$$

$$\int_0^\infty \cdots \int_0^\infty \frac{h(S_1', \ldots, S_k')}{S_1', \ldots, S_k'} \exp(-0.5\alpha^T \Sigma^{-1}\alpha)dS_1', \ldots, dS_k',$$

where

$$\alpha_i = (\sigma_i(T-t)^{1/2})^{-1}(\log \frac{S_i'}{S_i} - (r - \frac{\sigma^2}{2})(T-t)),$$

r is the risk-free interest rate and Σ is the covariance matrix, where the (i, j) entry is the covariance of dX_i and dX_j for the k assets. For this task, the multiple integrals can be estimated by using either pseudorandom or quasirandom sequences. Furthermore, if the integrands are smooth, lattice rules could also be used.

2   The Lattice Sequences

We will use this rank-1 lattice sequence (LS), defined by [11]:

$$\mathbf{x}_k = \left\{ \frac{k}{N} \mathbf{z} \right\}, \ k = 1, \ldots, N,$$

where $N \geq 2$ is an integer, $\mathbf{z} = (z_1, z_2, \ldots z_s)$ is the generating vector.

   The performance of the LS can be optimized through the Component-by-Component Fast Construction method (CBC) [1,9,10]. Now we construct several special generating vectors for the lattice sequence (GV) up to 2^{24} points following the idea in [4,7]. The first GV will be the CBC construction of rank-1 lattice rules with prime number of points and with product weights, denoted by **1PT**. The second GV will be the CBC construction of rank-1 lattice sequences with prime power of points and with product weights, denoted by **1EXPT**. The third GV will be the CBC construction of polynomial rank-1 lattice sequences in base 2 and with product weights, denoted by **2POLY**. The three GVs will be used in the lattice sequence, defined by the three steps below. At the first step of the algorithm the s dimensional GVs

$$\mathbf{z} = (z_1, z_2, \ldots, z_s) \tag{4}$$

is generated by the CBC method. The second step of the algorithm is generating the points of lattice rule by formula

$$\mathbf{x}_k = \left\{ \frac{k}{N} \mathbf{z} \right\}, \ k = 1, \ldots, N. \tag{5}$$

And at the third step of the algorithm an approximate value I_N of the multidimensional integral is evaluated by the formula:

$$I_N = \frac{1}{N} \sum_{k=1}^{N} f\left(\left\{ \frac{k}{N} \mathbf{z} \right\} \right). \tag{6}$$

Now we will use the three GVs to another non-standard lattice sequence (NP1), which applies the polynomial transformation function $\varphi(t) = 3t^2 - 2t^3$ to a nonperiodic integrand to make it suitable for applying a lattice rule [7]. The transformation must satisfy the following conditions

$$\varphi(0) = 0, \ \varphi(1) = 1, \ \varphi'(t) > 0.$$

Thus φ is a continuous bijection from $[0, 1]$ to $[0, 1]$.

3 Computational Results

We will be interested in the following integrals:
Example 1. s = 3.

$$\int\limits_{[0,1]^3} \exp(x_1 x_2 x_3) dx \approx 1.14649913323497. \tag{7}$$

Example 2. s = 4.

$$\int\limits_{[0,1]^4} x_1 x_2^2 e^{x_1 x_2} \sin(x_3) \cos(x_4) dx \approx 0.10897491798381. \tag{8}$$

Example 3. s = 5.

$$\int\limits_{[0,1]^5} \exp(-100 x_1 x_2 x_3)\big(\sin(x_4) + \cos(x_5)\big) dx \approx 0.185429894674946. \tag{9}$$

Example 4. s = 7.

$$\int\limits_{[0,1]^7} \exp\left(1 - \sum_{i=1}^{3} \sin\left(\frac{\pi}{2} \cdot x_i\right)\right) \cdot \arcsin\left(\sin(1) + \frac{\sum\limits_{j=1}^{7} x_j}{200}\right) dx \approx 0.481088487600829. \tag{10}$$

Example 5. s = 10.

$$\int\limits_{[0,1]^{10}} \frac{4 x_1 x_3^2 e^{2 x_1 x_3}}{(1 + x_2 + x_4)^2} e^{x_5 + \cdots + x_{10}} dx \approx 14.8087092095592. \tag{11}$$

Example 6. s = 15.

$$\int\limits_{[0,1]^{15}} \left(\sum_{i=1}^{10} x_i^2\right) \cdot (x_{11} - x_{12}^2 - x_{13}^3 - x_{14}^4 - x_{15}^5)^2 dx \approx 1.9644304795203. \tag{12}$$

Example 7. s = 25.

$$\int\limits_{[0,1]^{25}} \frac{4 x_1 x_3^2 e^{2 x_1 x_3}}{(1 + x_2 + x_4)^2} e^{x_5 + \cdots + x_{20}} x_{21} \ldots x_{25} dx \approx 103.987049568116. \tag{13}$$

Example 8. s = 30.

$$\int\limits_{[0,1]^{30}} \frac{4 x_1 x_3^2 e^{2 x_1 x_3}}{(1 + x_2 + x_4)^2} e^{x_5 + \cdots + x_{20}} x_{21} \ldots x_{30} dx \approx 3.242993405561. \tag{14}$$

Table 1. REER for the 3-MI (7).

N	CRU	t, [s]	1PT	t, [s]	1EXPT	t, [s]	2POLY	t, [s]	NP1-1PT	t, [s]	NP1-1EXPT	t, [s]	NP1-2POLY	t, [s]
2^{16}	8.2520e−6	1.1115	1.6540e−4	1.5851	2.2422e−6	1.5538	2.3206e−5	1.7170	1.8338e−6	1.5956	5.8922e−7	1.7839	**7.5484e−8**	1.2876
2^{18}	3.5285e−5	2.8969	1.3999e−4	5.1728	5.1882e−6	4.9803	5.4321e−6	5.5039	2.1316e−6	5.4525	7.9061e−8	5.0267	**7.2904e−8**	4.8263
2^{20}	2.4507e−4	11.5537	3.3369e−5	19.6834	3.2069e−7	20.1002	7.1570e−7	19.6523	9.5421e−8	19.4327	**5.4408e−8**	19.2552	6.3327e−8	20.2402
2^{22}	1.5882e−4	48.9150	2.4903e−6	79.3680	3.4554e−7	81.0995	1.7195e−6	79.9336	5.3142e−8	80.0544	**5.3063e−8**	81.0517	5.4980e−8	67.6634
2^{24}	6.0769e−5	146.682	5.2066e−7	248.106	**2.4282e−8**	247.874	2.2979e−6	244.432	5.2969e−8	255.356	5.2961e−8	254.319	5.4890e−8	255.073

Table 2. REER for the 4-MI (8).

N	CRU	t, [s]	1PT	t, [s]	1EXPT	t, [s]	2POLY	t, [s]	NP1-1PT	t, [s]	NP1-1EXPT	t, [s]	NP1-2POLY	t, [s]
2^{16}	1.7184e−3	0.9951	2.4189e−3	1.1905	2.8077e−4	1.4903	2.6567e−4	1.3684	3.3105e−5	1.2442	**1.0432e−6**	1.3997	2.2867e−6	1.2982
2^{18}	3.2513e−3	2.8285	2.5389e−3	4.9391	4.0130e−5	5.0288	1.0032e−4	4.8216	2.0125e−6	5.2619	**9.0844e−7**	5.1619	9.3055e−7	4.9553
2^{20}	2.4597e−3	11.9121	4.8114e−4	19.5007	**3.6981e−7**	19.9723	2.8624e−5	20.1365	5.7980e−7	20.1600	4.7743e−7	20.2610	5.6390e−7	20.0602
2^{22}	1.4415e−3	47.9877	3.9253e−5	78.4301	4.8984e−6	80.7908	9.9289e−6	79.3939	**4.5249e−7**	79.3673	5.0459e−7	79.5096	5.2205e−7	68.8742
2^{24}	1.0926e−4	150.424	5.0941e−6	254.927	7.3840e−7	255.908	1.4987e−5	253.659	5.0474e−7	262.829	**5.0431e−7**	261.353	5.0822e−7	257.837

Table 3. REER for the 5-MI (9).

N	CRU	t, [s]	1PT	t, [s]	1EXPT	t, [s]	2POLY	t, [s]	NP1-1PT	t, [s]	NP1-1EXPT	t, [s]	NP1-2POLY	t, [s]
2^{16}	2.1318e−3	0.8202	1.0302e−3	1.2695	1.1708e−4	1.3277	5.7686e−4	1.2896	1.2713e−4	1.2387	**2.5915e−5**	1.3992	1.1893e−4	1.2587
2^{18}	1.0794e−3	3.5397	1.3191e−4	5.1498	3.5492e−5	5.3695	2.5841e−4	4.9974	3.9269e−5	5.0156	1.4441e−5	5.3613	**8.2446e−7**	5.2271
2^{20}	1.4381e−3	12.5479	1.3128e−4	19.7372	4.2077e−5	20.2525	7.3197e−5	20.2551	5.6335e−6	20.9560	**4.3773e−7**	21.1565	1.7998e−6	21.3167
2^{22}	8.3316e−4	48.8885	1.0408e−4	82.5369	3.6274e−6	84.1442	1.1004e−5	82.5330	7.4122e−7	83.4617	**1.2205e−7**	83.2092	1.9671e−6	68.9935
2^{24}	1.6885e−4	154.888	7.8327e−6	261.027	2.5809e−6	261.745	3.8090e−6	256.256	8.0564e−7	267.656	1.4409e−7	267.110	**1.4122e−7**	261.935

Table 4. REER for the 7-MI (10).

N	CRU	t, [s]	1PT	t, [s]	1EXPT	t, [s]	2POLY	t, [s]	NP1-1PT	t, [s]	NP1-1EXPT	t, [s]	NP1-2POLY	t, [s]
2^{16}	4.0062e−3	1.0892	7.9528e−5	1.5475	5.7737e−5	1.5443	3.1105e−4	1.5698	4.4713e−3	1.4692	**3.6700e−6**	1.4930	1.8256e−4	1.4658
2^{18}	3.2923e−4	3.8354	2.0853e−4	5.6514	**4.4756e−6**	5.9641	5.5603e−5	6.3398	2.9318e−5	5.9375	2.6771e−5	5.7091	9.1966e−6	5.9079
2^{20}	1.1771e−3	15.7454	5.0568e−5	23.0960	3.2016e−6	23.6369	1.4754e−5	23.3675	1.9180e−5	23.7671	2.7513e−6	23.2645	**4.8911e−7**	23.9874
2^{22}	5.1123e−4	63.5753	4.8218e−7	92.2730	1.6344e−6	94.9988	7.3069e−6	93.3058	**1.3506e−7**	92.8146	2.6264e−7	92.6743	1.0977e−6	77.6263
2^{24}	9.8318e−5	198.696	**1.4746e−8**	301.270	2.7572e−7	302.984	5.4474e−6	294.750	6.8725e−8	308.547	1.2724e−7	306.979	2.8228e−6	302.676

Table 5. REER for the 10-MI (11).

N	CRU	t, [s]	1PT	t, [s]	1EXPT	t, [s]	2POLY	t, [s]	NP1-1PT	t, [s]	NP1-1EXPT	t, [s]	NP1-2POLY	t, [s]
2^{16}	8.0933e−4	0.8815	**3.2254e−4**	1.3792	3.2208e−3	1.4607	9.1079e−4	1.4327	3.1766e−3	1.7431	1.0477e−3	1.4855	2.7315e−3	1.3024
2^{18}	7.0237e−5	3.4401	6.8270e−4	5.6438	5.5938e−4	5.3956	4.3233e−4	6.0597	1.6734e−3	5.5246	**6.4146e−5**	5.5521	6.1373e−4	5.5534
2^{20}	1.5885e−4	14.6252	5.2104e−4	22.1159	1.9895e−4	23.4919	2.6133e−4	22.5770	2.6305e−3	23.7152	**1.4761e−4**	22.9228	2.0503e−4	22.8717
2^{22}	2.1825e−5	60.2465	1.6552e−4	89.2836	1.3072e−4	92.5986	3.3908e−4	91.2479	**1.4725e−5**	92.6444	3.3354e−5	91.4595	1.8942e−4	75.6532
2^{24}	6.9192e−5	183.000	3.0497e−5	296.741	6.5162e−5	297.498	1.1223e−4	286.668	**1.7202e−5**	302.023	2.4351e−5	301.697	1.8961e−4	293.508

Table 6. REER for the 15-MI (12).

N	CRU	t, [s]	1PT	t, [s]	1EXPT	t, [s]	2POLY	t, [s]	NP1-1PT	t, [s]	NP1-1EXPT	t, [s]	NP1-2POLY	t, [s]
2^{16}	1.3391e−3	1.0546	5.9489e−4	1.5435	1.5183e−4	1.6086	**6.2742e−7**	1.5151	1.3563e−3	1.6729	2.0261e−2	1.5365	2.0999e−2	1.6106
2^{18}	2.8104e−3	3.9360	1.4530e−4	6.4288	**3.6509e−5**	6.6034	1.4915e−4	6.0089	7.3551e−4	6.3738	4.0287e−3	6.3738	2.7643e−3	6.1582
2^{20}	9.3890e−5	16.2590	1.9395e−4	24.2053	**1.7698e−6**	25.2711	1.2229e−4	24.1438	7.3551e−4	25.5167	4.0287e−3	25.4622	4.4813e−3	25.7828
2^{22}	8.6161e−4	65.5093	1.0548e−4	97.6555	**1.6627e−5**	100.0253	8.1993e−5	96.7782	3.6237e−4	100.8739	1.1845e−4	100.6483	4.5135e−3	81.8601
2^{24}	2.3334e−4	199.483	2.5671e−5	320.376	8.2552e−6	320.167	**5.8471e−6**	308.633	6.2646e−5	327.654	9.6375e−5	328.142	4.5153e−3	316.546

Table 7. REER for the 25-MI (13).

N	CRU	t, [s]	1PT	t, [s]	1EXPT	t, [s]	2POLY	t, [s]	NP1-1PT	t, [s]	NP1-1EXPT	t, [s]	NP1-2POLY	t, [s]
2^{16}	6.4246e−2	0.9761	**2.5713e−3**	1.6864	1.0591e−2	1.8600	9.4390e−2	1.7199	7.8887e−2	1.6904	2.3464e−1	1.6435	9.2559e−2	1.6894
2^{18}	2.2972e−2	3.9291	1.6482e−2	7.1900	**2.5971e−3**	6.7375	4.5965e−2	6.3649	2.0303e−2	7.1896	3.7904e−2	7.1043	8.1092e−2	6.6853
2^{20}	6.7567e−3	15.9719	1.0504e−2	26.9373	**7.4885e−5**	27.7567	1.8766e−1	25.9933	1.1957e−2	27.2861	1.7907e−2	27.1209	5.3147e−2	26.5250
2^{22}	9.7312e−3	64.5971	1.6456e−3	106.7073	**1.3889e−3**	107.8308	2.7217e−1	104.2942	2.2592e−2	109.1403	6.5981e−3	108.4354	5.3299e−2	88.8773
2^{24}	8.8542e−2	205.605	2.6623e−4	348.144	**1.9956e−4**	347.559	2.9660e−1	330.280	7.7893e−3	356.124	4.3892e−3	357.497	5.3279e−2	339.151

Table 8. REER for the 30-MI (14).

N	CRU	t, [s]	1PT	t, [s]	1EXPT	t, [s]	2POLY	t, [s]	NP1-1PT	t, [s]	NP1-1EXPT	t, [s]	NP1-2POLY	t, [s]
2^{16}	3.7043e−2	1.0575	**4.6815e−3**	2.0073	6.3343e−3	2.0583	1.4197e−1	1.6880	2.9312e−1	1.9883	7.1441e−1	1.7511	1.6467e−1	1.7770
2^{18}	4.2678e−2	4.2779	**4.9931e−3**	6.9416	6.1085e−3	7.3514	4.9720e−1	6.4858	1.6821e−1	7.2369	1.4231e−1	7.5274	3.7326e−2	6.9677
2^{20}	**5.1293e−3**	16.9705	6.9007e−3	28.7444	1.5035e−2	29.6477	4.8006e0	27.8997	1.5613e−1	29.4081	7.6153e−2	29.4541	1.0346e−1	27.9230
2^{22}	2.0267e−3	67.4620	**1.1173e−3**	114.2573	8.5930e−3	116.0081	7.3876e0	110.9041	1.0566e−1	114.2560	5.5677e−2	114.4987	1.0169e−1	92.9773
2^{24}	2.3749e−3	209.214	**2.3364e−3**	360.710	2.8792e−3	359.820	8.1599e0	338.904	1.3316e−2	369.607	5.4988e−3	371.880	1.0182e−1	347.467

The following observations can be done. Regarding the computational complexity of the algorithms, as to be expected, the fastest method is the Crude Monte Carlo. The computational time for the non standard lattice sequence is slightly bigger than the computational time for the standard lattice sequence. Regarding the relative error the following implications can be done.

For the 3-dimensional integrals for a number of samples $N = 2^{24}$ the best approach is the 1EXPT – it gives a relative error 2.4282e−8 – see Table 1. For the 4-dimensional integrals for a number of samples $N = 2^{24}$ the best approach is NP1-EXPT – it gives a relative error 5.0431e−7 – see Table 2. For the 5-dimensional integrals for the maximum number of samples the best approach is the NP1-2POLY giving a relative error of 1.4122e−7 – see Table 3. For the 7-dimensional integrals for the maximum number of samples $N = 2^{24}$ the best approach is the 1PT - it gives a relative error 1.4746e−8 – see Table 4. For the 10-dimensional integrals for the maximum number of samples $N = 2^{24}$ the best approach is the NP1-1PT - it gives a relative error 1.7202e−5 – see Table 5. For the 15-dimensional integrals for a number of samples $N = 2^{24}$ the best approach is 2POLY - it gives a relative error 5.8471e−6 – see Table 6 and it at least 1 magnitude better than all the other algorithms. For the 25-dimensional integrals for a number of samples $N = 2^{24}$ the best approach here is 1EXPT - it gives a relative error 1.9956e−4 – see Table 7. Note that the result produced by the lattice sequence is with 1 order better magnitude than all the others algorithms. This multidimensional integral should be a hard task, because of its larger reference value compared to the other integrals in this study. For the 30-dimensional integrals for a number of samples $N = 2^{24}$ the best approach is the 1PT - it gives a relative error 2.3364e−3 – see Table 8, and the result is very close to this produced by the 1EXPT.

The convergence of all compared stochastic methods can be seen on Fig. 1. It can be seen that for a given GV the difference in the relative error between the standard and non-standard lattice sequence are of the same magnitude.

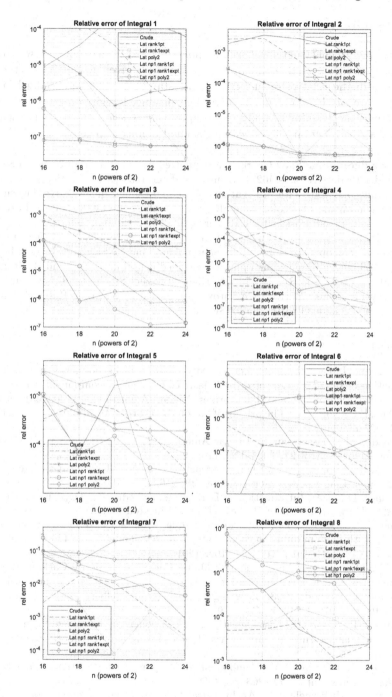

Fig. 1. Convergence of the stochastic methods for the integrals (7)-(14).

4 Conclusion

In this paper we construct standard and non-standard lattice sequence with three different generating vectors. It can be seen that for a given generating vector, the difference in the relative error between the standard and non-standard lattice sequence are of the same magnitude. The produced relative errors are much better than the standard Monte Carlo approach. The obtained results will be important regrading large-scale computational finance problems.

Acknowledgements. Slavi Georgiev is supported by the Bulgarian National Science Fund under Project KP-06-M62/1 "Numerical deterministic, stochastic, machine and deep learning methods with applications in computational, quantitative, algorithmic finance, biomathematics, ecology and algebra" from 2022 and by the National Program "Young Scientists and Postdoctoral Researchers - 2" – Bulgarian Academy of Sciences. Venelin Todorov is supported by BNSF under Project KP-06-N52/5 "Efficient methods for modeling, optimization and decision making" and BNSF under Project KP-06-N62/6 "Machine learning through physics-informed neural networks" and BNSF under Bilateral Project KP-06-Russia/17 "New Highly Efficient Stochastic Simulation Methods and Applications". Borislav Chakarov is supported by Scientific Research Fund of Ruse University under Project No FNSE-04.

References

1. Baldeaux, J., Dick, J., Leobacher, G., Nuyens, D., Pillichshammer, F.: Efficient calculation of the worst-case error and (fast) component-by-component construction of higher order polynomial lattice rules. Numer. Algorithms **59**, 403–431 (2012)
2. Black, F., Scholes, M.: The pricing of options and corporate liabilities. J. Pol. Econ. **81**, 637–659 (1973)
3. Broadie, M., Glasserman, P.: Pricing American-style securities using simulation. J. Econ. Dyn. Control **21**, 1323–1352 (1997)
4. Cools, R., Kuo, F., Nuyens, D.: Constructing embedded lattice rules for multivariate integration. SIAM J. Sci. Comput. **28**, 2162–2188 (2006)
5. Chance, D.M.: An Introduction to Derivatives, 3rd edn. The Dryden Press (1995)
6. Duffie, D.: Dynamic Asset Pricing Theory, Princeton (1992)
7. Kuo, F.Y., Nuyens, D.: Application of quasi-Monte Carlo methods to elliptic PDEs with random diffusion coefficients - a survey of analysis and implementation. Found. Comput. Math. **16**, 1631–1696 (2016)
8. Lai, Y., Spanier, J.: Applications of Monte Carlo/Quasi-Monte Carlo methods in finance: option pricing. In: Niederreiter, H., Spanier, J. (eds.) Monte-Carlo and Quasi-MC Methods 1998, pp. 284–295. Springer, Heidelberg (2000). https://doi.org/10.1007/978-3-642-59657-5_19
9. Nuyens, D., Cools, R.: Fast algorithms for component-by-component construction of rank-1 lattice rules in shift-invariant reproducing kernel Hilbert spaces. Math. Comput. **75**, 903–920 (2006)
10. Sloan, I.H., Reztsov, A.V.: Component-by-component construction of good lattice rules. Math. Comput. **71**, 263–273 (2002)
11. Wang, Y., Hickernell, F.J.: An historical overview of lattice point sets. In: Fang, K.T., Niederreiter, H., Hickernell, F.J. (eds.) Monte Carlo and Quasi-Monte Carlo Methods 2000, pp. 158–167. Springer, Heidelberg (2002). https://doi.org/10.1007/978-3-642-56046-0_10

Circular Intuitionistic Fuzzy Knapsack Problem

Velichka Traneva$^{(\boxtimes)}$ ⓘ, Petar Petrov ⓘ, and Stoyan Tranev ⓘ

"Prof. Asen Zlatarov" University, "Prof. Yakimov" Blvd, 8000 Bourgas, Bulgaria
veleka13@gmail.com, tranev@abv.bg

Abstract. The Knapsack problem is an NP-hard combinatorial optimization problem whose objective is to select those items to put in the knapsack to reach the highest possible total value without exceeding its capacity. Nowadays, there is great uncertainty in the parameters of this problem. Traditional methods for solving this problem cannot account for the uncertainty in the environment. In 2020, Atanassov introduced an extension of the Intuitionistic Fuzzy Set called the Circular Intuitionistic Fuzzy Set (C-IFS) to model the greater uncertainty of the environment. This study suggests an index-matrix approach to a circular intuitionistic fuzzy knapsack problem (C-IFKP) by extending the classical dynamic optimization algorithm. The urgency and duration of request satisfaction are suggested by experts and their ratings are taken into account in the proposed algorithm. Software for performing the proposed C-IFKP is also developed. The efficiency of the algorithm is applied to optimize the execution of the requests that an Ambulance team must satisfy in a certain time, taking into account the urgency of the request and its duration for execution. Three scenarios are proposed to the decision maker for the final choice - pessimistic, optimistic, and average.

Keywords: Index matrices · Intuitionistic fuzzy logic · Knapsack problem

1 Introduction

The objective of the Knapsack problem (KP) is to optimize the total utility value of all chosen items by the decision-maker to the capacity of knapsack [16]. The name "Knapsack problem" dates back to the early works of George Dantzig [9] in the fifties of the last century. The dynamic programming approach to the KP was investigated by Gilmore and Gomory in 1966 [12]. In today's inflationary environment, the amount of weight and the profit of the parameters is vague.

Work on Sect. 3 and Sect. 4 is supported by the Asen Zlatarov University through project Ref. No. NIX-482/2023 "Application of Intuitionistic Fuzzy Logic in Decision Making". Work on Sect. 1 and Sect. 2 is supported by the Asen Zlatarov University through project Ref. No. NIX-486/2023 "Modeling Management Decisions with New Analysis Tools in a Fuzzy Environment".

I. Lirkov and S. Margenov (Eds.): LSSC 2023, LNCS 13952, pp. 279–287, 2024.
https://doi.org/10.1007/978-3-031-56208-2_28

Fuzzy sets of Zadeh (FSs) [25] is a means of describing unclear parameters using degrees of membership and non-membership. A more powerful toolkit for modeling KP with fuzzy parameters is that of the circular intuitionistic fuzzy sets (C-IFSs, [4]) proposed by Atanassov, which are one of the latest extensions of intuitionistic FSs (IFSs, [1]). C-IFSs are defined as sets where each element of the universe has degrees of membership and non-membership with a circle around them [4]. The classical KP transforms to the fuzzy (FKP), intuitionistic fuzzy KP (IFKP), or circular IFKP (C-IFKP). The term "fuzzy knapsack" participates in the titles of 37 articles in the Scopus, and "IF knapsack" participates only in the title of 1 article [11]. A multiple choice FKP is presented in [15] by an approximate algorithm for its solution. In [18], a type of FKP is presented. In [8,18,19], a dynamic programming approach has been given for solving FKP. Ant Colony Optimization algorithm on Multiple-Constraint Knapsack Problem using intuitionistic fuzzy (IF) pheromone updating is presented in [11]. This study suggests an index-matrix approach [2] to a C-IFKP by extending the classical dynamic optimization algorithm and IFKP approach from [14,23,26]. The urgency coefficients and the expected duration of the requests are suggested by experts and their ratings are taken into account in the proposed algorithm. Software for performing the proposed C-IFKP is also developed. The efficiency of the algorithm is applied to optimize the execution of the requests that an Ambulance team must satisfy in a certain time. The originality of this paper comes from the development of the C-IFKP and its application to the Ambulance team problem. The rest of this study is organized as follows: In Sect. 2, the preliminaries of the C-IF triples and the IMs are given. In Sect. 3, we suggest a type of 0–1 C-IFKP and the software for its performance, and then apply it to the Ambulance team problem. Finally, the work ends with Sect. 4, which outlines the conclusions and suggestions for further research.

2   Basic Definitions of the Concepts of IMs and C-IF Triples

2.1   Operations and Relations on Circular Intuitionistic Fuzzy Triples (C-IFTs)

C-IFSs are introduced by Atanassov in 2020 as an extension of the IFSs and have a different from IFS by including a circle with radius r of the number consisting of membership degree and non-membership degree [4]. The C-IFT has the form as $\langle a(p), b(p); r \rangle = \langle \mu(p), \nu(p); r \rangle$, where $a(p), b(p) \in [0,1]$ and $a(p) + b(p) \leq 1$ that is used as an evaluation of a proposition p (see [4,6]). $\mu(p)$ and $\nu(p)$ respectively determine the "truth degree" and "falsity degree" of the proposition p. $r \in [0, \sqrt{2}]$ is a radius of the circle around $\langle a(p), b(p) \rangle$. Let us have two C-IFTs $x = \langle a, b; r_1 \rangle$ and $y = \langle c, d; r_2 \rangle$. Let us define an operation $* \in \{\min, \max\}$. The following operations are defined in [4,6,13]:

$$\neg x = \langle b, a \rangle; x \wedge_{1*} y = \langle \min(a,c), \max(b,d); *(r_c h11, r_2) \rangle;$$
$$x \vee_{1*} y = \langle \max(a,c), \min(b,d); *(r_1, r_2) \rangle; x \wedge_{2*} y = x + y = \langle a + c - a.c, b.d; *(r_1, r_2) \rangle;$$
$$x \vee_{2*} y = x.y = \langle a.c, b + d - b.d; *(r_1, r_2) \rangle; x @_* y = \langle \tfrac{a+c}{2}, \tfrac{b+d}{2}; *(r_1, r_2) \rangle$$
$$x -_* y = \langle \max(0, a - c), \min(1, b + d, 1 - a + c); *(r_1, r_2) \rangle$$

Let us expand the definition of the operation "division" between two IFPs [7, 10] to include the C-IFTs.

$$x :_* y = \begin{cases} \langle \min(1, a/c), \min(\max(0, 1 - a/c), \\ \max(0, (b - d)/(1 - d))); *(r_1, r_2) \text{ if } c \neq 0 \ \&d \ \neq 1 \\ \langle 0, 1; *(r_1, r_2) \rangle \text{ otherwise} \end{cases}$$

Since they produce outcomes with minimal and maximum levels of uncertainty, respectively, the operations presented here are based on the minimum and maximum of the radius individually. In [6], the definition of Szmidt and Kacprzyk's form of IF Hamming distance [20] is extended for the C-IFSs. Using the formula of the C-IFSs distance, we suggest the following relation for comparison of two C-IFTs extending the definition for IFPs measure $R = 0.5(1 + \pi_x)\text{distance}(\langle 1, 0; \sqrt{2} \rangle, x)$ that can be used for ranking the alternatives [20, 21]:

$$x \geq_{R^{circ}} y \text{ iff } R^{circ}_{\langle a, b \rangle} \leq R^{circ}_{\langle c, d \rangle} \tag{1}$$

where $R^{circ}_{\langle a, b; r \rangle} = 0.25(2 - a - b)\left(\frac{|\sqrt{2} - r|}{\sqrt{2}} + 1 - a\right)$ is a distance from the ideal positive alternative $\langle 1, 0; \sqrt{2} \rangle$ to x.

We say that the C-IFT x is in α-proximity with C-IFT y according to the Szmidt and Kacprzyk's form of the distance [6] if the circular IF distance between them $d(x, y) = 0.5\left(\frac{|r_1 - r_2|}{\sqrt{2}} + 0.5(|a - c| + |b - d| + |c + d - a - b|)\right) \leq \alpha$, where $\alpha \in [0; 1]$.

2.2 Definition, Operations, and Relations with Circular IF IMs (C-IFIMs)

The theory of index matrices (IMs) arose in 1987 in [2]. Different operations, relations, and operators over IMs are defined [3, 22]. Let $\mathcal{I}$ be a fixed set of indices. By two-dimensional C-IFIM $A = [K, L, \{\langle \mu_{k_i, l_j}, \nu_{k_i, l_j}; r_{k_i, l_j} \rangle\}]$ with index sets K and L ($K, L \subset \mathcal{I}$), we denote the object analogous to the definition of IFIM [3]:

$$A = \begin{array}{c|cccccc} & l_{ch11} & \cdots & l_j & \cdots & l_n \\ \hline k_{ch11} & \langle \mu_{k_{ch11}, l_{ch11}}, \nu_{k_{ch11}, l_{ch11}}; r_{k_{ch11}, l_{ch11}} \rangle & \cdots & \langle \mu_{k_{ch11}, l_j}, \nu_{k_{ch11}, l_j}; r_{k_{ch11}, l_j} \rangle & \cdots & \langle \mu_{k_{ch11}, l_n}, \nu_{k_{ch11}, l_n}; r_{k_{ch11}, l_n} \rangle \\ \vdots & \vdots & & \vdots & & \vdots \\ k_m & \langle \mu_{k_m, l_{ch11}}, \nu_{k_m, l_{ch11}}; r_{k_m, l_{ch11}} \rangle & \cdots & \langle \mu_{k_m, l_j}, \nu_{k_m, l_j}; r_{k_m, l_j} \rangle & \cdots & \langle \mu_{k_m, l_n}, \nu_{k_m, l_n}; r_{k_m, l_n} \rangle \end{array}$$

The definition of a 3-D C-IFIM is similar to the given in [3] and extends the previous definition of 2-D C-IFIM. In [2, 3], a lot of operations are defined over the IMs.

Transposition [3]: A' is the transposed IM of A.

Reduction [3]: The operations $(k, \bot)$-reduction of an IM A is defined by: $A_{(k, \bot)} = [K - \{k\}, L, \{c_{t_u, v_w}\}]$, where $c_{t_u, v_w} = a_{k_i, l_j}(t_u = k_i \in K - \{k\}, v_w = l_j \in L)$.

Projection [3]: Let $M \subseteq K$ and $N \subseteq L$. Then, $pr_{M, N} A = [M, N, \{b_{k_i, l_j}\}]$, where for each $k_i \in M$ and each $l_j \in N$, $b_{k_i, l_j} = a_{k_i, l_j}$.

Substitution [3]: $\left[\frac{p}{k}; \bot\right] A = [(K - \{k\}) \cup \{p\}, L, \{a_{k, l}\}]$

Internal subtraction of IMs' components [22]: $IO_{-(\max, \min)}(\langle k_i, l_j, A \rangle, \langle p_r, q_s, B \rangle) = [K, L, \{\langle \gamma_{t_u, v_w}, \delta_{t_u, v_w} \rangle\}]$.

Index type operations [22]:

$AGIndex_{(\max_R^{circ}),(\nleq)}(A) = \langle k_i, l_j \rangle$, where $\langle k_i, l_j \rangle$ (for $1 \leq i \leq m, 1 \leq j \leq n$) is the index of the maximum C-IFT of A in the sense of the relation (1) that has no empty value.

$Index_{(\max_R^{circ}),k_i}(A) = \{\langle k_i, l_{v_1} \rangle, \ldots, \langle k_i, l_{v_x} \rangle, \ldots, \langle k_i, l_{vV} \rangle\}$, where $\langle k_i, l_{v_x} \rangle$ (for $1 \leq i \leq m, 1 \leq x \leq V$) are the indices of the maximum element of k_i-th row of A.

Let us introduce some operations over C-IFIMs $A = [K, L, \{\langle \mu_{k_i,l_j}, \nu_{k_i,l_j}; r_{k_i,l_j} \rangle\}]$ and $B = [P, Q, \{\langle \rho_{p_r,q_s}, \sigma_{p_r,q_s} \rangle; \delta_{k_i,l_j}\}]$ analogous to those in [3,22].

Addition-$(\circ_1, \circ_2, *)$: $A \oplus_{(\circ_1, \circ_2, *)} B = [K \cup P, L \cup Q, \{\langle \phi_{t_u,v_w}, \psi_{t_u,v_w} \rangle; \eta_{t_u,v_w}\}]$, where
$\langle \circ_1, \circ_2 \rangle \in \{\langle \max, \min \rangle, \langle \min, \max \rangle, \langle \text{ average, average} \rangle\}$ and $* \in \{\langle \max, \min \}$.
$\langle \phi_{t_u,v_w}, \psi_{t_u,v_w}; \eta_{t_u,v_w} \rangle = \langle \circ_1(\mu_{k_i,l_j}, \rho_{p_r,q_s}), \circ_2(\nu_{k_i,l_j}, \sigma_{p_r,q_s}); *(r_{t_u,v_w}, \delta_{t_u,v_w}) \rangle$.

Termwise subtraction-(max,min): $A -_{(\max,\min,*)} B = A \oplus_{(\max,\min,*)} \neg B$.

Termwise   **mul-**
tiplication: $A \otimes_{(\circ_1, \circ_2, *)} B = [K \cap P, L \cap Q, \{\langle \phi_{t_u,v_w}, \psi_{t_u,v_w}; \eta_{t_u,v_w} \rangle\}]$, where
$\langle \phi_{t_u,v_w}, \psi_{t_u,v_w}; \eta_{t_u,v_w} \rangle = \langle \circ_1(\mu_{k_i,l_j}, \rho_{p_r,q_s}), \circ_2(\nu_{k_i,l_j}, \sigma_{p_r,q_s}); *(r_{t_u,v_w}, \delta_{t_u,v_w}) \rangle$.

Aggregation operations: Let us extend the operations $\#_q, (q \leq i \leq 3)$ from [24] for scaling aggregation operations over C-IFTs $x = \langle a, b; r_1 \rangle$ and $y = \langle c, d; r_2 \rangle$:
$x \#_1, * y = \langle min(a,c), max(b,d); *(r_1, r_2) \rangle$;
$x \#_2, * y = \langle average(a,c), average(b,d); *(r_1, r_2) \rangle$;
$x \#_3, * y = \langle max(a,c), min(b,d); *(r_1, r_2) \rangle$ and $* \in \{min, max\}$.

Let $k_0 \notin K$ will be the fixed index. The definition of the aggregation operation by the dimension K over 3-D C-IFIM A is [3,24]:

$$\alpha_{K,\#_q,*}(A, k_0) = \begin{array}{c|ccc} h_g \in H & l_1 & \cdots & l_n \\ \hline k_0 & \#_{q,*}^{m} \langle \mu_{k_i,l_1,h_g}, \nu_{k_i,l_1,h_g}; r_{k_i,l_1,h_g} \rangle & \cdots & \#_{q,*}^{m} \langle \mu_{k_i,l_n,h_g}, \nu_{k_i,l_n,h_g}; r_{k_i,l_n,h_g} \rangle \\ & {}_{i=1} & & {}_{i=1} \end{array},$$

where $1 \leq q \leq 3$.

Aggregate global internal operation [22]: $AGIO_{\oplus_{(\#_q,*)}}(A)$. If $q = 1, q = 2$ or $q = 3$ then we obtain respectively a pessimistic, averaged, or optimistic scenario.

Operation "Purge" of IM A Here, we define a new operation "Purge" by a dimension K as follows: $Purge_K(A)$ reduces each k_x-th row of A, if $a_{k_x,l_j} \leq a_{k_y,l_j}$, but $a_{k_x,l_e} \geq a_{k_y,l_e}$ for $1 \leq x \leq m, 1 \leq y \leq m, 1 \leq j \leq n$ and $1 \leq e \leq n$.

3   A 0-1 Circular Intuitionistic Fuzzy KP

In this section, a 0–1 C-IFKP for optimal scheduling of an Ambulance team is defined and suggest an IM approach for its solution (C-IFKP) extending IFKP [23] and the dynamic programming method [14,16,26]. The problem is: An Ambulance team has a certain time $T = \langle \rho, \sigma; r_T \rangle$ to serve the requests received on the emergency telephones. Incoming m requests $\{k_1, \ldots, k_i, \ldots, k_m\}$ must be evaluated by the experts $\{d_1, \ldots, d_s, \ldots, d_D\}$ (for $s = 1, \ldots, D$) with the specified IFP rating $re_s = \langle \delta_s, \epsilon_s \rangle$ ($1 \leq s \leq D$) by criteria c_1 and c_2: for urgency in terms of saving the patient's life a_{k_i,c_1} (for $i = 1, ..., m$) and expected duration of the request a_{k_i,c_2} (for $i = 1, ..., m$). It is crucial to properly assess the time to service the request and its urgency in terms of endangering the patient's life. The C-IFSs are a suitable tool for modeling the optimization problem. The purpose is to determine the optimal schedule of the Emergency team per their working day with three decision-making scenarios - optimistic, averaged, and pessimistic.

Let us expand an IM approach for the solution of C-IFKP extending IFKP [23].

Step 1. 3-D evaluation IFIM $EV[K, C, E, \{ev_{k_i,c_j,d_s}\}]$ is formed in accordance with the above problem, where $K = \{k_1, k_2, \ldots, k_m\}$, $C = \{c_1, c_2\}$, $E = \{d_1, \ldots, d_s, \ldots, d_D\}$ and the element $\{ev_{k_i,c_j,d_s}\} = \langle \mu_{k_i,c_j,d_s}, \nu_{k_i,c_j,d_s} \rangle$ (for $1 \le i \le m, 1 \le j \le n, 1 \le s \le D$) is the estimate of the d_s-th expert for the k_i-th request by the c_j-th criterion ($j = 1, 2$). The expert is not sure about its evaluation according to a given criterion due to changes in some uncontrollable factors and his evaluations are in the form of IFPs. Let the score (rating) re_s of the s-th expert ($s \in E$) be specified by an IFP $\langle \delta_s, \epsilon_s \rangle . \delta_s$ and ϵ_s are interpreted respectively as his degree of competence and incompetence. Then we create $EV^*[K, C, E, \{ev^*_{k_i,c_g,d_s}\}] = re_1 pr_{K,C,d_1} EV \oplus_{(o_{ch11},o_2)} \cdots \oplus_{(o_{ch11},o_2)} re_D pr_{K,C,d_D} EV$.
$EV := EV^*(ev_{k_i,l_j,d_s} = ev^*_{k_i,l_j,d_s}, \forall k_i \in K, \forall l_j \in L, \forall d_s \in E)$.

The averaged IF value of the k_i-th request against the c_j-th criterion is determined in the present moment $h_f \notin E$ using the α_E-th aggregation operation:

$$PI[K, C, h_f, \{pi_{k_i,c_g,h_f}\}] = \alpha_{E,\#_2}(EV^*, h_f) = \begin{array}{c|cc} h_f & c_1 & c_2 \\ \hline k_1 & \overset{D}{\underset{s=1}{\#_2}} \langle \mu_{k_1,c_1,d_s}, \nu_{k_1,c_1,d_s} \rangle & \overset{D}{\underset{s=1}{\#_2}} \langle \mu_{k_1,c_2,d_s}, \nu_{k_1,c_2,d_s} \rangle \\ \vdots & \vdots & \vdots \\ k_m & \overset{D}{\underset{s=1}{\#_2}} \langle \mu_{k_m,c_1,d_s}, \nu_{k_m,c_1,d_s} \rangle & \overset{D}{\underset{s=1}{\#_2}} \langle \mu_{k_m,c_2,d_s}, \nu_{k_m,c_2,d_s} \rangle \end{array} \mid h_f \notin E.$$

Now, we determine C-IFIM A, which comprises of the assessments of the requests at this time h_f based on the urgency and duration of the request:

$$A[K, C\{a_{k_i,c_g}\}] = \begin{array}{c|cc} & c_1 & c_2 \\ \hline k_{ch11} & \langle \mu^a_{k_{ch11},c_1}, \nu^a_{k_{ch11},c_1}; r^a_{k_{ch11},c_1} \rangle & \langle \mu^a_{k_{ch11},c_2}, \nu^a_{k_{ch11},c_2}; r^a_{k_{ch11},c_2} \rangle \\ \vdots & \vdots & \vdots \\ k_m & \langle \mu^a_{k_m,c_1}, \nu^a_{k_m,c_1}; r^a_{k_m,c_1} \rangle & \langle \mu^a_{k_m,c_2}, \nu^a_{k_m,c_2}; r^a_{k_m,c_2} \rangle \end{array},$$

where $K = \{k_1, \ldots, k_i, \ldots, k_m\}, i = 1, \ldots, m; C = \{c_1, c_2\}$, a_{k_i,c_g} (for $i = 1, \ldots, m; g = 1, 2$) are obtained as C-IFTs by transforming the IFPs pi_{k_i,c_j,d_s} in a C-IFTs by

for $g = 1$ to 2, $i = 1$ to m $\{\mu^a_{k_i,c_g} = \mu^{pi}_{k_i,c_g,h_f}; \nu^a_{k_i,c_g} = \nu^{pi}_{k_i,c_g,h_f}$ and

$$r^a_{k_i,c_g} = \max | \sqrt{(\mu^{ev}_{k_i,c_g,d_s} - \mu^{pi}_{k_i,c_g,h_f})^2 + (\nu^{ev}_{k_i,c_g,d_s} - \nu^{pi}_{k_i,c_g,h_f})^2} | \}.$$
$$1 \le s \le D$$

A similar approach for constructing C-IFTs is developed in [4,13]. Then, a check is made on the input data for the duration of the requests for not exceeding the given time of the Ambulance team T.

for $i = 1$ to m $\{$If $a_{k_i,c_2} > T$ then $A_{(k_i,\perp)} \}$

Let us denote by $|K| = m$ the number of the elements of the set K, then $|C| = 2$. We also define C-IFIM $X[K, C]$ with elements x_{k_i, c_g} (for $1 \leq i \leq m$, $\leq g \leq 2$) and: $\{x_{k_i, c_g}\} \in \begin{cases} \langle 1, 0; 0 \rangle, & \text{if the request } k_i \text{ is selected} \\ \langle 0, 1; 0 \rangle & \text{otherwise} \end{cases}$

Let in the beginning of the algorithm, all elements of IM X are equal to $\langle 0, 1; 0 \rangle$.

Construct IM $S^0[u_0, L] = \begin{array}{c|cc} & c_1 & c_2 \\ \hline u_0 & s^0_{u_0, c_1} & s^0_{u_0, c_2} \end{array} = \begin{array}{c|cc} & c_1 & c_2 \\ \hline u_0 & \langle 0, 1; \sqrt{2} \rangle & \langle 0, 1; \sqrt{2} \rangle \end{array}$

Step 2. for $i = 1$ to m do {Create IMs $R_i[k_i, C] = pr_{k_i, C} A$; $SH^{i-1}_{ch11} = \left[\frac{u_i}{u_{i-1}}; \perp \right] S^{i-1}$

for $h = 1$ to $i + 1$ do $\{SH^{i-1}_{ch11} = SH^{i-1}_{ch11} \oplus_{(\circ_1, \circ_2, *)} \left[\frac{u_h}{k_i}; \perp \right] R_i \}$

$S^i[U^i, L] = S^{i-1} \oplus_{((\circ_1, \circ_2, *))} SH^{i-1}_{ch11}$; for $h = 1$ to $i + 1$ do { If $s^i_{h, w} > T$ then $S^i_{(h, \perp)} \}$

The "Purge" operation is running by $S^i = Purge_{U^i} S^i$. } Go to *Step 3*.

Step 3. This step finds the index of the most urgent request by $\{Index_{(\max_{Rcirc}), c_1}(A) = \langle u_g, c_1 \rangle$

for $i = m$ to 1 do

{Find the α-nearest elements of $s^i_{u_g, c_1}$ (or $s^i_{u_g, c_2}$) ($\alpha = 0.5$) of S^i and choose the closest element from them - $s^i_{u_{g*}, c_1}$ (or $s^i_{u_{g*}, c_2}$).

If $\{s^i_{u_g *, c_1}$ (or $s^i_{u_g *, c_2})\} \in S^i$ and $\{s^i_{u_g *, c_1}$ (or $s^i_{u_g *, c_2})\} \notin S^{i-1}$ then

$\{x_{k_i, p} = \langle 1, 0; \sqrt{2} \rangle$ and $x_{k_i, w} = \langle 1, 0; \sqrt{2} \rangle$; $s^i_{u_g, c_1} = s^i_{u_g *, c_1} -_* a_{k_i, c_1}$; $s^i_{u_g, c_2} = s^i_{u_g *, c_2} -_* a_{k_i, c_2} \} \}$ Go to *Step 4*.

Step 4. The best outcome and response time for emergency requests are

$AGIO_{\oplus_{(\#_q, *)}} \left(pr_{K, c_1} A \otimes_{(\circ_1, \circ_2, *)} pr_{K, c_1} X \right)$;

$AGIO_{\oplus_{(\#_q, *)}} \left(pr_{K, c_2} A \otimes_{(\circ_1, \circ_2, *)} pr_{K, c_2} X \right)$.

If $q = 1$, $q = 2$, or $q = 3$, we get the pessimistic, averaged, or optimistic estimate of the optimal benefit, respectively. If the algorithm uses $\langle \circ_1, \circ_2 \rangle = \langle \min, \max \rangle$ in all operations, the pessimistic scenario has been chosen. If $\langle \circ_1, \circ_2 \rangle = \langle \max, \min \rangle$ is utilized, the optimistic scenario has been chosen. In the case of greater vagueness, the operation "$* = \max$", else "$* = \min$."

The suggested approach's time complexity is comparable to that of the conventional dynamic programming algorithm [16, 26] - $O(m.C)$.

We have created a command-line tool, written in C++, to automate the algorithm's calculations. The basic IM operations are implemented using the IM structure we created for this purpose, which contains std::tuple types from the STL [17]. The requests' C-IFTs and the knapsack capacity are inputs to the program. When the program is finished, a suggested solution is displayed on the computer screen along with a thorough output of each algorithm iteration. The algorithm's representational software has been created for private use and won't be made available to the general public.

Here, the C-IFKP that is being proposed in this section is made clear by an application to streamline the fulfillment of the requests that an ambulance crew must fulfill in a specific amount of time $T = \langle 0.99, 0.0; 1 \rangle$ to respond to emergency phone requests. A team of experts $\{d_1, d_2, d_3\}$ with given IFP rating $re_s = \langle \delta_s, \epsilon_s \rangle$ $(1 \le s \le D)$ is required to evaluate incoming m requests $\{k_1, k_2, k_3, k_4\}$ by criteria c_1 and c_2: for urgency in terms of saving the patient's life a_{k_i,c_1} (for $i = 1, ..., m$) and expected duration of the request a_{k_i,c_2} (for $i = 1, ..., m$). The purpose is to determine the optimal schedule of the Emergency team per their working day with three decision-making scenarios.

Solution of the problem:

Step 1. 3-D evaluation IFIM $EV[K, C, E, \{ev_{k_i,c_g,d_s}\}]$ is formed.

$$EV = \left\{ \begin{array}{c|cc|c|cc|c|cc} & d_1 & & & d_2 & & & d_3 & \\ & c_1 & c_2 & & c_1 & c_2 & & c_1 & c_2 \\ \hline k_{ch11} & \langle 0.1, 0.89 \rangle & \langle 0.05, 0.85 \rangle & k_{ch11} & \langle 0.2, 0.79 \rangle & \langle 0.1, 0.9 \rangle & k_{ch11} & \langle 0.15, 0.85 \rangle & \langle 0.15, 0.85 \rangle \\ k_2 & \langle 0.27, 0.73 \rangle & \langle 0.05, 0.94 \rangle & k_2 & \langle 0.28, 0.70 \rangle & \langle 0.02, 0.97 \rangle & k_2 & \langle 0.24, 0.76 \rangle & \langle 0.08, 0.92 \rangle \\ k_3 & \langle 0.36, 0.63 \rangle & \langle 0.26, 0.73 \rangle & k_3 & \langle 0.4, 0.6 \rangle & \langle 0.3, 0.7 \rangle & k_3 & \langle 0.32, 0.66 \rangle & \langle 0.22, 0.76 \rangle \\ k_4 & \langle 0.47, 0.52 \rangle & \langle 0.4, 0.6 \rangle & k_4 & \langle 0.5, 0.49 \rangle & \langle 0.36, 0.63 \rangle & k_4 & \langle 0.44, 0.55 \rangle & \langle 0.32, 0.66 \rangle \end{array} \right\},$$

$\{ev_{k_i,c_j,d_s}\}$ (for $1 \le i \le 4, 1 \le g \le 2, 1 \le s \le 3$) is the estimate of the d_s-th expert for the k_i-th request by the c_g-th criterion. Let the experts have the following rating coefficients respectively $\{r_1, r_2, r_3\} = \{\langle 0.7, 0.1 \rangle, \langle 0.6, 0.08 \rangle, \langle 0.8, 0.07 \rangle\}$. We create $EV^*[K, C, E, \{ev^*_{k_i,c_g,d_s}\}] = re_1 pr_{K,C,d_1} EV \oplus_{(\circ_{ch11}, \circ_2)} \cdots \oplus_{(\circ_{ch11}, \circ_2)} re_D pr_{K,C,d_D} EV$.
$EV := EV^*$. Then, we calculate C-IFIM A, which consists of the evaluations in a moment h_f of the requests by 2 criteria:

$$A[K,C] = \begin{array}{c|cc} & c_1 & c_2 \\ \hline k_{ch11} & \langle 0.10, 0.86; 0, 06 \rangle & \langle 0.07, 0.88; 0, 05 \rangle \\ k_2 & \langle 0.18, 0.75; 0, 03 \rangle & \langle 0.04, 0.95; 0, 04 \rangle \\ k_3 & \langle 0.25, 0.66; 0, 03 \rangle & \langle 0.18, 0.753; 0, 03 \rangle \\ k_4 & \langle 0.33, 0.56; 0, 04 \rangle & \langle 0.25, 0.66; 0, 04 \rangle \end{array}$$

where $K = \{k_1, k_2, k_3, k_4\}$, $C = \{c_1, c_2\}$ and $\{a_{k_i,c_1}, a_{k_i,c_2}\}$ are respectively C-IF urgency and duration of the k_i-th request. $X[K, C]$ with elements equal to $\langle 0, 1; \sqrt{2} \rangle$ is created. Go to *Step 2*.

Step 2. The algorithm calculates sequentially the IMs. The final IM after the operation "Purge" is the following":

$$S^4[U_4, C] = \begin{array}{c|cc} & c_1 & c_2 \\ \hline u_0 & \langle 0, 1; \sqrt{2} \rangle & \langle 0, 1; \sqrt{2} \rangle \\ u_1 & \langle 0.10, 0.86; 0.06 \rangle & \langle 0.07, 0.88; 0.05 \rangle \\ u_3 & \langle 0.27, 0.64; 0.03 \rangle & \langle 0.11, 0.83; 0.04 \rangle \\ u_5 & \langle 0.33, 0.57; 0.03 \rangle & \langle 0.24, 0.66; 0.03 \rangle \\ u_6 & \langle 0.45, 0.43; 0.03 \rangle & \langle 0.27, 0.63; 0.03 \rangle \\ u_9 & \langle 0.51, 0.36; 0.03 \rangle & \langle 0.33, 0.55; 0.04 \rangle \\ u_{10} & \langle 0.55, 0.32; 0.03 \rangle & \langle 0.43, 0.44; 0.03 \rangle \\ u_{11} & \langle 0.63, 0.24; 0.03 \rangle & \langle 0.45, 0.41; 0.03 \rangle \end{array}$$

Go to *Step 3*.

Step 3. At this step, the index of the largest C-IF urgency is found. $\langle 0.33, 0.56; 0.04 \rangle$ (or $\langle 0.25, 0.66; 0.04 \rangle$) $\in S^4$, but they $\notin S^3$, therefore $x_{k_4,c_1} = \langle 1, 0; \sqrt{2} \rangle$ and $x_{k_4,c_2} = \langle 1, 0; \sqrt{2} \rangle$. Then $c^*_{ch11} = \langle 0.63, 0.24; 0.03 \rangle -_* \langle 0.33, 0.56; 0.04 \rangle = \langle 0.27, 0.64; 0.03 \rangle$ and $c^*_2 = \langle 0.45, 0.41; 0.03 \rangle_* - \langle 0.25, 0.66; 0.04 \rangle = \langle 0.11, 0.83; 0.04 \rangle$. Find the the 0.5-nearest elements of c^*_{ch11} and c^*_2 in S^3 - $\langle 0.180.750.03 \rangle$ and $\langle 0.040.950.04 \rangle$. These elements do not belong to S^2, therefore $x_{k_2,c_1} = \langle 1, 0; \sqrt{2} \rangle$ and $x_{k_2,c_2} = \langle 1, 0; \sqrt{2} \rangle$. Then $c^*_{ch11} = \langle 0.27, 0.64; 0.03 \rangle -_* \langle 0.18, 0.75; 0.03 \rangle = \langle 0.10, 0.86; 0.06 \rangle$ and

$c_2^* = \langle 0.11, 0.83; 0.04 \rangle -_* \langle 0.04, 0.95; 0.04 \rangle = \langle 0.07, 0.88; 0.05 \rangle$. Analogously $x_{k_1,c_1} = \langle 1, 0; \sqrt{2} \rangle$ and $x_{k_1,c_2} = \langle 1, 0; \sqrt{2} \rangle$. After that is calculated: $x_{k_2,c_1} = \langle 0, 1; \sqrt{2} \rangle$ and $x_{k_2,c_2} = \langle 0, 1; \sqrt{2} \rangle$. In this example, the team can service the 4th, 2nd, and 1st requests.

Step 4. The following table presents the results for the optimistic, pessimistic, and average scenarios for the optimal benefit (Table 1):

Table 1. The results for the optimistic, pessimistic, and average scenarios.

Scenario	c_1	c_2
Optimistic	$\langle 0.51, 0.36; 0.03 \rangle$	$\langle 0.14, 0.79; 0.04 \rangle$
Pessimistic	$\langle 0.10, 0.86; 0.03 \rangle$	$\langle 0.04, 0.95; 0.04 \rangle$
Average	$\langle 0.20, 0.72; 0.04 \rangle$	$\langle 0.12, 0.83; 0.04 \rangle$

4   Conclusion

In this research, we proposed an extension of the IF dynamic method to the KP [23] that enables the search for an optimal solution of the type 0–1 C-IFKP. An application to optimize the execution of the requests that an ambulance team must fulfill in a limited amount of time, taking into account the urgency of the request and its execution time, clarifies the developed IMs technique for the solution of the C-IFKP. Pessimistic, optimistic, and average scenarios are put out to the decision-maker for consideration before making a final decision. The software tool for C-IFKP implementation is created and used with a numerical example. In the future, we will expand the approach, so that it can be applied over elliptic IF data [5] as well as the software for its implementation.

References

1. Atanassov, K.T.: Intuitionistic fuzzy sets. In: VII ITKR Session, Sofia, 20-23 June 1983 (Deposed in Centr. Sci.-Techn. Library of the Bulg. Acad. of Sci., 1697/84) (1983). (in Bulgarian)
2. Atanassov, K.: Generalized index matrices. Comptes rendus de l'Academie Bulgare des Sciences **40**(11), 15–18 (1987)
3. Atanassov, K.: Index matrices: towards an augmented matrix calculus. In: Studies in Computational Intelligence, vol. 573. Springer, Cham (2014). https://doi.org/10.1007/978-3-319-10945-9
4. Atanassov, K.: Circular intuitionistic fuzzy sets. J. Intell. Fuzzy Syst. **39**(5), 5981–5986 (2020)
5. Atanassov, K.: Elliptic intuitionistic fuzzy sets. C. R. Acad. Bulg. Sci. **74**(65), 812–819 (2021)
6. Atanassov, K., Marinov, E.: Four distances for circular intuitionistic fuzzy sets. Mathematics **9**(10), 11–21 (2021). https://doi.org/10.3390/math9101121

7. Atanassov, K.: Remark on an intuitionistic fuzzy operation "division". Issues IFSs GNs **14**, 113-116 (2018-2019)
8. Chakraborty, D., Singh, V.: On solving fuzzy knapsack problem by multistage decision making using dynamic programming. AMO **16**(3), 575–585 (2014)
9. Dantzig, G.: Linear Programming and Extensions. Princeton University Press, Princeton (1963)
10. De, S.K., Bisvas, R., Roy, R.: Some operations on IFSs. Fuzzy sets Syst. **114**(4), 477–484 (2000)
11. Fidanova, S., Atanassov, K.: ACO with intuitionistic fuzzy pheromone updating applied on multiple-constraint Knapsack problem. Mathematics **9**(13), 1456 (2021)
12. Gilmore, P., Gomory, R.: The theory and computation of knapsack functions. Oper. Res. **14**, 1045–1074 (1966)
13. Kahraman, C., Alkan, N.: Circular intuitionistic fuzzy TOPSIS method with vague membership functions: supplier selection application context. NIFS **27**(1), 24–52 (2021)
14. Kellerer, H., Pferschy, U., Pisinger, D.: Knapsack Problems. Springer, Berlin (2004). https://doi.org/10.1007/978-1-4613-0303-9_5
15. Kuchta, D.: A generalisation of an algorithm solving the fuzzy multiple-choice knapsack problem. Fuzzy Sets Syst. **127**(2), 131–140 (2002)
16. Martello, S., Toth, P.: Knapsack Problems: Algorithms and Computer Implementations. John Wiley & sons, Hoboken (1990)
17. Mavrov, D.: An application for performing operations on two-dimensional index matrices. Ann. "Informatics" Sect. Union Sci. Bulgaria **10**, 66–80 (2019-2020)
18. Singh, V.: an approach to solve fuzzy knapsack problem in investment and business model. In: Nogalski, B., Szpitter, A., Jaboski, A., Jaboski, M. (eds.). Networked Business Models in the Circular Economy (2020). https://doi.org/10.4018/978-1-5225-7850-5.ch007
19. Singh, V.P., Chakraborty, D.: A dynamic programming algorithm for solving bi-objective fuzzy knapsack problem. In: Mohapatra, R.N., Chowdhury, D.R., Giri, D. (eds.) Mathematics and Computing. SPMS, vol. 139, pp. 289–306. Springer, New Delhi (2015). https://doi.org/10.1007/978-81-322-2452-5_20
20. Szmidt, E., Kacprzyk, J.: Amount of information and its reliability in the ranking of Atanassov's intuitionistic fuzzy alternatives. In: Rakus-Andersson, E., Yager, R., (eds.). Recent Advances in Decision Making, SCI, vol. 222, pp. 7–19. Springer, Heidelberg (2009). https://doi.org/10.1007/978-3-642-02187-9_2
21. Traneva, V., Tranev, S.: Intuitionistic fuzzy two-factor variance analysis of movie ticket sales. J. Intell. Fuzzy Syst. **42**(1), 563–573 (2022)
22. Traneva, V., Tranev, S.: Index Matrices as a Tool for Managerial Decision Making. Publ, House of the Union of Scientists, Bulgaria (2017). (in Bulgarian)
23. Traneva, V., Petrov, P., Tranev, S.: Intuitionistic fuzzy knapsack problem trough the index matrices prism. In: Georgiev, I., Datcheva, M., Georgiev, K., Nikolov, G. (eds.) NMA 2022. LNCS, vol. 13858, pp. 314–326. Springer, Cham (2023). https://doi.org/10.1007/978-3-031-32412-3_28
24. Traneva, V., Tranev, S., Stoenchev, M., Atanassov, K.: Scaled aggregation operations over two- and three-dimensional index matrices. Soft. Comput. **22**, 5115–5120 (2019)
25. Zadeh, L.: Fuzzy Sets. Inf. Control **8**(3), 338–353 (1965)
26. Knapsack problem using dynamic programming. https://codecrucks.com/knapsack-problem-using-dynamic-programming/. Accessed 18 Mar 2023

Large-Scale Models: Numerical Methods, Parallel Computations and Applications

Clouds Formed by Thermals Arising and Evolving Under the Influence of the Coriolis Force

Hristo Chervenkov[(⊠)] [iD] and Valery Spiridonov

National Institute of Meteorology and Hydrology, Tsarigradsko Shose blvd. 66,
1784 Sofia, Bulgaria
hristo.tchervenkov@meteo.bg
http://www.meteo.bg

Abstract. In the presented study, a 'mushroom-like' cloud structure in the atmosphere of the Northern Hemisphere has been studied. Examples of such a structure are shown and compared with the result of a simple analytic model of thermals developing in a gravitational field of the rotating Earth. Most generally, this structure can be described as a consequence of the interaction of the Coriolis force at a certain degree of atmospheric instability. The conditions for the occurrence and development of single thermal have been analyzed. For this purpose, the parcel theory of thermal formation is utilized. The studied cloud structures are relatively rare, which makes it difficult to determine the conditions for their appearance. These clouds are mostly observed in the cool part of the year. Subsequently, the contrast between the temperature in the tropical part of the ocean and that in the polar regions favors the occurrence of horizontal convection caused by the Coriolis force. It was found that the instability index, introduced by the authors in another study, should have values below minus one. The outcomes of the study suggest that there is an optimal range of temperature difference between the thermals and environment, as well as suitable wind and a level of instability for the development of the considered phenomena.

Keywords: Instability Index · Coriolis Thermals · Mushroom-like Clouds

1 Introduction and Problem Formulation

In our work [7] an index was obtained characterizing the instability of atmospheric processes in a horizontal surface (hereinafter referred to as the index). It was derived from the hydrostatic system in spherical coordinates. This index defines a necessary condition for atmospheric instability. In our next study [2] a statistical model (relationship) between the index and the registered temperature anomalies in two regions of Europe has been proposed and tested. The average monthly values of temperature and geopotential during this period were utilized to calculate the index and anomalies (defined as the difference between the monthly means concerning the reference period 1981–2010) in these regions. The mean values of this index, calculated at isobaric levels of 850 hPa and 500 hPa (approximately 1.5 km and 5.8 km above the mean sea level), were used as predictors in the modelling of temperature anomalies by months for the period 2011–2020. For the anomalies, data from the ERA5 reanalysis [5] in regular $0.25° \times 0.25°$ grid downloaded from the Copernicus Climate Data Store (CCDS) [3] was used. Due to the applied methods and the utilized amount of environmental records, the performed task could be regarded from a methodological point of view as big data analytics [4]. In this study, it was shown how to interpret the instability of the atmosphere in terms of the index for practical tasks.

The considered objects are clearly distinguishable on satellite images. The used visualisation, which is organized as seamless animation, is comprised of imagery from the geostationary satellites of the European Organisation for the Exploitation of Meteorological Satellites (EUMETSAT), U.S. National Oceanic and Atmospheric Administration (NOAA), the meteorological administrations of China (CMA), and Japan (JMA) and is combined with data from EUMETSAT's polar orbiting Metop satellites. It shows an entire year of weather across the globe throughout 2019. This ultra-high resolution (4k) visualisation has been produced by EUMETSAT's data visualisation team and is composed of a satellite infrared data layer, provided by MeteoFrance, superimposed over NASA's 'Blue Marble Next Generation' ground maps, which change with the seasons. It is freely available at https://youtu.be/8w3o6_cn-O8. In some subsets of this visualisation, the development of peculiar mushroom-shaped cloud formations is detected. Figure 1 shows three examples of 'mushroom-like' thermals in the Northern Hemisphere. They are apparently only for a few hours a year. Under certain conditions 'mushroom-like' thermals can be observed in laboratory experiments of modelling of convection. Such structures are reported in [6]. In the study, insight into the qualitative nature of similar thermals is afforded by observation of flow patterns made visible by electrochemical techniques. The thermals displayed in Fig. 2 were generated along a line that is more or less parallel to the edge of heated surface. In the atmosphere, such thermals appear during strong convection and can be seen after a powerful explosion [9]. The development of 'Mushroom-like' thermals is also considered by other authors [1]. Here the thermals caused by the Coriolis force will be investigated. A comparison of Figs. 1 and 2 gives reason to assume that the observed cloudiness is a consequence of

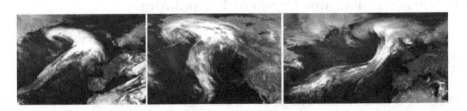

Fig. 1. Three examples of Coriolis thermal occurrence (see definition below).

Fig. 2. Photographs of thermals rising from a heated horizontal surface according [6]

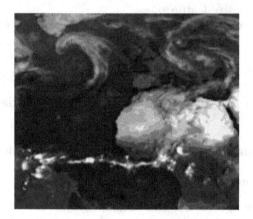

Fig. 3. Thermals occurring in the Equator region

convection. Convection in this case is driven by the Coriolis force instead of gravity force. Therefore, cloudiness with a specific shape is formed in the horizontal plane. Hereinafter, these two types will be referred as Coriolis and gravitational convection. The Coriolis thermals also arise with a symmetrical 'mushroom-like' shape. Figure 3 provides again a rationale for this terminology. It shows the thermals forming in the region of the Equator. They rise gravitationally to the condensation level and do not go to the poles because the Coriolis force is zero in this region. At latitudes where the Coriolis force is significant, traces of thermals can be observed again. Although the conditions in nature are different from those in the experiment, the developed forms are similar. In the Northern Hemisphere, the prevailing circulation is from west to east, which determines the tilt of the Coriolis thermals in that direction. To study this phenomenon, the so-called 'Parcel theory' will be applied as shown in the next section.

2 Methodology

The conditions for the occurrence and development of a single thermal will be analyzed here. For this purpose, the well-known parcel theory of thermals will be applied. This is a rather simple approach, but useful for a qualitative understanding of parcels evolving in the gravity field. If the temperature of the rising parcel remains higher than the surrounding atmosphere (despite its cooling), the parcel, being less dense than the surrounding environment, will continue to rise [8]. A similar mechanism can be assumed for Coriolis thermals. Before that, let us point out the main assumptions of the parcel theory applied to the gravitational rise of thermals. The main assumptions of the gravity parcel theory are:

- Buoyant convection leads to the formation of cumulus clouds (though other clouds have convective features).
- Parcel and environment are in dynamic equilibrium—the internal (of the parcel) and external (of the surrounding environment) pressure are equal: $P = P'$.
- Parcel maintains its identity; no mixing with environment occurs.
- The atmosphere is in hydrostatic equilibrium.
- No compensating motion of the atmosphere is caused as the parcel moves.

Neglected processes are:

- Drag (the friction between the parcel and environment during the movement)
- Mixing
- Effects of condensed water

For Coriolis thermals, some of these assumptions are self-fulfilling, while others are not relevant to them. For example, the processes in Coriolis convection develop at or above the level of condensation, so there are no effects of condensed water.

Detailed determination of the formula for gravitational acceleration can be found in many books (see, for example, [8]). The way in which this formula is found is described as follows. First, using the hydrostatic equilibrium of the environment, the expression for the pressure change as a function of height can be replaced by the density multiplied by $g = 9.81 \text{ m.s}^{-1}$ which is the acceleration of gravity:

$$a = g(\rho - \rho')/\rho', \qquad (1)$$

where a is the vertical acceleration and ρ is the air density. As in the most literature sources, the prime symbol denotes the properties of the air parcel in order to distinguish them from the properties of the environment. If the parcel's density is less than the surrounding air then $\rho - \rho' > 0$, and the acceleration is positive (i.e. upward motion). Substituting for density from the ideal gas law, and remembering that $P = P'$, we can obtain:

$$(\rho - \rho')/\rho' = (T' - T)/T, \qquad (2)$$

where T' is the internal and T the external temperature. Then the acceleration of the parcel in Eq. 1 can be written in terms of temperature as:

$$a = \frac{T' - T}{T} g \qquad (3)$$

Obviously, the necessary condition in Eq. 3 for lifting (i.e. upward motion) is $T' - T > 0$. Replacing the gravitational acceleration g with the Coriolis force acceleration fc, where f is the Coriolis parameter and c is the wind vector module, the acceleration of the Coriolis thermals is determined by:

$$a = \frac{T'(y) - T(y)}{T(y)} fc \qquad (4)$$

Here, the notation (y) marks the latitudinal dependence of the acceleration of the Coriolis thermals. The following transformations matched the instability condition in [7]. The geostrophic wind is taken without its meridional component, although instability is the reason for its appearance. Here the condition *ante factum* (before the event) is imposed. The aim is to determine the value of the instability index under the assumptions by which it was determined, namely:

$$fc = fu_g \text{ and } v_g = 0, \qquad (5)$$

where u_g and v_g are the zonal and meridional component of the geostrophic wind correspondingly. In the transformations below the requirement is introduced that the acceleration does not change the sign along the thermal trajectory. Another requirement is a continuous steady convective flow while the thermals form a symmetrical 'mushroom-like' shape. This flow is seen as a trail along its path, similar to those observed in Fig. 2 (see also Fig. 1).

From the equation of state and the condition of dynamic equilibrium follows $\rho RT = \rho' RT'$, or $\rho/\rho' = T'/T$.

Subsequent transformations exploit the fact that in the one-dimensional case the partial and total derivatives are equivalent. If H represents the geopotential in isobaric coordinates, the following transformations are using the formula for the geostrophic wind (see [7] again):

$$fu_g = -g\frac{dH}{dy}, \quad a = \frac{T'(y) - T(y)}{T(y)} \left[-g\frac{dH}{dy} \right] \qquad (6)$$

Given that:

$$\Gamma \doteq \frac{g}{C_p} = 0.0097 \text{ K.m}^{-1}, \qquad (7)$$

where $C_p = 1012 \text{ m}^2.\text{s}^{-2}.\text{K}^{-1}$ is the specific heat capacity of dry air at constant pressure. Replacing g with ΓC_p in Eq. 6 for the Archimedean acceleration is obtained:

$$a = \frac{T'(y) - T(y)}{T(y)} \left[-\Gamma C_p \frac{dH}{dy} \right] \qquad (8)$$

Considering H as an implicit function of T, dH/dy can be represented as:

$$\frac{dH}{dy} = \frac{dH}{dT}\frac{dT}{dy} \qquad (9)$$

Trough the instability index:

$$I \doteq \Gamma\frac{dH}{dT} - 1, \qquad (10)$$

is reached to:

$$\frac{dH}{dy} = \frac{I+1}{\Gamma}\frac{dT}{dy} \quad (11)$$

Thus, Eq. 8 can be rewritten as:

$$a = \frac{T'(y) - T(y)}{T(y)}\left[-\Gamma C_p \frac{I+1}{\Gamma}\frac{dT}{dy}\right] \quad (12)$$

or finally:

$$a = \frac{T'(y) - T(y)}{T(y)}\left[-C_p(I+1)\frac{dT}{dy}\right] \quad (13)$$

The acceleration a is positive if $I+1$ is negative (i.e. $I+1 < 0$) and if the thermal is warmer than the environment $(T'(y) - T(y) > 0)$. In this case, acceleration directs thermals from high to low temperatures, i.e. where $dT/dy > 0$, from the tropics to the midlatitudes (see Fig. 1). The convective instability considered in this way it imposes the stronger requirement $I < -1$ instead of the general condition for instability $I < 0$. This level of instability is a necessary condition for transporting air from low to high latitudes in such a way.

3   Results and Discussion

Despite the stronger requirement for the instability condition, it remains only necessary but not sufficient. The question of the optimal conditions at which the phenomenon under consideration is observed remains open. These thermals in their symmetrical shape are not often observed. Thermals dissipate relatively quickly or remain with only their cyclonic part. Well developed symmetrical thermals are more often observed during the colder period of the year. This may be due to more appropriate conditions for their formation.

The mechanism of formation of such structures can be assumed as follows. As a result of evaporation in the tropical zone of the ocean, clouds are formed above and at the level of condensation (due to gravitational convection). Coriolis convection then takes effect. The greater the temperature difference between the thermal and the environment, the greater the probability that the thermal will develop and maintain its symmetrical, 'mushroom-like', shape. In the tropics, the ocean has a high temperature that does not change significantly throughout the year. In the north (during the period October–March) the temperatures decrease significantly and their contrast with that of the thermal increases (the thermal does not mix and exchange with the surrounding air, according to the assumptions of the parcel theory). The assumptions for the 'Coriolis variant' of parcel theory are:

- Parcel and environment are in dynamic equilibrium: $P = P'$.
- Parcel maintains its identity; no mixing with the environment occurs.
- The acceleration does not change the sign.
- No compensating motion of the atmosphere is caused as the parcel moves.

4 Conclusion

The convection considered so far is related to the instability studied in [7] and is determined by the index introduced there. This leads to the development of different types of atmospheric circulation and associated cloud systems. The studied cloud system is relatively rare, which makes it difficult to determine the conditions for its appearance. It has been established that the instability index should have values below minus one, but the question remains what is its lower limit. This phenomenon is observed mainly in the cooler part of the year. During this period, the contrast between the temperature in the tropical part of the ocean and of the level of condensation in the polar regions is greater than that in the warmer part. This can be considered the second necessary condition for the occurrence of this phenomenon. Whether the thermals will develop a symmetrical shape or will arise only with its cyclone part seems to be determined by the difference in heat and ambient temperature. The appearance of the thermals in question probably requires a period of conducive conditions, but how long that period should be cannot be said at this time. Perhaps there is an optimal range of temperature difference between the thermals, environment, suitable wind, and level of instability. This is probably the reason for the rare occurrence of such thermal development. Thermals that do not develop a symmetrical shape probably form under more common environmental conditions.

Acknowledgment. This work has been accomplished with the financial support by the Grant №BG05M2OP001-1.001-0003, financed by the Science and Education for Smart Growth Operational Program (2014–2020) and co-financed by the European Union through the European structural and Investment funds.

References

1. Atkinson, J.W., Davidson, P.A.: The evolution of laminar thermals. J. Fluid Mech. **878**, 907–931 (2019). https://doi.org/10.1017/jfm.2019.690
2. Chervenkov, H., Spiridonov, V.: Statistical model of temperature anomalies using a new instability index. C. R. Acad. Bulg. Sci. **75**(11), 1621–1627 (2022). https://doi.org/10.7546/crabs.2022.11.09
3. CCDS Homepage. https://cds.climate.copernicus.eu/#!/home. Accessed 1 Mar 2023
4. Fathi, M., Haghi Kashani, M., Jameii, S.M., et al.: Big data analytics in weather forecasting: a systematic review. Arch. Comput. Methods Eng. **29**, 1247–1275 (2022). https://doi.org/10.1007/s11831-021-09616-4
5. Hersbach, H., Bell, B., Berrisford, P., Hirahara, S., Horányi, A., et al.: The ERA5 global reanalysis. Q. J. R. Meteorol. Soc. **146**, 1999–2049 (2020). https://doi.org/10.1002/qj.3803
6. Sparrow, E.M., Husar, R.B., Goldstein, R.J.: Observations and other characteristics of thermals. J. Fluid Mech. **41**(4), 793–800 (1970). https://doi.org/10.1017/S0022112070000927

7. Spiridonov, V., Chervenkov, H.: A new criterion for the stability of the atmospheric processes and its relationships with NAO. C. R. Acad. Bulg. Sci. **74**(9), 1363–1369 (2021). https://doi.org/10.7546/CRABS.2021.09.12

8. Stull, R.: Practical Meteorology: An Algebra-Based Survey of Atmospheric Science, 1st edn. University of British Columbia, Vancouver (2017)

9. Yun, T.Q.: Motion equation and solution of mushroom cloud. Atmos. Clim. Sci. **11**, 86–97 (2021). https://doi.org/10.4236/acs.2021.111006

A Nonstandard Finite Difference Method for a General Epidemic Model

István Faragó[(⊠)] [ID], Gabriella Svantnerné Sebestyén[ID],
and Bálint Máté Takács[ID]

Budapest University of Technology and Economics, Egry József u. 1.,
Budapest 1111, Hungary
faragois@cs.elte.hu

Abstract. This paper aims to investigate the qualitative properties of a general SIR model. The model includes a general function that describes the effect infectious individuals have on susceptible individuals. We use the nonstandard finite difference method to preserve the properties of the continuous model. Then, the theoretical results are confirmed by some numerical experiments.

Keywords: nonstandard finite difference method · epidemic model SIR · positivity

1 Introduction

Epidemic models have an important role from mathematical and biological points of view. These models help to analyze the behaviour of diseases. Their importance became apparent in the previous years during the Covid-19 pandemic. The classical SIR model was created by Kermack and McKendrick [7] in 1927, which is a compartmental model, meaning that the population is divided into three compartments:

- $S(t)$ is the number of susceptible individuals at time t,
- $I(t)$ is the number of infectious individuals at time t,
- $R(t)$ is the number of removed or deceased individuals at time t.

The assumptions of the model are the following:

- The births and natural deaths are omitted meaning that there are no vital dynamics in this model.
- Susceptible individuals leave the class S at rate μSI per unit time.
- Infectious individuals leave the class I at rate γI per unit time.

I. Lirkov and S. Margenov (Eds.): LSSC 2023, LNCS 13952, pp. 299–306, 2024.
https://doi.org/10.1007/978-3-031-56208-2_30

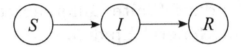

Fig. 1. The flow chart of the SIR model.

The flow chart of the SIR model is shown in Fig. 1.

The model has the form of the following system of ordinary differential equations

$$\frac{dS(t)}{dt} = -\mu S(t) \cdot I(t),$$

$$\frac{dI(t)}{dt} = \mu S(t) \cdot I(t) - \gamma I(t), \tag{1}$$

$$\frac{dR(t)}{dt} = \gamma I(t),$$

where μ and γ are given positive parameters.

This model has numerous variations. The stability of these models was also investigated [1,2]. Our aim is to generalize the aforementioned model, and then to create and analyze appropriate numerical schemes which preserve the qualitative properties of the continuous model e.g. positivity, boundedness, and conservation laws. In general, when we use standard numerical methods we get an upper bound for the step–size [4]. In this case, the nonstandard finite difference method is a good strategy to guarantee the qualitative properties without any restriction for the choice of the step–size [6].

2   The Mathematical Model and Its Properties

Consider the following generalization of model (1):

$$\frac{dS(t)}{dt} = -S(t) \cdot g(I(t)),$$

$$\frac{dI(t)}{dt} = S(t) \cdot g(I(t)) - \gamma I(t), \tag{2}$$

$$\frac{dR(t)}{dt} = \gamma I(t),$$

in which function $g(I(t))$ describes the effect infectious people have on healthy people. In epidemiology, this function is called the force of infection, which expresses the rate at which susceptible individuals acquire an infectious disease. The classical choice for the function g is when

$$g(I) = I. \tag{3}$$

In this case, we get the classical SIR model [7]. There are other forms of the function $g(I)$, e.g.

$$g(I) = \frac{I^p}{c_1 + c_2 I^q}, \quad c_1, c_2, p, q > 0 \tag{4}$$

when the force of infection is a nonlinear function [8,11]. The term I^p denotes the infection for of an infectious disease and the function $\frac{1}{c_1+c_2 I^q}$ measures the inhibition effect of the susceptible individuals when their number increases. The interested reader may consult [5].

In the following, we assume that function $g(x) : \mathbb{R} \to \mathbb{R}$ has the following properties

(C1) $g(x) \geq 0$ for all $x \in \mathbb{R}$,
(C2) $g(0) = 0$,
(C3) $g(x)$ is bounded in the sense that there exists a constant $\alpha \in \mathbb{R}^+$ such that $|g(x)| \leq \alpha|x|$ for every $x \in \mathbb{R}$.

Note that choices (3) and (4) do not have the property (C1) - however, we can modify them in a way that they do suffice this condition: e.g. $g(I) = |I|$. In this case, this function only differs from the usual choice when $I < 0$, but that case is not interesting from a biological standpoint, meaning that it never occurs when the initial conditions are chosen in a biologically reasonable way.

It is also worth mentioning that condition (C2) is only assumed because of its biological meaning (if there are no infectious people, then there should be no epidemic), but all the following results are true even if condition (C2) does not hold.

We analyze the model (2) and its attributes.

Theorem 1. *System* (2) *has a unique, continuously differentiable solution for all* $t \geq 0$.

Proof. The right-hand side of Eq. (2) has the local Lipschitz property, meaning that if we write Eq. (2) in the form

$$\frac{dU(t)}{dt} = F(U(t)),$$

where $U(t) = (S(t), I(t), R(t))^T$ and $F : \mathbb{R}^3 \to \mathbb{R}^3$, then there exist constants $L, d \in \mathbb{R}^+$ such that

$$\|F(U_1) - F(U_2)\| \leq L\|U_1 - U_2\|$$

holds for any U_1, U_2 for which $\|U_1\|, \|U_2\| \leq d$. Here $\|.\|$ denotes an appropriate norm of $\mathbb{R}^3$. It can also be proved that function F is bounded, meaning that $\|F(U)\| \leq c\|U\|$ holds for some appropriate constant $c \in \mathbb{R}^+$. Then, by the usual arguments (see e.g. [12] Theorem 4.2.1), it has a unique global solution.

Theorem 2. *Let* $N(t) = S(t) + I(t) + R(t)$ *be the sum of the three compartments. Then* $N(t) = N$ *($N > 0$ constant) for all* $t \geq 0$, *which is called mass–preservation property.*

Proof. We sum the right–hand side of (2), and we get

$$\frac{dN(t)}{dt} = \frac{dS(t)}{dt} + \frac{dI(t)}{dt} + \frac{dR(t)}{dt} = 0. \tag{5}$$

That means the total population $N(t)$ is constant in time.

Theorem 3. *Let* $S(0), I(0), R(0) \geq 0$ *be the initial conditions of system* (2) *at* $t = 0$. *Then, the solution of the problem* (2) *is non–negative, that is* $S(t), I(t), R(t) \geq 0$, *for all* $t > 0$.

Proof. The first equation can be written as

$$\frac{S'(t)}{S(t)} = -g(I(t)) \tag{6}$$

By integrating equation (6) we get

$$S(t) = S(0) \exp\left(-\int_0^t g(I(s))\, ds\right) \geq 0. \tag{7}$$

That means $S(t)$ is non–negative for all t.

From the second equation, we obtain

$$\frac{dI(t)}{dt} \geq -\gamma I(t) \tag{8}$$

from which we get $I(t)$ is also non–negative, since

$$I(t) \geq I(0) \exp\left(-\int_0^t \gamma\, ds\right) \geq 0. \tag{9}$$

It can be also shown that $R(t)$ is non–negative: by integrating the third equation the solution has the form

$$R(t) = R(0) + \gamma \int_0^t I(s)\, ds \geq 0. \tag{10}$$

Theorem 4. *Let* $S(0), I(0), R(0) \geq 0$ *be the initial conditions of system* (2) *at* $t = 0$. *Then* $S(t)$ *is a monotonically decreasing and* $R(t)$ *is a monotonically increasing function.*

Proof. By using condition (C1) for $g(I(t))$ we get $S'(t) \leq 0$, since $I(t)$ is non–negative, and also $R'(t) \geq 0$.

3   Discrete Model and Its Properties

Since the calculation of the exact solution of Eq. (2) is hard, we solve the system by using a nonstandard finite difference method. Let us denote the approximations of $S(t), I(t)$, and $R(t)$ at point $t_n = nh$, $n = 0, 1, 2, \ldots$ by S^n, I^n, and R^n, respectively. Our main goal is the construction of such numerical schemes which preserve the qualitative properties of the solution of system (2).

Let us apply the following nonstandard scheme [9,10] to system (2):

$$\frac{S^{n+1} - S^n}{\Phi(\Delta t)} = -S^n g(I^n),$$

$$\frac{I^{n+1} - I^n}{\Phi(\Delta t)} = S^n g(I^n) - \gamma I^n, \tag{11}$$

$$\frac{R^{n+1} - R^n}{\Phi(\Delta t)} = \gamma I^n,$$

with initial conditions $S^0 = S(0), I^0 = I(0), R^0 = R(0) \geq 0, \forall n \in \mathbb{N}$.

Function $\Phi(\Delta t) : \mathbb{R}^+ \to \mathbb{R}$ is chosen in a way that the left-hand side of the scheme above is an approximation of the given derivative. The conditions function $\Phi(\Delta t)$ should meet for a first-order approximation are [4]:

(P1) it is sufficiently smooth (it should be continuously differentiable twice)
(P2) it is positive,
(P3) $\Phi(0) = 0$,
(P4) $\Phi'(0) = 1$.

Note that the previous conditions also hold if the usual assumption $\Phi(\Delta t) = \Delta t + O\left((\Delta t)^2\right)$ is made.

Theorem 5. *Assume that we apply scheme* (11) *to solve system* (2). *Then, the following statements hold:*

(a) For every choice of $\Phi(\Delta t)$ we get

$$S^n + I^n + R^n =: N^n = N^{n+1}$$

for every $n > 0$. This means that the size of the total population is constant.
(b) Assume that $S^0, I^0, R^0 \geq 0$. Then, if

$$\Phi(\Delta t) \leq \min\left\{\frac{1}{g(I^n)}, \frac{1}{\gamma}\right\}, \tag{12}$$

then $S^n, I^n, R^n \geq 0$ also holds.
(c) Assume that $S^0, I^0, R^0 \geq 0$ and condition (12) *also holds. Then*

$$S^{n+1} \leq S^n, \quad R^n \leq R^{n+1} \quad \forall n \in \mathbb{N} \tag{13}$$

also holds.

Proof. We can prove the statements by induction. Statement (a) can be proved by observing the difference $N^{n+1} - N^n$ in the following way

$$N^{n+1} - N^n = (S^{n+1} + I^{n+1} + R^{n+1}) - (S^n + I^n + R^n) = S^n - \Phi(\Delta t) \cdot S^n g(I^n)$$
$$+ I^n + \Phi(\Delta t) \cdot (S^n g(I^n) - \gamma I^n) + R^n + \Phi(\Delta t) \cdot \gamma I^n - S^n - I^n - R^n$$
$$= 0.$$

Statement (b) can be proved by rewriting the first line of scheme (11) as

$$S^{n+1} = S^n(1 - \Phi(\Delta t)g(I^n)).$$

If condition (12) is fulfilled, then S^{n+1} is non-negative. From the second equation of scheme (11), we get

$$I^{n+1} = (1 - \gamma \cdot \Phi(\Delta t))I^n + \Phi(\Delta t)S^n g(I^n).$$

If condition (12) is fulfilled, then I^{n+1} is non-negative too. By similar arguments, one can also prove the non-negativity of R^{n+1}.

If statement (b) is true, then statement (c) can also be seen from the form of the numerical scheme.

Remark 1. It is worth mentioning here that condition (12) cannot be checked a priori. However, by statement (a) of Theorem 5, condition (12) holds if

$$\Phi(\Delta t) \le \min \left\{ \frac{1}{\max\limits_{x \in [0,N]} g(x)}, \frac{1}{\gamma} \right\}, \tag{14}$$

where $N := S(0) + I(0) + R(0)$. Condition (14) is now an a priori condition. The term $\max\limits_{x \in [0,N]} g(x)$ is well-defined because of condition (C3).

Remark 2. If we use the notation

$$C := \min \left\{ \frac{1}{\max\limits_{x \in [0,N]} g(x)}, \frac{1}{\gamma} \right\}, \tag{15}$$

then we get the following conditions for different step–size functions:

- If $\Phi(\Delta t) = \Delta t$ then we get an upper bound for the step–size, namely

$$\Delta t \le C. \tag{16}$$

- Another possible choice for function Φ for which the conditions (P1)–(P4) and condition (14) also hold for every choice of Δt is the following:

$$\Phi(x) = \frac{2C}{\pi} \arctan \left(\frac{\pi}{2C} x \right). \tag{17}$$

It can be seen that this is an appropriate choice. Then, since condition (14) is satisfied for every choice of $\Delta t \in \mathbb{R}^+$, the numerical scheme preserves the desired properties for any positive time step.

Theorem 6. *Let* $\Phi(\Delta t) = \dfrac{2C}{\pi} \arctan \left(\dfrac{\pi}{2C} \Delta t \right)$ *be a step–size function. Then the approximation* (11) *preserves the nonnegativity property for any non-negative step–size.*

Note that another possible choice for function $\Phi(\Delta t)$ is $\Phi(\Delta t) = \dfrac{1 - e^{-\beta \Delta t}}{\beta}$ (see e.g. [3]). In this case condition (14) is satisfied if $\beta =: \dfrac{1}{C}$.

4 Numerical Experiments

In this section, we conduct some numerical experiments to confirm the previous theoretical results.

Let us choose parameter γ to be $\gamma = 0.01$, function $g(I)$ to be

$$g(I) := \frac{2I}{1 + 0.1I}$$

(see [11]) and assume that the initial conditions are $S(0) = 0.9$, $I(0) = 0.1$ and $R(0) = 0$.

We are going to solve Eq. (2) using the numerical scheme (11) on the time interval $[0, 10]$. The bound given by (14) can be calculated as

$$\min\left\{ \frac{1}{\max\limits_{x \in [0,N]} g(x)}, \frac{1}{\gamma} \right\} = \min\left\{ \frac{1}{\max\limits_{x \in [0,1]} \dfrac{2x}{1 + 0.1x}}, \frac{1}{0.01} \right\} = \frac{1.1}{2} = 0.55$$

This essentially means that if we apply the choice $\Phi(\Delta t) = \Delta t$ and $\Delta t \geq 0.55$, then the numerical scheme might not preserve the qualitative properties of the continuous solution. However, if we use $\Phi(x) = \dfrac{2C}{\pi} \arctan\left(\dfrac{\pi}{2C}x\right)$, then the numerical solution behaves as expected. In Fig. 2 the numerical solutions can be seen in the first and second cases with choice $\Delta t = 0.8$. As we can see, in the first case the solution becomes negative, while in the second case, it behaves as expected. This showcases the advantage of the non-standard method: in this case, the solution remains positive for any choice of Δt, while in the case of the 'standard' Euler scheme, the time step should be smaller than a given bound for the required properties to hold.

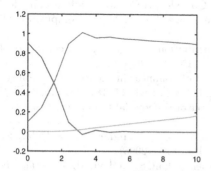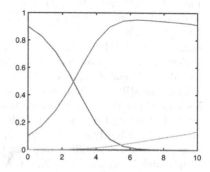

Fig. 2. The numerical solutions for $\Delta t = 0.8$ in the case of the first (left) and second (right) choice of the function $\Phi(\Delta t)$. The blue, red, and yellow curves are the numerical approximations of functions $S(t)$, $I(t)$, and $R(t)$, respectively. (Color figure online)

5 Summary

We have investigated an SIR type epidemics model with general incidence function. The continuous problem has the following properties: the solution is non-negative for non-negative initial conditions and the sum of the compartments is constant. Our aim was to use such numerical methods which preserve the above properties of the continuous problem. We used the nonstandard difference method and analyzed the properties of the discrete model. We have proved that when we use an appropriate choice of $\Phi(\Delta t)$ the discrete model preserves the properties of the continuous model without any restriction for the step size. We have made numerical experiments too. These simulations show that by using an adequate step-size function the size of the compartments remains non-negative. Moreover, since we are able to use bigger step-sizes compared to the ones of the standard finite difference method, the proposed non-standard method is more efficient.

References

1. Brauer, F., Castillo-Chavez, C.: Mathematical Models in Population Biology and Epidemiology. Springer, New York (2001). https://doi.org/10.1007/978-1-4757-3516-1
2. Capasso, V.: Mathematical Structures of Epidemic Systems. Springer, Heidelberg (1993). https://doi.org/10.1007/978-3-540-70514-7
3. Chapwanya, M., Lubuma, J., Mickens, R.: From enzyme kinetics to epidemiological models with Michaelis-Menten contact rate: design of nonstandard finite difference schemes. Comput. Math. Appl. **64**(3), 201–213 (2012)
4. Farago, I., Mosleh, R.: Reliable numerical modeling of malaria propagation. Appl. Math. **63**(3), 259–271 (2018)
5. Holling, C.S.: The components of predation as revealed by a study of small-mammal predation of the European pine sawfly. Can. Entomol. **91**(5), 293–320 (1959)
6. Jodar, L., Villanueva, R.J., Arenas, A.J., Gonzalez, G.C.: Nonstandard numerical methods for a mathematical model for influenza disease. Math. Comput. Simul. **79**(3), 622–633 (2008)
7. Kermack, W.O., McKendrick, A.G.: A contribution to the mathematical theory of epidemics. I. Proc. Roy. Soc. Lond. Ser. A **115**, 700–721 (1927)
8. Liu, W.M., Levin, S.A., Iwasa, Y.: Influence of nonlinear incidence rates upon the behavior of SIRS epidemiological models. J. Math. Biol. **23**, 187–204 (1986)
9. Mickens, R.E.: Nonstandard Finite Difference of Differential Equations. World Scientific Publishing Co., Pte. Ltd. (1994)
10. Mickens, R. E.: Advances in the Applications of Nonstandard Finite Difference Schemes. World Scientific Publishing Co., Pte. Ltd. (2005)
11. Mehdizadeh Khalsaraei, M., Shokri, A., Noeiaghdam, S., Molayi, M.: Nonstandard finite difference schemes for an SIR epidemic model. Mathematics **9**(23), 3082 (2021)
12. Schaeffer, D.G., Cain, J.W.: Ordinary Differential Equations: Basics and Beyond. Springer, New York (2016). https://doi.org/10.1007/978-1-4939-6389-8

Minimization of Energy Functionals via FEM: Implementation of hp-FEM

Miroslav Frost[1], Alexej Moskovka[2(✉)], and Jan Valdman[3,4]

[1] Institute of Thermomechanics, Czech Academy of Sciences, Dolejškova 5, 18200 Prague, Czech Republic
mfrost@it.cas.cz

[2] Department of Mathematics, Faculty of Applied Sciences, University of West Bohemia, Technická 8, 30100 Pilsen, Czech Republic
alexmos@kma.zcu.cz

[3] Institute of Information Theory and Automation, The Czech Academy of Sciences, Pod Vodárenskou věží 4, 18208 Prague, Czech Republic
jan.valdman@utia.cas.cz

[4] Faculty of Information Technology, Czech Technical University in Prague, Thákurova 9, 16000 Prague, Czech Republic

Abstract. Many problems in science and engineering can be rigorously recast into minimizing a suitable energy functional. We have been developing efficient and flexible solution strategies to tackle various minimization problems by employing finite element discretization with P1 triangular elements [1,2]. An extension to rectangular hp-finite elements in 2D is introduced in this contribution.

Keywords: hp finite elements · energy functional · trust-region methods · p-Laplace equation · hyperelasticity · MATLAB code vectorization

1 Introduction

The finite element method (FEM) can be efficiently used to minimize energy functionals appearing in various types of problems. The simplest P1 finite elements were implemented in MATLAB for the discretization of p-Laplace energy functional in [1]. We introduced several vectorization techniques in [2] for an efficient evaluation of the discrete energy gradient and, additionally, applied these techniques for the minimization of hyperelasticity in 2D and 3D. Recently, our approach has been successfully applied to 2D/3D problems in solid mechanics, namely the resolution of elastoplastic deformations of layered structures [3] or superelastic and pseudoplastic deformations of shape-memory alloys [4].

A. Moskovka and J. Valdman announce the support of the Czech Science Foundation (GACR) through the grant 21-06569K. M. Frost acknowledges the support of the Czech Science Foundation (GACR) through the grant 22-20181S.

I. Lirkov and S. Margenov (Eds.): LSSC 2023, LNCS 13952, pp. 307–315, 2024.
https://doi.org/10.1007/978-3-031-56208-2_31

The hp-FEM is an advanced numerical method based on FEM dating back to the pioneering works of I. Babuška, B. A. Szabó and co-workers in 1980s, e.g. [5] and [6]. It provides increased flexibility and convergence properties compared to the "conventional" FEM. There are recent MATLAB implementations including triangular elements [12] and rectangular elements [7].

In this paper, we combine energy evaluation techniques of [2] and the hp-FEM implementation [7]. The trust-region (TR) method [8] is applied for the actual minimization of energies. It is available in the MATLAB Optimization Toolbox and was found to be very efficient in the comparison performed in [2]. It requires the gradient of a discrete energy functional and also allows to specify a sparsity pattern of the corresponding Hessian matrix which is directly given by a finite element discretization. We employ two different options:

- option 1: the TR method with the gradient evaluated directly via its explicit form and the specified Hessian sparsity pattern.
- option 2: the TR method with the gradient evaluated approximately via central differences and the specified Hessian sparsity pattern.

We demonstrate the capabilities of our implementation on two particular problems in 2D: the scalar p-Laplace problem and the vector hyperelasticity problem. The underlying MATLAB code is available at

https://www.mathworks.com/matlabcentral/fileexchange/125465

for free download and testing. Running times were obtained on a MacBook Pro (M2 Pro processor, 2023) with 16 GB memory running MATLAB R2023a.

2   Hierarchical Shape Basis Functions

Given a reference element $T_{ref} = [-1,1]^2$ and $p \in \mathbb{N}$ we define by $S^p(T_{ref})$ a space of all (local) shape basis functions of polynomial degrees less or equal to p defined on T_{ref} and we denote by $n_{p,ref}$ their number. The construction of these functions is based on Legendre polynomials and is described in detail in [6]. Generally, there are three types of shape basis functions shown in Fig. 1:

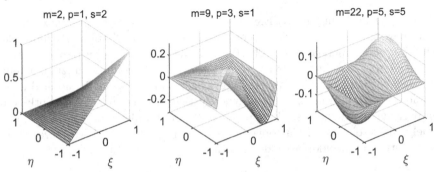

Fig. 1. Example of nodal, edge, and bubble shape basis functions on T_{ref}. Generated by a modification of the 'Benchmark 2' from [7].

- the nodal (Q1) shape basis function of the first order is nonzero in one particular node and vanishes in all other nodes;
- the edge shape basis function of the p-th order is nonzero on one particular edge and vanishes on all other edges;
- the bubble shape basis function of the p-th order vanishes on the whole boundary of T_{ref}.

All shape basis functions on T_{ref} are sorted by the polynomial degree p and the type of shape function. Every local basis function is assigned a unique index m. It is determined by the polynomial order p and an additional index s given by the local index of a node, edge or bubble.

We assume a computational domain Ω and its decomposition T into quadrilaterals in the sense of Ciarlet [9]. We denote by $|\mathcal{N}|$, $|\mathcal{E}|$ and $|T|$ the number of nodes, edges and elements, respectively. The local shape basis functions are used for the construction of global ones defined on the whole T. We denote by $S^p(T)$ the space of all global basis functions of polynomial degrees less or equal to p and by n_p their number. In [7] we introduced several key matrices providing the relation between the topology of T and the corresponding global basis functions. The first matrix collects the indices of all global basis functions and their type, the second collects for every element the indices of global basis functions that are nonzero on that element, and the third stores for every element the signs of local basis functions (necessary for edge functions of an odd degree).

Example 1. Quadrilaterals of the L-shape domain are shown in Fig. 2 (left) in which $|\mathcal{N}| = 21$, $|\mathcal{E}| = 32$, $|T| = 12$. For $p = 2$ we have $n_p = 53$ global basis functions (21 nodal and 32 edge) and the Hessian sparsity pattern (right) can be extracted directly.

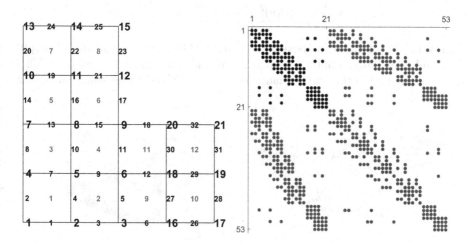

Fig. 2. A rectangular mesh (left) and the corresponding Hessian sparsity pattern for $p = 2$ hierarchical elements (right). The upper diagonal submatrix highlighted by the black color corresponds to Hessian sparsity for $p = 1$.

3   Models and Implementation

3.1   p-Laplace Equation

We are interested in a (weak) solution of the p-Laplace equation [10]:

$$\Delta_\alpha u = f \qquad \text{in } \Omega,$$
$$u = g \qquad \text{on } \partial\Omega, \tag{1}$$

where the p-Laplace operator is defined as $\Delta_\alpha u = \nabla \cdot \left(\|\nabla u\|^{\alpha-2} \nabla u \right)$ for some power $\alpha > 1$ (the integer p denotes the polynomial degree of $S^p(T_{ref})$). The domain $\Omega \in \mathbb{R}^d$ is assumed to have a Lipschitz boundary $\partial\Omega$, $f \in L^2(\Omega)$ and $g \in W^{1-1/\alpha,\alpha}(\partial\Omega)$, where L and W denote standard Lebesque and Sobolev spaces. It is known that (1) represents an Euler-Lagrange equation corresponding to a minimization problem

$$J(u) = \min_{v\in V} J(v), \qquad J(v) := \frac{1}{\alpha} \int_\Omega \|\nabla v\|^\alpha \, \mathrm{d}\mathbf{x} - \int_\Omega f\, v \, \mathrm{d}\mathbf{x}, \tag{2}$$

where $V = W_g^{1,\alpha}(\Omega) = \{v \in W^{1,\alpha}, v = g \text{ on } \partial\Omega\}$ includes Dirichlet boundary conditions on $\partial\Omega$. The minimizer $u \in V$ of (2) is known to be unique for $\alpha > 1$. It corresponds to the classical Laplace operator for $\alpha = 2$. The analytical handling of (1) is difficult for general f. The Eq. (1) in 2D ($d = 2$) takes the form

$$\nabla \cdot \left(\left((\partial_1 u)^2 + (\partial_2 u)^2 \right)^{\frac{\alpha-2}{2}} \nabla u \right) = f \qquad \text{in } \Omega \tag{3}$$

and the corresponding energy reads

$$J(v) := \frac{1}{\alpha} \int_\Omega \left((\partial_1 v)^2 + (\partial_2 v)^2 \right)^{\frac{\alpha}{2}} \, \mathrm{d}\mathbf{x} - \int_\Omega f\, v \, \mathrm{d}\mathbf{x}. \tag{4}$$

The evaluation of integrals above is based on the Gaussian quadrature. We apply the number of Gauss points corresponding to the quadrature of order $p+1$. This does not guarantee the exact quadrature, but it proved to be sufficient in our

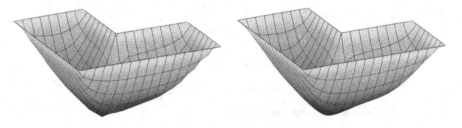

Fig. 3. Numerical solutions of the p-Laplacian with $\alpha = 3$ for a computational mesh with $|T| = 192$ and polynomial bases for $p = 1$ (left) and $p = 4$ (right).

numerical tests. Figure 3 illustrates numerical solutions for the L-shape domain from Fig. 2, for a constant $f = -10$, $\alpha = 3$ and zero Dirichlet boundary condition on $\partial\Omega$.

Table 1 shows the performance for Q2 elements (corresponding to the choice $p = 2$). We notice that computations using the explicitly evaluated gradient are faster than using the numerical gradient (via central differences). A comparison to Q1 elements (corresponding to the choice $p = 1$) or P1 (triangular) elements of [2] is depicted in Fig. 4. We observe a lower number of needed degrees of freedom (dofs) for Q2 elements and slightly lower running times to achieve the same accuracy. Since the exact energy is not known in this example, we use J_{ref} as the smallest of all achieved energy values $J(u)$ of Table 1 decreased by 10^{-4}.

Table 1. Performance of p-Laplacian for $\alpha = 3$ and Q2 elements.

level	$|\mathcal{T}|$	dofs	explicit gradient			numerical gradient		
			time [s]	iters	$J(u)$	time [s]	iters	$J(u)$
1	48	113	0.06	7	−7.9209	0.07	7	−7.9209
2	192	513	0.14	8	−7.9488	0.18	8	−7.9488
3	768	2177	0.49	10	−7.9562	0.67	10	−7.9562
4	3072	8961	1.73	12	−7.9587	2.38	12	−7.9587
5	12288	36353	8.31	13	−7.9596	10.60	13	−7.9596
6	49152	146433	80.81	13	−7.9600	136.92	14	−7.9600

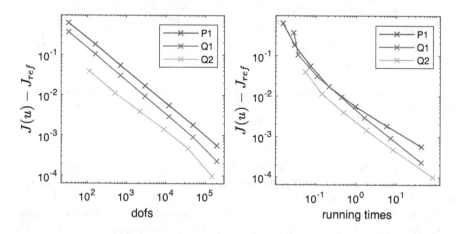

Fig. 4. Performance of p-Laplacian for $\alpha = 3$: comparison of elements.

3.2 Hyperelasticity

Boundary value problems in (non-linear) elastostatics provide examples of vector problem which can be directly dealt with our approach, see [2]. Deformation, $\mathbf{v}(\mathbf{x})$, of a (hyper)elastic body spanning the domain $\Omega \in \mathbb{R}^d$ subjected to volumetric force, $\mathbf{f}(\mathbf{x})$, can be obtained by minimization of the corresponding energy functional, J, which takes the form:

$$J(\mathbf{v}(\mathbf{x})) = \int_\Omega W\big(\mathbf{F}(\mathbf{v}(\mathbf{x}))\big)\,d\mathbf{x} - \int_\Omega \mathbf{f}(\mathbf{x}) \cdot \mathbf{v}(\mathbf{x})\,d\mathbf{x}, \tag{5}$$

where $\mathbf{F}(\mathbf{v}(\mathbf{x})) = \nabla\mathbf{v}(\mathbf{x})$ is deformation gradient and

$$W(\mathbf{F}) = C_1\big(I_1(\mathbf{F}) - \dim - 2\log(\det \mathbf{F})\big) + D_1(\det \mathbf{F} - 1)^2, \tag{6}$$

is so-called compressible Neo-Hookean energy density with C_1, D_1 being material constants and $I_1(\mathbf{F}) = \|\mathbf{F}\|^2$ denotes the squared Frobenius norm; see [11] for details on the underlying continuum mechanics theory and its mathematically rigorous formulation.

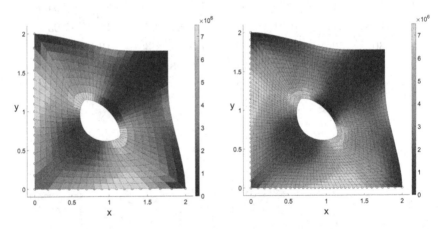

Fig. 5. Deformation and the corresponding Neo-Hookean density distributions for the 2D hyperelastic problem: a mesh with $|\mathcal{T}| = 512$ and Q4 elements (left) and a mesh with $|\mathcal{T}| = 2048$ and Q3 elements (right).

We assume the same benchmark problem as in [2]: a 2D hyperelastic domain given by a square $[0, 2] \times [0, 2]$ perforated by a disk with radius $r = 1/3$ in its center is subjected to a constant volumetric vector force $\mathbf{f} = (-3.5 \cdot 10^7, -3.5 \cdot 10^7)$ acting in a bottom-left direction; zero Dirichlet boundary conditions are applied on the left and bottom edge. We assume the Young modulus $E = 2 \cdot 10^8$ and the Poisson ratio $\nu = 0.3$. We consider arbitrary, although mutually consistent physical units. For illustration, Fig. 5 shows examples of the corresponding deformed mesh together with the underlying Neo-Hookean density distribution. Figure 6 depicts a comparison of P1, Q1 and Q2 elements. Similarly to Fig. 4, Q2 elements are superior to Q1 and P1 in accuracy with respect to the number of dofs, however, we observe only a little improvement with respect to the evaluation times.

3.3 Remarks on 2D Implementation

As an example, we introduce the following block that describes the evaluation of the p-Laplace energy:

```
1  function e = energy(v,mesh,params)
2  v_elems = v(e2d_elems);        % values on elements in hp basis
3  F_elems = evaluate_F_scalar_2D(v_elems,Dphi_elems);  % grads
4  densities_elems = density_pLaplace_2D(F_elems,w,alpha);
5  e = sum(areas_elems.*densities_elems) - b_full'*v;  % energy
6  end
```

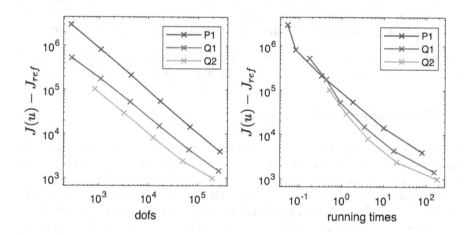

Fig. 6. Performance of hyperelasticity: comparison of elements.

The structure corresponds exactly to the code of [2]. The main difference is that some objects inside 'evaluate_F_scalar_2D' and 'density_pLaplace_2D' are higher-dimensional. This is because more Gauss points (we denote by n_{ip} their number) are needed for the integration of higher polynomial functions in the hierarchical basis. The function 'evaluate_F_scalar_2D' is provided below

```
1  function F_elems = evaluate_F_scalar_2D(v_elems,Dphi)
2  v_elems3D = reshape(v_elems',size(v_elems,2),1,size(v_elems,1));
3  v_x_elems = Dphi{1}.*v_elems3D;  % np_ref x nip x ne
4  v_y_elems = Dphi{2}.*v_elems3D;  % np_ref x nip x ne
5  F_elems = cell(1,2);
6  F_elems{1,1} = squeeze(sum(v_x_elems,1));  % nip x ne
7  F_elems{1,2} = squeeze(sum(v_y_elems,1));  % nip x ne
8  end
```

and evaluates the following objects:

– (lines 3–4) 3D matrices 'v_x_elems' and 'v_y_elems' of size $n_{p,ref} \times n_{ip} \times |\mathcal{T}|$ store for every element the partial derivatives of local basis functions in all n_{ip} Gauss points multiplied by the corresponding values of v.
– (lines 6–7) the final matrices of size $n_{ip} \times |\mathcal{T}|$ storing the partial derivatives of v in all Gauss points on every element.

The function '**density_pLaplace_2D**' evaluating energy densities on every element is introduced below

```
1  function densities = density_pLaplace_2D(F,w,alpha)
2  densities = (1/alpha)*sqrt(F{1,1}.^2 + F{1,2}.^2).^alpha;
3  densities = densities'*w;
4  end
```

and evaluates the following objects:

– (line 2) the matrix of size $n_{ip} \times |\mathcal{T}|$ storing the p-Laplace energy densities in all Gauss points of every element.
– (lines 3) vector of length $|\mathcal{T}|$ containing energy densities on all elements.

Similarly, the majority of the original P1 functions related to the evaluation of the hyperelastic energy are extended to the higher dimension in the same way.

4 Conclusions and Outlook

The hp-FEM for 2D rectangular elements was successfully incorporated into our vectorized MATLAB code and its improved convergence performance was demonstrated on two examples.

This work contributes to our long-term effort in developing a vectorized finite element-based solvers for energy minimization problems. Since many such problems emerge in science and engineering, the code is designed in a modular way so that various modifications (e.g., in functional types or boundary conditions) can be easily adopted. Our future research directions include implementing the hp-FEM in 3D or tuning the applied minimization algorithms.

References

1. Matonoha, C., Moskovka, A., Valdman, J.: Minimization of p-Laplacian via the finite element method in MATLAB. In: Lirkov, I., Margenov, S. (eds.) LSSC 2021. LNCS, vol. 13127, pp. 496–503. Springer, Cham (2022). https://doi.org/10.1007/978-3-030-97549-4_61
2. Moskovka, A., Valdman, J.: Fast MATLAB evaluation of nonlinear energies using FEM in 2D and 3D: nodal elements. Appl. Math. Comput. **424**, 127048 (2022)
3. Drozdenko, D., Knapek, M., Kruvzík, M., Máthis, K., et al.: Elastoplastic deformations of layered structures. Milan J. Math. **90**, 691–706 (2022)

4. Frost, M., Valdman, J.: Vectorized MATLAB implementation of the incremental minimization principle for rate-independent dissipative solids using FEM: a constitutive model of shape memory alloys. Mathematics **10**, 4412 (2022)
5. Szabó, B., Babuška, I.: Finite Element Analysis. Wiley, New York (1991)
6. Szabó, B., Babuška, I.: Introduction to Finite Element Analysis. Wiley, New York (2011)
7. Moskovka, A., Valdman, J.: MATLAB implementation of HP finite elements on rectangles using hierarchical basis functions. In: Wyrzykowski, R., Deelman, E., Dongarra, J., Karczewski, K. (eds.) PPAM 2022. LNCS, vol. 13827, pp. 287–299. Springer, Cham (2023). https://doi.org/10.1007/978-3-031-30445-3_24
8. Conn, A.R., Gould, N.I.M., Toint, P.L.: Trust-Region Methods. SIAM, Philadelphia (2000)
9. Ciarlet, P.G.: The Finite Element Method for Elliptic Problems. SIAM, Philadelphia (2002)
10. Lindqvist, P.: Notes of the p-Laplace equation (sec. ed.), report 161 of the Department of Mathematics and Statistics, University of Jyväskylä, Finland (2017)
11. Marsden, J.E., Hughes, T.J.: Mathematical Foundations of Elasticity. Prentice-Hall, Englewood Cliffs (1983)
12. Innerberger, M., Praetorius, D.: MooAFEM: an object oriented MATLAB code for higher-order adaptive FEM for (nonlinear) elliptic PDEs. Appl. Math. Comput. **442**, 127731 (2023)

Influence of the Grid Resolutions on the Computer Simulated Transport and Transformation Atmospheric Composition Processes over the Territory of Bulgaria

Georgi Gadzhev$^{(\boxtimes)}$ (iD), Ivelina Georgieva, Kostadin Ganev, Vladimir Ivanov (iD), and Nikolay Miloshev

National Institute of Geophysics, Geodesy and Geography, Bulgarian Academy of Sciences, Acad. G. Bonchev str., bl. 3, 1113 Sofia, Bulgaria
ggadjev@geophys.bas.bg

Abstract. It should be expected that the computational grid resolution will have a significant impact on the computer simulation results. Therefore, a numerical experiment is made in order to evaluate the horizontal grid resolution impact on the simulated transport and transformation atmospheric composition processes over the territory of Bulgaria. The computer simulations are performed with a set of models used worldwide - US EPA Models 3 System. Using the "nesting" capabilities of the models, a resolution of 9 km was achieved for the territory of Bulgaria, by sequentially solving the task in several consecutive, nested areas. Three cases are considered in this paper: – The computer simulations results from the domain with a horizontal resolution (both of the emission source description and the grid) of 27 km; – The computer simulations results from the domain with a horizontal resolution (both of the emission source description and the grid) of 9 km; – Hybrid case with the computer simulations performed with a grid resolution of 9 km, but with emissions like in the 27×27 km domain. The simulations were performed, for all the three cases, for the period 2007–2014 year, thus creating an ensemble large and comprehensive enough, as to reflect the most typical atmospheric conditions with their typical recurrence. The CMAQ "Integrated process rate analysis" option is applied to obtain the spatial/temporal distribution of the different transport and transformation atmospheric composition processes. Comparing them for the above 3 cases makes it possible to evaluate the grid resolution impact.

Keywords: air pollution modeling · emission scenarios · numerical simulation · atmospheric composition · grid resolution

© The Author(s), under exclusive license to Springer Nature Switzerland AG 2024
I. Lirkov and S. Margenov (Eds.): LSSC 2023, LNCS 13952, pp. 316–324, 2024.
https://doi.org/10.1007/978-3-031-56208-2_32

1   Introduction

The main factors in air quality modeling are meteorology and emissions. The atmospheric composition pattern results from a complex interaction between atmospheric circulation and the physical and chemical processes of air pollutants, in both gaseous and aerosol forms. Many studies have investigated the role of meteorology on air quality [8–10]. The detailed and accurate distribution of emissions sources (location, quantity and strength) is a key factor for obtaining qualitative and representative results in atmospheric composition modeling. Combining the mentioned two factors in multi-scale models and creating new specialized numerical approaches leads to a better understanding of air pollution formation on the regional-local scale.

The improvement of the ability in atmospheric composition modeling is related to the quality of the input data of the two factors (meteorology and emissions) and an adequate description of their interaction. The computational grid refinement always leads to more detailed simulations of meteorological fields and pollution transport and transformation processes. Having in mind the uncertainties and errors in the raw meteorological and emission input data, this will not necessarily lead to better simulations. Nevertheless, it is useful to investigate the role of computational grid resolution on the simulated atmospheric composition properties.

A statistically significant ensemble of the atmospheric composition over Bulgaria was created, taking into account interactions between different processes (dynamic and chemical), and the main pathways and contributions were studied. To analyze the contribution of different dynamic and chemical processes in the formation of air pollution over Bulgaria, the CMAQ "Integrated Process Rate Analysis" (PA) option was used. This option presents the concentration change for each of the compounds as a sum of the contribution of each of the processes, which determine the concentration of pollutants in the atmosphere. Results from modeling the atmospheric composition over Bulgaria have already been published in [6] and [7]. This paper presents a detailed study of the grid resolution impact on the simulated processes determining the behavior of Nitrogen oxides in the atmosphere of Bulgaria by comparing the different emission sources and grid descriptions.

2   Methods

Based on 3D modelling tools, an extensive database was created and used for this paper. The simulations were based on the following models - WRF v.3.2.1 [11] and [12] Weather Research and Forecasting Model used as meteorological pre-processor; CMAQ v.4.6 - Community Multi-Scale Air Quality model [1,2,4] the Chemical Transport Model (CTM), and SMOKE - the Sparse Matrix Operator Kernel Emissions Modelling System [3] - the emission pre-processor. The models were adapted and validated for Bulgaria. This gave us the opportunity to conduct extensive studies on a fully competitive modern level of the climate of atmospheric composition in the country.

TNO inventory [5] is exploited for the territories outside Bulgaria. The National inventory provided by Bulgarian Executive Environmental Agency is used for the Bulgarian domain.

The "NCEP Global Analysis Data" with a horizontal resolution of $1° \times 1°$ is used as background meteorological data in the study.

The PA option allows for determining the contributions of individual processes to a particular pollutant and a group of pollutants. One group of gas pollutants, Nitrogen oxides (GNOY), is considered in this paper:

$$GNOY = NO + NO_2 + NO_3 + 2N_2O_5 + (HONO + HNO_3 + PNA)$$

The dependence of the concentration change Δc_i^1 for the respective time step Δt from the different processes can be obtained as follows:

$$\Delta c_i^1 = (\Delta c_i^1)_{horizontaldiffusion} + (\Delta c_i^1)_{verticaldiffusion} +$$
$$(\Delta c_i^1)_{horizontaladvection} + (\Delta c_i^1)_{verticaladvection} + (\Delta c_i^1)_{drydeposition} +$$
$$(\Delta c_i^1)_{emissions} + (\Delta c_i^1)_{chemical} + (\Delta c_i^1)_{cloud} + (\Delta c_i^1)_{aerosol}$$

where the c_i is the concentration of the i-th admixture, and the Δc_i^1 is the mean concentration change of i-th admixture in the first model layer (1) for the time t to Δt.
or:

$$\Delta c_i^1 = HDIF + VDIF + HADV + ZADV + DDEP + EMIS + CHEM + CLDS + AERO$$

where the calculated processes are: HDIF - horizontal diffusion; VDIF - vertical diffusion; HADV - horizontal advection; ZADV - vertical advection; DDEP - dry deposition; EMIS - emissions; CHEM – chemical transformations; CLDS - cloud processes; AERO – aerosol processes. Based on the simulations of the atmospheric composition, the PA for GNOY was calculated for the following Cases:

Case 1 (C1) – The computer simulation results from the domain with a horizontal resolution (both of the emission source description and the grid) of 9 km;

Case 2 (C2) – The computer simulation results from the domain with a horizontal resolution (both of the emission source description and the grid) of 27 km;

Case 3 (C3) – Hybrid case with the computer simulations performed with a grid resolution of 9 km, but with emissions like in the 27×27 km domain.

The differences between Cases C1 and C2 (C1–C2) and Cases C1 and C3 (C1–C3) was also evaluated.

All the simulations are performed day by day for 7 years period from 2008 to 2014. The outputs were averaged over the whole 7-year period (ensemble), and so the "typical" annual evaluations were obtained. In the current paper, the annual behavior of the processes contribution to the surface concentrations of Nitrogen oxides for Bulgaria is presented.

3   Results

In the first place, when consider the results, should be clarifed the locations of the most intensive and elevated emission sources. They are present in Fig. 1.

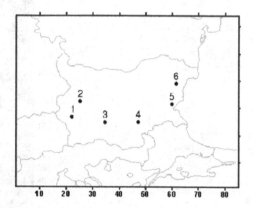

Fig. 1. Location of the zones with the highest emissions in Bulgaria: 1 – TPP Bobovdol, 2 – Sofia city, 3 – Plovdiv city, 4 – TPP Maritsa Iztok, 5 – Burgas city, 6 – Devnya Industrial Area.

The horizontal advection contribution plots for C1, C2, and C3 in Fig. 2 show that its contribution over the mountains (low emission areas) is positive, over the rest of the area it is almost zero, and only over the significant emission sources in the area is negative. The contribution of vertical advection is also positive over the mountains but generally opposite to the horizontal, and this is better observed for the two cases with the finer meteorology C1 and C3, while in C2, due to the coarser resolution, this effect is not as well manifested. The plots for the difference between the two cases with finer and coarser meteorology and emissions C1–C2 show that for HADV, there is almost no difference between the two cases except for some of the locations with more detailed emissions. The same applies when comparing the case with fine emissions and meteorology and the hybrid case (C1–C3). The differences are mainly over the sites with modi-fied emissions. The last two conclusions also apply to the emission contribution (EMIS). Regarding the vertical advection, the difference between the two main cases (C1–C2) is very well expressed, especially over the mountains, which means that the advection contribution at 9 km resolution is greater than that at 27 km. On the plots for the horizontal diffusion contribution C1, C2, and C3 in Fig. 3 can be seen that the contribution over the whole region is almost entirely negative, tending to zero, and only for C1 and C3 near the large sources, the contribution becomes slightly negative. This effect is well seen on the plot for the difference between 9 and 27 km resolutions (C1–C2), where the difference between the con-tributions at the two resolutions is almost identical to the contribution at 9 km resolution. On the plot of the difference between the 9 km resolution simulation

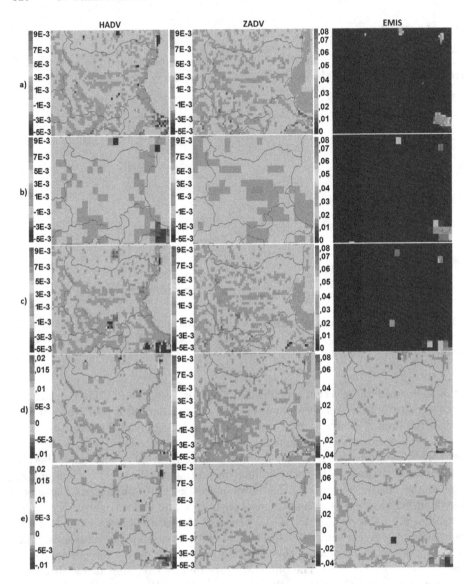

Fig. 2. Averaged over the ensemble annual contribution [ppmv] of HADV, ZADV and EMIS to the formation of surface GNOY at 12:00 UTC for (a) C1, (b) C2, (c) C3, (d) C1–C2 and (e) C1–C3.

and the hybrid simulation, it can be seen that in the horizontal diffusion fields, the difference is entirely positive, indicating that the contribution of horizontal diffusion at 9 km resolution is larger than that of the hybrid configuration, and means that for more detailed emissions, horizontal diffusion is more intense over the whole area. The plots for the vertical diffusion contribution C1, C2, and C3

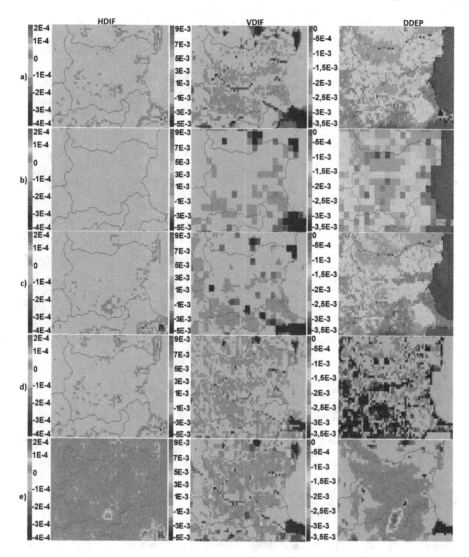

Fig. 3. Averaged over the ensemble annual contribution [ppmv] of HDIF, VDIF and DDEP to the formation of surface GNOY at 12:00 UTC for (a) C1, (b) C2, (c) C3, (d) C1–C2 and (e) C1–C3.

show that the contribution over the region is negative, tending to zero, and over the mountains, it is positive, tending to zero. Also, the positive contribution of VDIF in the case of C1 under each of the elevated large point sources is well highlighted, as well as the higher negative contribution around these point sources and also over the more congested sections of the road network. In C3, the high negative contribution under the point sources stands out clearly. The plot for the difference between the 9 and 27 km resolution configurations (C1–C2) shows

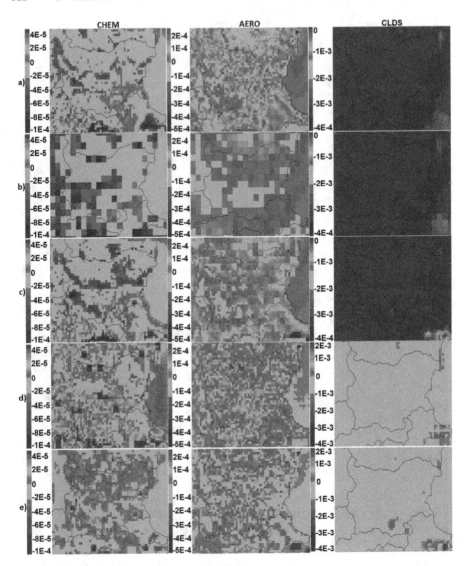

Fig. 4. Averaged over the ensemble annual contribution [ppmv] of CHEM, AERO and CLDS to the formation of surface GNOY at 12:00 UTC for (a) C1, (b) C2, (c) C3, (d) C1–C2 and (e) C1–C3.

that the difference is almost identical to that of the 9 km resolution configuration (C1). The differences are mainly near and under the elevated large point sources on large territories around them. On the plot for the difference between the 9 km configuration and the hybrid (C1–C3) in the vertical diffusion contribution fields, the contribution over the country is almost entirely positive and tends to zero, and under the elevated large point sources are very well distinguished. The large positive difference is near Thermal Power Plant (TPP) Maritsa East, in particular is clearly visible. This effect can be easily explained by the fact that when the elevated large point source emissions are not concentrated in one point but spread over several points, it will result in smaller concentrations aloft, hence smaller vertical gradients and less intensive vertical diffusion. The plot for the contribution of dry deposition in C1 shows that it is, as expected, entirely negative and larger over the mountains. At C2, no remarkable features stand out, which is a consequence of the coarse resolution. The results for C3 are pretty similar to those of C1, as only near the Maritsa East TPP the contribution is significantly larger. The results for the difference between 9 km and hybrid cases show very well the location of large elevated large point sources as spots with the most negative values (much smaller surface concentrations in C3 compared to C1). On the plots for the contribution of chemical processes C1, C2, and C3 in Fig. 4, it can be seen that over most of the area, it is negative, and only in places in the southern part of the area it becomes positive. The plots for the differences C1–C2 and C1–C3 highlight very well the differences in source locations in the 27 km and hybrid simulations as locations with high positive differences. This is a demonstration of the nonlinearity of the chemical processes, which rate also depends on pollution concentrations. On the plots for the contribution of aerosol processes in C1 and C3, most of the area is predominantly positive. Only in the area of Sofia and some point sources does the contribution becomes negative, tending to zero. In plot C2, the western part of the domain and southern Bulgaria also stand out with a negative contribution, and above the rest parts, it is positive and tends to be zero. Plot C1–C2 showing the difference between the 9 and 27 km cases, closely resembles C1 due to the near to zero values in C2. The main conclusion that can be outdrawn is that a more detailed description of emissions and meteorology leads to a better description of aerosol processes. On the plot for the differences between the 9 km and the hybrid simulations, the location of the large sources is well highlighted by large positive values. The cloud contribution plots for C1, C2, and C3 show that over most of the region is negative, close to zero, and fairly similar for the three cases. This is natural, keeping in mind that the clouds are large, and their description is probably good enough even with a 27 km resolution. From the plots for the differences C1–C2 and C1–C3, it can be seen that for the entire area, the differences of the contributions are negative and very close to zero, and only near the Maritsa East TPP there is an area with slightly positive differences, which also tend to zero.

4   Conclusion

Generally it can be seen that in the fields of the different processes contribution, except from some point/small areas connected to large sources, the results for case C1 (fine grid and fine source resolution) are pretty close to the results for case C3 (fine grid and coarse source resolution) and significantly differ from the results for case C2 (coarse grid and coarse source resolution). This means that perhaps, at least in this case and for this region, the grid resolution has stronger impact on the simulation results than the emission description resolution.

Acknowledgments. This work has been carried out in the framework of the National Science Program "Environmental Protection and Reduction of Risks of Adverse Events and Natural Disasters", approved by the Resolution of the Council of Ministers No 577/17.08.2018 and supported by the Ministry of Education and Science (MES) of Bulgaria (Agreement No D01D01-271/09.12.2022). This work has also been accomplished with the financial support of Grant No BG05M2OP001-1.001-0003, financed by the Science and Education for Smart Growth Operational Program (2014–2020) and co-financed by the European Union through the European Structural and Investment funds. Special thanks are due to US EPA and US NCEP for providing free-of-charge data and software and to the Netherlands Organization for Applied Scientific research (TNO) for providing the high-resolution European anthropogenic emission inventory.

References

1. Byun, D., Ching, J.: Science algorithms of the EPA models-3 community multiscale air quality (CMAQ) modeling system. EPA report 600/R-99/030, Washington, D.C., EPA/600/R-99/030 (NTIS PB2000-100561) (1999). https://cfpub.epa.gov/si/si_public_record_report.cfm?Lab=NERL&dirEntryId=63400

2. Byun, D.: Dynamically consistent formulations in meteorological and air quality models for multiscale atmospheric studies. Part I: governing equations in a generalized coordinate system. J. Atmos. Sci. **56**, 3789–3807 (1999)

3. CEP: Sparse Matrix Operator Kernel Emission (SMOKE) Modeling System. University of Carolina, Carolina Environmental Programs, Research Triangle Park, North Carolina (2003)

4. CMAS Homepage. http://www.cmaq-model.org. Accessed 22 Feb 2023

5. Denier van der Gon, H., Visschedijk, A., van de Brugh, H., Droge, R.: A high resolution European emission data base for the year 2005. TNO-report TNO-034-UT-2010-01895 RPT-ML, Apeldoorn, The Netherlands (2010)

6. Gadzhev, G., Ganev, K., Mukhtarov, P.: Influence of the grid resolutions on the computer simulated surface air pollution concentrations in Bulgaria. Atmosphere **13**, 774 (2022). https://doi.org/10.3390/atmos13050774

7. Gadzhev, G., Ganev, K., Miloshev, N., Syrakov, D., Prodanova, M.: Numerical study of the atmospheric composition in Bulgaria. Comput. Math. Appl. **65**, 402–422 (2013)

8. Gilliam, R.C., Hogrefe, C., Godowitch, J.M., Napelenok, S., Mathur, R., Rao, S.T.: Impact of inherent meteorology uncertainty on air quality model predictions. J. Geophys. Res. Atmos. **120**(23), 12259–12280 (2015)

9. McNider, R.T., Pour-Biazar, A.: Meteorological modeling relevant to mesoscale and regional air quality applications: a review. J. Air Waste Manag. **70**, 2–43 (2020)

10. Rao, S.T., Luo, H., Astitha, M., Hogrefe, C., Garcia, V., Mathur, R.: On the limit to the accuracy of regional-scale air quality models. Atmos. Chem. Phys. **20**, 1627–1639 (2020)

11. Shamarock, W.C., et al.: A description of the advanced research WRF version 2 (2007). http://www.dtic.mil/dtic/tr/fulltext/u2/a487419.pdf

12. UCAR/ NCAR Homepage. https://ncar.ucar.edu/what-we-offer/models/weather-research-and-forecasting-model-wrf. Accessed 22 Feb 2023

Development of New High Performance Computer Architectures and Improvements in Danish Eulerian Model for Long Range Transport of Air Pollutants

Krassimir Georgiev[1], Zahari Zlatev[2], and Ivan Lirkov[1(✉)]

[1] Institute of Information and Communication Technologies,
Bulgarian Academy of Sciences, Sofia, Bulgaria
georgiev@parallel.bas.bg, ivan.lirkov@iict.bas.bg
[2] Department of Environmental Science, Aarhus University, Roskilde, Denmark
zz@envs.au.dk

Abstract. The paper is devoted to an analysis and comparison in the development of new high - performance computers and the improvements and development of new more reliable versions of the Danish Eulerian model for computer studying of the transport of the air pollutants over Europe and surrounding areas, studying some economical and agricultural problems, regional and global climate changing, etc.

Keywords: High-performance computer architectures · mathematical and computer modelling in environmental studies · speed-up and efficiency of parallel algorithms · partial and ordinary differential equations

1 Introduction

The computers are becoming faster and faster, more and more different for the computers knowing in the end of the 20th Century. Their capabilities to deal with very large data sets are steadily increasing. Problems that require a lot of computing time and rely on the use of huge files of input data can now be solved on PCs (such problems had to be treated on powerful mainframes only some years ago). Let's go back to the paper "Ordering the universe: The role of mathematics" written by Arthur Jaffe [28] which was published in 1984 - "Although the fastest computers can execute millions of operations in one second they are always too slow. This may seem a paradox, but the heart of the matter is: the bigger and better computers become, the larger are the problems scientists and engineers want to solve". The relevance of this statement can be illustrated by the difference between the requirements to the air pollution problems studied in 1984 and those studied at present and particularly comparing the stage of the

I. Lirkov and S. Margenov (Eds.): LSSC 2023, LNCS 13952, pp. 325–334, 2024.
https://doi.org/10.1007/978-3-031-56208-2_33

presented here Danish Euleran model at that time and in now days. In 1984 it was difficult to solve models containing, after the discretization, more than 2048 equations. Some of the biggest problems solved at present contain more than 200 000 000 ordinary differential equations. Moreover, in the case of studying air pollution problems, one can solve such systems of equations hundred thousand time steps and even bigger computational tasks in a comprehensive study related to future climate changes and high pollution levels. Therefore, the scientists and engineers will formulate and trying to solve bigger and bigger tasks, with requirements for handling huge data sets, because the solutions obtained in this way: (i) will be closer to the reality; (ii) will contain more useful details; and (iii) will give more reliable answers to questions which are at present open. These and some other discussions show why even more and more problems can successfully be handled on the nowadays workstations and PCs, it is necessary **to run highly optimized codes on big modern supercomputers** (Table 1).

Table 1. Performance in solving a system of linear equation

Computer	LINPACK Benchmark $n = 100$ OS/Computer Mflop/s	TPP Best effort $n = 1000$, Mflop/s	Theoretical Peack Mflop/s
Cray C90 1proc. 10 ns	CF77 5.0-Zp-Wd-e68 387	902	952
Fujitsu VPP/500 1proc. 10 ns	FORTRAN77 EX/VP 203	1048	1250

2   Development of New High Performance Computer Architectures Since 1997 and Till 2017

2.1   Vector Computers (CRAY Y-MP C90A, Fujitsu, etc.)

The Cray C90 series (initially named the Y-MP C90) was a vector processor supercomputer launched by Cray Research in 1991. The C90 was a development of the Cray Y-MP architecture. Compared to the Y-MP, the C90 processor had a dual vector pipeline and a faster 4.1 ns clock cycle (244 MHz), which together gave three times the performance of the Y-MP processor. The maximum number of processors in a system was also doubled from eight to 16. The C90 series used the same Model E IOS (Input/Output Subsystem) and UNICOS operating system as the earlier Y-MP Model E The C90 series included the C94, C98 and C916 models (configurations with a maximum of four, eight, and 16 processor respectively) and the C92A and C94A (air-cooled models). Maximum SRAM memory was between 1 and 8 GB, depending on model. According to J. Dongara [10,31,32] the performance of the Cray machine with one processor and Fujitsu using the standard linear equations software is as follows: where n is the number of the equations in the system and "TPP" means Toward Peak Performance.

Parallel Computers with Distributed Memory (IBM SP, CRAY T3E, Beowulf Clusters, Macintosh G4 Clusters, etc.)

Scalable POWERparallel (SP) is a series of supercomputers from IBM. SP systems were part of the IBM RISC System/6000 (RS/6000) family. The first model, the SP1, was introduced in February 1993. The SP is a distributed memory system, consisting of multiple RS/6000-based nodes interconnected by an IBM-proprietary switch called the High Performance Switch (HPS). IBM SP/2, which was first use by the authors of this study, has the following main characteristics [20,27]: 8 "Winter-Hawk II" IBM RS/6000 SP 375MHz POWER3 SMP Wide Nodes; Processors: 4×375 Mhz POWER3-II 64-bit superscalar CPUs; Up to four double precision floating-point operations per cycle; Memory: 4 Gigabytes of RAM (128-bit data path, 93.75Mhz system bus); Network: IBM SPSwitch (MX2) Interconnect (intranode) 10/100 Integrated Ethernet (internet) Gigabit (1000SX) Ethernet (not used); Disk storage (internal): Two (2) 18.2 GB Ultra SCSI-2 Disk Drives (Mirrored) 4 Gigabytes Swap (paging) Space; Disks: 16×36.4 GB 10,000 RPM Serial Storage Architecture Drives; Software: Operating System: AIX 5.1 Maintenance Level 2; Compilers: XL Fortran 7.1, C For AIX 5.0, VisualAge C++ for AIX 5.0, etc.

Cray T3E architecture are microprocessor for super-computing system. Cray T3E is a RISC(Reduced Instruction Set Computer) architecture which is very powerful microprocessors. T3E systems contain a large number of processing elements (PE). Each PE consists of a DEC Alpha EV5 RISC microprocessor. Some of the main features of Cray T3E architecture are as follows [20,25,44]: the CRAY T3E is a scalable shared-memory multiprocessor; the system architecture is designed to tolerate latency and enhance scalability; the T3E system was fully self-hosted and ran the UNICOS/mk distributed operating system; Cray T3E scalability can handle added processors and memory as well as larger I/O and interconnection bandwidths; Cray T3E system has their own local memory with capacity from 64 megabytes to 2 gigabytes; Cray T3E architecture uses multiple ports to perform I/O. These multiple ports have one or more I/O channels, which are integrated into the network that interconnects the processing nodes.

In the beginning of the 21 Century in IICT - BAS a Macintosh G4 cluster was installed. The cluster was based on the AppleSeed paradigm. Purportedly, the cluster is capable of delivering over 20 GFlops of peak power, and it is based on eight Dual PowerPC G4/450, 4 GB of RAM, 0.5 terabyte of hard disk space, running Mac OS 9, over 100 Mb/s Fast Ethernet, and is switched by one Asanté Intracore 8000. PowerPC G4 series of microprocessors, the Power Mac G4 was marketed by Apple as the first "personal supercomputers" [45], reaching speeds of 4 to 20 gigaFLOPS.

2.2 Parallel Computers with Shared Memory (SGI Origin, SUN, etc.)

The SGI Origin 2000 is a family of mid-range and high-end server computers developed and manufactured by Silicon Graphics (SGI). They were introduced in 1996 to succeed the SGI Challenge and POWER Challenge. At the time of introduction, these ran the IRIX operating system, originally version 6.4 and later, 6.5 and supported Fortran 90, Fortran 77, C, and C++ compilers [18,46].

Sun products included computer servers and workstations built on its own RISC-based SPARC processor architecture, as well as on x86-based AMD Opteron and Intel Xeon processors. Sun also developed its own storage systems and a suite of software products, including the Solaris operating system, developer tools, Web infrastructure software, and applications.

2.3 Parallel Computers with Two Level of Parallelism (IBM SMP, IBM BlueGene/P, Clusters of Multiprocessor Nodes, etc.)

Symmetric multiprocessing (**SMP**) is the most popular form of multiprocessor system available. In SMP environment, multiple processors share other hardware resources [5,47]. Multiple processors share the same memory and disk space, but use a single operating system. The workload for a parallel job is distributed across the processors in the system. The shared resources might limit the actual speed at which the job completes in the system. To scale the system, one can increase the number of processors, add memory, or increase storage. The scalability strategy that you implement depends on how your job is limited within your current system. In most SMP machines, all processors are connected in a shared backplane. The characteristics of an SMP are the following [29,30]: (a) two or more similar (or often identical) processors are employed in a stand-alone system; (b) all processors share the same memory and I/O devices via one or more shared buses with similar access time; (c) all processors are capable of performing the same functions; (d) the distribution of workload between processors is performed by the operating system in such a way that multiple independent or dependent tasks are shared between processors without any special consideration in the application program.

IBM BG/P compute nodes are connected in a three-dimensional (3D) torus topology [19,20]. Each compute node is also connected to a collective network, rooted at the I/O nodes. BG/P nodes are now four way symmetric multiprocessors (SMPs). The torus network logic has been enhanced with a direct memory access (DMA) engine, thereby offloading communication. The BG/P system currently offers 2 GB of main memory per node. Many other important features and characteristics of the IBM Blue Gene series can be found in [48]. The IBM Blue Gene/P computer which was used by the authors was installed in Sofia (Bulgarian Supercomputer Center, National Center for Supercomputing Applications) and consisted of two racks, 2048 Power PC 450 based compute nodes, 8192 processor cores, and a total of 4TB random access memory. Each processor core had a double-precision, dual pipe floating-point core accelerator. 16 I/O nodes were connected via fibre optics to a 10 Gb.s Ethernet switch. The smallest partition size, available currently, was 128 compute nodes (512 processor cores). The theoretical performance of the computer was 27.85 Tflops while the maximum LINPACK performance achieved was 23.42 Tlops.

3   Improvements in Danish Eulerian Model for Long Range Transport of Air Pollutants Since 1997 According to the New High Performance Computers in Use

The Unified Danish Eulerian Model (UNI-DEM) is an appropriate model for studying the transport on long distances and the transformations in the atmosphere of pollutants, which are potentially harmful. The mathematical model and the related software were developed at the National Environmental Research Institute (NERI, Roskilde, Denmark) [49]. The space domain of the model covers Europe and some neighboring parts of this continent belonging to the Atlantic Ocean, Asia and Africa. Large-scale models, as UNI-DEM, can efficiently be run only on high-performance parallel computer architectures [2–4,6,8,12–14,20]. The main idea in the parallel version of DEM is the domain partitioning approach. For maximum portability, only standard MPI routines [15–17] and/or OpenMP primitives are used in the UNI-DEM code. The parallelization is based on domain partitioning of the horizontal grid: - domain overlapping of the advection-diffusion subproblem (requires communication on each time step (communication stage), and - nonoverlapping subdomains in chemistry-deposition subproblem (improving the data locality for more efficient cache utilization by using chunks to group properly the small tasks). In addition pre-processing and post-processing stages are needed for scattering the input data and gathering the results (cheap and reduce the communications during actual computations, but their relative weight grows up with increasing the number of MPI tasks (affects the total speed-up and efficiency). The parallel code is highly portable and performs very efficiently on different kind of supercomputers; parallel and vector computers. Some important application areas of UNI-DEM are: Ecosystem mutual relations; Forestry and wild life protection; Human health preservation; Agricultural production; Global climate changes, etc. Using different kind of high performance computing systems UNI-DEM was used in the following important applications: (a) Studying the phenomenon itself (better understanding physical and chemical processes either in scientific studies treatment of tasks required by policy makers) [1,11,21,23,24,26,33–36,38,39,42]; (b) Improving as much as possible the reliability of the control that are to be used for keeping the pollution the prescribed acceptable limits; (c) Studying the ozone episodes; (d) Source-receptor relations [7,9,41]; (e) Calculating losses of crops due to high ozone levels [8]; (f) Influence of the future climate changes on air pollutants [22,37,40], etc.

In the following four tables, the main output results achieved when UNI-DEM was running on different kind of supercomputers can be found. Enough references are add in order to see more results obtained in the period after year 1997. From the Table 2 one can see the size of the computational tasks and reason to create numerical methods, algorithms and software for high speed parallel computers. The most often case which was used in the computer experiments was with a n umber of chemical species equal to 35 (Table 3).

Table 2. Size of the computational task

Grid => No. of the chemical species	32 × 32	96 × 96	288 × 288	480 × 480
1	1024	9216	82944	230 400
2	2048	18432	165 888	460 800
10	10240	92160	839 440	2 304 000
35	35 840	322 560	2 903 040	8 064 000
56	57 344	516 096	4 644 864	12 902 400
168	172 032	1 548 288	13 934 592	38 707 200

Table 3. High performance runs of 2D version of UNI-DEM on 50 km × 50 km grid (96 × 96 mesh)

Computer type	Number of processors	Communications	CPU time (in hours)	Efficiency
Cray C90 A	1	vector	9	80
Sun	16	Open MP	1.43	79
Cray T3E	16	MPI	1.69	93
MAC G4	8	MPI	2.69	84
SunFire 6800	8	MPI	2.16	85
IBM SMP	16	MPI/Open MP	1.41	87
SGI Origin 2000	32	Open MP	0.74	61

It was important to obtain output results over more and more fine grids. Starting in 1997 with mesh size of 150 km. We improve the UNI-DEM to run it on 10 km. Mesh size when the using computer is fat enough (Tables 4 and 5).

From all reported output data and parameters of the parallel runs, it is well seen that:

- By using high performance parallel computers to run the variable grid-size code UniDEM, detailed air pollution results for a large region (whole Europe) and for a very long period (one or several years) can be obtained within a reasonable time.

Table 4. High performance runs of 2D version of U61NI-DEM on 16.7 km × 16.7 km grid (288 × 288 mesh)

Computer type	Number of processors	Communications	CPU time (in hours)	Efficiency
Mac G4	8	MPI	48.63	96
SunFire 6800	8	MPI	1.43	89
IBM SMP	16	MPI/Open MP	20.70	83

Table 5. High performance runs of 2D version of UNI-DEM on 10 km × 10 km grid (480 × 480 mesh)

Computer type	Number of processors	Communications	CPU time (in hours)	Efficiency
Cray T3E	32	MPI	61.02	92
IBM SP	32	MPI	53.33	86
IBM BlueGene/P	32	MPI	31.60	83

– The parallel code, created by using MPI standard library, appears to be highly portable and shows good efficiency and scalability on both distributed-memory and shared-memory parallel machines.

Acknowledgement. This work is accomplished with the support by the Grant No BG05M2OP001-1.001-0003, financed by the Science and Education for Smart Growth Operational Program (2014–2020) and co-financed by the European Union through the European structural and Investment funds. The work is partially supported by the Bulgarian National Science Fund under grant No. KP-06-N52/5.

References

1. Antonov, A., Georgiev, K., Komsalova, E., Zlatev, Z.: Implementation of bilinear nonconforming finite elements in an Eulerian air pollution model: results obtained by using the rotational test. In: Dimov, I., Lirkov, I., Margenov, S., Zlatev, Z. (eds.) NMA 2002. LNCS, vol. 2542, pp. 379–386. Springer, Heidelberg (2003). https://doi.org/10.1007/3-540-36487-0_42
2. Brandt, J., Christensen, J., Dimov, I., Georgiev, K., Uria, I., Zlatev, Z.: Treatment of large air pollution models. In: Vulkov, L., Waśniewski, J., Yalamov, P. (eds.) WNAA 1996. LNCS, vol. 1196, pp. 66–77. Springer, Heidelberg (1997). https://doi.org/10.1007/3-540-62598-4_80
3. Brandt, J., Dimov, I., Georgiev, K., Uria, I., Zlatev, Z.: Numerical algorithms for the three-dimensional version of the Danish Eulerian Model. In: Geernaert, G., Hansen, A., Zlatev, Z. (eds.) Regional Modelling of Air Pollution in Europe, p. 249 (1997)
4. Brandt, J., Dimov, I., Georgiev, K., Wasniewski, J., Zlatev, Z.: Coupling the advection and the chemical parts of large air pollution models. In: Waśniewski, J., Dongarra, J., Madsen, K., Olesen, D. (eds.) PARA 1996. LNCS, vol. 1184, pp. 65–76. Springer, Heidelberg (1996). https://doi.org/10.1007/3-540-62095-8_8
5. Cheung, P., Luk, W.: The Electrical Engineering Handbook (2005)
6. Dimov, I., Georgiev, K., Waśniewski, J., Zlatev, Z.: Three-dimensional version of the Danish Eulerian Model. In: Dongarra, J., Madsen, K., Waśniewski, J. (eds.) PARA 1995. LNCS, vol. 1041, pp. 151–157. Springer, Heidelberg (1996). https://doi.org/10.1007/3-540-60902-4_18
7. Dimov, I., Georgiev, K., Zlatev, Z.: Long-range transport of air pollutants and source-receptor relations. Notes Numer. Fluid Mech. **62**, 155–166 (1998)

8. Dimov, I., Georgiev, K., Ostromsky, T., van der Pas, R.J., Zlatev, Z.: Computational challenges in the numerical treatment of large air pollution models. Ecol. Model. **179**(2), 187–203 (2004)

9. Dimov, I., Georgiev, K., Zlatev, Z.: Some source-receptor relations studied by the Danish Eulerian Model. In: Proceedings of 8th International Conference on Harmonisation within Atmospheric Dispersion Modelling for Regulatory Purposes, Sofia, Bulgaria, pp. 424–428 (2002)

10. Dongara, J.: Performance of various computers using standard linear equations software. ACM SIGARCH Comput. Archit. News **20**(3) (1992). https://doi.org/10.1145/141868.141871

11. Zlatev, Z., Ebel, A., Georgiev, K.: Environmental modelling for security: future needs and development of computer networking, numerics and algorithms. In: Ebel, A., Davitashvili, T. (eds.) Air, Water and Soil Quality Modelling for Risk and Impact Assessment. NASTC, pp. 351–356. Springer, Dordrecht (2007). https://doi.org/10.1007/978-1-4020-5877-6_33

12. Farago, I., Georgiev, K., Thomsen, P.G., Zlatev, Z.: Numerical and computational issues related to applied mathematical modelling. Appl. Math. Model. **32**(8), 1475–1476 (2008). https://doi.org/10.1016/j.apm.2007.06.035

13. Farago, I., Georgiev, K., Havasi, A., Zlatev, Z.: Efficient numerical methods for scientific applications: introduction. Comput. Math. Appl. **65**(3), 297–300 (2013). https://doi.org/10.1016/j.camwa.2013.01.001

14. Farago, I., Georgiev, K., Havasi, A., Zlatev, Z.: Efficient algorithms for large scale scientific computations: introduction. Comput. Math. Appl. **67**(12), 2085–2087 (2014). https://doi.org/10.1016/j.camwa.2014.05.021

15. Georgiev, K., Zlatev, Z.: Running an advection-chemistry code on message passing computers. In: Alexandrov, V., Dongarra, J. (eds.) EuroPVM/MPI 1998. LNCS, vol. 1497, pp. 354–363. Springer, Heidelberg (1998). https://doi.org/10.1007/BFb0056595

16. Georgiev, K., Zlatev, Z.: Parallel sparse matrix algorithm for air pollution models. Parallel Distrib. Comput. Pract. **2**(4), 429–442 (1999)

17. Georgiev, K., Zlatev, Z.: Some numerical experiments with the parallel version of the Danish Eulerian Model. Notes Numer. Fluid Mech. **73**, 283–291 (2000)

18. Georgiev, K., Zlatev, Z.: Fine-grid resolution in Danish Eulerian model and an implementation on SGI origin 2000 computer. In: Margenov, S., Waśniewski, J., Yalamov, P. (eds.) LSSC 2001. LNCS, vol. 2179, pp. 272–280. Springer, Heidelberg (2001). https://doi.org/10.1007/3-540-45346-6_28

19. Georgiev, K., Zlatev, Z.: Runs of UNI-DEM model on IBM Blue Gene/P computer and analysis of the model performance. In: Lirkov, I., Margenov, S., Waśniewski, J. (eds.) LSSC 2009. LNCS, vol. 5910, pp. 188–1963. Springer, Heidelberg (2010). https://doi.org/10.1007/978-3-642-12535-5_21

20. Georgiev, K., Zlatev, Z.: Studying an Eulerian computer model on different high-performance computer platforms and some applications. In: AIP Conference Proceedings, vol. 1301, p. 476 (2010)

21. Georgiev, K., Zlatev, Z.: Specialized sparse matrices solver in the chemical part of an environmental model. In: Dimov, I., Dimova, S., Kolkovska, N. (eds.) NMA 2010. LNCS, vol. 6046, pp. 158–166. Springer, Heidelberg (2011). https://doi.org/10.1007/978-3-642-18466-6_18

22. Georgiev, K., Zlatev, Z.: Studying air pollution levels in the Balkan Peninsula area by using an IBM Blue Gene/P computer. Int. J. Environ. Pollut. **46**(1–2), 97–114 (2011). https://doi.org/10.1504/IJEP.2011.042611

23. Georgiev, K., Zlatev, Z.: Implementation of sparse matrix algorithms in an advection-diffusion-chemistry module. J. Comput. Appl. Math. **236**(3), 342–353 (2011). https://doi.org/10.1016/j.cam.2011.07.026
24. Georgiev, K., Zlatev, Z.: Application of parallel algorithms in an air pollution model. In: Zlatev, Z., et al. (eds.) Large Scale Computations in Air Pollution Modelling. NATO Science Series, vol. 57, pp. 173–184. Kluwer Academic Publishers (1999)
25. Georgiev, K., Zlatev, Z.: Running large-scale air pollution models on parallel computers. In: Air Pollution Modelling and its Application XIII, pp. 223–232. Kluwer Academic/Plenum Press, London - New York (2000)
26. Georgiev, K.: Two-dimensional computer modelling of long-range transport of air pollutants. In: Proceedings of 9th International Symposium "Ecology 2000", Bourgas, Bulgaria, pp. 74–80 (2000)
27. IBM Blue Gene team: IBM J. Res. Dev. **52**(1/2) (2008)
28. Jaffe, A.: Ordering the universe: the role of mathematics. SIAM Rev. **26**, 473–500 (1984). https://doi.org/10.1137/1026103
29. Ostromsky, Tz., Georgiev, K., Zlatev, Z.: An efficient highly parallel implementation of a large air pollution model on an IBM Blue Gene supercomputer. In: AIP Conference Proceedings, vol. 1487, pp. 135–142 (2012). https://doi.org/10.1063/1.4758951
30. Patterson, D., Hennessy J.: Computer Organization and Design (1998)
31. Zlatev, Z., Dimov, I., Georgiev, K.: Modeling the long-range transport of air pollutants. IEEE Comput. Sci. Eng. **1**(3), 45–52 (1994)
32. Zlatev, Z., Dimov, I., Georgiev, K.: Three-dimensional version of the Danish Eulerian Model. Zeitung fuer Augewandte Mathematik und Mechanik **76**, 473–476 (1996)
33. Zlatev, Z., Dimov, I., Georgiev, K., Waśniewski, J.: Parallel computations with large atmospheric models. In: Dongarra, J., Madsen, K., Waśniewski, J. (eds.) PARA 1995. LNCS, vol. 1041, pp. 550–560. Springer, Heidelberg (1996). https://doi.org/10.1007/3-540-60902-4_58
34. Zlatev, Z., Georgiev, K.: Treatment of large scientific problems: an introduction. In: Dongarra, J., Madsen, K., Waśniewski, J. (eds.) PARA 2004. LNCS, vol. 3732, pp. 828–830. Springer, Heidelberg (2006). https://doi.org/10.1007/11558958_99
35. Zlatev, Z., Ebel, A., Georgiev, K.: Large-scale computations in environmental modelling. Int. J. Environ. Pollut. **32**(2), 135–138 (2008)
36. Zlatev, Z., Ebel, A., Georgiev, K., Jose, R.S.: Large scale computations in environmental modelling: editorial introduction. Ecol. Model. **217**(3–4), 207–208 (2008). https://doi.org/10.1016/j.ecolmodel.2008.06.001
37. Zlatev, Z., Georgiev, K., Dimov, I.: Influence of climatic changes on pollution levels in the Balkan Peninsula. Comput. Math. Appl. **65**(3), 544–562 (2013). https://doi.org/10.1016/j.camwa.2012.07.006
38. Zlatev, Z., Georgiev, K., Dimov, I.: Parallel computations in a large-scale air pollution model. In: Faragó, I., Havasi, Á., Zlatev, Z. (eds.) Advanced Numerical Methods for Complex Environmental Models: Needs and Availability, pp. 169–207. Bentham Science Publishers (2013). https://doi.org/10.2174/97816080577881130101. eISBN: 978-1-60805-778-8. ISBN: 978-1-60805-777-1
39. Zlatev, Z., Georgiev, K., Dimov, I.: Part C: Sensitivity of European pollution levels to changes of human-made emissions. In: Faragó, I., Havasi, Á., Zlatev, Z. (eds.) Advanced Numerical Methods for Complex Environmental Models: Needs and Availability, pp. 307–333. Bentham Science Publishers (2013). https://doi.org/10.2174/97816080577881130101. eISBN: 978-1-60805-778-8. ISBN: 978-1-60805-777-1

40. Zlatev, Z., Dimov, I., Georgiev, K., Blaheta, R.: Using advanced mathematical tools in complex studies related to climate changes and high pollution levels. In: Lirkov, I., Margenov, S. (eds.) LSSC 2017. LNCS, vol. 10665, pp. 552–559. Springer, Cham (2018). https://doi.org/10.1007/978-3-319-73441-5_61

41. Zlatev, Z., Dimov, I., Georgiev, K., Wasnievski, J.: Long-Range Transport of Air Pollutants and Source-Receptor Relations, Report UNIC-94-04. Technical University of Denmark (1994)

42. Zlatev, Z., Dimov, I., Georgiev, K.: Optimizing large air pollution models on high speed computers. In: Dimov, I., Sendov, Bl., Vassilevski, P. (eds.) Advances in Numerical Methods & Applications, pp. 309–320. World Scientific, Singapore (1994)

43. Zlatev, Z., Dimov, I., Georgiev, K.: Experiments with the Danish Eulerian model. In: Candev, M. (ed.) Lecture Notes on Biomathematics and Bioinformatics, pp. 76–83. DATECS Publication, Sofia (1995)

44. https://home.cscamm.umd.edu/facilities/computing/sp2/

45. https://www.geeksforgeeks.org/cray-t3e-architecture/

46. https://www.sfgate.com/business/article/Apple-Unveils-Personal-Supercomputer-2909963.php/

47. http://www.computinghistory.org.uk/det/11264/SGI-Origin-2000/

48. https://www.ibm.com/support/knowledgecenter/SSZJPZ_11.7.0com.ibm.swg.im. iis.productization.iisinfsv.install.doc/topics/wsisinst_pln_engscalabilityparallel. html

49. https://www.dmu.dk/AtmosphericEnvironment/DEM/ . Atmosphere, vol. 2, pp. 201–221 (2011). https://doi.org/10.3390/atmos2030201

Evaluation of the Effects of the National Emission Reduction Strategies for Years 2020–2029 and After 2030 on the AQI on the Territory of Bulgaria

Ivelina Georgieva[(✉)], Georgi Gadzhev[ID], Kostadin Ganev, Vladimir Ivanov[ID], and Nikolay Miloshev

National Institute of Geophysics, Geodesy and Geography, Bulgarian Academy of Sciences, Acad. G. Bonchev str., bl. 3, 1113 Sofia, Bulgaria
iivanova@geophys.bas.bg

Abstract. The present paper presents the results obtained in the frame of an extensive study of the effects of different emission reduction scenarios on the air quality in Bulgaria. The set of models applied for atmospheric composition simulations is the same used in the operational Bulgarian Chemical Weather Forecast and Information System. Thus, the results are fully compatible with the operational chemical weather forecast. The models are also widely used in air pollution modelling, so the obtained computer simulation results are in harmony with evaluations made for other regions. Based on 3D modelling tools, an extensive database was created and used for different studies of atmospheric composition. The models in the system were adapted and validated for Bulgaria. It gave the opportunity to conduct extensive studies on a fully competitive modern level of the climate of atmospheric composition in the country. The provided model simulations are with horizontal resolution 9 km for the region of Bulgaria. Complying the EU Directive 2016/2284, Bulgaria developed national emission reduction strategies for years 2020–2029 and after 2030. Evaluation of the effects of these strategies on the AQI on the territory of Bulgaria is the objective of the present study. Five emission scenarios are considered in the paper for two different periods (2020–2029 and after 2030) with existing measures (WEM) and with additional measures (WAM), and they compare the results with the reference period (2005).

Keywords: air pollution modeling · emission scenarios · numerical simulation · atmospheric composition · emission reduction strategies

1 Introduction

Complying with the EU Directive 2016/2284 [6], Bulgaria developed national emission reduction strategies for the years 2020–2029 and after 2030. Evaluation of the effects of these strategies on the Air Quality Index (AQI) on the territory

© The Author(s), under exclusive license to Springer Nature Switzerland AG 2024
I. Lirkov and S. Margenov (Eds.): LSSC 2023, LNCS 13952, pp. 335–342, 2024.
https://doi.org/10.1007/978-3-031-56208-2_34

of Bulgaria is the objective of the present study. The studies are performed by applying computer simulations with the US EPA Models-3 System. The NCEP Global Analysis Data meteorological background with $1° \times 1°$ resolution is used as a meteorological background. The models' nesting capabilities were applied to downscale the simulations to 9 km resolution for the territory of Bulgaria. TNO inventory [5] is exploited for the territories outside Bulgaria. The National inventory provided by Bulgarian Executive Environmental Agency is used for the Bulgarian domain. Every EU member state (According to the legislation) has to report a set of emission projections scenarios [7]: projections scenario 'with existing measures' (WEM) means projections of anthropogenic Greenhouse Gas (GHG) or air pollutant emissions by sources that encompass the effects of currently implemented or adopted policies and measures; projections scenario 'with additional measures' (WAM) means projections of anthropogenic GHG or air pollutant emissions by sources that encompass the effects of policies and measures which have been adopted and implemented, as well as planned policies that are judged to have a realistic chance to be adopted and implemented in the future. In the paper, five emission scenarios are considered: 2005 emissions (reference period), 2020–2029 emissions projected with existing measures (WEM), and with additional measures (WAM), projected after 2030 WEM and WAM emissions. Only the Bulgarian emissions for 2020–2029 and 2030 have been modified according to the forecast scenarios. The considerations and conclusions in the paper are based on simulated AQI for the territory of Bulgaria. Comparing the simulated AQI with the different scenarios makes it possible to evaluate the effect of the national emission reduction strategies [6].

2   Methods

Based on 3D modelling tools, an extensive database was created and used for this paper. The simulations were based on the US EPA Model-3 system; WRF v.3.2.1 [12,15] Weather Research and Forecasting Model, used as meteorological pre-processor; CMAQ v.4.6 - Community Multi-Scale Air Quality model [1,2,4] is the Chemical Transport Model (CTM), and SMOKE - the Sparse Matrix Operator Kernel Emissions Modelling System [3] is the emission pre-processor. The models in the system were adapted and validated for Bulgaria. This gave us the opportunity to conduct extensive studies on a fully competitive modern level of the climate of atmospheric composition in the country. Previous results from air pollution modeling for Bulgaria have been published in many papers [8,9,13,14]. The AQI gives an integrated assessment of the impact of pollutants and directly measures the effects of air quality on human health. AQI evaluations are based on air pollutants concentrations (from numerical modelling) and make it possible to reveal the AQI status spatial/temporal distribution and behavior. The index is calculated to take into account 5 main pollutants: Ozone (O_3), Nitrogen dioxide (NO_2), Sulphur dioxide (SO_2), Carbon oxide (CO), and Particle Matter (PM_{10} - Particulate Matter, with a diameter of less or equal to $10\,\mu m$) [11]. Each pollutant's concentration falls into one of the 10 bands, and the AQI for each pollutant is determined. The overall AQI is one of the pollutants which individual index is in the highest band. The pollutant, which determines the overall AQI for the given particular case, is referred as to the "dominant pollutant". The calculated AQI in that way and compared for different scenarios according to the national emission reduction strategies [6], gives a good assessment of the measures, whether they are in effect or planned compared to the results of 2005, and generally shows the impact of the strategy itself on the territory of the country.

3   Results

Based on the simulations of atmospheric air quality, the AQI was calculated for the following emission scenarios:

Scenario 1: Emissions in 2005 (reference period) (Emission data for Bulgaria, according to the 2005 inventory);
Scenario 2: Emission data for Bulgaria for 2020–2029, emissions for 2020–2029, projected with existing measures (WEM);
Scenario 3: Emission data for Bulgaria for 2020–2029, emissions for 2020–2029, projected with additional measures (WAM);
Scenario 4: Emission data for Bulgaria after 2030, emissions projected after 2030 with existing measures (WEM);

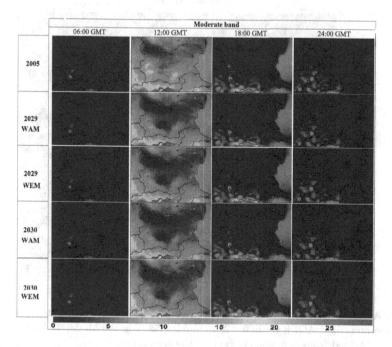

Fig. 1. Summarized annual recurrence of the AQI [%] for all emission scenarios in Moderate (indexes 4, 5, 6) band.

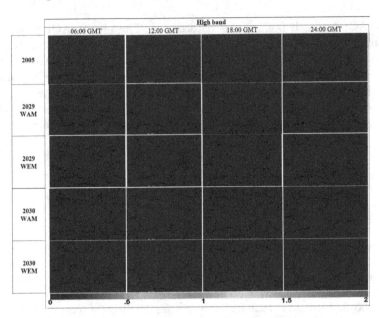

Fig. 2. Summarized annual recurrence of the AQI [%] for all emission scenarios in High (7, 8, 9, 10) band.

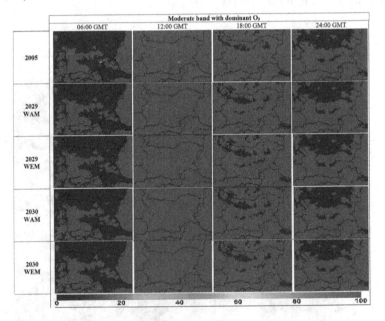

Fig. 3. Annual recurrence [%] of the cases where the dominant pollutants in determining the summarized AQI is O_3 in 6, 12, 18 and 24 GMT in the Moderate band (indexes 4, 5, 6), obtained from emissions in each of the scenarios.

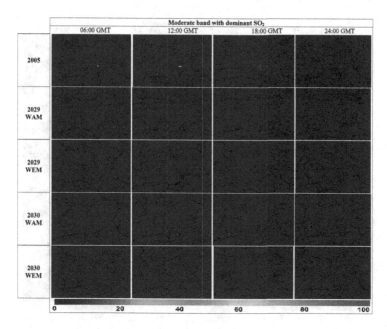

Fig. 4. Annual recurrence [%] of the cases where the dominant pollutant in determining the summarized AQI is SO_2 in 6, 12, 18 and 24 GMT in the he Moderate band (indexes 4, 5, 6), obtained from emissions in each of the scenarios.

Scenario 5: Emission data for Bulgaria after 2030, emissions projected after 2030 with additional measures (WAM);

Figure 1 and 2 shows the daily variations of the annual recurrence of pollution indexes for all emission scenarios. For simplicity, the pollution indexes are grouped into two bands: Moderate band (indexes 4, 5, 6) and High band (indexes 7, 8, 9, 10). It is seen that with emissions from 2005, the recurrence of the indexes in the two bands is higher compared to scenarios with different applied measures.

The differences are most noticeable in the midday hours (12:00 GMT), where in the Moderate band, the recurrence is about 15% for the whole country and 20% for Sofia, Bobov Dol TPP, and Maritsa Iztok TPP for the 2005 reference, while the recurrence decreases to 10–13% in these areas when measures are applied. For the High band, again in the midday hours, a recurrence of about 1% above Maritsa Iztok TPP is found, which recurrence is zeroed when different measures are applied in the scenarios. This shows that both Moderate and High bands in the mentioned areas decreased when the respective measures were applied. The dominant pollutants for each AQI category are defined. The values of the index are determined by the pollutants that reach a certain threshold for a given category, which are shown in Fig. 3 and 4. The recurrence maps show that O_3 and SO_2 are the dominant pollutants, as the other pollutants have a negligible, almost zero recurrence in determining the index. For all emission scenarios, the dominant pollutant in the Moderate band (Fig. 3) O_3 is 100% recurrence over

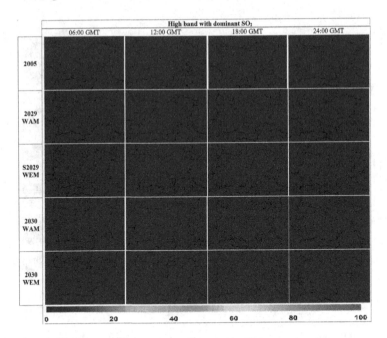

Fig. 5. Annual recurrence [%] of the cases where the dominant pollutant in determining the summarized AQI is SO_2 in 6, 12, 18 and 24 GMT in the High band (indexes 7, 8, 9, 10), obtained from emissions in each of the scenarios.

the entire country during the day. During the other hours, there are "holes" with zero ozone recurrence, which means that another pollutant dominates. Due to the fact that ozone over Bulgaria is mainly due to transboundary transport [10], there is no special difference in the maps during the different scenarios, which shows that ozone impacts on AQI are not affected by different measures applied, but its complex distribution is maintained. It should be noted, however, that in one of the "holes", Maritsa Iztok TPP area, an 80% recurrence of SO_2 (Fig. 4) domination emissions from 2005 is found, and this recurrence disappears when considering scenarios with different applied additional measures. The analysis shows that there is a high level of recurrence of SO_2 in the Maritsa Iztok TPP in the afternoon hours (Fig. 5), around 90% for 2005. This recurrence is nullified when measures are applied.

4   Conclusion

It can be concluded that with emissions from 2005, the recurrence of the indexes in the bands Moderate and High, is higher compared to scenarios with different applied measures.

- The highest recurrence of AQI from the Moderate band is in the afternoon, about 15% for the whole country and 20% for Sofia, Bobov Dol TPP, and

Maritsa Iztok TPP for the 2005 reference, and this recurrence decreases to 10–13% when measures are applied.

- For all emission scenarios, the dominant pollutant in the Moderate band is O_3 with 100% recurrence over the entire country. During the other hours, there are "holes" with zero ozone recurrence, which means that another pollutant dominates. The ozone being the dominant pollutant in the Moderate band.
- For the High band, in the midday hours, a recurrence is about 1% above Maritsa Iztok TPP and it is zeroed when different measures are applied in the scenarios.
- The recurrence of the index in the High band is low, but Sulfur dioxide is definitely the dominant pollutant in the formation of the indices in this category, the other pollutants have a negligible, almost zero recurrence in determining the index.
- An 80% recurrence of SO_2 domination emissions from 2005 is found in Maritsa Iztok TPP area (in the afternoon hours, around 90%), and this recurrence is nullified when measures are applied.

This shows that both Moderate and High bands in the mentioned areas decreased when the respective measures were applied.

Applying different measures in different scenarios leads to a definite positive effect on recurrence for most pollutants and bands.

For most of the higher AQI categories, the recurrence is almost zero due to the reduction of pollutant concentrations when measures are applied (WAM scenario).

Acknowledgments. This work has been carried out in the framework of the National Science Program "Environmental Protection and Reduction of Risks of Adverse Events and Natural Disasters", approved by the Resolution of the Council of Ministers No 577/17.08.2018 and supported by the Ministry of Education and Science (MES) of Bulgaria (Agreement No D01D01-271/09.12.2022). This work has also been accomplished with the financial support of Grant No BG05M2OP001-1.001-0003, financed by the Science and Education for Smart Growth Operational Program (2014–2020) and co-financed by the European Union through the European Structural and Investment funds. Special thanks are due to US EPA and US NCEP for providing free-of-charge data and software and to the Netherlands Organization for Applied Scientific research (TNO) for providing the high-resolution European anthropogenic emission inventory.

References

1. Byun, D., Ching, J.: Science Algorithms of the EPA Models-3 Community Multiscale Air Quality (CMAQ) Modeling System. EPA Report 600/R-99/030, Washington, D.C., EPA/600/R-99/030 (NTIS PB2000-100561) (1999). https://cfpub.epa.gov/si/si_public_record_report.cfm?Lab=NERL&dirEntryId=63400
2. Byun, D.: Dynamically consistent formulations in meteorological and air quality models for multiscale atmospheric studies part I: governing equations in a generalized coordinate system. J. Atmos. Sci. **56**, 3789–3807 (1999)

3. CEP: Sparse Matrix Operator Kernel Emission (SMOKE) Modeling System, University of Carolina, Carolina Environmental Programs, Research Triangle Park, North Carolina (2003)
4. CMAS. http://www.cmaq-model.org. Accessed 22 Feb 2023
5. Denier van der Gon, H., Visschedijk, A., van de Brugh, H., Droge, R.: A high resolution European emission data base for the year 2005, TNO-report TNO-034-UT-2010-01895 RPT-ML, Apeldoorn, The Netherlands (2010)
6. Directive (EU) 2016/2284 of the European Parliament and of the Council of 14 December 2016 on the reduction of national emissions of certain atmospheric pollutants, amending Directive 2003/35/EC and repealing Directive 2001/81/EC. https://www.eea.europa.eu/themes/air/air-pollution-sources-1/national-emission-ceilings
7. EEA Technical report No 4/2015, Projections in hindsight an assessment of past emission projections reported by Member States under EU air pollution and GHG legislation. https://www.eea.europa.eu/publications/projections-in-hindsight/download
8. Gadzhev, G., Ganev, K., Prodanova, M., Syrakov, D., Atanasov, E., Miloshev, N.: Multi-scale atmospheric composition modelling for Bulgaria. In: Steyn, D., Builtjes, P., Timmermans, R. (eds.) Air Pollution Modeling and its Application XXII. NAPSC, pp. 381–385. Springer, Dordrecht (2014). https://doi.org/10.1007/978-94-007-5577-2_64
9. Gadzhev, G.K., Ganev, K.G., Prodanov, M., Syrakov, D.E., Miloshev, N.G., Georgiev, G.J.: Some numerically studies of the atmospheric composition climate of Bulgaria. In: AIP Conference Proceedings, vol. 1561, pp. 100–111 (2013). https://doi.org/10.1063/1.4827219
10. Gadzhev, G., Ganev, K., Miloshev, N., Syrakov, D., Prodanova, M.: Numerical study of the atmospheric composition in Bulgaria. Comput. Math. Appl. 65, 402–422 (2013)
11. de Leeuw, F., Mol, W.: Air Quality and Air Quality Indices: a world apart. ETC/ACC Technical Paper 2005/5 (2005). http://acm.eionet.europa.eu/docs/ETCACC_TechnPaper_2005_5_AQ_Indices.pdf
12. Shamarock, W.C., et al.: A Description of the Advanced Research WRF Version 2 (2007). http://www.dtic.mil/dtic/tr/fulltext/u2/a487419.pdf
13. Syrakov, D., Prodanova, M., Georgieva, E., Etropolska, I., Slavov, K.: Impact of NOx emissions on air quality simulations with the Bulgarian WRF-CMAQ modelling system. Int. J. Environ. Pollut. 57(3–4), 285–296 (2015)
14. Syrakov, D., et al.: Application of WRF-CMAQ model system for analysis of sulfur and nitrogen deposition over Bulgaria. In: Nikolov, G., Kolkovska, N., Georgiev, K. (eds.) NMA 2018. LNCS, vol. 11189, pp. 474–482. Springer, Cham (2019). https://doi.org/10.1007/978-3-030-10692-8_54
15. UCAR/NCAR. https://ncar.ucar.edu/what-we-offer/models/weather-research-and-forecasting-model-wrf. Accessed 22 Feb 2023

Mathematical and Computational Modeling of a Nonlinear Elliptic Problem in Divergence Form

Sergey Korotov[1]([✉])[iD] and Michal Křížek[2][iD]

[1] Division of Mathematics and Physics, UKK Mälardalen University,
Box 883, 721 23 Västerås, Sweden
sergey.korotov@mdu.se
[2] Institute of Mathematics, Czech Academy of Sciences,
Žitná 25, 115 67 Prague 1, Czech Republic
krizek@math.cas.cz

Abstract. We introduce our main results on solving a nonlinear steady-state heat conduction problem in anisotropic and nonhomogeneous media and its finite element approximation. In particular, we concentrate on temperature distribution profile typical e.g. for large transformers and rotary machines. Several theorems on the existence and uniqueness of weak and discrete solutions and their convergence are presented.

Keywords: Heat conduction problem · Radiation · Uniqueness · Finite element method · Convergence

1 Motivation

We shall deal with a nonlinear steady-state heat conduction problem in large transformers and rotary machines whose material is anisotropic and also nonhomogeneous. This problem is nonpotential [15] (i.e., it cannot be transformed to the minimization of some functional). Moreover, it is typically of nonmonotone type [7]. Hence, the classical theory of monotone operators cannot be applied. Thus, every class of such nonlinear problems requires developing new investigation techniques. In this paper we present our own approach for proving the existence and uniqueness of a weak solution and to establish convergence of the corresponding finite element approximations.

A typical example of an orthotropic and nonlinear material is the magnetic circuit of a large oil-immersed transformer with iron sheets (see Fig. 1a). The height of such transformers can be up to 5 m and diameters of its legs 2 m. Since the total power is up to 400 MW and the efficiency is 97 %, about 12 MW changes into residual heat every second. Therefore, the knowledge of temperature distribution is very important to design a suitable cooling. If the temperature exceeds the prescribed limits (usually in the upper concave corners of the transformer on Fig. 1a), then the oil starts to boil, which may cause a short circuit and can completely destroy the entire transformer, which cannot be easily repaired if a demage occurs.

I. Lirkov and S. Margenov (Eds.): LSSC 2023, LNCS 13952, pp. 343–353, 2024.
https://doi.org/10.1007/978-3-031-56208-2_35

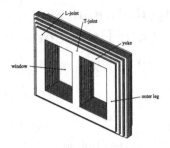
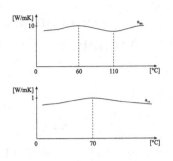

Fig. 1. a) Magnetic circuit of a large transformer with iron sheets represents an orthotropic nonlinear material. b) Temperature dependence of heat conductivities across and along iron sheets.

The associated 3×3 matrix $A = A(u)$ of heat conductivities (see problem (P) below) is diagonal and such that $a_{11} \neq a_{22} = a_{33}$. Note that the well-known Kirchhoff transformation (see [15, p. 162]) cannot be applied in the case of anisotropic nonlinear media, since the nonlinear temperature dependence of heat conductivities is different across and along the iron sheets (see Fig. 1b).

Another typical example which we have calculated is the temperature distribution in a stator of large rotary machines, see Fig. 2. Their material is again anisotropic and nonlinear. Moreover, it is highly nonhomogeneous. For instance, heat conductivities of copper wires (Cu) substantially differ from heat conductivities of insulation (Is), namely, $a_{ii}^{\mathrm{Cu}} = 332$ W/(mK) for $i = 1, 2, 3$, whereas $a_{11}^{\mathrm{Is}} = 0.2$ W/(mK) and $a_{22}^{\mathrm{Is}} = 0.5$ W/(mK). This shows that there are large jumps in the corresponding heat conduction coefficients. The power of such rotary machines is up to 40 MW and its diameter about 2 m. If the temperature in copper wires exceeds the prescribed limits, then their insulation starts to burn, which can destroy the entire machine. Therefore, a detailed knowledge of temperature distribution is again very important.

Throughout this paper we use the standard Sobolev space notation [15]. In the next section, we give classical and weak formulations of a nonlinear steady-state heat conduction problem in anisotropic and nonhomogeneous media. Then we concentrate on the uniqueness and nonuniqueness of their solutions and introduce a comparison principle. We also present theorems on the existence of weak and finite element solutions and their convergence.

2   Classical and Weak Formulation

A classical formulation of a nonlinear steady-state heat conduction in anisotropic and nonhomogeneous media (see [14]) consists of finding a function $u \in C^1(\overline{\Omega}) \cap C^2(\Omega)$ (called the *classical solution*) such that

$$-\mathrm{div}(A(x, u)\,\mathrm{grad}\,u) = f \quad \text{in} \quad \Omega, \tag{P}$$

$$\alpha u + n^{\top} A(s, u)\,\mathrm{grad}\,u = g \quad \text{on} \quad \partial\Omega.$$

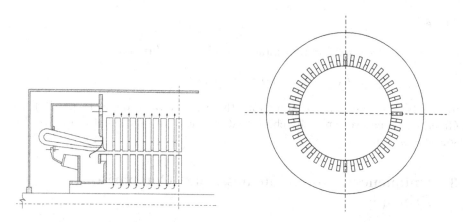

Fig. 2. Left: Packs of the stator of a large motor are separated by radial ventilation channels. Right: Cross-section of an individual pack.

Here $\Omega \subset \mathbb{R}^d$, $d \in \{1,2,3\}$, is a bounded domain with Lipschitz continuous boundary $\partial\Omega$, the closure $\overline{\Omega} = \Omega \cup \partial\Omega$, n is the outward unit normal to $\partial\Omega$, u is the temperature distribution, $f \in L^2(\Omega)$ is the density of volume heat sources, $g \in L^2(\partial\Omega)$ is the density of surface heat sources, $\alpha \in L^\infty(\partial\Omega)$ is the heat transfer coefficient for which there exists a constant $C > 0$ such that

$$\alpha(s) \geq C > 0 \quad \forall s \in \Gamma, \tag{C}$$

where $\Gamma \subset \partial\Omega$ has a positive measure and $A = (A_{ij})_{i,j=1}^d$ is a bounded measurable matrix function of heat conductivities such that there exist positive constants C_E and C_L for which

$$\eta^\top A(x,\xi)\eta \geq C_E\|\eta\|^2 \quad \forall x \in \Omega \quad \forall \xi \in \mathbb{R}^1 \quad \forall \eta \in \mathbb{R}^d \tag{E}$$

and

$$|A_{ij}(x,\zeta) - A_{ij}(x,\xi)| \leq C_L|\zeta - \xi| \quad \forall x \in \Omega \quad \forall \zeta, \xi \in \mathbb{R}^1. \tag{L}$$

Here $\|\cdot\|$ stands for the standard Euclidean norm.

Theorem 1. *Let* (C), (E), (L) *hold and let A be a diagonal matrix such that A_{ii} are continuously differentiable on $\overline{\Omega} \times \mathbb{R}^1$ for all arguments. Then there exists at most one solution of* (P).

For the proof see [6]. Since the classical solution may not exist (for example, when A has jumps or when Ω has concave corners), we now introduce a weak solution of the problem (P).

A function $u \in V$ is said to be a *weak solution* of the problem (P) if the following variational equality holds

$$a(u;u,v) = F(v) \quad \forall v \in V := H^1(\Omega), \tag{V}$$

where

$$a(y; w, v) = (A(y)\operatorname{grad} w, \operatorname{grad} v)_{0,\Omega} + \langle \alpha w, v \rangle_{0,\partial\Omega},$$

$$F(v) = (f, v)_{0,\Omega} + \langle g, v \rangle_{0,\partial\Omega}$$

for $v, w, y \in V$ and $\langle \cdot, \cdot \rangle_{0,\partial\Omega}$ stands for the usual scalar product in $L^2(\partial\Omega)$. By Green's theorem we can easily check that the classical solution of (P) is also the solution of (V).

3   Uniqueness, Nonuniqueness, and Comparison Principles

First we present sufficient conditions for the uniqueness of the weak solution.

Theorem 2. *Let* (C), (E), (L) *hold and let* $u_1, u_2 \in V$ *be two solutions of the problem* (V). *Then* $u_1 = u_2$ *a.e. in* Ω.

For the proof see [7]. Another uniqueness theorems for nonlinear elliptic problems with Dirichlet boundary conditions are given e.g. in Boccardo, Gallouët, Murat [1], and Jensen [8].

If a nonlinear elliptic equation is not in the divergence form, there exist examples of nonunique solutions, see e.g. Gilbarg, Trudinger [5] or Meyers [17]. We can also get nonunique solutions of problem (P) which is in the divergence form if the Lipschitz condition (L) is violated. Namely, in [7] we define

$$A(x, \xi) \equiv \frac{1}{u_i'(x)} \quad \text{for} \quad x \in [0, 1], \quad \xi = u_i(x), \quad i = 1, 2,$$

and show that one-dimensional nonlinear heat conduction problem (P) can surprisingly have two different solution $u_1 \neq u_2$ for the same boundary conditions (see [15, p. 171]). By Tietze's extension theorem (see [19]) there exists a continuous extension (still denoted by A) so that $A(\cdot, \cdot) : \Omega \times \mathbb{R}^1 \to \mathbb{R}^1$ is bounded and that (C) and (E) hold.

We see that

$$-(A(x, u_i)u_i')' = 0 \quad \text{for } i = 1, 2,$$

i.e., u_1 and u_2 are solutions of (P) with nonhomogeneous boundary conditions and $f = 0$.

The Lipschitz condition (L) can be weakened and the uniqueness can still be achieved. To this end we assume that there exists a positive constant C_H such that

$$|A_{ij}(x, \zeta) - A_{ij}(x, \xi)| \leq C_H |\zeta - \xi|^e \quad \forall \zeta, \xi \in \mathbb{R}^1, \; i, j = 1, \dots, d, \qquad \text{(H)}$$

where $e \in \left[\frac{1}{2}, 1\right]$ is a given Hölder exponent, i.e., $A(\cdot, \cdot)$ is e-Hölder continuous with respect to the last variable.

Theorem 3. *Let* (C), (E), (H) *hold and let* $u_1, u_2 \in V$ *be two solutions of the problem* (V). *Then* $u_1 = u_2$ *a.e. in* Ω.

Its proof in [12] involves also a nonlinear dependence of f and g on u, and mixed Dirichlet-Newton boundary conditions.

Comparison and maximum principles are important features of second order elliptic equations that distinguish them from higher order equations and systems of elliptic equations.

Theorem 4. *Let* (C), (E), (L) *hold and let* $u_1, u_2 \in V$ *be two solutions of the problem* (V) *corresponding to* $f_1, f_2 \in L^2(\Omega)$ *and* $g_1, g_2 \in L^2(\partial\Omega)$, *respectively. Assume that*

$$f_1 \geq f_2 \quad a.e. \quad in \quad \Omega \quad and \quad g_1 \geq g_2 \quad a.e. \quad on \quad \partial\Omega.$$

Then $u_1 \geq u_2$ *a.e. in* Ω.

For the proof see [13]. The comparison principle for linear problems is a direct consequence of the weak maximum principle. Note that the comparison principle immediately implies the uniqueness of the weak solution. It also yields a natural assertion:

Any rise of the density of heat sources always causes that the temperature will not decrease in almost all points. This shows that the nonlinear steady-state heat conduction mathematical model (P) has reasonable physical properties.

4   Existence of Weak and Discrete Solutions

Let $V_h \subset C(\overline{\Omega})$ be a nonempty finite-dimensional subspace. Then a Galerkin solution $u_h \in V_h$ of (V) is defined by the following relation

$$a(u_h; u_h, v_h) = F(v_h) \quad \forall v_h \in V_h. \tag{G}$$

The space V_h is usually constructed by means of the finite element method, see [2, p. 132].

Theorem 5. *If* (C), (E), (L) *hold, then there exists a Galerkin solution* $u_h \in V_h$ *satisfying* (G).

For the proof see [7]. It is based on the Brouwer fixed-point theorem [15]. Note that our technique to prove the uniqueness of the solution of the problem (V) cannot be modified to Galerkin approximations, since the corresponding test functions (see [7, p. 184]) do not belong to finite element spaces, in general. However, the uniqueness was proved recently in [18] by a different method.

Remark. We used the classical iteration method of freezing coefficients (called also the Picard iterations or the method of secant modules or Kačanov's method, cf. (K) below) to approximate the Galerkin solution $u_h \in V_h$ corresponding to weak nonlinearities of Fig. 2. When the space V_h occurring in (G) was generated by the standard bilinear elements, only 5 iterations were enough to reach a "numerical fixed point" starting from various initial temperatures.

Let $\{V_h\}_{h \to 0}$ be a family of finite-dimensional subspaces of $H^1(\Omega) \cap C(\overline{\Omega})$ such that the following density requirement holds

$$\forall v \in C^\infty(\overline{\Omega}) \quad \exists \{v_h\}_{h \to 0}: \quad v_h \in V_h, \quad \|v - v_h\|_{1,\Omega} \to 0 \quad \text{as} \quad h \to 0. \qquad \text{(D)}$$

Then we have the following statement.

Theorem 6. *Let* (C), (D), (E), (L) *hold and let* $\{u_h\}_{h \to 0}$, $u_h \in V_h$, *be a sequence of Galerkin solutions satisfying* (G). *Then there exists an element* $u \in H^1(\Omega)$ *such that*

$$u_h \rightharpoonup u \quad \text{(weakly) in} \quad H^1(\Omega) \quad \text{as} \quad h \to 0, \qquad \text{(W)}$$

and u *is a solution of the problem* (V).

For the proof see [7]. It is based on the Eberlein-Schmulyan theorem.

5   Convergence of Finite Element Approximations

From the weak convergence (W) and the compactness of the imbedding operator $H^1(\Omega) \hookrightarrow L^2(\Omega)$ we can prove the convergence of the Galerkin solutions in the $\| \cdot \|_{0,\Omega}$-norm. To prove the (strong) convergence (S) in the $\| \cdot \|_{1,\Omega}$-norm, we shall, in addition, require the following boundedness condition

$$V_h \subset W_4^1(\Omega), \quad \|v_h\|_{1,4,\Omega} \le C(v) \quad \forall h \in (0, h_0), \qquad \text{(B)}$$

where $v \in C^\infty(\overline{\Omega})$, v_h satisfies the density condition (D), $h_0 > 0$, and $C(v)$ is a constant independent of h. For instance, functions v_h can be defined as the V_h-interpolant of v.

The next theorem from [15] establishes convergence of the sequence $\{u_h\}_{h \to 0}$ without any additional regularity assumptions on the solution u.

Theorem 7. *Let* (B), (C), (D), (E) *and* (L) *hold. Then the convergence* (W) *is strong, i.e.,*

$$\|u - u_h\|_{1,\Omega} \to 0 \quad \text{as} \quad h \to 0. \qquad \text{(S)}$$

Further, we indroduce some a priori error estimate for the following Dirichlet problem

$$-\text{div}(A(x, u)\,\text{grad}\,u) = f \quad \text{in} \quad \Omega,$$
$$u = 0 \quad \text{on} \quad \partial\Omega,$$

where $f \in L^2(\Omega)$ and A is an L^∞ matrix function which is Lipschitz continuous with respect to the last variable and is uniformly positive definite, see (L) and (E). This problem can be converted by the Kirchhoff transformation [15, p. 162] to a linear problem provided $A(x,\xi) = a(\xi)I$, where I is the identity matrix. To the authors' knowledge, it is an open problem how to modify the Kirchhoff transformation to nonlinear anisotropic materials (see Fig. 1b).

The weak formulation of the above Dirichlet problem consists of finding $u \in V$ such that

$$a(u; u, v) = F(v) \quad \forall v \in V := H_0^1(\Omega), \tag{U}$$

where $a(y; w, v) = (A(y) \operatorname{grad} w, \operatorname{grad} v)_{0,\Omega}$ and $F(v) = (f, v)_{0,\Omega}$ for $y, w, v \in H^1(\Omega)$. By [7], $u \in V$ exists uniquely. Let its Galerkin approximation u_h satisfying (G) belong to the space

$$V_h = \{v_h \in H_0^1(\Omega) \mid v_h|_K \in P_K \quad \forall K \in T_h\},$$

where $P_K \supset P_k(K)$ and $k \geq 1$ (integer). Consider the adjoint problem: Find $\phi \in H_0^1(\Omega)$ such that

$$-\operatorname{div}(A^\top(x, u) \operatorname{grad} \phi) + (\operatorname{grad} u)^\top B^\top(x, u) \operatorname{grad} \phi = \zeta,$$

where u is the solution of (U), $\zeta \in L^2(\Omega)$, $B = (B_{ij})$, $B_{ij}(x,\xi) = \partial A_{ij}(x,\xi)/\partial \xi$ and, moreover, we assume the following regularity

$$\|\phi\|_{2,\Omega} \leq C\|\zeta\|_{0,\Omega}. \tag{R}$$

Theorem 8. *Let $u \in H^{k+1}(\Omega)$, $k \geq 1$, be the solution of* (U), *let* (E), (L) *and* (R) *hold, let the derivatives $\partial A_{ij}/\partial \xi$ and $\partial^2 A_{ij}/\partial \xi^2$ be bounded and continuous on $\overline{\Omega} \times \mathbb{R}^1$ and let $\{T_h\}_{h \to 0}$ be a regular family of triangulations. Then there exists $h_0 > 0$ such that for any $h \in (0, h_0)$ we have*

$$\|u - u_h\|_{0,\Omega} + h\|u - u_h\|_{1,\Omega} \leq Ch^{k+1},$$

where C depends on the norm $\|u\|_{k+1,\Omega}$.

For the proof see [16].

6   Variational Crimes

In practical implementation of the finite element method, the so-called *variational crimes* are often committed, see [2]. They include that: 1) integrals are evaluated numerically, 2) the domain Ω with a piecewise curved boundary is approximated by a polygonal or polyhedral domains, and 3) boundary conditions are approximated by finite element functions. Since quadrature formulae are used, we need stronger smoothness assumptions: $A_{ij} = A_{ij}(x,\xi)$, $\partial A_{ij}/\partial x_k$ and $\partial A_{ij}/\partial \xi$ are continuous and bounded in $\overline{\Omega} \times \mathbb{R}^1$ for all $i, j, k = 1, \ldots, d$, i.e. there exists a constant C such that

$$|A_{ij}(x,\xi)| \leq C, \quad \left|\frac{\partial A_{ij}}{\partial x_k}(x,\xi)\right| \leq C, \quad \left|\frac{\partial A_{ij}}{\partial \xi}(x,\xi)\right| \leq C \quad \forall x \in \overline{\Omega} \quad \forall \xi \in \mathbb{R}^1. \tag{A}$$

To approximate the Galerkin solution we apply the following numerical integration formula (see [2]) over a simplex $K \in T_h$

$$\int_K \varphi(x) \, dx \approx \text{meas } K \sum_{m=1}^{M_K} \omega_{K,m} \varphi(x_{K,m}),$$

where the weights $\omega_{K,m} \in \mathbb{R}^1$ are such that

$$\omega_{K,m} > 0 \quad \text{and} \quad \sum_{m=1}^{M_K} \omega_{K,m} = 1, \tag{N}$$

and all nodes $x_{K,m} \in K$ for $k = 1, \ldots, M_K$. Define the spaces of linear elements

$$X_h = \{v_h \in H^1(\Omega) \mid v_h|_K \in P_1(K) \quad \forall K \in T_h\}, \qquad V_h = H_0^1(\Omega) \cap X_h.$$

For any $v_h \in X_h$ and any simplex $K \in T_h$ put $v_K = v_h|_K$.

A function $u_h^* \in V_h$ is said to be an *approximate solution* of the problem (U) if

$$a_h(u_h^*; u_h^*, v_h) = F_h(v_h) \quad \forall v_h \in V_h, \tag{Q}$$

where the corresponding quadrature rules are defined for $y_h, w_h, v_h \in X_h$ by

$$a_h(y_h; w_h, v_h) = \sum_{K \in T_h} \text{meas } K \sum_{m=1}^{M_K} \omega_{K,m} (\text{grad } w_K)^\top A(x_{K,m}, y_K(x_{K,m})) \text{grad } v_K,$$

$$F_h(v_h) = \sum_{K \in T_h} \text{meas } K \sum_{m=1}^{M_K} \omega_{K,m} f(x_{K,m}) v_K(x_{K,m}).$$

Theorem 9. *Let $f \in W_\infty^1(\Omega)$, let (A), (E) and (N) hold. Then for any $V_h \neq \emptyset$ there exists the corresponding approximate solution $u_h^* \in V_h$ of the problem (Q). Moreover, if the family $\{T_h\}_{h \to 0}$ of triangulations satisfies the following inverse inequality inequality*

$$\|v_h\|_{1,\Omega} \leq C h^{-1} \|v_h\|_{0,\Omega} \qquad \forall v_h \in V_h, \tag{I}$$

we have

$$\|u - u_h^*\|_{1,\Omega} \to 0 \quad \text{as} \quad h \to 0.$$

For the proof see [4,11]. Notice that no additional regularity assumptions on u are required.

The problem (Q) can be approximately solved by the Kačanov method, see [15, p. 138]. This method enables us to transform the discrete nonlinear problem to an infinite sequence of linear finite-dimensional problems as follows:

Given $y_h^0 \in V_h$ arbitrarily, we define $y_h^{j+1} \in V_h$ recurrently by

$$a_h(y_h^j; y_h^{j+1}, v_h) = F_h(v_h) \quad \forall v_h \in V_h, \quad j = 0, 1, \ldots \tag{K}$$

Further, we briefly introduce the convergence analysis of the finite element method in the case when an approximation of a curved boundary as well as numerical integration are used.

Let $\Omega \subset \mathbb{R}^2$ be a bounded domain with Lipschitz continuous boundary which is piecewise of the class $\mathcal{C}^3$. Let us consider a family $\{\Omega_h\}_{h \to 0}$ of polygonal approximations of Ω, and a family $\{T_h\}_{h \to 0}$ of triangulations of $\overline{\Omega}_h$ with $h \in (0, h_0)$, $h_0 > 0$ sufficiently small. We assume that all vertices of all triangles from any T_h belong to $\overline{\Omega}$ and the corners of $\overline{\Omega}_h$ are lying on $\partial \Omega$. Now let $\tilde{\Omega} \subset \mathbb{R}^2$ be a bounded domain such that $\Omega \cup \Omega_h \subset \tilde{\Omega}$ for all $h \in (0, h_0)$ and let $A = A(x, \xi)$, $x \in \tilde{\Omega}$, $\xi \in \mathbb{R}^1$, satisfy (A) and (E), where Ω is replaced by $\tilde{\Omega}$. Let $V_h \subset H_0^1(\Omega_h)$ be the space of linear elements. Then by [4] we have:

Theorem 10. *Let $f \in W_\infty^1(\tilde{\Omega})$ and let (A), (E), (N) hold on $\tilde{\Omega}$. Then for any V_h there exists the corresponding approximate solution $u_h^* \in V_h$ of the problem (Q). Moreover, if the family $\{T_h\}_{h \to 0}$ of triangulations satisfies the inverse inequality (I) on Ω_h, we have*

$$\|\tilde{u} - u_h^*\|_{1,\Omega_h} \to 0 \quad \text{as} \quad h \to 0,$$

where $\tilde{u} \in H^1(\mathbb{R}^2)$ is the standard Calderon extension of the weak solution u onto $\mathbb{R}^2$.

7   Concluding Remarks

Each body loses heat energy from its surface by electromagnetic waves. This effect is called radiation. Losses of energy are proportional to the 4th power of the surface temperature (the Kirchhoff law). Radiation is small at room temperature, but it should not be neglected if the surface temperature is high. It is described by the nonlinear Stefan-Boltzmann boundary condition

$$\alpha u + n^\top A \operatorname{grad} u + \beta u^4 = g,$$

on some part of the boundary $\partial \Omega$, where $\beta = \sigma f_{\text{em}}$, $\sigma = 5.669 \times 10^{-8} \text{ Wm}^{-2}\text{K}^{-4}$ is the Stefan-Boltzmann constant, $0 \le f_{\text{em}} \le 1$ is the relative emissivity function.

If A is independent of the solution u, the heat radiation problem can be transformed to the minimization of a nonquadratic functional over a nonempty convex set. The existence and uniqueness of the weak solution u is guaranteed [15]. A discrete problem then consists of minimizing the same functional over V_h. For a linear rate of convergence in the $H^1(\Omega)$-norm for linear elements and $u \in H^2(\Omega)$ see [15].

There are various generalizations of the above methods and theorems in the literature. Continuous and discrete maximum principles for mixed boundary conditions and some other types of nonlinearities are investigated e.g. in [3,9,10]. As an example consider the problem

$$-\Delta u + cu = f \text{ in } \Omega \quad \text{and} \quad \alpha u + \frac{\partial u}{\partial n} = g \text{ on } \partial \Omega,$$

In [3], the following maximum principle in the form of an a priori two-sided estimate is proved:

Let $c, f \in C(\overline{\Omega})$, $\alpha, g \in C(\partial\Omega)$, and let there exist positive constants c_0 and α_0 such that

$$c(x) \geq c_0 \quad \forall x \in \overline{\Omega} \quad \text{and} \quad \alpha(s) \geq \alpha_0 \quad \forall s \in \partial\Omega.$$

Then

$$\min\left\{0, \min_{x \in \overline{\Omega}} \frac{f(x)}{c(x)}, \min_{s \in \partial\Omega} \frac{g(x)}{\alpha(s)}\right\} \leq u(x) \leq \max\left\{0, \max_{x \in \overline{\Omega}} \frac{f(x)}{c(x)}, \max_{s \in \partial\Omega} \frac{g(x)}{\alpha(s)}\right\}.$$

In [15], partial differential equations of parabolic and hyperbolic type that model time dependent heat conduction are presented. A detailed proof of the convergence of semidiscrete finite element approximations to a linear parabolic problem is given. The speed of heat conduction is infinite for the parabolic type equations whereas it is finite for the hyperbolic type equations.

Acknowledgements. Supported by GrantS No. 23-06159S and 24-10586S of the Grant Agency of the Czech Republic and the Czech Academy of Sciences (RVO 67985840).

References

1. Boccardo, L., Gallouët, T., Murat, F.: Unicité de la solution de certaines équations elliptiques non linéaires. C. R. Acad. Sci. Paris Ser. I Math. **315**, 1159–1164 (1992)
2. Brandts, J., Korotov, S., Křížek, M.: Simplicial Partitions with Applications to the Finite Element Method. Springer, Cham (2020)
3. Faragó, I., Korotov, S., Szabó, T.: On continuous and discrete maximum principles for elliptic problems with the third boundary condition. Appl. Math. Comput. **219**, 7215–7224 (2013)
4. Feistauer, M., Křížek, M., Sobotíková, V.: An analysis of finite element variational crimes for a nonlinear elliptic problem of a nonmonotone type. East-West J. Numer. Math. **1**, 267–285 (1993)
5. Gilbarg, D., Trudinger, N.S.: Elliptic Partial Differential Equations of Second Order. Springer, Berlin (1977)
6. Hlaváček, I., Křížek, M.: On a nonpotential and nonmonotone second order elliptic problem with mixed boundary conditions. Stability Appl. Anal. Contin. Media **3**, 85–97 (1993)
7. Hlaváček, I., Křížek, M., Malý, J.: On Galerkin approximations of a quasilinear nonpotential elliptic problem of a nonmonotone type. J. Math. Anal. Appl. **184**, 168–189 (1994)
8. Jensen, R.: The maximum principle for viscosity solutions of fully nonlinear second order partial differential equations. Arch. Rational Mech. Anal. **101**, 1–27 (1988)
9. Karátson, J., Korotov, S.: Discrete maximum principles for finite element solutions of nonlinear elliptic problems with mixed boundary conditions. Numer. Math. **99**, 669–698 (2005)
10. Karátson, J., Korotov, S., Křížek, M.: On discrete maximum principles for nonlinear elliptic problems. Math. Comput. Simulation **76**, 99–108 (2007)

11. Korotov, S., Křížek, M.: Finite element analysis of variational crimes for a quasi-linear elliptic problem in $3D$. Numer. Math. **84**, 549–576 (2000)
12. Křížek, M.: The uniqueness of the solution of a nonlinear heat conduction problem under Hölder's continuity condition. Appl. Math. Lett. **103**, 1–6, Article no. 106214 (2020)
13. Křížek, M., Liu, L.: On a comparison principle for a quasilinear elliptic boundary value problem of a nonmonotone type. Appl. Math. (Warsaw) **24**, 97–107 (1996)
14. Křížek, M., Liu, L.: Finite element approximation of a nonlinear heat conduction problem in anisotropic media. Comput. Methods Appl. Mech. Eng. **157**, 387–397 (1998)
15. Křížek, M., Neittaanmäki, P.: Mathematical and Numerical Modelling in Electrical Engineering: Theory and Applications. Kluwer Academic Publishers, Dordrecht (1996)
16. Liu, L., Křížek, M., Neittaanmäki, P.: Higher order finite element approximation of a quasilinear elliptic boundary value problem of a non-monotone type. Appl. Math. **41**, 467–478 (1996)
17. Meyers, N.G.: An example of non-uniqueness in the theory of quasi-linear elliptic equations of second order. Arch. Rational Mech. Anal. **14**, 177–179 (1963)
18. Pollock, S., Zhu, Y.: Uniqueness of discrete solutions of nonmonotone PDEs without a globally fine mesh condition. Numer. Math. **139**, 845–865 (2018)
19. Rudin, W.: Real and Complex Analysis. McGraw-Hill, New York (1966)

Two Approaches for Identifying Epidemiological Parameters Illustrated with COVID-19 Data for Bulgaria

Tchavdar T. Marinov[1], Rossitza S. Marinova[2,3(✉)], and Nicci Shelby[1]

[1] Southern University at New Orleans, 6801 Press Drive,
New Orleans, LA 70126, USA
tmarinov@suno.edu
[2] Concordia University of Edmonton, 7128 Ada Blvd, Edmonton,
AB T5B-4E4, Canada
rossitza.marinova@concordia.ab.ca
[3] Department Computer Science, Varna Free University, Varna, Bulgaria

Abstract. This work focuses on special numerical techniques for identifying time-dependent epidemics parameters such as infection and recovery rates, and the corresponding basic and effective reproduction numbers. The first method is based on solving the inverse problem for coefficients' identification in an Adaptive Susceptible-Infected-Removed (A-SIR) epidemic model with time-dependent infection and recovery rates. The second approach is based on direct integration of the SIR system of ordinary differential equations and the least square method. This approach is limited to estimation of the basic and effective reproduction numbers. The developed methods are used for identifying the parameters with real data for COVID-19 for Bulgaria. The results demonstrate how well the methods simulate the dynamics of the infectious disease.

Keywords: Inverse problem · Infectious disease · Infection and recovery rates · Time-dependent parameters · COVID-19

1 Introduction

According to the COVID-19 data, published in [12], there are 1,296,009 cases with 38,198 deaths in Bulgaria as of February 14, 2023 [1]. The total population of Bulgaria is reported to be 6,844,597. The percentage of the Bulgarian population with at least one doze COVID-19 vaccine being 35.8%, is unquestionably the lowest vaccination rate in the European Union (EU) for the age group 18 years and older, according to [2]. The following publications implement various SIR-type models to investigate the infectious disease dynamics in Bulgaria [3–5,7,8]. The parameters are identified from formulated inverse problems as in [11].

The present work considers a SIR-type model. The model groups individuals as **S**usceptible, **I**nfectious, and **R**ecovered (SIR). The notations in the SIR

I. Lirkov and S. Margenov (Eds.): LSSC 2023, LNCS 13952, pp. 354–362, 2024.
https://doi.org/10.1007/978-3-031-56208-2_36

model include: $S(t)$ – the number of *susceptible*, $I(t)$ – *infectious*, $R(t)$ – *removed* individuals, and t is the time. The total population $N = S(t) + I(t) + R(t)$ is considered constant in the equations below:

$$L_1(S, I, \beta) = \frac{dS(t)}{dt} + \beta(t)S(t)I(t) = 0 \tag{1}$$

$$L_2(S, I, \beta, \gamma) = \frac{dI(t)}{dt} - \beta(t)S(t)I(t) + \gamma(t)I(t) = 0 \tag{2}$$

$$L_3(I, R, \gamma) = \frac{dR(t)}{dt} - \gamma(t)I(t) = 0. \tag{3}$$

The classical SIR model, proposed in [6], assumes constant infection β and recovery γ rates. Here, $\beta = \beta(t)$ and $\gamma = \gamma(t)$ are time-dependent. Therefore, the *basic reproduction number (ratio, rate)* $R_0 = R_0(t)$, and the *effective reproduction number* $R_e = R_e(t)$, are also time-dependent functions,

$$R_0(t) = \frac{\beta(t)\, N}{\gamma(t)}, \quad R_e(t) = \frac{\beta(t)S(t)}{\gamma(t)}. \tag{4}$$

2   Identifying Problem Parameters

With proper initial conditions, and the coefficients $\beta(t)$ and $\gamma(t)$ given, one can construct a numerical scheme to solve the system of ordinary differential equations (1)–(3). In case of new disease, like COVID-19, the values of the coefficients are unknown. Additional information is needed for identifying $\beta(t)$ and $\gamma(t)$.

We assume that the values of $S(t)$ and $I(t)$ at some time moments ν_1, ν_2, ..., ν_m, are known and given by

$$S(\nu_l) \approx \sigma_l, \quad I(\nu_l) \approx \lambda_l, \tag{5}$$

for $l = 1, 2, \ldots, m$, as seen in Fig. 1.

The inverse problem approach to solving the adaptive SIR model (A-SIR) considers the time-dependent infection and recovery rates as piece-wise constant functions of time [9, 10]

$$\beta(t) = \beta_k \text{ and } \gamma(t) = \gamma_k, \text{ or } \frac{\beta(t)}{\gamma(t)} = \frac{\beta_k}{\gamma_k} \text{ for } \nu_{k-1} < t < \nu_k.$$

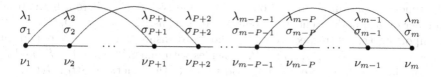

Fig. 1. The time moments ν_l, where σ_l and λ_l are given, and the subsets of fixed length of P days for identifying the problem parameters.

We propose two approaches for identifying the parameters of the inverse problem for the A-SIR epidemic model. Both approaches consider the coefficients β and γ as constants in each of the sub-interval from ν_l to ν_{l+P}. We identify those constants (or their ratio β/γ) and assign the values as β_{l+P} and γ_{l+P}, or β_{l+P}/γ_{l+P}. The first approach is to solve the inverse problem following the idea of Method of Variational Imbedding (MVI). We will call this approach MVI for brief. The second approach identifies the ratio β_{l+P}/γ_{l+P} using direct integration of the classical SIR system and the Least Square Method (LSM).

3 Solving the Inverse Problem Using MVI

Let $[0, P]$ be a time sub-interval (Fig. 1), where P is a number of days. The approximate values of the susceptible $S(t)$ and currently infectious $I(t)$ are known at specified time moments, given in (5). Following the idea of MVI, we construct a functional using the original Eqs. (1)–(2), and the available data values (5). We define the functional over the sub-interval $[0, P]$ as

$$\mathcal{F} = \int_0^P \left[L_1^2 + L_2^2 + \sum_{i=1}^{P-1} \delta(t - \nu_i) \, \mu_i \left((S(t) - \sigma_i)^2 + (I(t) - \lambda_i)^2 \right) \right] dt, \quad (6)$$

where μ_i are the weights prescribed for the i-th node and $\delta(t)$ is the *Dirac delta function* $\delta(t - \nu_i)$ defined as:

$$\delta(t - \nu_i) = \begin{cases} \infty, & t = \nu_i \\ 0, & t \neq \nu_i \end{cases} \quad \text{and} \quad \int_{-\infty}^{\infty} \delta(t - \nu_i) dt = 1.$$

The absolute minimum of $\mathcal{F}$ is equal to zero with the functional becoming zero if and only if Eqs. (1), (2) and (5) are satisfied. We want to emphasize here that the functions (S, I) and the parameters (β, γ) are unknown and one have to identify them simultaneously by solving a minimization problem. Finding $R(t)$ is straightforward after knowing the values of $I(t)$, $S(t)$, $\beta(t)$, and $\gamma(t)$.

3.1 Discretization of the Minimization Sub-problem

We find the minimum of the functional $\mathcal{F}$ numerically over a sub-interval of the entire period assuming constant infection and recovery rates. In order to do this, we first approximate the derivatives and integrals in (6). Let $\tau = \frac{P}{n}$ be the time-step of a uniform grid on the finite interval $[0, P]$, where n is the total number of grid nodes. The grid nodes are defined as: $t_k = k\tau$, $k = 0, 1, \ldots, n$. It is important that τ (respectively n) is chosen to ensure that every time moment $\nu_l, \nu_{l+1}, \ldots, \nu_{l+P}$, $l = 1, \ldots, m - P$, coincides with one grid node. Let S_k and I_k be notations for the corresponding grid values of the functions $S(t_k)$, $I(t_k)$.

The problem under consideration is non-linear and it requires linearization at some stage of the solution process. Let $\hat{S}_k$, $\hat{I}_k$ be known approximate values (say from the previous iteration), used in the non-linear term of the differential equations (1), (2). In order to secure $O(\tau^2)$ errors of approximation of the operators in (6), we discretize the derivatives in L_1 and L_2 at the grid nodes $t_{k-1/2}$. We introduce the notations for the approximations of $S(t)$ and $I(t)$ at the midpoints of $[t_{k-1}, t_k]$, where $k = 1, 2, \ldots, n$:

$$\hat{S}_{k-1/2} = 0.5(\hat{S}_{k-1} + \hat{S}_k), \quad \hat{I}_{k-1/2} = 0.5(\hat{I}_{k-1} + \hat{I}_k).$$

Consequently, the differential operators L_1 and L_2 are approximated by the following linear difference operators Λ_1 and Λ_2

$$\Lambda_{1,k} = \frac{S_k - S_{k-1}}{\tau} + \beta \hat{S}_{k-1/2} \hat{I}_{k-1/2}$$

$$\Lambda_{2,k} = \frac{I_k - I_{k-1}}{\tau} - \beta \hat{S}_{k-1/2} \hat{I}_{k-1/2} + \gamma \hat{I}_{k-1/2},$$

for $k = 1, 2, \ldots, n$. We approximate $\delta(t_k - \nu_i)$ as

$$\delta(t_k - \nu_i) = \begin{cases} \dfrac{1}{\tau}, & t_k = \nu_i \\ 0, & t_k \neq \nu_i \end{cases} \quad \text{for} \quad k = 1, 2, \ldots, n, \quad i = 1, 2, \ldots, P - 1.$$

To simplify the presentation, let us introduce the notations

$$\bar{\mu}_k = \begin{cases} \mu_i, & t_k = \nu_i, \\ 0, & \text{otherwise} \end{cases} \quad \bar{\sigma}_k = \begin{cases} \sigma_i, & t_k = \nu_i, \\ 0, & \text{otherwise} \end{cases} \quad \bar{\lambda}_k = \begin{cases} \lambda_i, & t_k = \nu_i, \\ 0, & \text{otherwise} \end{cases}.$$

This way, the discretized version of the functional $\mathcal{F}$ becomes

$$\Phi = \sum_{k=1}^{n} \left[(\Lambda_{1,k}^2 + \Lambda_{2,k}^2)\tau + \bar{\mu}_k \left((S_k - \bar{\sigma}_k)^2 + (I_k - \bar{\lambda}_k)^2 \right) \right].$$

3.2 Equations for S and I

The necessary conditions for minimization of the function Φ with respect to its arguments S_k and I_k require the derivatives to be zero. For the case under consideration, we obtain the following equations in the internal grid nodes t_k, where $k = 1, 2, \ldots, n - 1$:

$$S_{k-1} - (2 + \tau\mu_k)S_k + S_{k+1}$$
$$= -\tau\beta(\hat{S}_{k+1/2}\hat{I}_{k+1/2} - \hat{S}_{k-1/2}\hat{I}_{k-1/2}) - \tau(u_{k+1/2} - u_{k-1/2}) - \tau\mu_k\sigma_k, \tag{7}$$

$$I_{k-1} - (2 + \tau\mu_k)I_k + I_{k+1}$$
$$= \tau\beta(\hat{S}_{k+1/2}\hat{I}_{k+1/2} - \hat{S}_{k-1/2}\hat{I}_{k-1/2}) - \tau\gamma(\hat{I}_{k+1/2} - \hat{I}_{k-1/2}) - \tau\mu_k\lambda_k. \tag{8}$$

The equations at t_0 and at t_n are

$$S_1 - S_0 = \tau(-\beta \hat{S}_{1/2}\hat{I}_{1/2} - u_{1/2}), \tag{9}$$

$$I_1 - I_0 = \tau(\beta \hat{S}_{1/2}\hat{I}_{1/2} - \gamma \hat{I}_{1/2}), \tag{10}$$

$$S_n - S_{n-1} = \tau(-\beta \hat{S}_{n-1/2}\hat{I}_{n-1/2} - u_{n-1/2}), \tag{11}$$

$$I_n - I_{n-1} = \tau(\beta \hat{S}_{n-1/2}\hat{I}_{n-1/2} - \gamma \hat{I}_{n-1/2}). \tag{12}$$

This way, if approximate values of β and γ are known, there are two well-posed systems of $(n + 1)$ linear equations: (7), (9), and (11) for the unknown set of new values $(S_0, S_1, S_2, \ldots, S_n)$; and (8), (10), and (12) for $(I_0, I_1, I_2, \ldots, I_n)$. Due to the early linearization, these two systems are with constant matrices. Hence, we need to invert them once and use the inverse matrices during the entire iterative process, which helps reduce the computational time of this large scale problem.

3.3   Equations for β and γ

We rewrite the function Φ in the form

$$\Phi = \alpha_{00} + \alpha_{10}\beta + \alpha_{01}\gamma + \alpha_{20}\beta^2 + \alpha_{11}\beta\gamma + \alpha_{02}\gamma^2,$$

where

$$\alpha_{00} = \sum_{k=1}^{n} \left[(I_k - I_{k-1})^2 + (S_k - S_{k-1})^2 + \tau^2 u_{k-1/2}^2 \right]$$
$$+ \bar{\mu}_k \left[(S_k - \bar{\sigma}_k)^2 + (I_k - \bar{\lambda}_k)^2 \right],$$

$$\alpha_{10} = \sum_{k=1}^{n} -2\hat{I}_{k-1/2}\hat{S}_{k-1/2}(I_k - I_{k-1} - S_k + S_{k-1} - \tau u_{k-1/2})\tau,$$

$$\alpha_{01} = \sum_{k=1}^{n} 2\hat{I}_{k-1/2}(I_k - I_{k-1})\tau,$$

$$\alpha_{20} = \sum_{k=1}^{n} 2\hat{I}_{k-1/2}^2 \hat{S}_{k-1/2}^2 \tau^2, \quad \alpha_{11} = \sum_{k=1}^{n} -2\hat{I}_{k-1/2}^2 \hat{S}_{k-1/2}\tau^2, \quad \alpha_{02} = \sum_{k=1}^{n} \hat{I}_{k-1/2}^2 \tau^2.$$

The necessary conditions for minimization of the function Φ with respect to β and γ give a system of two linear equations with solution:

$$\beta = -\frac{2\alpha_{02}\alpha_{10} - \alpha_{01}\alpha_{11}}{-\alpha_{11}^2 + 4\alpha_{02}\alpha_{20}}, \quad \gamma = -\frac{\alpha_{10}\alpha_{11} - 2\alpha_{01}\alpha_{20}}{\alpha_{11}^2 - 4\alpha_{02}\alpha_{20}}.$$

3.4   Numerical Algorithm

We have constructed two systems of linear equations, one system for (S, I) under the condition that (β, γ) are given, and another system for (β, γ) under the condition that (S, I) are given. This allows us to build an algorithm for finding a solution to the entire non-linear problem, by means of an iterative procedure, replacing (β, γ) (when calculating (S, I)), or (S, I) (when calculating (β, γ)) with their values from the previous iteration. If the iterations converge, then they will give one of the possible solutions of the problem. Thus, the existence of the solution to the problem can be established *a-posteriori*. If the iterative process diverges, this will mean that there exists no solution to the problem.

4 Identifying the Reproduction Numbers by LSM

As we mentioned, we assume that the parameters β and γ are constant in every sub-interval $[0, P]$, where P is a number of days, see Fig. 1. Equations (1) and (2), after excluding dt, give an ordinary differential equation connecting the functions I and S:

$$\frac{dI}{dS} = \frac{\beta SI - \gamma I}{-\beta SI}. \tag{13}$$

The general solution of Eq. (13) represents I as a function of S:

$$I = -S + \frac{\gamma}{\beta} \ln S + C.$$

The straightforward way to identify the γ/β and C is to take $P = 1$ and to use the values (5) as boundary conditions. This approach does not work well in practice because the "noise" in the datasets $\{\sigma_i\}$ and $\{\lambda_i\}$ propagates to γ/β. For this reason we use subsets of fixed length of $P > 1$ days, as shown in Fig. 1. Following the LSM, we construct the functions

$$\Psi\left(\frac{\beta_k}{\gamma_k}, C_k\right) = \sum_{l=k-P}^{k} \mu_l \left(\frac{\gamma_k}{\beta_k} \ln(\sigma_l) + C_k - \sigma_l - \lambda_l\right)^2$$

for $k = P+1, P+2, \ldots, m$. Here μ_l are the weight coefficient of the corresponding nodes. From the necessary conditions for minimization of the function Ψ with respect to $\frac{\gamma_k}{\beta_k}$ and C_k we receive a system of two equations with solution for $\frac{\gamma_k}{\beta_k}$

$$\frac{\gamma_k}{\beta_k} = \frac{A_{22}B_1 - A_{12}B_2}{A_{11}A_{22} - A_{12}^2},$$

where

$$A_{11} = \sum_{l=k-P}^{k} \mu_l \left[\ln(\sigma_l)\right]^2, \quad A_{12} = \sum_{l=k-P}^{k} \mu_l \ln(\sigma_l), \quad A_{22} = \sum_{l=k-P}^{k} \mu_l,$$

$$B_1 = \sum_{l=k-P}^{k} \mu_l(\sigma_l + \lambda_l)\ln(\sigma_l), \quad B_2 = \sum_{l=k-P}^{k} \mu_l(\sigma_l + \lambda_l).$$

The solution C_k is not essential. Substituting $\frac{\gamma_k}{\beta_k}$ in (4) gives $R_0 = R_0(t)$ and $R_e = R_e(t)$ as functions of time.

5   Results and Conclusions

We applied the methods to real COVID-19 data, published in [12], for Bulgaria, for a period of two years ending on February 19, 2023. The parameters are identified over three different time periods: 7 days; 14 days; and 21 days. Figure 2a–2b show the time-dependent infection and recovery rates identified by the MVI, Fig. 2c–2d – the effective reproduction numbers identified by MVI and LSM, respectively. The results obtained by these different methods are almost identical. The infection and recovery rates also agree with the results obtained in [8].

The obtained results from our simulations with data from other countries (e.g. Canada and USA) and USA states (Georgia, Louisiana, and Texas) show an interesting relation between the number of currently infectious individuals and the effective reproduction number, namely that the local maximum values of the effective reproduction number R_e occur approximately three weeks before the local maximum values of the number of currently infectious individuals. The COVID-19 data for Bulgaria, however, does not follow such behavior.

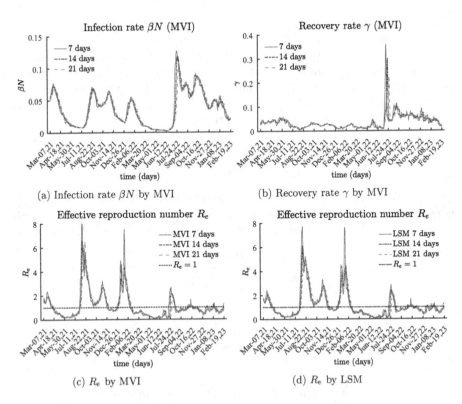

(a) Infection rate βN by MVI

(b) Recovery rate γ by MVI

(c) R_e by MVI

(d) R_e by LSM

Fig. 2. Infection and recovery rates, and reproduction numbers

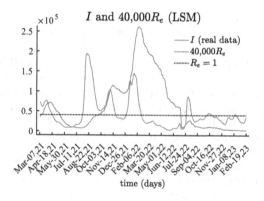

Fig. 3. R_e by LSM and the reported number of infectious people in Bulgaria.

Figure 3 compares the reported COVID-19 data for the daily infectious individuals $I(t)$ and the scaled computed effective reproduction number $R_e(t)$, based on 21-day period. The above mentioned observation for Bulgaria is seen only during a relatively short period of time between Fall 2021 and spring 2022. This effect can help to predict the upcoming peak of the current wave. Further work is required to find out why the data for Bulgaria does not follow this pattern. On the other hand, the observed differences display the complexity of the problem and the sensitivity of the epidemic development depending on the environment.

Acknowledgments. The work of T.M. and N.S. was partially supported from the Grant LA Dept. of Health, Office of Public Health, Bureau of Community Preparedness of the State of LA (LDH), contract LaGov PO#: 2000620243.

References

1. Bulgarian Ministry of Health: Bulgarian COVID-19 Unified Information Portal (2020). https://coronavirus.bg/
2. European Centre for Disease Prevention and Control: COVID-19 vaccine tracker (2020). https://www.ecdc.europa.eu/en
3. Georgiev, S., Vulkov, L.: Numerical coefficient reconstruction of time-depending integer- and fractional-order SIR models for economic analysis of COVID-19. Mathematics **10**(22) (2022). https://doi.org/10.3390/math10224247
4. Georgiev, S.G., Vulkov, L.G.: Coefficient identification for SEIR model and economic forecasting in the propagation of COVID-19. In: Georgiev, I., Kostadinov, H., Lilkova, E. (eds.) Advanced Computing in Industrial Mathematics, pp. 34–44. Springer, Cham (2023). https://doi.org/10.1007/978-3-031-20951-2_4
5. Ivanova, M., Dospatliev, L.: Data analytics and SIR modeling of COVID-19 in Bulgaria. Int. J. Appl. Math. **33**(6), 1099–1114 (2020). https://doi.org/10.12732/ijam.v33i6.10
6. Kermack, W., McKendrick, A.: A contribution to the mathematical theory of epidemics. **115**(772), 700–721 (1927)

7. Margenov, S., Popivanov, N., Ugrinova, I., Harizanov, S., Hristov, T.: Mathematical and computer modeling of COVID-19 transmission dynamics in Bulgaria by time-depended inverse SEIR model. In: AIP Conference Proceedings, vol. 2333, p. 090024. AIP Publishing (2021). https://doi.org/10.1063/5.0041868

8. Margenov, S., Popivanov, N., Ugrinova, I., Hristov, T.: Mathematical modeling and short-term forecasting of the COVID-19 epidemic in Bulgaria: SEIRS model with vaccination. Mathematics 10(15) (2022). https://doi.org/10.3390/math10152570

9. Marinov, T., Marinova, R.: Dynamics of COVID-19 using inverse problem for coefficient identification in SIR epidemic models. Chaos Solitons Fractals: X 5(5), 100041 (2020). https://doi.org/10.1016/j.csfx.2020.100041

10. Marinov, T., Marinova, R.: Inverse problem for adaptive SIR model: application to COVID-19 in Latin America. Infect. Dis. Model. 7(1), 134–148 (2022). https://doi.org/10.1016/j.idm.2021.12.001

11. Marinov, T., Marinova, R., Omojola, J., Jackson, M.: Inverse problem for coefficient identification in SIR epidemic models. Comput. Math. Appl. 67(12), 2218–2227 (2014). https://doi.org/10.1016/j.camwa.2014.02.002

12. Worldometer: COVID-19 Coronavirus Pandemic (2020). https://www.worldometers.info/coronavirus/

Improved Stochastic Lattice Methods for Large-Scale Air Pollution Model

Venelin Todorov[1,2]([✉]) [ID], Slavi Georgiev[1,3][ID], Ivan Dimov[2], Rayna Georgieva[2], and Tzvetan Ostromsky[2]

[1] Department of Information Modeling, Institute of Mathematics and Informatics, Bulgarian Academy of Sciences, Acad. G. Bonchev Str., Bl. 8, 1113 Sofia, Bulgaria
{vtodorov,sggeorgiev}@math.bas.bg

[2] Department of Parallel Algorithms, Institute of Information and Communication Technologies, Bulgarian Academy of Sciences, Acad. G. Bonchev Str., Bl. 25A, 1113 Sofia, Bulgaria
{venelin,rayna,ceco}@parallel.bas.bg, ivdimov@bas.bg

[3] Department of Applied Mathematics and Statistics, Faculty of Natural Sciences and Education, Angel Kanchev University of Ruse, 8 Studentska Str., 7004 Ruse, Bulgaria
sggeorgiev@uni-ruse.bg

Abstract. In the current study, a large-scale air pollution model is adopted, focusing on the Sobol' approach for sensitivity analysis. In this paper we will use the advanced stochastic approach based on component by component construction methods. Optimized algorithms based on lattice rules have been designed and implemented, while their performance has been compared to the best available stochastic approaches, applied for multidimensional sensitivity analysis. Numerical results show a significant improvement over the current stochastic methods. The obtained results would have an important multi-sided role in the area of air pollution modeling.

Keywords: Multidimensional integrals · Sensitivity analysis · Air pollution modeling

1 Introduction

It is well known that the adaptation to the climate changes is an important global problem. It threatens whole ecosystems on Earth, having a vast impact on health and economy. A lot of effort is put into resolving the issue. There is still abundant interest in theoretical research too. One of the most powerful tools to do it is mathematical modeling. An important issue when large-scale mathematical models are used to support decision makers is their reliability. Sensitivity analysis (SA) of model outputs to variation or natural uncertainties of model inputs is very significant for improving the reliability of these models. By definition, sensitivity analysis is a procedure for studying how sensitive are the output results of large-scale mathematical models to some uncertainties of

© The Author(s), under exclusive license to Springer Nature Switzerland AG 2024
I. Lirkov and S. Margenov (Eds.): LSSC 2023, LNCS 13952, pp. 363–371, 2024.
https://doi.org/10.1007/978-3-031-56208-2_37

the input data [4,6]. The sensitivity analysis studies the level of uncertainty at which the model input data impacts the model output in terms of accuracy. Various factors as instrumental error, data approximation and compression, and many others might cause the perturbation in the input. The real-world experiments are very costly, time-consuming and they require highly trained crew to conduct them. Due to this fact, the mathematical modeling is very useful and important when describing natural phenomena [2,4]. The mathematical formulation of the task for sensitivity analysis leads to estimating multidimensional integrals [6]. Stochastic methods, and particularly Monte Carlo methods, are one of the best tools for multidimensional integrals' evaluation due to their simplicity and applicability for various integrand functions of any dimension defined in complicated domains [2,5].

2   Improved Stochastic Approaches for Sensitivity Analysis

We will use the rank-1 lattice sequence (LS) defined by [7]: $\mathbf{x}_k = \left\{ \dfrac{k}{N} \mathbf{z} \right\}$, $k = 1, \ldots, N$, where $N \geq 2$ is an integer, $\mathbf{z} = (z_1, z_2, \ldots z_s)$ is the generating vector (GV). More about the widely used Fibonacci LS can be found in [7]. Now we will improve the LS by using the Component by Component Fast Construction method (CBC) [1]. Now we generate several special generating vector for up to 2^{24} points for the LS following the idea in [3]. The first GV will be the CBC construction of rank-1 lattice rule with prime number of points and with product weights, we will call it **1PT**. The second GV will be the CBC construction of rank-1 lattice rule with prime number of points and with order dependent weights, we will call it **1OD**. The third GV will be the CBC construction of rank-1 LS with prime power of points and with product weights, we call it **1EXPT**. The fourth GV will be the CBC construction of rank-1 LS with prime power of points and with order dependent weights, denoted by **1EXOD**. On the first stage of generation of LS, $\mathbf{z} = (z_1, z_2, \ldots z_s)$ is obtained by the CBC. In particular, in the beginning $z_1 := 1$. Then, z_1 is hold fixed, and $z_2 \in \mathbb{U}^N := \{z \in \mathbb{N} : 1 \leq z \leq N - 1, \gcd(z, N) = 1\}$ is chosen in such a way that the predefined error criterion, such as worst function errors or Zaremba index, is minimized in two dimensions. Then, iteratively for $i = 3, \ldots, s$, z_i is chosen from $\mathbb{U}^N$ in such a way to minimize the predefined error criterion in i dimensions.

The current sensitivity study has been based on the ANOVA-representation of the model function, and the Sobol' approach using sensitivity measures/sensitivity indices (SI) S_i and total sensitivity indices (TSI) S_i^{tot}, defined via the corresponding partial and total variances [6]. The latter are measures of the overall influence (full effect) of an input parameter on variations in the output. In this way, the main problem from mathematical point of view is the approximate evaluation of Sobol' sensitivity indices presented by integrals of various dimensions defined by the terms $g_d(\mathbf{x}), d = 1, \ldots$ from the model function $f(\mathbf{x})$, $\mathbf{x} \in \mathbb{U}^n$ decomposition: $I^d = \int_{\mathbb{U}^d} g_d(\mathbf{x}) d\mathbf{x}$, $d = 1, \ldots$.

One should notice that $f_0 = \int_{\mathbb{U}^n} f(\mathbf{x})d\mathbf{x}$. Finally, the approximate value I_N^d of the corresponding integral is obtained by the following Monte Carlo estimate:

$$I_N^d = \frac{1}{N} \sum_{k=1}^{N} func\left(\left\{\frac{k}{N}\mathbf{z}\right\}\right),$$ where $func$ is g_d or f and N is the number of samples.

3   Case Study and Numerical Results

The stochastic algorithms under consideration have been applied to a particular sensitivity study via the Uniform Danish Eulerian model for a remote transport of air pollutants [8]. The current study has been provided in two main directions: sensitivity studies with respect of emission levels, and sensitivity studies according to chemical reaction rates. The following notation has been used inside the tables: EQ - estimated quantity, RV - reference value, CRU - crude (plain) Monte Carlo algorithm using pseudorandom numbers.

3.1   Sensitivity Studies with Respect to Emission Levels

The main four groups of air pollutants are presented here $\mathbf{E} = (\mathbf{E}^A, \mathbf{E}^N, \mathbf{E}^S, \mathbf{E}^C)$, where $\mathbf{E}^A$–ammonia (NH_3), $\mathbf{E}^S$–sulphur dioxide (SO_2), $\mathbf{E}^N$–nitrogen oxides $(NO + NO_2)$, $\mathbf{E}^C$ – anthropogenic hydrocarbons.

The relative errors (the absolute difference between the exact/reference value and the obtained value, divided by the exact value) related to sensitivity studies with respect to emission levels are presented in Table 1, Table 2, Table 3, Table 4, and Table 5. The computational times of the algorithms under consideration are comparable (Fig. 1).

Table 1. Relative error for the evaluation of $f_0 \approx 0.048$.

N	CRU	1PT	1OD	1EXPT	1EXOD	FIBO
	Relative error	Relative error	Relative error	Relative error	Relative error	Relative error
2^{10}	1.02e−02	2.36e−04	**1.92e−04**	4.28e−04	6.43e−04	2.09e−04
2^{12}	3.42e−03	1.35e−04	6.80e−05	7.91e−05	1.29e−04	**4.32e−05**
2^{14}	2.51e−03	1.82e−05	3.01e−05	**1.64e−05**	4.07e−05	2.25e−05
2^{16}	1.73e−03	**4.66e−06**	8.56e−06	5.21e−06	7.78e−06	8.70e−06
2^{18}	4.32e−04	3.26e−06	1.78e−06	2.57e−06	**1.61e−06**	1.79e−06
2^{20}	6.72e−05	8.44e−07	5.73e−07	5.31e−07	**3.21e−07**	4.21e−07
2^{22}	6.46e−05	7.68e−08	1.08e−07	1.44e−07	4.84e−07	**5.44e−08**
2^{24}	1.63e−05	**1.87e−08**	3.17e−08	2.85e−08	4.71e−08	1.51e−07

Table 2. Relative error for the evaluation of the total variance $\mathbf{D} \approx 0.0002$.

N	CRU	1PT	1OD	1EXPT	1EXOD	FIBO
	Relative error	Relative error	Relative error	Relative error	Relative error	Relative error
2^{10}	1.15e−01	9.59e−02	3.38e−02	**1.57e−02**	3.25e−02	1.63e−01
2^{12}	2.87e−02	1.13e−02	5.07e−03	**1.24e−04**	6.05e−03	2.39e−02
2^{14}	4.30e−02	5.46e−03	1.14e−02	7.70e−04	**1.80e−04**	2.90e−03
2^{16}	1.76e−02	6.13e−04	6.39e−04	**2.09e−04**	2.26e−04	2.65e−04
2^{18}	1.16e−02	2.39e−04	9.43e−05	1.19e−04	**9.19e−05**	3.01e−04
2^{20}	5.80e−03	5.79e−05	1.35e−05	3.38e−05	**8.12e−06**	1.19e−04
2^{22}	7.36e−04	2.58e−05	3.26e−05	**7.81e−06**	3.01e−05	2.59e−05
2^{24}	1.99e−03	6.71e−06	1.73e−06	**9.92e−08**	2.15e−06	4.91e−06

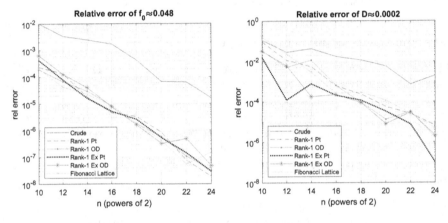

Fig. 1. Relative errors for the calculation of $f_0 \approx 0.048$ (left) and $\mathbf{D} \approx 0.0002$ (right).

Table 3. Relative error for estimation of sensitivity indices of input parameters using various Monte Carlo and quasi-Monte Carlo approaches ($N = 2^{16}$).

EQ	RV	CRU	1PT	1OD	1EXPT	1EXOD	FIBO
S_1	9e−01	8.23e−03	1.28e−04	**2.18e−05**	1.30e−04	4.52e−04	3.62e−04
S_2	2e−04	2.29e+00	1.99e−01	4.46e−01	**2.28e−02**	1.23e−01	1.74e−01
S_3	1e−01	5.32e−02	2.05e−03	2.79e−03	**6.40e−04**	1.32e−03	3.22e−03
S_4	4e−05	3.02e+00	**4.46e−03**	1.41e+00	4.07e−01	6.38e−01	4.87e−01
S_1^{tot}	9e−01	6.97e−03	1.94e−04	2.57e−04	**6.98e−05**	2.26e−04	4.61e−04
S_2^{tot}	2e−04	1.67e+00	4.61e−01	1.08e−01	**2.96e−02**	1.31e−01	3.45e−01
S_3^{tot}	1e−01	6.25e−02	2.27e−03	**9.30e−04**	1.09e−03	3.19e−03	1.96e−03
S_4^{tot}	5e−05	1.14e+00	7.05e−01	4.30e−01	**3.73e−02**	4.88e−01	5.06e−01

Table 4. Relative error for estimation of sensitivity indices of input parameters using various Monte Carlo and quasi-Monte Carlo approaches ($N = 2^{20}$).

EQ	RV	CRU	1PT	1OD	1EXPT	1EXOD	FIBO
S_1	9e−01	7.87e−04	2.13e−06	4.73e−05	6.29e−06	6.13e−05	**5.29e−08**
S_2	2e−04	5.55e−01	5.98e−02	2.46e−02	**6.82e−04**	5.00e−03	3.17e−03
S_3	1e−01	3.42e−03	2.08e−04	1.92e−04	**5.60e−05**	2.38e−04	6.88e−05
S_4	4e−05	2.16e−01	3.03e−01	8.10e−02	**1.59e−02**	3.48e−02	1.88e−01
S_1^{tot}	9e−01	5.27e−04	2.84e−05	2.84e−05	**6.70e−06**	3.29e−05	2.14e−05
S_2^{tot}	2e−04	4.43e−01	8.72e−02	**5.62e−04**	4.29e−03	1.46e−02	4.56e−03
S_3^{tot}	1e−01	5.33e−03	2.45e−04	3.47e−04	**4.53e−05**	4.71e−04	4.69e−05
S_4^{tot}	5e−05	1.47e−01	2.47e−02	6.82e−03	3.89e−03	**1.90e−03**	6.08e−02

Table 5. Relative error for estimation of sensitivity indices of input parameters using various Monte Carlo and quasi-Monte Carlo approaches ($N = 2^{24}$).

EQ	RV	CRU	1PT	1OD	1EXPT	1EXOD	FIBO
S_1	9e−01	2.07e−04	2.18e−06	2.17e−06	6.69e−07	**4.63e−07**	3.43e−06
S_2	2e−04	3.34e−03	1.27e−04	1.05e−03	1.37e−04	**8.60e−05**	1.57e−03
S_3	1e−01	1.36e−03	3.89e−06	1.66e−05	5.35e−06	**3.20e−07**	1.15e−05
S_4	4e−05	1.68e−01	**4.13e−04**	5.83e−03	4.73e−04	1.27e−03	1.24e−02
S_1^{tot}	9e−01	1.67e−04	5.15e−07	2.43e−06	5.75e−07	**7.62e−08**	2.08e−06
S_2^{tot}	2e−04	6.33e−02	3.33e−04	5.85e−04	1.61e−04	9.65e−05	**4.16e−05**
S_3^{tot}	1e−01	1.76e−03	1.76e−05	1.73e−05	5.92e−06	**2.75e−06**	2.46e−05
S_4^{tot}	5e−05	1.53e−02	1.40e−03	9.66e−04	**9.28e−04**	9.84e−04	1.57e−02

The general conclusion in this case is that our constructed LS - 1PT, 1OD, 1EXPT and 1EXOD significantly outperform standard LS and standard Monte Carlo, and 1EXPT has a slightly advantage over the 1EXOD and it is significantly better than 1PT and 1OD especially in the case of quantities with smaller values (for example, see Table 2).

3.2 Sensitivity Studies with Respect to Chemical Reaction Rates

The following six chemical reactions have been recognized as important during previous studies [8,9]:

[#1] $NO_2 + h\nu \Longrightarrow NO + O;$ [#22] $HO_2 + NO \Longrightarrow OH + NO_2;$
[#3] $O_3 + NO \Longrightarrow NO_2;$ [#27] $HO_2 + HO_2 \Longrightarrow H_2O_2;$
[#7] $NO_2 + O_3 \Longrightarrow NO_3;$ [#28] $OH + CO \Longrightarrow HO_2.$

The results (in particular, relative errors) related to sensitivity studies with respect to the rates of several chemical reactions are presented in Table 6, Table 7,

Table 6. Relative error for the evaluation of $f_0 \approx 0.27$.

N	CRU Relative error	1PT Relative error	1OD Relative error	1EXPT Relative error	1EXOD Relative error	FIBO Relative error
2^{10}	6.97e−04	5.44e−03	3.68e−04	**2.45e−04**	6.74e−04	2.08e−03
2^{12}	8.06e−04	1.55e−03	**7.22e−06**	1.83e−04	8.40e−05	1.40e−04
2^{14}	2.80e−03	3.46e−04	4.85e−05	**8.47e−06**	3.21e−05	3.98e−04
2^{16}	5.95e−04	4.32e−05	1.28e−05	**4.54e−06**	1.43e−05	2.61e−04
2^{18}	7.66e−04	2.63e−05	1.76e−06	9.69e−07	**9.46e−07**	7.29e−06
2^{20}	3.45e−04	6.70e−06	6.56e−07	**2.14e−07**	2.49e−07	4.57e−07
2^{22}	4.70e−05	7.02e−07	1.51e−07	1.55e−07	**1.10e−07**	5.67e−07
2^{24}	8.62e−06	2.78e−07	1.15e−07	**3.92e−08**	4.58e−08	1.19e−06

Table 7. Relative error for the evaluation of the total variance $\mathbf{D} \approx 0.0025$.

	CRU Relative error	1PT Relative error	1OD Relative error	1EXPT Relative error	1EXOD Relative error	FIBO Relative error
2^{10}	7.22e−02	3.97e−01	**5.10e−03**	4.31e−02	5.27e−02	6.73e+00
2^{12}	9.54e−02	9.31e−02	**8.03e−03**	1.00e−02	1.22e−02	5.27e−01
2^{14}	6.40e−02	1.90e−02	4.32e−03	1.94e−03	**1.25e−03**	1.02e−01
2^{16}	2.97e−02	8.16e−04	1.40e−04	3.21e−04	**8.28e−05**	1.97e−03
2^{18}	7.42e−03	2.49e−03	3.66e−04	**2.22e−04**	8.04e−04	4.53e−03
2^{20}	8.92e−03	8.46e−04	2.06e−04	**7.24e−05**	1.09e−04	9.33e−03
2^{22}	2.21e−03	4.30e−05	5.95e−05	2.65e−05	**8.92e−06**	2.21e−02
2^{24}	1.29e−03	3.26e−06	3.43e−05	**2.65e−06**	3.80e−06	5.03e−04

Table 8, Table 9, and Table 10. The computational times of the algorithms under consideration are comparable (Fig. 2).

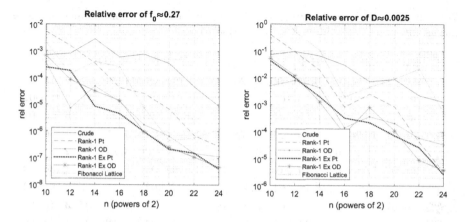

Fig. 2. Relative errors for the calculation of $f_0 \approx 0.27$ (left) and $\mathbf{D} \approx 0.0025$ (right).

Table 8. Relative error for estimation of sensitivity indices of input parameters using various MC and QMC approaches ($N \approx 65536$).

EQ	RV	CRU	1PT	1OD	1EXPT	1EXOD	FIBO
S_1	4e−01	5.61e−03	3.01e−03	3.36e−03	2.51e−03	**2.19e−03**	3.82e−02
S_2	3e−01	4.73e−02	1.24e−03	5.96e−03	**5.39e−04**	4.13e−03	1.03e−02
S_3	5e−02	1.38e−01	3.32e−02	**4.07e−03**	7.61e−03	9.16e−03	5.48e−01
S_4	3e−01	1.05e−02	6.41e−03	4.87e−04	1.78e−03	**1.13e−04**	1.07e−02
S_5	4e−07	8.73e+02	**2.32e+01**	2.01e+02	1.09e+02	1.52e+02	3.40e+03
S_6	2e−02	2.69e−01	2.12e−02	2.58e−02	2.27e−02	**1.55e−03**	1.32e+00
S_1^{tot}	4e−01	2.83e−02	8.29e−03	7.00e−03	**1.93e−03**	3.66e−03	7.92e−02
S_2^{tot}	3e−01	3.63e−02	6.84e−03	2.17e−03	**1.43e−04**	4.99e−03	3.06e−02
S_3^{tot}	5e−02	1.62e−01	3.07e−02	2.24e−02	5.99e−03	**5.49e−03**	1.31e+00
S_4^{tot}	3e−01	2.47e−02	8.06e−03	**9.62e−04**	1.08e−03	1.84e−03	3.84e−01
S_5^{tot}	2e−04	2.44e+00	**1.79e−01**	1.28e+00	5.23e−01	4.77e−01	8.85e+01
S_6^{tot}	2e−02	2.72e−01	4.71e−03	**3.04e−03**	6.19e−02	2.05e−02	2.15e+00
S_{12}	6e−03	6.31e−01	3.66e−01	3.19e−01	**3.37e−02**	7.56e−02	3.21e+00
S_{14}	5e−03	8.28e−01	**1.40e−02**	2.81e−02	4.43e−02	3.05e−02	8.64e+00
S_{24}	3e−03	2.23e−01	4.6330e−02	5.27e−02	**4.17e−02**	2.3106e−01	1.37e+01
S_{45}	1e−05	1.89e+01	3.37e+00	1.29e+00	**2.15e−01**	3.49e−01	4.25e+01

The general conclusion in this case is that our constructed LS - 1PT, 1OD, 1EXPT and 1EXOD significantly outperform the standard LS and the standard Monte Carlo, and 1EXPT has slightly advantage over 1EXOD and significantly better for most of the cases than 1PT and 1OD.

Table 9. Relative error for estimation of sensitivity indices of input parameters using various MC and QMC approaches ($N \approx 1048576$).

EQ	RV	CRU	1PT	1OD	1EXPT	1EXOD	FIBO
S_1	4e−01	2.55e−03	**5.54e−05**	2.02e−04	1.75e−04	1.23e−04	9.21e−03
S_2	3e−01	1.91-02	3.61e−04	2.70e−04	**1.08e−04**	3.22e−04	1.47e−02
S_3	5e−02	3.50e−03	5.44e−04	1.30e−03	**3.36e−04**	4.56e−04	6.50e−01
S_4	3e−01	1.79e−02	2.18e−03	4.57e−04	3.24e−04	**1.41e−04**	1.53e−01
S_5	4e−07	1.87e+02	2.41e+02	6.70e+01	**6.19e+00**	4.55e+01	2.68e+03
S_6	2e−02	5.38e−02	**1.77e−04**	2.12e−03	2.29e−04	8.70e−04	1.13e+00
S_1^{tot}	4e−01	7.95e−03	8.42e−04	**2.30e−05**	1.27e−04	4.24e−04	9.69e−03
S_2^{tot}	3e−01	1.97e−02	7.81e−04	2.30e−04	2.22e−04	**2.51e−05**	3.01e−02
S_3^{tot}	5e−02	3.10e−03	5.38e−03	2.20e−03	2.75e−04	**8.18e−05**	1.37e+00
S_4^{tot}	3e−01	2.01e−02	1.67e−03	9.30e−05	**8.39e−05**	1.76e−04	3.67e−01
S_5^{tot}	2e−04	5.80e−01	1.49e−01	1.0619e−01	2.88e−01	**2.79e−02**	3.90e+01
S_6^{tot}	2e−02	7.27e−02	6.15e−04	**7.35e−05**	1.06e−03	1.46e−03	1.76e+00
S_{12}	6e−03	1.33e−01	5.98e−02	7.30e−03	**6.65e−03**	1.41e−02	8.40e−02
S_{14}	5e−03	2.71e−01	3.99e−03	1.18e−03	1.70e−02	**9.30e−04**	1.85e−01
S_{24}	3e−03	3.07e−01	2.29e−02	3.0215e−02	8.44e−03	**4.76e−03**	1.41e+01
S_{45}	1e−05	3.15e+00	1.43e−01	2.96e−01	**9.87e−02**	2.62e−01	2.60e+01

Table 10. Relative error for estimation of sensitivity indices of input parameters using various QMC approaches ($N \approx 2^{24}$).

EQ	RV	CRU	1PT	1OD	1EXPT	1EXOD	FIBO
S_1	4e−01	2.43e−03	1.03e−05	1.16e−05	**3.11e−06**	4.95e−06	1.33e−03
S_2	3e−01	1.86e−03	2.41e−05	1.55e−05	**9.76e−06**	1.44e−05	6.09e−04
S_3	5e−02	3.93e−03	6.97e−04	4.51e−04	**8.87e−06**	3.13e−05	3.47e−03
S_4	3e−01	1.73e−03	6.66e−05	3.95e−05	2.19e−06	**1.21e−06**	6.25e−04
S_5	4e−07	6.30e+01	1.01e+00	2.29e+00	2.12e+00	**2.09e−01**	8.95e+00
S_6	2e−02	1.39e−02	4.33e−05	**7.91e−06**	3.84e−05	1.99e−05	2.88e−03
S_1^{tot}	4e−01	1.97e−03	1.19e−05	1.38e−05	**1.26e−06**	1.5792e−05	3.14e−04
S_2^{tot}	3e−01	2.60e−03	3.99e−06	9.30e−06	6.92e−06	**8.77e−07**	2.27e−04
S_3^{tot}	5e−02	2.07e−03	3.19e−04	4.99e−05	4.94e−05	**2.51e−05**	3.10e−03
S_4^{tot}	3e−01	1.75e−03	1.13e−04	1.39e−05	1.09e−05	**2.44e−06**	4.26e−04
S_5^{tot}	2e−04	4.30e−02	3.72e−03	6.18e−03	**2.45e−03**	5.25e−03	2.20e−02
S_6^{tot}	2e−02	5.34e−03	3.91e−05	1.13e−04	**7.30e−06**	8.68e−05	1.99e−02
S_{12}	6e−03	2.49e−02	**6.28e−05**	7.74e−04	4.39e−04	2.83e−04	3.91e−02
S_{14}	5e−03	1.21e−02	4.02e−05	7.56e−04	**2.55e−05**	4.81e−04	5.45e−03
S_{24}	3e−03	2.66e−02	1.47e−03	6.41e−04	4.04e−04	**1.48e−04**	9.25e−03
S_{45}	1e−05	1.52e−01	1.23e−02	3.96e−02	1.44e−03	**4.94e−04**	5.84e−02

4   Conclusion

In this work we propose advanced stochastic methods based on lattice sequence with improved generating vectors. The proposed methods show significant advantages over the standard Monte Carlo method and the well-known Fibonacci lattice sequence. The results obtained here will be important for environmental safety and human health.

Acknowledgements. Slavi Georgiev is supported by the Bulgarian National Science Fund (BNSF) under Project KP-06-M62/1 "Numerical deterministic, stochastic, machine and deep learning methods with applications in computational, quantitative, algorithmic finance, biomathematics, ecology and algebra" from 2022. Venelin Todorov is supported by BNSF under Project KP-06-N52/5 "Efficient methods for modeling, optimization and decision making" and BNSF under Project KP-06-N62/6 "Machine learning through physics-informed neural networks". The work is also supported by BNSF under Bilateral Project KP-06-Russia/17 "New Highly Efficient Stochastic Simulation Methods and Applications".

References

1. Baldeaux, J., Dick, J., Leobacher, G., Nuyens, D., Pillichshammer, F.: Efficient calculation of the worst-case error and (fast) component-by-component construction of higher order polynomial lattice rules. Numer. Algor. **59**, 403–431 (2012)
2. Dimov, I.T.: Monte Carlo Methods for Applied Scientists, p. 291. World Scientific, Singapore (2008)
3. Kuo, F.Y., Nuyens, D.: Application of quasi-Monte Carlo methods to elliptic PDEs with random diffusion coefficients - a survey of analysis and implementation. Found. Comput. Math. **16**, 1631–1696 (2016)
4. Saltelli, A., Tarantola, S., Campolongo, F., Ratto, M.: Sensitivity Analysis in Practice: A Guide to Assessing Scientific Models. Halsted Press, New York (2004)
5. Sobol, I.M.: Numerical Methods Monte Carlo. Nauka, Moscow (1973)
6. Sobol, I.M.: Sensitivity estimates for nonlinear mathematical models. Math. Modeling Comput. Experiment **1**(4), 407–414 (1993)
7. Wang, Y., Hickernell, F.J.: An historical overview of lattice point sets. In: Fang, K.T., Niederreiter, H., Hickernell, F.J. (eds.) Monte Carlo and Quasi-Monte Carlo Methods 2000, pp. 158–167. Springer, Heidelberg (2002). https://doi.org/10.1007/978-3-642-56046-0_10
8. Zlatev, Z., Dimov, I.T., Georgiev, K.: Three-dimensional version of the Danish Eulerian model. Z. Angew. Math. Mech. **76**(S4), 473–476 (1996)
9. Zlatev, Z., Dimov, I.T.: Computational and Numerical Challenges in Environmental Modelling. Elsevier, Amsterdam (2006)

HPC and HPDA: Algorithms and Applications

Parallel Solution of the Schrödinger-Poisson Equation on GPUs

Johann Cervenka[(✉)] ⓘ, Robert Kosik ⓘ, and Felipe Ribeiro ⓘ

Institute for Microelectronics, TU Wien, Vienna, Austria
cervenka@iue.tuwien.ac.at
http://www.iue.tuwien.ac.at

Abstract. Quantum mechanical effects exhibited in carrier transport must often be accounted for in the development of future electronic devices. To achieve physically reasonable results, the transport equation of the quantum mechanical system and the electrical problem (Poisson equation) have to be solved self-consistently.

To calculate IV-characteristics the Newton method has to be applied on a coupled Schrödinger-Poisson system for each bias point, requiring the assembly of the Jacobian with respect to the unknowns. In a typical simulation several millions of Schrödinger-type equations need to be solved for the assembly and a parallelization of the procedure is essential.

Special care has to be taken because of the memory limitation of the GPU. To prevent a parallel storage of the system matrices, the discretization is carried out by a reformulation of the problem in terms of one-sided boundary conditions. An explicit scheme can be employed and no individual system matrices need to be assembled.

Traditional CPUs are utilized for reference. Benchmarks study the scalability of the approach when using up to several thousands of CUDA cores in parallel.

Keywords: Schrödinger equation · Poisson equation · self-consistent · GPU · CUDA

1 Introduction

In the study of solid state physics and electronics, the Schrödinger-Poisson equation is used to describe the behavior of electrons in semiconductor devices. It is defined by a system of two partial differential equations which have to be solved self-consistently, meaning that the solution for the wave function and the electric potential must be related in such a way that satisfies both the Schrödinger equation for the wave function and the Poisson equation for the electric potential.

Robert Kosik was funded by FWF Austrian Science Fund, project number P33151. Some of the computational results presented here have been achieved using the Vienna Scientific Cluster (VSC5).

© The Author(s), under exclusive license to Springer Nature Switzerland AG 2024
I. Lirkov and S. Margenov (Eds.): LSSC 2023, LNCS 13952, pp. 375–382, 2024.
https://doi.org/10.1007/978-3-031-56208-2_38

The method iteratively updates the density function and electric potential until a converged solution is obtained. Solving this system is computationally expensive.

The problem structure shows a variety of similar independent calculations which is feasible for parallelization and transferable to GPUs. For the implementation we rely on the CUDA platform. Simulations are carried out on general purpose GPUs.

2   Simulation Setup

In the described method a one-dimensional semiconductor device with one carrier type is modeled. The device behavior is defined by its implanted doping profiles at given operating conditions. On the left and right sides, a bias voltage is applied to the semiconductor device.

Caused by the miniaturization of the device, quantum effects have to be accounted for, which may be described by the Schrödinger equation. In the stationary case, the time-independent Schrödinger equation describes the wave functions of the electrons of different wave vectors in the semiconductor regions. Summing up the associated density distributions weighted by the based Fermi-Dirac distribution of particles [5], the distribution function of the carriers can be obtained.

The electrical system is described by the Poisson equation. The two equations are coupled via the carrier concentration and the electric potential. Both systems have to be solved self-consistently.

2.1   The Electrical System

The Poisson equation is the basis of the electrical system

$$\triangle U(x) = -\frac{q}{\varepsilon} \left(n(x) - N(x) \right) \tag{1}$$

Here a constant electric permittivity ε is assumed. $N(x)$ is the net-doping in the semiconductor, $n(x)$ is the density of electrons, $q = -e$ the electron charge and $U(x)$ the electrostatic potential. In the one-dimensional case, the Laplace operator is given by the second order spatial derivative $\mathrm{d}^2/\mathrm{d}x^2$. The boundary conditions are set at the contacts according to the bias voltage.

2.2   The Quantum System

In the stationary case a particle in the semiconductor can be described by the stationary Schrödinger equation

$$-\frac{\hbar^2}{2\,m} \triangle \psi(x) + (V(x) - E)\, \psi(x) = 0 \tag{2}$$

$\psi(x)$ describes the wave function of a particle with effective mass m, energy E and the imposed potential energy $V(x)$, which is related to the electrostatic potential $U(x)$ by

$$V(x) = q\,U(x). \tag{3}$$

This partial differential equation describes the wave function of each particle with wave number k in the system with defined energy $E(k) = \frac{(\hbar k)^2}{2m}$. Transparent boundary conditions are applied which are discussed later in Sect. 3.1.

The overall carrier densities in the semiconductor can be obtained by the sum of the carriers weighted by their statistics (Fermi-Dirac distribution) [5]

$$\tilde{n}(x) = \int\limits_{k} w(k)\,\psi(x,k)\,\psi^*(x,k)\,\mathrm{d}k \tag{4}$$

which closes the loop to the electrical system.

2.3 Coupling of the Equations

Even without a self-consistent coupling of the equations the solving procedure is a demanding process which is shown in Fig. 1. From an initial guess of the carrier distribution the full simulation process has to applied, which starts with the calculation of the electric potential, then solves the Schrödinger equations and finally by summation delivers a new carrier distribution.

The calculation of the Poisson equation requires only solving one sparse linear system (independent of k) which rank grows with the number of spatial points, whereas for solving the Schrödinger system one equation has to be solved for each value in k-space. Hence the computational costs for calculation of $\tilde{n}$ in Eq. 4 scale like $\mathcal{O}(N_x \times N_k)$.

Both the Poisson and the Schrödinger equation are linear partial differential equations. Nevertheless, self-consistent coupling results in a nonlinear partial differential equation system.

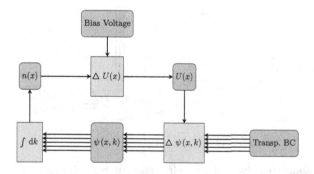

Fig. 1. Coupling of the Poisson and Schrödinger equations. Dark colored boxes denote the data flow, light colored boxes denote numerical calculations. (Color figure online)

2.4 Newton Method

Unfortunately, naive approaches with the use of already calculated values as the new input, do not converge. Therefore a multivariate Newton method is chosen. The linear approximation of the whole system is defined by

$$\mathbf{n}^{(i)} + \Delta\mathbf{n}^{(i)} = \tilde{\mathbf{n}}^{(i)} + \mathbf{J}\Delta\mathbf{n}^{(i)} \tag{5}$$

where $\tilde{\mathbf{n}}^{(i)}$ is a function of $\mathbf{n}^{(i)}$ (see Eq. 4). This defines the update of the carrier densities from one iteration $\mathbf{n}^{(i)}$ to the next one $\mathbf{n}^{(i+1)} = \mathbf{n}^{(i)} + \Delta\mathbf{n}^{(i)}$.

The Newton method introduces the Jacobian $\mathbf{J}$ of the output to the input carrier densities in the feedback loop.

The adapted process flow can bee seen in Fig. 2.

In the closed loop this procedure requires solving a linear system for which the calculation of the Jacobian $\mathbf{J}$ is needed. The linearization leads to modified Poisson equations and inhomogeneous Schrödinger-type equations. To calculate one column $\mathbf{J}_j$ of the Jacobian, one Poisson-type equation has to be solved followed by many subsequent Schrödinger-type simulations.

The computational costs for the numerical calculation of the Jacobian scale like $\mathcal{O}(N_x^2 \times N_k)$. For the assembly several millions of Schrödinger equations have to be solved routinely. This mechanism requires high demands of computational power.

For the calculation of IV-characteristics, subsequent runs of the Newton method for each bias update have to be performed.

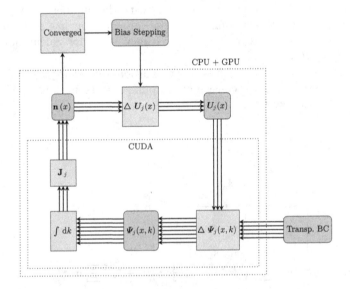

Fig. 2. Final simulation loop of the self-consistent Schrödinger-Poisson system. For each column j of the Jacobian one Poisson-type equation needs to be solved. In the inner CUDA loop for each column j a set of Schrödinger-type equations is solved.

3 The CUDA Implementation

The existing work flow from parallelization on the cluster has been adapted to be capable of GPU calculations using CUDA [2]. Solving the Poisson equation and the solution of the Jacobian system incurs only relatively small computational costs. For the first implementation only the most demanding part, the assembly of the Jacobian, has been ported to the GPU (see Fig. 2). In the developed method the columns of the Jacobian are calculated in parallel as they can be calculated independently of each other. Calculation of a column requires to solve a large set of Schrödinger-type equations, one for each k-value.

3.1 Memory Limitations

Usually GPUs are more limited in memory than CPU systems. To lower the amount of Schrödinger equation data which has to be stored in parallel during the calculation, a numerical method has been chosen which avoids the storage of the system matrices.

For our application an incoming plane wave is assumed in the left boundary which is partially reflected [4]. An outgoing plane wave is assumed as a solution at the right boundary

$$\tilde{\psi}'_{\text{out}} = ik_{\text{out}}\tilde{\psi}_{\text{out}} \qquad (6)$$

which can be fulfilled while simultaneously imposing $\tilde{\psi}_{\text{out}} = 1$. This Schrödinger equation for $\tilde{\psi}$ with one-sided boundary conditions can be solved by an explicit scheme which delivers the solution in one run. A solution ψ fulfilling the transparent boundary condition also at the inflow side

$$\psi'_{\text{in}} + ik_{\text{in}}\psi_{\text{in}} = 2ik_{\text{in}} \qquad (7)$$

can then be calculated by a rescaling of $\tilde{\psi}$.

3.2 Interface Between CPU and GPU

GPU programming requires the provision of input data to the GPU and also reverse, output data back to the CPU. The memory areas for the GPUs have to be reserved by CPU code and provided with the data on the CPU.

Special care has to be taken on the memory reservations of the GPU. Each allocation on the device is highly expensive in terms of computation times. As many device memory allocations as possible have been pulled outside of the inner device loops [1].

Already in the sequential algorithm the Eigen library had been chosen to handle the input data as matrices. The Library shows a convenient C++ interface to handle the input vectors and their operations, also conserving the performance [3]. Fortunately, the Eigen library is also capable of CUDA code for the GPU calculations.

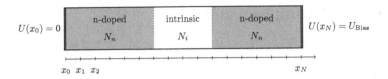

Fig. 3. Simulated one-dimensional semiconductor structure.

It works excellent in the device code areas of the CUDA code. Restrictions have to be taken on the interfaces between CPU code and GPU code. High performance device memory usually cannot be accessed by CPUs and vice versa. Here only limited C-style interfaces are available to allocate memory and copy memory to or from the device. Fortunately, the Eigen library shows the ability to provide user allocated arrays to their data types. This allows to include memory, allocated on the GPUs, within the Eigen data structures.

4 Simulation Results

The applicability of the method was evaluated by self-consistently solving the carrier and potential distributions of an nin-structure at different bias points.

The one-dimensional simulated structure, which is shown in Fig. 3, consists of a silicon block which is highly n-doped at the left and right regions connected by an intrinsically doped area in between. At the left and right boundaries a varying bias voltage is applied. For the carriers a Fermi-Dirac distribution at room temperature can be assumed.

The device lengths are 60 nm for each n block and 10 nm for the intrinsic region. The carrier concentrations are chosen as $N_n = 2 \times 10^{18}$cm^{-3} for the n-doped area and $N_i = 0.2 \times 10^{18}$cm^{-3} for the intrinsic area. To achieve accurate results a resolution of 1000 spatial points and a number of 20000 k-values in the energy range from 0 to 0.4 eV are chosen. In Fig. 4 several distributions of the wave functions are shown.

5 Benchmarks

In our benchmarks we study the scalability of the approach when using up to several thousands of CUDA cores in parallel. We use double precision floating point arithmetic. The calculations are applied on NVIDIA RTX 3070, GTX 1080 Ti, Tesla T4, and A100 with 5888, 3584, 2560, and 6912 CUDA cores, respectively. For a comparison the CPU version has also been benchmarked using a single i7 core.

In our first benchmark we calculated N columns of the Jacobian matrix when using N CUDA cores, i.e., each core calculates exactly one column of the Jacobian. The speedup for the parallel calculation is depicted in the left Fig. 5. Here the speedup has been scaled with the runtime on one i7 core. It is calculated as $\frac{N \cdot t_{i7}(1)}{t(N)}$, where $t(N)$ is the run time for N parallel jobs.

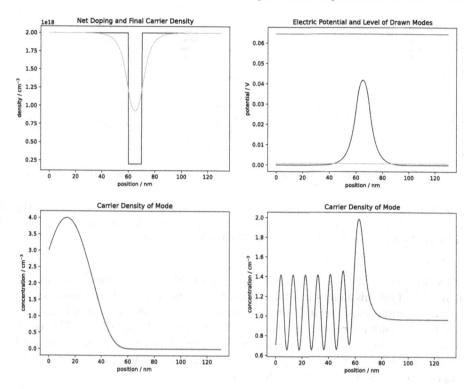

Fig. 4. Result of the simulated nin-structure. The net-doping, carrier concentrations and resulting potential distribution are plotted for zero bias. Additionally two separate modes are shown whose energy levels are plotted with the potential.

A single thread using one CUDA core on the RTX 3070 is about 60 times slower than the same calculation using a single thread on an i7 CPU, which appears relatively slow. However, this still implies that for our application the 5888 CUDA cores of the RTX 3070 can achieve a parallel performance corresponding to approximately 20 i7 cores. The A100 shows double the speedup (factor 40).

For our second benchmark we took a fixed workload and distributed it to an increasing number N of CUDA cores in parallel. This scenario corresponds to the full calculation of the Jacobian. The speedup relative to the A100 is calculated as the ratio of the execution times $\frac{t_{A100}(1)}{t(N)}$ and is shown in Fig. 5 right.

The Tesla T4 which is a general purpose GPU shows similar performance to the 1080 Ti. The newer generation A100 and RTX 3070 show a significantly better performance due to nearly twice the number of CUDA cores and can achieve a maximum speedup of a factor 800 and 400, respectively.

Both benchmarks show a similar scaling behavior, as should be expected. Compared with the RTX 3070 the A100 achieves double the speedup, but as it is much more expensive it is not cost effective for this type of simulations. The

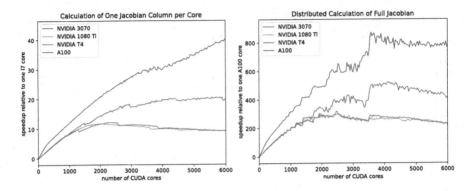

Fig. 5. In the left picture a comparison of the speedup for calculating N columns of the Jacobian using N CUDA cores is shown. In the right picture the speedup of a problem with fixed total size when distributed to N CUDA cores is shown.

RTX 3070 is an affordable consumer graphics card and for an end-user with a limited budget this makes it a fair option to be run on a single workstation.

In our application the individual threads are completely separated from each other and do not share any memory except for input and output data. Yet, there appears to be an associated overhead increasing with the number of threads and we do not see perfect scalability. The causes for this overhead and measures to improve on it are a topic of on-going research.

6 Conclusion and Outlook

In this paper a work flow for handling a mixed calculation on CPUs and GPUs of a Schrödinger-Poisson System was demonstrated. The method has been validated by an example of an nin-structure.

Future investigations will be taken in also implementing the Poisson part on the GPUs. This will significantly reduce the amount of memory which has to be transferred from/to the GPUs.

References

1. CUDA - Memory Model. https://medium.com/analytics-vidhya/cuda-memory-model-823f02cef0bf. Accessed 20 May 2023
2. CUDA C++ Programming Guide - NVIDIA Documentation Center. https://docs.nvidia.com/cuda/cuda-c-programming-guide/. Accessed 20 May 2023
3. Eigen: A C++ Linear Algebra Library. https://eigen.tuxfamily.org. Accessed 20 May 2023
4. Lent, C.S., Kirkner, D.J.: The quantum transmitting boundary method. J. Appl. Phys. **67**(10), 6353–6359 (1990). https://doi.org/10.1063/1.345156
5. Lundstrom, M.: Fundamentals of Carrier Transport. Cambridge University Press, Cambridge (2000). https://doi.org/10.1017/CBO9780511618611. Accessed 20 May 2023

Anastylosis of Frescos – A Web Service in an HPC Environment

Dimo Dimov$^{(\boxtimes)}$, Todor Gurov, Sofiya Ivanovska, and Svetlozar Yordanov

Institute of Information and Communication Technologies (IICT) at Bulgarian
Academy of Sciences (BAS), Sofia, Bulgaria
dimo.dimov@iict.bas.bg, {gurov,sofia,svetlozar}@parallel.bas.bg

Abstract. The parallelization in a High Performance Computing
(HPC) environment of a method, algorithm and software called RINC-
CAS (Rotation-Invariant NCC for 2D Color Matching of Arbitrary
Shaped Fragments of a Fresco) is presented. RINCCAS is designed for
virtual restoration of frescoes from their ruins, a known problem of
national and world heritage conservation. The method was developed
at IICT-BAS for participation in the DAFNE computer competition,
Italy, 2019. Prepared in the MATLAB environment, RINCCAS uses a
classic NCC approach for positioning of the fragments (rectangular coordinates and accidental rotation angle). It extends NCC to color input
images (frescoes and fragments) and to arbitrary shapes of the fragments. RINCCAS is a cubic complexity method. To reduce the execution time, we use parallelization through independent subtasks whose
optimal distribution in consecutive sessions we call *sequential concatenation* within one or more HPC nodes. A description and comparative
analysis of the experiments with Avitohol-HPC at IICT-BAS are presented. Upper limits are set for the amount of input data relative to the
amount of available RAM, not meeting which significantly slows performance due to system memory swapping. An improvement called *parallel
concatenation* is proposed to significantly alleviate memory constraints
using the same theoretical formulation. In conclusion, an invitation is
extended to the recently opened website of the RINCCAS-HPC service
for Anastylosis of Frescos' enthusiasts and specialists.

Keywords: Pattern Recognition · Normalized Cross Correlation
(NCC) · Digital Anastylosis of Frescos from Ruins · Cultural Heritage
Preservation on HPC · Optimal HPC resource planning

1 Introduction

Anastylosis of Frescos, or virtual restoration of frescoes from their ruins, is the
application area of the RINCCAS method which was initially developed at IICT-
BAS for participation in DAFNE (Digital Anastylosis of Frescos challeNgE)
computer competition, June-July 2019, [1–3].

RINCCAS-HPC is a service based on the RINCACAS method [1] and is
developed to run in a HPC environment.

I. Lirkov and S. Margenov (Eds.): LSSC 2023, LNCS 13952, pp. 383–393, 2024.
https://doi.org/10.1007/978-3-031-56208-2_39

The input fresco is expected to be represented by a painting or sketch relatively well preserved over time. The fragments collected from the ruins, of random shapes and sizes, possibly mixed with other, spurious fragments, are also expected to be given by their images. This problem statement is well known in the practice of World Heritage conservation, [2–4].

RINCCAS uses the Normalized Cross Correlation (NCC), analyzed by Lewis, [5], and implemented by MATLAB (R2010b, or higher) in a function called *normxcorr2*. Modifying it to speed up, RINCCAS uses it to check enough digital situations (positions and rotations at RGB color cannels) of each fragment compared to the fresco. The NCC approach is extended to arbitrary curvilinear 2D shapes of fragments – each fragment being approximated by one or more non-overlapping so-called Maximal & Axes-Collinear Inner Rectangle(s) (MACIR).

RINCCAS mainly consists of two phases. Phase-1 positions the fragments with high accuracy on the given fresco, e.g. enough so that in Phase-2 the false fragments can be effectively recognized. Phase-1 is of cubic complexity while Phase-2 is much faster (linear complexity) and allows interactive expert optimization in false fragment detection, [1]. The high accuracy of the results (positioning and recognition) characteristic of RINCCAS requires significant computational resources, even for medium-sized tasks of the given type, cf. [2].

To reduce the heavy computational load of Phase-1, a resizing procedure can be applied to the input images (fresco and fragments). For example, reducing them 2× gives about 8× acceleration. But to count on satisfactory positioning, the resolution of the given images must be high enough, which is a well-known trade-off between scaling and precision. Thus, the main goal of the RINCCAS-HPC was to accelerate Phase-1 using HPC, namely – Avitohol of IICT-BAS, [6].

The paper has the following structure: Sect. 2 presents recent related works; Sect. 3 addresses RINCCAS software parallelization, first describing the parallelization setup, and second – a method and algorithm for optimal scheduling the parallel computation in one or more HPC nodes. Section 4 comments on the experiments conducted, the results obtained, and the directions for future work as well. A brief conclusion, acknowledgements, and list of references follow.

2   Related Works

Out of about a dozen candidates and/or actual participants in the DAFNE competition, only 5 computer programs were rated high enough. Their authors were invited to describe their approaches in the DAFNE Special Edition of Pattern Recognition Letters [3]. Each of the articles [1,7–9] gives its analytical overview on the DAFNE problem until 2019. These reviews can be considered a recent state-of-the-art description, although they are specifically oriented to support of the respectively proposed method and algorithm. RINCCAS [1] builds on and extends the classic NCC [5], emphasizing its guarantees of high recovery accuracy, i.e. positioning fragments of arbitrary shapes on the fresco, albeit at the expense of speed. It is the precise positioning that ultimately allows the complete isolation of the false fragments from the initial set of fragments (ruins).

The other 3 approaches [7–9] are based on a modern method of *key points* (KP), the first and most famous of which is SIFT, [10]. So in [7] they focus on the FAST and BRISK versions, while in [9] – on SIFT and SURF. To determine the random rotation under sufficient textural/color similarity, in a next step, for each fragment, the optimal correspondence of its KP-set with some KP-subset of the fresco is sought. The latter is an (almost) exponential problem, which is why all three teams focus on improved heuristics using chromaticity by regions, several histograms, and then versions of the *k*-NN method, $k = 1$ (pairwise KPs, i.e. via 'biangles'), $k = 2$ (via triangles), etc. Isolation of the spurious fragments is carried out in the meantime (or finally) according to statistical approaches. Thus, the accuracy of the recovery largely depends on the density of the KPs in the different regions, the luck in finding KP matches, the competences of the human operator in determining the set of preliminary ones, etc.

As a best match approach (or an NCC one) it is worth to mention [4] used in the so called Mantegna's project, which RINCCAS was compared with, [1]. See also [11], where a FFT extension of NCC works with partially covered shapes (in our case the fresco and a fragment) and/or curvilinear shapes. And both lead to a higher complexity for the DAFNE problems.

3 Parallelization of the Software Structure of RINCCAS

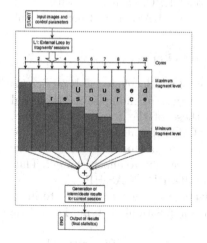

Fig. 1. RINCCAS-HPC flow chart, Phase-1 running on a node with 32 (=16+16) cores. For illustrative purposes, the core assignments are arranged in descending order by area of the loaded fragments.

Phase-1 of RINCCAS consists of 3 nested loops:

L1. External loop by fragments, their number is K.

L2. Intermediate loop of p_i per number of rotations of $2\pi/p_i$ [rad] of each current fragment $i, i = 1, \ldots, K$, where p_i is its perimeter (i.e. number of the enclosing pixels).

L3. Internal loop for the RGB color channels as well as for the number of non overlapping MACIRs, partially covering each fragment.

Figure 1 shows a general block diagram of the RINCCAS software, where the inner (nested) loops, L2 and L3 are represented in parallel columns in the body of L1.

Asymptotically, the computation time *per fragment* is of cubic order [1]:

$$T_{frag} \sim c(N + n)^2 n \log(N + n), \qquad (1)$$

where $N^2 = N_x N_y$ reflects the input fresco image area (in pixels), and $n^2 \approx 0.5 A_{frag}$ is about half the real area A of the fragment, since its curvilinear shape is approximated by a geometrical chain of its MACIRs, c is a constant depending on the computer used.

MACIR is an innovation, [1] allowing to position a curvilinear fragment using an NCC method via positioning a few MACIRs of the fragment instead of itself. If the 1st MACIR is not enough to cover half of the fragment, more may be added, but no more than 5 by default.

An additional advantage of MACIRs and chains of them is that during rotation they are almost independent in size, which contributes to speed-up.

The memory requirements M depend only quadratically on fresco and fragments' sizes:

$$M_{\underline{frag}} \sim c_1 + c_2(N + n)^2, \tag{2}$$

where c_1 and c_2 are computable constants, usually $c_1 \approx 0$.

The two main HPC resources we are interested in, see Eq. 1 and Eq. 2, are defined per fragment. They can be considered by analogy for MACIRs as well as for arbitrary groups of fragments and/or their MACIRs, since the virtual positioning of fragments, MACIRs and/or groups thereof are associative and commutative operations. The interdependence between them arises in the interpretation of positioning results, e.g. with overlapping fragments in RINCCAS Phase-2, and is not considered here.

Here we consider the optimal use of the provided computing resource for Phase-1, for example, up to 4 nodes of Avitohol HPC, each node with w=32 (=16+16) cores, with total random access memory (RAM) 64GB per node, [6].

3.1   Preferred RINCCAS Parallelization Scheme

Our goal is to provide a significant acceleration of the developed web service for Anastylosis, for example, up to 4800 × 3800 (color fresco) and up to 2–3 thousand fragments, each with sizes up to 5–10% of those of the fresco, rarely more. An additional constraint for now is that RINCCAS parallelization should not significantly change the basic Phase-1 calculation algorithm to avoid variations in positioning (beyond the popular $\pm 10^{-9}$ tolerance) and to affect Phase-2. Thus we save running new accuracy tests on the parallel RINCCAS-HPC.

The task parallelisms is the chosen parallel computing scheme, due to independence among individual tasks of a 'fragment positioning' type, i.e. K the number of tasks for a given fresco, see Fig. 1.

Even early experiments running RINCCAS on one only HPC node (32 cores) found that the trivial 'as it is' scheme results in a significant underloading of the node, up to 70–80%. This naturally led us to a task parallelisms approach (a task ⇔ a fragment), which is relatively simple to implement and does not require communications among individual tasks.

Thus, in order to ensure the maximum speed of the RINCCAS-HPC process, it is enough *to observe the following 2 principles*: **P1**) Full load (up to 100%) of each core of the node, and **P2**) Compliance with the given limit of RAM, at any moment of the operation of the given node.

The optimal load of the computing resource is considered here as a reasonable balance between the above two principles to achieve a minimum time to complete a full experiment (all the given tasks) of the 'fresco anastylosis' type.

A characteristic feature of *task parallelism* is that the computation depends largely on the order in which tasks are submitted to the cores. If the cores are 'fully loaded' (at 100%), then the computation in each core depends only on the tasks sent to it and no 'mutual help' between the cores is possible except in the intervals between sessions, i.e. of *unused resource* that we minimize. Only then there is time for some system-side optimization related to preventive swapping of memory. But then requests for new memory drop sharply, i.e. the need to swap memory is eliminated (at least according to our observations). Thus, our (ideal) goal is to ensure a load of up to 100% for all cores on a given node, which we can influence (manage) only with a *pre-planning of the experiment* by appropriate distribution the tasks (i.e. fragments) to the cores.

Two main cases of optimal design of experiments:

i) Optimal scheduling of a (single) CPU-node, in the case of Avitohol, up to 16 CPU-cores, each hyper-threaded, or 32 cores in total.

ii) Optimal scheduling of several (more than one) Avitohol CPU-nodes.

Due to the status quo of RINCCAS, we introduce one more principle here: **P3**) The order of fragments in operations is fixed at the beginning of Phase-1 in descending order by size of area. This is necessary to simplify the software there and also in Phase-2, where this allows the combinatorial problem of overlapping the positioned fragments to be solved in linear time.

This fragments arrangement allows us to treat the above two cases equally, as for a single *concatenated node*, with 32 concatenated cores, where the concatenations are among the respective cores in the nodes involved.

On the other hand, the fixed order of the fragments does not allow the optimization to directly use methods and results from the field of 'knapsack filling problem', [12]. More details on optimal planning a node can be found in [13].

Here we propose a heuristic method for preliminary optimal planning for a concatenated node, with V number of nodes, $V \geq 1$, each node with W number of cores, $W \geq 1$, observing the 3 principles formulated above: $P1$, $P2$ and $P3$.

3.2   Preliminary Optimal Planning for RINCCAS-HPC

Planning (or scheduling) is *preliminary* because the only effective control in task parallelisms is start-stop of computations. We describe here the proposed planning incrementally, for a single node, and generalize for a concatenated node.

To begin with, we consider that all the cores of the node are loaded each one with an elementary task (fragment), i.e. that the number of fragments K is much larger than the number of cores W, now $W = 32$ (otherwise choose less powerful HPC). The processing time of the fragments according to Eq. 1 is a monotonic function of their size, i.e. the largest fragment finishes last, and the corresponding time T_{max}, is assigned to the core w executing it, $1 \leq w \leq W$.

Thus, if they wait it for to finish, the remaining cores would be idle, see also Fig. 1. So, for the total unused HPC time, we have:

$$T_{\underline{unUs}} \sim W(N + n_{max})^2 n_{max} - \sum_{i=1}^{W}(N + n_i)^2 n_i \; , \; W = 32, \qquad (3)$$

where n_{max} is the size of the longest fragment. In order not to waste this resource, we apply a simple strategy – after the first finished core, we add a new fragment to the node, which fragment is naturally loaded into the only free core, i.e. the one just finished. We proceed similarly for the next finishing core, and so on, every time a core is freed, including the one we have temporarily called 'max'. Thus we exhaust all fragments and claim that the resulting natural schedule is the best one if ignoring the last value of $T_{\underline{unUs}}(k)$, i.e. for $k = K$. The selected fixed order P3 of the fragments is monotonic, therefore, the value $T_{\underline{unUs}}(K)$ is known in advance, and is usually quite small, because it is formed mainly from the last fragments of the base-list P3, see also Fig. 2.

The only problem here is the limit for available RAM, $\underline{Lim} \leq 64$ GB.

In parallel computing environment, RAM is not freed during execution in a core, even if the core continues with another task (fragment). This is managed by the host OS, in this case Linux, as a prevention against RAM partitioning.

The RAM is released at the end of the interval, we call it a *session*, when all tasks submitted to the cores in the given node have finished. Thus, under a 'session' we understand the processing of the row of fragments that are distributed as sub-rows on the cores of the given node, starting from some moment (the beginning of the session) when no RAM has yet been requested. The session ends when all the fragments to it are exhausted, but the release of the used RAM starts after the completion of the last, the longest time working core w, $T_{max}(w)$, $1 \leq w \leq W$ of the node. According to Eq. 2, the total required memory for the session is $M_{\underline{Ses}} = \sum_{i \in \underline{Ses}} M(i)$, which should not exceed $\underline{Lim}$.

If so far we have considered only an initial (full) session, and if the overall task is not very small, then according to the $\underline{Lim}$ constraint, we will most likely have to truncate it to some sequential fragment for which the inequality $M_{\underline{Ses}} \leq \underline{Lim}$ starts running again. In the diagrams of Fig. 2, this is fragment #257. Of course, to ensure the optimum, it is recommended the session to be shortened a little further to the left, where the last minimum for $T_{\underline{unUs}}$ is located (if there is one), Eq. 3. We declare the next session by starting the next fragment of the list, then resetting the main quantities: T_{max}, $T_{\underline{unUs}}$ and $M_{\underline{Ses}}$, and continue the calculation at the given node, until the next reason to stop of $\underline{Lim}$ type, etc. until the list of fragments is exhausted.

Thus, the modelling of the RICCASS-HPC Phase-1 for 'minimum execution time' on one concatenated node, aggregated over a set J of sessions, is given by:

$$\sum_{j \in J} T_{max}(\underline{Ses}(j)) = \min_{J}; \begin{cases} T(w, j) = \sum_i \{T(i)|F(i) \in \underline{Ses}(w, j) \subset \underline{Ses}(j) \subseteq F\}; \\ T_{max}(\underline{Ses}(j)) = \max_{w} \{T(w, j)|w \leq W, \underline{Ses}(j) \subseteq F\}; \end{cases}$$

$$(4)$$

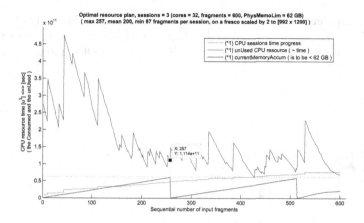

Fig. 2. Plots of the principal quantities: T_{max} (in progress, in green), T_{unUs}(initial status, in red) and M_{Ses} (the progress & limit, in blue) for the Pic-1 experiment (the first competition task from DAFNE). Obviously, the first session should be limited to fragment #257 $\gg$ 32 $= W$, hence a *regular* session, while the last one is shortened because of end. The 3 peaks (in cyan) at fragments #256, #513, and #600 represent the only non-zero values of T_{unUs} (at the ends of sessions).

$$\sum_{F(i)\in \underline{Ses}(j)} M(i) \leq \underline{Lim}, \ \forall Ses(j) \subseteq F; \ \bigcup_{j\in J} \underline{Ses}(j) \equiv F; \ \underline{Ses}(j) \bigcap_{j\neq k} \underline{Ses}(k) \equiv \emptyset.$$

$$(5)$$

where F is the set of fragments (understand $|F| = K$, to relate to Eq. 1–3); W is the number of cores in a concatenated node, $W = 32$ for Avitohol, $W = 8$ or less for a conventional PC. The input series of fragments is split into J, a set of sessions $j \in J$, $\underline{Ses}(j) \in F$. Each session is deployed on all W cores, i.e. $\underline{Ses}(j) = \cup_{w=1}^{W} \underline{Ses}(w, j)$, where $\underline{Ses}(w, j)$ are its parallel sub-sessions on each of the cores $w \leq W$. $T(i)$ is the time according to Eq. 1 for a processed fragment $F(i), 1 \leq i \leq |F|$. $T(w, j)$ is the total processing time of the fragments of the sub-session $\underline{Ses}(w, j)$, and $T_{max}(\underline{Ses}(j))$ is the maximum among them in the session $\underline{Ses}(j)$, which equals the real time for processing the session.

Sessions are non-intersecting rows of fragments and exhaust F. The goal is the time for all fragments, i.e. the total time by sessions to be minimal, where each session is limited according to the condition $\underline{Lim}$ for the required RAM.

Work on several HPC nodes, V, $1 \leq V \leq 4$, can be considered according to the model from Eq. 1–5 as a split of the initial node of *concatenated sessions* among the new V nodes, via one of the criteria: 1) a minimum of the total execution time of the all V nodes, 2) a minimum of the maximum time among them, which is similar to asking nodes to finish at the same time, 3) others.

Figure 3 illustrates the model diagrams (MD) for the main model quantities for Pic-1, a fresco from the DAFNE competition, also given in the RINCCAS-HPC website, see next Sect. 4. A concatenated node of fragments ordered by area is divided among $V = 4$ (physical) nodes by the 1st criterion, where the start-end fragments are illustrated via vertical dot-lines (in blue). The respective values, T_{max}, T_{unUs}, and M_{Ses}, cf. Fig. 2 are to have a similar behavior in the 4 nodes split, see Fig. 3, namely T_{max} in (TR), T_{unUs} in (DL), and M_{Ses} in (DR).

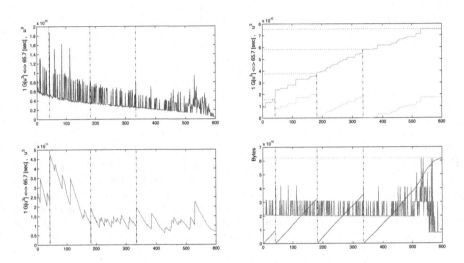

Fig. 3. (TopLeft) MD of processing time T_{frag} by fragments, see Eq. 1. Peaks correspond to additional MACIRs, 2nd to 5th, where available. Vertical dot-lines split T_{frag} among V=4 nodes, see also next (TR). **(TopRight)** MD of accumulated T_{max} processing time by fragments in a given node of concatenated sessions. Jumps correspond to the peaks in (TL). Vertical dot-lines split T_{max} progress while respective horizontal lines show T_{max} for each node. And the latter should be (almost) equal. **(DownLeft)** MD of the initial status of T_{unUs} (Eq. 3, CPU time) when processing, fragment by fragment in the given node. The best places to end the current session are the local minima here. This is the rule used to split the concatenation into 4 separate processes (HPC nodes), according to the *minimum total time* criterion. **(DownRight)** MD for the RAM need of each fragment in the concatenation (multiplied here by 100 for visibility). The jumps correspond to the peaks in (DL). The diagram is slightly tilted down to the right, which is natural but difficult to notice since of fresco dominates the fragments in area. Respective M_{Ses} for the 4 nodes and Lim have been also added.

4   Experiments and Discussion

The memory consumption $M_{Ses}(j) = \sum_{F(i)\in Ses(j)} M(i)$ is relatively balanced between sessions $Ses(j), j \in J$, and the differences are due to the differences in the participating fragments (in area and in number of MACIRs).

The main limitation of the considered *sequentially concatenated node* approach is that for the given Lim, there exists a border frame F_{lim} above which the accumulated work area $(N + n)^2$ for the fresco and its current MACIR, see Eq. 1–2, will cause the optimal planning to fail. Thus some sessions in the concatenation (or even all sessions) will always be underloaded, i.e. the number of tasks (fragments processed) will be less than the number of cores $W = 32$, which contradicts principle $P1$ of the optimization idea. Obviously, no increase in the number V of nodes, and/or improvements in fragment ordering, can significantly change these loads $M_{Lim}(j)$, $j \in J$ per session. The underloading of a node (by sessions), i.e. when

$$\exists (j \in J) \, \{ |\underline{Ses}(j)| < W \} \, , \tag{6}$$

is usually best expressed in the first session j, $\#j = 1$, although a set of shorter sessions also appear, as a result of the overall optimization.

In Table 1, the Pic-3 fresco is shrunk by a factor of $s = 1.5$ to *enter* the acceptable borders for underloading the nodes (Eq. 6), which is evident from the jump in execution time, compared to the original Pic-3. Of course, increasing the number of nodes leads to an (almost) proportional decrease in execution time (regardless of all the negatives of possible underloading with large frescoes).

Table 1. Comparison of times for the 3 competition tasks from DAFNE, processed on three different computer configurations.

#	Fresco (N^2)	Fragments number $|F|$	MACIR size (n^2), max, (mean)	configuration (W, $\underline{Lim}$)		
				(4, 8)	(32, 64)	(4×32, 64)
Pic-1	2400 × 1983	600	$349^2, (151^2)$	84 h	17 h	3–8 h
Pic-2	3400 × 2459	1188	$274^2, (130^2)$	284 h	57 h	16–21 h
Pic-3	4851 × 3816	2045	$336^2, (159^2)$	n/a	571 h	148–219 h
Pic-3 (scaled)	3234 × 2546	2045	$224^2, (106^2)$	n/a	101 h	27–41 h

System Environment: The operating environment of Avitohol-HPC is Linux Red Hat, with RT library v83 for MATLAB R2014a installed. RINCCAS-HPC executables are prepared in advance on a conventional PC in Windows-7 environment, compiled in Linux Mint environment on another PC and finally transferred to Avitohol for execution in batch mode. All necessary scripts for the implementation were written according to the instructions for working with Avitohol, [6].

A *web site* has been developed: https://rinccas.avitohol.acad.bg/.

Directions for Future Work: One promising approach to the nodal under-loading problem, see Eq. 4–6, is to split the fresco into 2 or more rectangular parts, then do anastylosis on each part separately. Thus the underloading problem weakens quadratically, i.e. up to 4 times, as well as the total time for anastylosis, but we concentrate primarily on underloading, Eq. 6. Of course, the individual parts have to overlap in border strips, each with a width equal to $1/2$ the size of the largest MACIR. This preserves the accuracy of the NCC positioning of MACIRs after the simple union (by maxima) of the NCC results by parts. Splitting up the larger fragments would have produced considerably less effect, due to the generally dominant area of the fresco. However, this can be necessary for reducing the width of the strip when splitting the fresco.

We can process each of these rectangular parts of the fresco according to the *sequentially concatenated node* approach already described; and if not, then we need to split the fresco into more parts. The definition of the strip for overlapping the neitbouring parts ensures independence of their treatments, i.e. parallel work on them according to the same task parallelism scheme. The latter gives reasons to call this extension a *parallel concatenated node* (of physical and/or other serially concatenated nodes). Obviously, this continuation can be described by a theoretical formulation, a simple extension of the one presented here.

5   Conclusions

The described development RINCCAS-HPC realizes an optimal parallelization of the RINCCAS method [1], more precisely, its Phase-1 for NCC virtual positioning of fragments, i.e. colored images supposed to be parts of a given image (a fresco). For relatively small dimensions of the fresco image, i.e. up to 3500×2500 [pix], successful experiments have been conducted on 1–4 nodes of HPC-Avitohol at IICT-BAS. Optimality consists in the maximum ($\sim 100\%$) load of all cores in the given HPC node, according to the principle *'the earliest released resource is loaded again immediately'*. To speed up the execution of a given anastylosis, it is suggested to use more nodes, considered here as *sequential concatenated node* which has its restrictions about the input data volume. For optimal use of the provided HPC resource, for larger fresco images, it is suggested that the available HPC nodes be organized into a *parallel concatenated node* (of physical and/or other serially concatenated nodes). The latter is still under development.

Through the announced site of RINCCAS-HPC, the full RINCCAS service is now available, i.e. the result of HPC-Avitohol can be refined through Phase-2 of the RINCCAS method at the customer's request to purify any spurious fragments that may have entered the input. Since it is much faster, Phase-2 is usually run on the same PC that supports the site and can be successfully used for interactive expert optimization when detecting false (spurious) fragments.

Acknowledgements. This work is supported by the project NI4OS-Europe, National Initiatives for Open Science in Europe, H2020, contract no. 857645, as well as by the National Geoinformation Center (part of National Roadmap of RIs) under grant D01-164/28.07.2022.

References

1. Dimov, D.T.: Rotation-invariant NCC for 2D color matching of arbitrary shaped fragments of a fresco. Pattern Recognit. Lett. **138**, 431–438 (2020)
2. Cantoni, V., Foresti, G.L., Sebe, N.: Editorial for the special issue on the DAFNE project (Digital Anastylosis of Frescoes challeNgE). Pattern Recognit. Lett. **147**, 179–180 (2021)
3. Cantoni, V., Lombardi, L., Giovanna, M., Alessandro, S., Setti, A.: Digital Anastylosis of Frescoes challeNgE (DAFNE). Firenze Univ. Press, 2019, pp. 24–31 (2019)

4. Fornasier, M., Toniolo, D.: Fast, robust and efficient 2D pattern recognition for re-assembling fragmented images. Pattern Recognit. **38**, 2074–2087 (2005)
5. Lewis, J.P.: Fast normalized cross-correlation. Indust. Light & Magic, p. 7 (1995)
6. Supercomputer System Avitohol at IICT-BAS. http://www.iict.bas.bg/avitohol
7. Lermé, N., Hégarat-Mascle, S.L., Zhang, B., Aldea, E.: Fast and efficient reconstruction of digitized frescoes. Pattern Recognit. Lett. **138**, 417–423 (2020)
8. Barra, P., Barra, S., Nappi, M., Narducci, F.: SAFFO: a SIFT based approach for digital anastylosis for fresco reconstruction. Pattern Recognit. Lett. **138**, 123–129 (2020)
9. Teixeira, T.D., Andrade, M.L., Luz, M.R.: Reconstruction of frescoes by sequential layers of feature extraction. Pattern Recognit. Lett. **147**, 172–178 (2021)
10. Lowe, D.G.: Distinctive image features from scale-invariant keypoints. Int. J. Comput. Vis. **60**, 91–110 (2004)
11. Padfield, D.R.: Masked object registration in the Fourier domain. IEEE Trans. Image Process. **21**, 2706–2718 (2012)
12. Martello, S., Pisinger, D., Toth, P.: Dynamic programming and strong bounds for the 0–1 knapsack problem. Manag. Sci. **45**, 414–424 (1999)
13. Dimov, D., Ivanovska, S., Alexiev, K., Hristov, A.: Anastylosis of Frescos – an optimised implementation on Avitohol HPC, BG-SIAM, Sofia, 2022 (to appear)

A Resolvent Quasi-Monte Carlo Method for Estimating the Minimum Eigenvalues Using the Error Balancing

Silvi-Maria Gurova$^{(\boxtimes)}$ [ID], Emanouil Atanassov [ID], and Aneta Karaivanova [ID]

Institute of Information and Communication Technologies, Bulgarian Academy of Sciences, Acad. G. Bonchev Str, bl. 2, Sofia, Bulgaria
{smgurova,emanouil,anet}@parallel.bas.bg

Abstract. There are iterative Monte Carlo (MC) methods that can be used for estimating the extreme eigenvalues of large dimensional matrices. The Power MC method allows for finding the approximate maximum eigenvalue of the considered matrix. In the case when we need to estimate the minimum eigenvalue, it is recommended to use the Resolvent MC method. The recent developments of quasi-random sequence generators and their successful application for solving large-scale problems motivate us to investigate the quasi-Monte Carlo (QMC) approaches for solving eigenvalue problems. In this work, we propose a Resolvent QMC algorithm to estimate the minimum eigenvalues of large-scale dimension symmetric matrices. To generate the quasi-random sequences we use BRODA's Sobol Randomized Sequence Generator (RSG). Numerical experiments were done to investigate the balance between both errors - systematic and stochastic errors, which depend on the power of the resolvent matrix, the parameter controlling the convergence in the iteration process of the Resolvent MC/QMC method, and the number of realizations of the MC/QMC estimator. Numerical results show good scalability in the case of using Sobol sequences on GPU accelerators.

Keywords: Quasi-Monte Carlo · eigenvalue · resolvent matrix

1 Introduction

Many real-world problems described in scientific fields such as image and signal processing, control systems, machine learning, and quantum mechanics require

The work of the first author has been partially supported by the National Geoinformation Center (part of National Roadmap of RIs) under grants No. D01-164/28.07.2022 and has also been accomplished with partial support by Grant No. BG05M2OP001-1.001-0003, financed by the Science and Education for Smart Growth Operational Program (2014–2020) and co-financed by the European Union through the European Structural and Investment funds. The work of the other authors was accomplished with the financial support of the MES by Grant No D01-168/28.07.2022 for providing access to e-infrastructure of the NCHDC - part of the Bulgarian National Roadmap on RIs and as well as by grant No 30 from CAF America.

I. Lirkov and S. Margenov (Eds.): LSSC 2023, LNCS 13952, pp. 394–403, 2024.
https://doi.org/10.1007/978-3-031-56208-2_40

estimating the largest and smallest eigenvalues of given matrices. For example, the energy levels of quantum systems are determined by the eigenvalues of the Hamiltonian operator. Knowledge of the largest and smallest eigenvalues can provide insights into the ground state energy and excited states of the system [1,2]. Some of the popular deterministic methods that are used to estimate eigenvalues are the Power method, Rayleigh quotient iteration, Resolvent Power method, and Lanczos algorithm [3,4]. In the last decades, Monte Carlo (MC) and Quasi-Monte Carlo (QMC) iterative algorithms for finding the extreme eigenvalues based on the Power method and on the Resolvent Power method were developed by scientists who were working in this area [5,6,8,9]. We note that QMC methods use low discrepancy sequences, such as Sobol or Halton sequences, which offer more uniform and deterministic coverage of the sample space instead of using random samples [12,13,16]. Compared with deterministic methods, the MC and QMC methods possess some advantages such as flexibility and generality, applicability to large matrices, accessibility, simplicity, and easy parallelization of the algorithm [7,10,11]. The convergence of QMC methods is typically $\mathcal{O}((logN)^k N^{-1})$ where k is the dimension of the problem, while the MC methods usually have a convergence $\mathcal{O}(N^{-\frac{1}{2}})$.

In this paper, we propose a Resolvent QMC method for estimating the minimum eigenvalues of large-scale dimension symmetric matrices and compare its complexity with Resolvent MC one. The error of the Resolvent QMC method depends on the power of the resolvent matrix, the number of iterations in the Markov chain, and the number of realization N. Numerical experiments were performed to investigate errors' balancing which permits us to achieve the desired accuracy with a smaller number of computations. To generate sequences with low discrepancy we use BRODA's Sobol RSG [17,18].

2   Formulation of the Problem

Let $A = \{a_{ij}\}_{i,j=1}^n \in \mathbb{R}^{n \times n}$ be a given (non-singular) symmetric matrix. Consider the following eigenvalue problem: finding $\lambda(A)$ such that:

$$A\boldsymbol{x} = \lambda(A)\boldsymbol{x}, \quad \boldsymbol{x} \in \mathbb{R}^n,$$

$$0 < \lambda_{min} = \lambda_n < \lambda_{n-1} \leq \lambda_{n-2} \leq \ldots \leq \lambda_2 < \lambda_1 = \lambda_{max}.$$

We denote a matrix polynomial of degree k by the equation:

$$p_k(A) = \sum_{i=0}^{k} c_i . A^i, \quad c_i \in \mathbb{R}, \quad k = 1,2,3,\ldots\ldots$$

The well-known Power method [3,4] gives an estimate for the dominant eigenvalue λ_1 by the following formula:

$$\lambda_1 \approx \lim_{k \to \infty} \frac{(\boldsymbol{h}, p_k(A)\boldsymbol{f})}{(\boldsymbol{h}, p_{k-1}(A)\boldsymbol{f})}, \tag{1}$$

where $h = \{h_i\}_{i=1}^n$ and $f = \{f_i\}_{i=1}^n$ two arbitrary vectors in $\mathbb{R}^n$. In the case when $p_k(A) = A^k$, the formula (1) turns into the Rayleigh quotient iteration approach for estimating the λ_1:

$$\lambda_1 \approx \lim_{k \to \infty} \frac{(h, A^k f)}{(h, A^{k-1} f)}. \tag{2}$$

Consider the case, when we have

$$p_k(A) = \sum_{i=0}^k q^i C_{i+m-1}^i A^i, \quad c_i \in \mathbb{R}, \tag{3}$$

where C_{i+m-1}^i are binomial coefficients and q is an acceleration parameter of the series (3). If $|q| \, ||A|| < 1$ and $k \to \infty$ we obtain the m-power of the resolvent matrix.

$$R_q^m = p_\infty = p(A) = \sum_{i=0}^\infty q^i C_{i+m-1}^i A^i = [I - qA]^{-m}, \tag{4}$$

where the $R_q = [I - qA]^{-1} \in \mathbb{R}^{n \times n}$ is the resolvent matrix of the equation $x = qAx + f$. Thus we obtain the Resolvent Power method to find the extreme eigenvalues, $\lambda_{\min}$ and $\lambda_{\max}$ [9,10], i.e.

$$\lambda = \frac{(h, Ap(A)f)}{(h, p(A)f)} = \frac{(h, AR_q^m f)}{(h, R_q^m f)}. \tag{5}$$

If $q > 0$,

$$\frac{(h, AR_q^m f)}{(h, R_q^m f)} \approx \lambda_{\max}.$$

If $q < 0$,

$$\frac{(h, AR_q^m f)}{(h, R_q^m f)} \approx \frac{1}{q}\left(1 - \frac{1}{\mu^{(m)}}\right) \approx \lambda_{\min},$$

where $\mu^{(m)}$ is the approximation to the dominant eigenvalue of R_q. Usually, the Monte Carlo methods are applied to estimate given functionals and we always obtain the approximate solution with certain stochastic errors. In our case, we must construct an MC method for estimating both functionals in the expression (5). Because the resolvent matrix R_q is presented as an infinite series (4) the corresponding method is called iterative Monte Carlo. Thus proposed Resolvent MC/QMC methods in the next two sections produce an approximation of the extreme eigenvalues in (5), whose accuracy we can control by the following parameters: (i) the number N of the realizations of the MC/QMC estimators; (ii) the number of jumps k in the Markov chain, which corresponds to the iterative process (4) with fixed k iterations; (iii) the power m of the resolvent matrix; and (iv) the fixed acceleration parameter q. The balance of the errors and consequently, the computational complexity, depends on all these parameters [9,10].

3   The Resolvent MC Method for Eigenvalue Problem

Before presenting the Resolvent MC method for estimating extreme eigenvalues by formula (5) we need to construct random variable $\theta^{(k)}$ for evaluating the scalar product $(\boldsymbol{h}, A^k \boldsymbol{f}) = \boldsymbol{h}^T A^k \boldsymbol{f}$. For this purpose, we consider the following Markov chain:

$$l_0 \to l_1 \to \cdots \to l_j \to \ldots \quad (1 \le l_j \le n) \tag{6}$$

with initial density vector $\boldsymbol{p} = \{p_\alpha\}_{\alpha=1}^n$, $Pr(l_0 = \alpha) = p_\alpha$ and the transition density matrix, $P = \{p_{\alpha\beta}\}_{\alpha,\beta=1}^n$, $Pr(l_j = \beta | l_{j-1} = \alpha) = p_{\alpha\beta}$, where

$$p_\alpha = \frac{|h_\alpha|}{\sum_{\alpha=1}^n |h_\alpha|} \quad \text{and} \quad p_{\alpha\beta} = \frac{|a_{\alpha\beta}|}{\sum_{\beta=1}^n |a_{\alpha\beta}|}.$$

Then the random variable $\theta^{(k)}$ has the form

$$\theta^{(k)} = \frac{h_{l_0}}{p_{l_0}} W_k f_{l_k}, \quad \text{where} \quad W_0 = 1, \quad W_j = W_{j-1} \frac{a_{l_{j-1} l_j}}{p_{l_{j-1} l_j}}, \quad j = \overline{1, k} \tag{7}$$

with mathematical expectation [10, 15]

$$E[\theta^{(k)}] = (\boldsymbol{h}, A^k \boldsymbol{f}) = \boldsymbol{h}^T A^k \boldsymbol{f} \approx \frac{1}{N} \sum_{s=1}^N (\theta^{(k)})_s. \tag{8}$$

Now using (7) we can construct random variables $\theta^{(1)}, ..., \theta^{(i)}, ...$ $i = 1, 2, ...$ It is easy to show that:

$$\sum_{i=0}^\infty q^i C_{i+m-1}^i E[\theta^{(i)}] = (h, [I - qA]^{-m} f), \quad m = 1, 2, ...$$

Taking into account the formula (5) we can estimate the extreme eigenvalues after k jumps in the Markov chain, i.e.

$$\lambda \approx \frac{\sum_{i=0}^k q^i C_{i+m-1}^i E[\theta^{(i+1)}]}{\sum_{i=0}^k q^i C_{i+m-1}^i E[\theta^{(i)}]}, \tag{9}$$

where $\theta^{(0)} = 1$ and the random variables $\theta^{(i)}$ are defined according (7). Let us rewrite (9) as follows:

$$\lambda \approx \frac{E[\theta^{(1)}] + q C_m^1 E[\theta^{(2)}] + q^2 C_{m+1}^2 E[\theta^{(3)}] + ... + q^k C_{k+m-1}^k E[\theta^{(k+1)}]}{1 + q C_m^1 E[\theta^{(1)}] + q^2 C_{m+1}^2 E[\theta^{(2)}] + ... + q^k C_{k+m-1}^k E[\theta^{(k)}]} \tag{10}$$

where

$$E[\theta^{(i)}] \approx \frac{1}{N} \sum_{s=1}^N (\theta^{(i)})_s \quad i = 1, 2, ..., k + 1. \tag{11}$$

We have two options for the value of the acceleration parameter q. In the case when $q > 0$, we find $\lambda = \lambda_1 = \lambda_{max}$ and in the case when $q < 0$ we estimate $\lambda_n = \lambda_{min}$ of the matrix A. We mention that when we compute the values of the random variables in (11) using pseudorandom generators to define jumps in the Markov chain (6) we have a Resolvent MC method. A Resolvent QMC method is obtained when we use low discrepancy sequences. The binomial coefficients in (10) can be computed in advance.

4   The Resolvent QMC Method and Error Balancing

The QMC methods use specially designed, deterministic sequences (called quasi-random or low discrepancy) in place of the (pseudo-) random numbers in MC methods. When these sequences are uniformly distributed, they can approximate integrals of functions similarly to MC methods, i.e., for a uniformly distributed sequence $\sigma = \{x_i\}_{i=0}^{\infty}$, and a continuous function g, defined over the k-dimensional unit cube,

$$\lim_{N \to \infty} \frac{1}{N} \sum_{i=0}^{N-1} g(x_i) = \int_{E^k} g(x)\, dx. \tag{12}$$

The sequence σ can be fully deterministic or randomized (the randomization process is usually called *scrambling*). In any case, the error in approximating such an integral depends on the uniformity of the distribution of the sequence and the smoothness of the function g under consideration. The Koksma-Hlawka inequality [19] bounds the error via the product of one particular measure of the uniformity of distribution of the sequence, the so-called star discrepancy, and one particular measure of the smoothness, the variation of the function in the sense of Hardy and Krause (see, e.g., [19] for the definitions). This estimate is valid also for functions that are not continuous, e.g., for functions that have a finite number of discontinuities along each dimension. This is exactly the case of an algorithm for computing scalar products of the type $(A^k f, h)$ (8). Note that the dimension, in this case, will be $k+1$. The variation of the function will depend on the particular choice of transition probabilities for the Markov chain, but in any case, it does not depend on the chosen sequence. Therefore it is beneficial to select a sequence with the lowest possible discrepancy. The Sobol sequences (see,e.g., [14]) are a class of uniformly distributed sequences that have a proven upper bound on their star-discrepancy in the order of $\mathcal{O}\left(N^{-1}(\log N)^k\right)$. Sequences with such upper bounds are called *low discrepancy sequences* and are the most popular uniformly distributed sequences. As the definition of the Sobol sequences includes computations in base 2, it is natural to use a power-of-two number of samples N, and this practice, which we also follow, has strong theoretical justification. When a randomization procedure is applied to the low discrepancy sequence, some additional theoretical and practical advantages can be obtained. For example, the estimate of the scalar product $(A^k f, h)$ will be guaranteed to be unbiased. On the other hand, even though the error will be stochastic, it will always be bounded by a constant multiplied by $\mathcal{O}\left(N^{-1}(\log N)^k\right)$. In our testing, we used a particular implementation of the Sobol sequence, provided by [18], which offers an optimized choice of the so-called direction numbers, application of full Owen scrambling, and use of GPU for faster generation. When replacing the use of pseudo-random numbers in the Resolvent MC method with the scrambled Sobol sequence, we essentially improve on the order of the error of the MC method for approximating the various scalar products of the type $(A^k f, h)$, while retaining the basic structure of the algorithm. This means that the stochastic error of each scalar product will be of the order of $\left(N^{-1}\log(N)^{k+1}\right)$, with a constant depending also on the matrix A and the power k. It is more difficult to manage the deterministic error that stems from the use of powers of the operator $(I - qA)^{-1}$, which are then approximated using the binomial coefficients. As the deterministic error of the power method is governed by the powers of the ratio $\frac{\mu_1}{\mu_2}$, where μ_1 and μ_2 are the two biggest eigenvalues ($\mu_1 > \mu_2$), we get that the deterministic error of the resolvent MC or QMC method will be governed by the powers of the ratio

$$\nu = \frac{1 - q\lambda_1}{1 - q\lambda_2} \tag{13}$$

where λ_1 and λ_2 will be either the smallest or the largest eigenvalues of A, depending on the acceleration parameter q.

In order for the algorithm to have good convergence, we need good separation between these two eigenvalues. It is also important to consider the tails in the binomial expansion, which definitely converge if the spectral radius of $|q| \, ||A|| < 1$. The binomial coefficients grow rather fast for a large m and easily become the dominant factor in the deterministic part of the error. Thus it is suggested to use a rather small number of transitions in the Markov chain, not more than 10. When one has some knowledge about the distribution of the eigenvalues of A and most importantly about λ_1 and λ_2, then one can define and solve numerically an optimization problem, providing the upper bound of the deterministic error of the algorithm. If the eigenvalues λ_1 and λ_2 are sufficiently separated and m is relatively small, then one can assume that the balancing problem is only between ν^m and $N^{-1}(\log N)^{m+1}$. In practice it has been frequently observed that the star-discrepancy frequently behaves as something like $N^{-1+\epsilon}$, we can expect that setting $N = \nu^{-1}$ will ensure a good balancing of the errors. If we do not have knowledge about ν, we can choose N based on the available computational budget and the complexity of the algorithm, which is roughly proportional to $N(m + 1)$. As we pointed out earlier, it is beneficial to always choose $N = 2^r$ for some r.

Table 1. Numerical results for $A \in \mathbb{R}^{500 \times 500}$ for different values N, k and m. The minimum eigenvalue is $\lambda_{\min} = \lambda_{500} = 0.152992$, $\lambda_{499} = 0.15452$ and the acceleration parameter $q = -0.157428$.

N	k	m	λ_{min} Resolvent MC	Absolute error Resolvent MC	λ_{min} Resolvent QMC	Absolute error Resolvent QMC
$512 * 2^0$	5	5	0.1585637	0.0055717	0.1561046	0.0031126
	5	10	0.1585163	0.0055243	0.1524884	**0.0005036**
	10	5	0.1556132	**0.0026212**	0.1560595	0.0030675
$512 * 2^1$	5	5	0.158395	0.005403	0.1596503	0.0066583
	5	10	0.1585518	0.0055598	0.1595946	0.0066026
	10	5	0.1580794	0.0050874	0.1596616	0.0066696
$512 * 2^5$	5	5	0.1572588	0.0042668	0.1591139	0.0061219
	5	10	0.1566524	0.0036604	0.1585388	0.0055468
	10	5	0.156875	0.003883	0.1591292	0.0061372
$512 * 2^6$	5	5	0.157129	0.004137	0.1589384	0.0059464
	5	10	0.1565796	0.0035876	0.1584563	0.0054643
	10	5	0.1570216	0.0040296	0.158961	0.005969

5   Numerical Results

The Resolvent QMC algorithm is realized by using formulas (10,11) for estimating the minimum eigenvalues of given symmetric matrices. The BRODA's Sobol RSG [18] is used to generate Sobol sequences with the needed dimension to construct the Markov chain. The results are compared with the corresponding Resolvent MC algorithm which uses a pseudo-random generator to construct the Markov chain. Both algorithms are written in the C++ programming language. As preprocessing, we use a subprogram to generate randomly the symmetric matrix A with a given dimension $n \times n$ in such a way that all eigenvalues are in the unit circle and satisfy inequalities:

$$0 < c < \lambda_n < \lambda_{n-1} < \dots < \lambda_2 < \lambda_1 < 1,$$

where the constant $c = 0.15$. The tests were conducted on an HPC cluster with NVIDIA Tesla V100 32 GB GPU cards located in the HPC center at the Institute of Information and Communication Technologies [20]. Numerical results for estimating of minimum eigenvalue for two testing matrices: (i) a matrix A with dimension $n = 500$ with $\lambda_{min} = \lambda_{500} = 0.152992$ and acceleration parameter $q = -0.157428$ and (ii) a matrix A with dimension $n = 1000$ with $\lambda_{min} = \lambda_{1000} = 0.151911$ and acceleration parameter $q = -0.2302458$ are shown in Table 1 and Table 2. The results are obtained for different numbers of realization N, a number of jumps k in the Markov chain, and m, the power of the Resolvent matrix.

Results confirm that when N increases, the MC/QMC solution does not improve and the absolute error does not decrease because this parameter only

Table 2. Numerical results for $A \in \mathbb{R}^{1000 \times 1000}$ for different values N, k and m. The minimum eigenvalue is $\lambda_{\min} = \lambda_{1000} = 0.151911$, $\lambda_{999} = 0.154138$ and the acceleration parameter $q = -0.230245$.

N	k	m	λ_{min} Resolvent MC	Absolute error Resolvent MC	λ_{min} Resolvent QMC	Absolute error Resolvent QMC
$512 * 2^0$	5	5	0.1556059	0.0045949	0.1705751	0.0186641
	5	10	0.168251	0.01634	0.1886959	0.0367849
	10	5	0.1593428	0.0074318	0.1701307	0.0182197
$512 * 2^1$	5	5	0.1504153	**0.0014957**	0.1533797	**0.0014687**
	5	10	0.1417805	0.0101305	0.1499974	0.0019136
	10	5	0.156739	0.004828	0.1534383	0.0015273
$512 * 2^5$	5	5	0.1550015	0.0030905	0.1573842	0.0054732
	5	10	0.1495483	0.0023627	0.1539357	0.0020247
	10	5	0.1559548	0.0040438	0.1575454	0.0056344
$512 * 2^6$	5	5	0.1557509	0.0038399	0.1576339	0.0057229
	5	10	0.1537008	0.0017898	0.1562911	0.0043801
	10	5	0.1562905	0.0043795	0.15764225	0.0057312

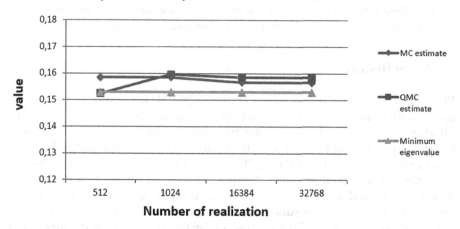

Fig. 1. The convergence to the exact minimum eigenvalue for matrix A using MC and QMC for $k = 5, m = 10$ for the different realizations N

affects the probable error. The absolute error when we use the QMC algorithm is the smallest when N = 512, k = 5, and m = 10 for results in Table 1 and when N = 1024, k = 5, and m = 5 for results in Table 2. These results are consistent with a theoretical estimate for the error balance, which depends on the parameters N, k, and m. Moreover, we see that the error is balanced when the number of realizations N is of the order of the dimension n of the matrix A.

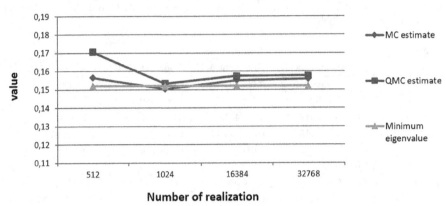

Fig. 2. The convergence to the exact minimum eigenvalue for matrix A using MC and QMC for $k = 5, m = 5$ for the different realizations N

In Fig. 1 and Fig. 2 the MC and QMC solutions are compared with the minimum eigenvalue for two matrices A in the case when we have a good balance of the errors. These figures show that When we increase the number of realizations N, an additional error appears caused by the binomial coefficients.

6   Conclusion

Finding the smallest eigenvalue by a Power MC/QMC method could be a tricky problem. One main option is to invert the original matrix and apply the Power MC/QMC method to the inverse matrix. The Resolvent MC/QMC overcomes inverting but the results show that it has got a limited applicability. We have developed an improved Resolvent MCM with a procedure for error balancing. Using the new algorithm, we can determine the smallest values of the algorithm's parameters m and k which leads to saving computational time. That means we are having a good approximation. This algorithm is developed for a hybrid implementation (GPU/CPU) that uses optimally the available resources thus obtaining fast estimation. The results from the previous section show that the algorithm implementation using Sobol's sequence is faster than using PRNs.

References

1. Rotter, I.: A non-Hermitian Hamilton operator and the physics of open quantum systems. J. Phys. A: Math. Theor. **42**(No15), 153001 (2009). https://doi.org/10. 1088/1751-8113/42/15/153001
2. Muminov, M.I., Rasulov, T.H.: Embedded eigenvalues of a Hamiltonian in bosonic fock space. Commun. Math. Anal. **17**(1), 1–22 (2014)

3. Golub, G.H., Van Loon, C.F.: Matrix Computations. The Johns Hopkins University Press, Baltimore (1996)
4. Isaacson, E., Keller, H.B.: Analysis of Numerical Methods. Dover Publications, Mineola; Wiley, New York (1996). ISBN 0-486-68029-0
5. Alexandrov, V., et al.: On the preconditioned quasi-Monte Carlo algorithm for matrix computations. In: Lirkov, I., Margenov, S., Waśniewski, J. (eds.) LSSC 2015. LNCS, vol. 9374, pp. 163–171. Springer, Cham (2015). https://doi.org/10.1007/978-3-319-26520-9_17
6. Dimov, I.T., et al.: Robustness and applicability of Markov chain Monte Carlo algorithms for eigenvalue problems. Appl. Math. Modelling **32**(8), 1511–1529 (2008). https://doi.org/10.1016/j.apm.2007.04.012
7. Mascagni, M., Karaivanova, A.: A Monte Carlo approach for finding more than one eigenpair. In: Dimov, I., Lirkov, I., Margenov, S., Zlatev, Z. (eds.) NMA 2002. LNCS, vol. 2542, pp. 123–131. Springer, Heidelberg (2003). https://doi.org/10.1007/3-540-36487-0_13
8. Mascagni, M., Karaivanova, A.: Matrix computations using quasirandom sequences. In: Vulkov, L., Yalamov, P., Waśniewski, J. (eds.) NAA 2000. LNCS, vol. 1988, pp. 552–559. Springer, Heidelberg (2001). https://doi.org/10.1007/3-540-45262-165
9. Dimov, I., Karaivanova, A.: A power method with monte Carlo iterations. In: Proceeding Recent Advances in Numerical Methods and Applications II, World Scientific, pp. 239–247 (1999). https://doi.org/10.1142/9789814291071_0022
10. Dimov, I.: Monte Carlo Methods For Applied Scientists. World Scientific, Singapore (2008). ISBN:9812779892
11. Dimov, I.T., Dimov, T.T., Gurov, T.V.: A new iterative Monte Carlo approach for inverse matrix problem. JCAM **92**(1), 15–36 (1998). https://doi.org/10.1016/S0377-0427(98)00043-0
12. Sobol, I.M., Shukhman, B.V.: Integration with quasirandom sequences: numerical experience. Int. J. Modern Phys. C **6**(2), 263–275 (1995)
13. Sobol, I.M., Asotsky, D., Kreinin, A., Kucherenko, S.: Construction and comparison of high-dimensional Sobol' generator. WILMOTT Mag. **64–79**, 2012 (2011)
14. Sobol, I.M.: Uniformly distributed sequences with an additional property of uniformity. USSR Comput. Math. Math. Phys. **16**(5), 236–242 (1976). https://doi.org/10.1016/0041-5553(76)90154-3
15. Gurova, S.-M., Karaivanova, A.: Monte Carlo method for estimating eigenvalues using error balancing. In: Lirkov, I., Margenov, S. (eds.) LSSC 2021. LNCS, vol. 13127, pp. 447–455. Springer, Cham (2022). https://doi.org/10.1007/978-3-030-97549-4_51
16. Faure, H., Lemieux, C.H.: Generalized Halton sequences in 2008: a comparative study. ACM Trans. Model. Comp. Simul. **19**(404), 1–31 (2009). https://doi.org/10.1145/1596519.1596520
17. Joe, S., Kuo, F.Y.: Constructing Sobol sequences with better two-dimensional projections. SIAM J. Sci. Comput. **30**(5), 2635–54 (2008)
18. BRODA's Sobol RSG. https://broda.co.uk/software.html. Accessed 15 Apr 2023
19. Basu, K., Owen, A.B.: Transformations and hardy-krause variation. SIAM J. Numer. Anal. **54**(3), 1946–1966 (2016). https://doi.org/10.1137/15M1052184
20. HPC center at IICT: https://ict.acad.bg/?page_id=1229. Accessed 15 Apr 2023

Influence of the Grid Resolutions on the Computer Simulated Air Quality Indices over the Territory of Bulgaria

Vladimir Ivanov$^{(\boxtimes)}$ ⓘ, Georgi Gadzhev ⓘ, Ivelina Georgieva, Kostadin Ganev, and Nikolay Miloshev

National Institute of Geophysics, Geodesy and Geography, Bulgarian Academy of Sciences, 1113 Acad. G. Bonchev str., bl. 3, Sofia, Bulgaria
vivanov@geophys.bas.bg

Abstract. The grid resolution would affect the pollution concentrations and thus on the Air Quality Indices (AQI) - a generalized assessment of the air quality impact on human health. Therefore, we made a numerical experiment for evaluation of the horizontal grid resolution impact on the simulated AQI over the territory of Bulgaria. We used a set of models used worldwide - WRF - the meteorological preprocessor, CMAQ - chemical transport model, SMOKE - emission model for performing computer simulations. The NCEP Global Analysis Data with a horizontal resolution of 1° × 1° are used as a background meteorological data used in the study. Using the "nesting" capabilities of the WRF and CMAQ models, a resolution of 9 km was achieved for the territory of Bulgaria, by sequentially solving the task in several consecutive, nested areas. The simulations were performed for three cases for grid and emission resolutions for the period 2008–2014, creating an ensemble large and comprehensive enough, reflects the most typical atmospheric conditions with their typical recurrence. The spatial/temporal distribution of the recurrence of the different AQI categories for Bulgaria are calculated. Comparing them for the above 3 cases makes it possible to evaluate the grid resolution impact.

Keywords: air quality index modelling · air pollution · modelling · emissions

1 Introduction

Air pollution is a health issue in many countries, including Bulgaria. It is one of the main factors for the quality of life, which provoke a wide variety of adverse effects on human health. According to the World Health Organization [1], in 2019, there were above 30% of premature deaths due to heart diseases, about 20% of deaths from pulmonary and respiratory diseases, and above 10% due to respiratory tract cancer. The EU legislation issue documents as the [2], and one

I. Lirkov and S. Margenov (Eds.): LSSC 2023, LNCS 13952, pp. 404–411, 2024.
https://doi.org/10.1007/978-3-031-56208-2_41

of the objectives is to make the public aware of the air quality information. Previous works deal with different aspects of air quality evaluation in Bulgaria [3–6]. It is known that the grid resolution would affect the pollution concentrations [7]. We deal with the Air Quality Index [8] based on a number of pollutants concentrations, used as a generalized assessment of the air quality impact on human health. The objective of our study is to evaluate the horizontal grid resolution impact on the simulated Air Quality Index over the territory of Bulgaria.

2   Methods

The numerical experiment was done to evaluate the horizontal grid resolution impact on the simulated AQI over the territory of Bulgaria. We use a set of models based on the US EPA Model-3 system used worldwide adapted and validated for Bulgaria – WRF v.3.4.1 - the meteorological preprocessor (UCAR/NCAR) [9,10], CMAQ v.4.6 - chemical transport model [11–13], and SMOKE - emission model for performing computer simulations [14]. The "NCEP Global Analysis Data" with a horizontal resolution of 1° x 1° are used as background meteorological data used in the study. Using the "nesting" capabilities of the WRF and CMAQ models, a resolution of 9 km was achieved for the territory of Bulgaria by sequentially solving the task in several consecutive, nested areas. The emission data for the territory of Bulgaria from the National Emission Inventory was used, and outside of the country, from the database of TNO [15]. Three cases are considered in this paper: Case 1: The computer simulations result from the domain with a horizontal resolution (both of the emission source description and the grid) of 27 km; Case 2: The computer simulations result from the domain with a horizontal resolution (both of the emission source description and the grid) of 9 km; Case 3: Hybrid case with the computer simulations performed with a grid resolution of 9 km, but with emissions like in the 27 × 27 km domain. The AQI estimates the effects of several air pollutants on human health (Table 1). The index is scaled into four ordered categories (Table 2). The AQI is determined from the pollutant with the highest category. For convenience, the results for the last two categories are combined in "High". The simulations were performed for all cases and for the period 2008–2014, creating an ensemble large and comprehensive enough to reflect the most typical atmospheric conditions with their typical recurrence. The spatial/temporal distribution of the annual recurrence of the different AQI categories for Bulgaria is calculated. Comparing them for the above 3 cases makes it possible to evaluate the grid resolution impact. All the simulations are performed day by day, the outputs were averaged over the whole period and ensemble, and so the "typical" annual evaluations were obtained. The calculations were implemented on the Supercomputer System "Avitohol" at the Institute of Information and Communication Technologies at the Bulgarian Academy of Sciences (IICT-BAS). The simulations for the selected domain were organized in different jobs [16,17].

Table 1. Air Quality Index categories definitions (hourly concentrations).

Band	Ozone(O_3) [$\mu g/m^3$]	Nitrogen dioxide(NO_2) [$\mu g/m^3$]	Sulfur dioxide(SO_2) [$\mu g/m^3$]	Carbon oxide(CO) [mg/m^3]	Particulate matter (PM) [$\mu g/m^3$]
Low	0–99	0–286	0–265	0–11.5	0–64
Moderate	100–179	287–572	266–531	11.6–17.3	65–96
High	180–359	573–763	532–1063	17.4–23.1	97–129
Very High	360 or more	764 or more	1064 or more	23.2 or more	130 or more

Table 2. Air Quality Index categories health effect.

Band	Health effect
Low	Effects are unlikely to be noticed even by individuals who know they are sensitive to air pollutants
Moderate	Mild effects, unlikely to require action, may be noticed amongst sensitive individuals
High	Significant effects may be noticed by sensitive individuals and action to avoid or reduce these effects may be needed (e.g. reducing exposure be spending less time in polluted areas outdoors). Asthmatics will find that their 'reliever' inhaler is likely to reverse the effects on the lung
Very High	The effects on sensitive individuals described for 'High' levels of pollution may worsen

3   Results

The first three rows of Fig. 1 show the geographic distributions of the considered combinations between grid resolution of the model simulations and emission data for the Low category recurrence of the AQI index. The first thing we see is that there are more locations with recurrences below 90% for the afternoon simulations. The morning C2–C1 differences in the annual recurrence of Low AQI in 07 GMT are between 0% and 10%, up to 20% in southwestern and central Bulgaria, and down to −5% for the Rila mountain. The morning ones of the C2–C3 are smaller. We suggest that the higher grid resolution is a factor that worsen the simulated air quality in the mountainous areas. In the afternoon, the C2–C1 difference is negative in the mountainous and adjacent areas and positive in the other ones, and the C2–C3 difference is negative only in Southwestern Bulgaria, to some extent on the Black Sea coast, and nearby a Thermal Power Plants (TPPs) in the south. There are Low category places very near the north border with recurrences between 80% and 90%, simulated only on C1 in 15 UTC. The C2–C1 difference at these is the highest one - above 10%. That feature is not recognizable in the other cases, which implies that it depends mainly on the resolution change of the grid conditions. According to another study [18,19], the dominant pollutant defining the Low AQI in Bulgaria is the O_3, and the NO_2 is in the second place. According

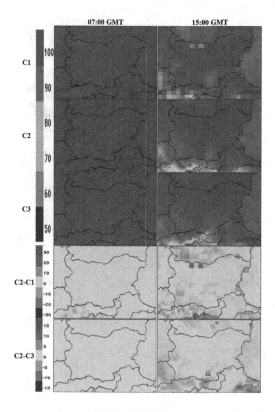

Fig. 1. Annual recurrence of the AQI [%] for all cases in Low category.

to the [19], higher concentrations of the NO_2 in the C2 case are simulated mainly over the big cities, and also that the main road network and the relative differences between C2 and C1, as well as C2 and C3 for the O_3 are very small, not directly connected to the sources of air pollution. Therefore, we suggest that the effects of the models' grid resolution and the emission grid change for the NO_2 and O_3 are the leading causes of the simulated differences between the three cases, more clearly manifested in the late afternoon.

The spatial distribution of the Moderate AQI recurrences in the three cases (Fig. 2) shows smaller frequencies. The morning simulations suggest higher recurrences in Southwestern Central Bulgaria for the C1 case. The recurrences of the C1 in the late afternoon are smaller, mainly in the north. The change of the emission grid and the process description resolution are evident from the more detailed results for the C2 in relation to the C1. C2–C1 in the morning and the late afternoon are opposite in the mountain and to the most northern and eastern parts. The C2–C3 are smaller in the mountain areas except in the southwest and not so patchy compared to the C2–C1. These patterns indicate the importance of the process description grid factor. The spatial distributions of the C2–C1 and the C2–C3 are opposed to the ones for the Low AQI, and that could be because the first two categories of the AQI largely determine the air quality in

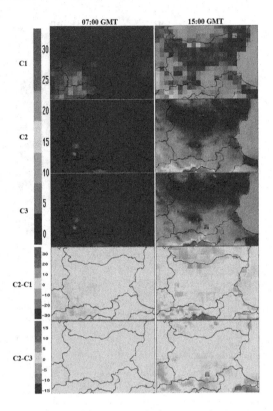

Fig. 2. Annual recurrence of the AQI [%] for all cases in Moderate category.

Bulgaria, and they complement each other. C2–C3 is connected with the emission resolution change mainly in Southwestern Bulgaria and around the TPPs in the south. The dominant pollutant of the Moderate AQI is the SO_2 [18,19]. The C2–C1 for that AQI category is bigger than the C2–C3 in the central mountain areas, where emission sources are missing. We suggest that the higher process description resolution intensifies the pollution transport from their sources to the other areas, and it is the key factor for these patterns. There could be a similar influence, but in the opposite direction at the most northern parts and the south central ones. We have to point out that the contribution of SO2 to the Moderate AQI is not 100% [18,19], and the influence of other species (O_3, PM) needs further investigation. As a whole, the southwestern region is characterized by the biggest negative health (positive concentrations) influence of the grid and emission resolutions changes on the Moderate AQI recurrences.

The annual recurrence of the High AQI (Fig. 3) is between 0% and 1%. The differences between cases are in the same interval and spatially homogeneous for the whole territory of Bulgaria. Therefore, the emission and grid resolutions have subtle or no influence on the spatial distribution of the High AQI recurrences in Bulgaria.

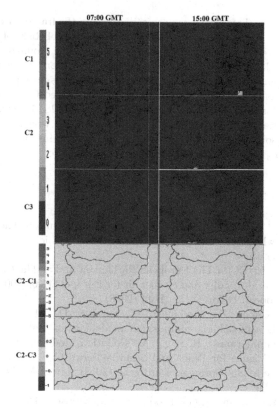

Fig. 3. Annual recurrence of the AQI [%] for all cases in High category.

4   Conclusion

In conclusion, we can make the following statements. The Low and Moderate AQI spatial differences between the cases are connected with the emission and grid resolutions. The spatial distributions of the Low AQI and Moderate AQI have opposite recurrence patterns on the territory of Bulgaria because they largely determine the health effect of the air quality in the country. The increase of the emission resolution for Moderate AQI is more evident in Southwestern Bulgaria. The higher process description resolution imposes an additional decrease in the Low AQI recurrences and increasing in the Moderate recurrences in the central mountainous areas. The High AQI recurrences are not significantly influenced by the grid and emission resolution changes. Generally, both the process description and the emission grid resolutions have an influence on the Low and Moderate AQI index recurrences in Bulgaria, which magnitudes and signs depend on the region of the country.

Acknowledgments. This work has been carried out in the framework of the National Science Program "Environmental Protection and Reduction of Risks of Adverse Events and Natural Disasters", approved by the Resolution of the Council of Ministers No 577/17.08.2018 and supported by the Ministry of Education and Science (MES) of Bulgaria (Agreement No D01-271/09.12.2022). This work has been partially supported by the National Center for High-performance and Distributed Computing (NCHDC), part of National Roadmap of RIs under grant No. D01-168/28.07.2022. Special thanks due to the Netherlands Organization for Applied Scientific research (TNO) for providing the high-resolution European anthropogenic emission inventory and to US EPA and US NCEP for providing free-of-charge data and software.

References

1. World Health Organization. Ambient (outdoor) air pollution [Fact sheet] (2022). https://www.who.int/news-room/fact-sheets/detail/ambient-(outdoor)-air-quality-and-health
2. Directive 2008/50/EC of The European Parliament and of The Council of 21 May 2008 on ambient air quality and cleaner air for Europe. (11.06.2008). Official Journal of the European Union L 152, 1–44 (2008)
3. Gadzhev, G., Ganev, K., Prodanova, M., Syrakov, D., Atanasov, E., Miloshev, N.: Multi-scale atmospheric composition modelling for Bulgaria. In: Steyn, D., Builtjes, P., Timmermans, R. (eds.) Air Pollution Modeling and its Application XXII. NATO Science for Peace and Security Series C: Environmental Security, vol. 137, pp. 381–385. Springer, Dordrecht (2013). https://doi.org/10.1007/978-94-007-5577-2_64
4. Gadzhev, G.K., Ganev, K.G., Prodanov, M., Syrakov, D.E., Miloshev, N.G., Georgiev, G.J.: Some numerically studies of the atmospheric composition climate of Bulgaria. In: AIP Conference Proceedings, vol.1561, pp. 100–111 (2013). https://doi.org/10.1063/1.4827219
5. Gadzhev, G., Ganev, K., Miloshev, N., Syrakov, D., Prodanova, M.: Numerical study of the atmospheric composition in Bulgaria. Comput. Math. Appl. **65**, 402–422 (2013)
6. Syrakov, D., Prodanova, M., Georgieva, E., Etropolska, I., Slavov, K.: Impact of NOx emissions on air quality simulations with the Bulgarian WRF-CMAQ modelling system. Int. J. Environ. Pollut. **57**(3–4), 285–296 (2015)
7. Gadzhev, G., Ganev, K., Mukhtarov, P.: Influence of the grid resolutions on the computer-simulated surface air pollution concentrations in Bulgaria. Atmosphere **13**(5), 774 (2022). https://doi.org/10.3390/atmos13050774
8. Leeuw, F. de, Mol, W.: Air Quality and Air Quality Indices: a world apart. ETC/ACC Technical Paper 2005/5 (2005). https://www.eionet.europa.eu/etcs/etc-atni/products/etc-atni-reports/etcacc_technpaper_2005_5_aq_indices/@@download/file/ETCACC_TechnPaper_2005_5_AQ_Indices.pdf
9. Skamarock, W.C., et al.: A Description of the Advanced Research WRF Version 3 (No. NCAR/TN-475+STR). University Corporation for Atmospheric Research (2008). https://doi.org/10.5065/D68S4MVH
10. UCAR/ NCAR. https://www2.mmm.ucar.edu/wrf/users/. Accessed 10 May 2022
11. CMAS. http://www.cmascenter.org. Accessed 10 May 2022
12. Byun, D., et al.: Description of the models-3 community multiscale air quality (CMAQ) modeling system. In: 10th Joint Conference on the Applications of Air

Pollution Meteorology with the A& WMA, pp. 264–268. 11–16 January 1998, Phoenix, Arizona (1998)

13. Byun, D., Ching, J.: Science Algorithms of the EPA Models-3 Community Multiscale Air Quality (CMAQ) Modeling System. EPA Report 600/R-99/030, Washington, D.C., EPA/600/R-99/030 (NTIS PB2000-100561). (1999). https://cfpub. epa.gov/si/si_public_record_report.cfm?Lab=NERL&dirEntryId=63400

14. CEP: Sparse Matrix Operator Kernel Emission (SMOKE) Modeling System, University of Carolina, Carolina Environmental Programs, Research Triangle Park, North Carolina (2003)

15. Denier van der Gon, H., Visschedijk, A., van de Brugh, H., Droge, R.: A high resolution European emission data base for the year 2005, TNO-report TNO-034-UT-2010-01895 RPT-ML, Apeldoorn, The Netherlands (2010)

16. Atanassov, E., Gurov, T., Ivanovska, S., Karaivanova, A.: Parallel Monte Carlo on intel MIC architecture. Proc. Comput. Sci. **108**, 1803–1810 (2017). https://doi. org/10.1016/j.procs.2017.05.149

17. Karaivanova, A., Atanassov, E., Gurov, T.: On the HPC/HPDA/AI competences in Bulgaria. In: 12th International Conference on Digital Presentation and Preservation of Cultural and Scientific Heritage, DiPP, vol. 12, 291–298 (2022). https:// doi.org/10.55630/dipp.2022.12.28

18. Georgieva, I., Ivanov, V.: Computer simulations of the impact of air pollution on the quality of life and health risks in Bulgaria. Int. J. Environ. Pollut. **64**(1–3), 35–46 (2018). https://doi.org/10.1504/IJEP.2018.099145

19. Georgieva, I., Gadzhev, G., Ganev, K., Melas, D., Wang, T.: High performance computing simulations of the atmospheric composition in Bulgaria and the city of Sofia. Cybern. Inf. Technol. **17**(5), 37–48 (2017). https://doi.org/10.1515/cait-2017-0053

Application of Active Subspaces for Model Reduction and Identification of Design Space

Sergei Kucherenko$^{(\boxtimes)}$ ⓘ, Nilay Shah ⓘ, and Oluyemi Zaccheus ⓘ

Imperial College London, London SW7 2AZ, UK
s.kucherenko@imperial.ac.uk

Abstract. The design space is defined as the combination of materials and process conditions which provides assurance of quality. Identification of the design space is a computationally demanding task especially in high dimensional settings. The active subspaces method is a technique that identifies the most important directions in the parameter space, enabling significant dimension reduction. We show how to apply the active subspaces method for model reductions and identification of design space. The results of constraint global sensitivity analysis match those obtained with the active subspaces method for the considered test case.

Keywords: Design Space · Active subspaces · Probability Map · Model Reduction

1 Introduction

The design space (DS) is defined as the combination of materials and process conditions which provides assurance of quality. The concept of DS is widely used in various industries, including pharmaceuticals, biotechnology, and chemical engineering. In high dimensional settings, identifying the design space can be a challenging computational task. The traditional approach based on exhausting sampling requires costly computations [2]. Advanced methods such as those based on adaptive sampling and metamodeling [7] may still require significant computational costs in high dimensional settings.

The active subspaces method (ASM) is a technique that identifies the most important directions in the parameter space, thus allowing for substantial reduction in dimensionality [1]. The ASM achieves this by performing the eigenvalue decomposition of the covariance-like matrix of response gradients. Unlike traditional global sensitivity analysis methods that can only rank variables and reduce dimensions for models with parameters aligned along the axes of the parameter space, the ASM allows for dimension reduction regardless of the orientation of the important directions. Each direction corresponds to a linear combination of the original input variables, forming what is called the active subspace. When applied to high-dimensional problems, ASM allows for the construction of metamodels in low-dimensional active subspaces, reducing computational costs.

I. Lirkov and S. Margenov (Eds.): LSSC 2023, LNCS 13952, pp. 412–418, 2024.
https://doi.org/10.1007/978-3-031-56208-2_42

We show that the active subspaces method can be applied for model reductions and identification of DS. Once DS is identified it is necessary to present it graphically. We demonstrate that the application of AS can reduce the number of required 2D projections for visualization of DS. Results of constraint global sensitivity analysis (cGSA) developed in [5] match those obtained with the active subspaces method for the considered synthetic test case.

We note that constrained sensitivity analysis was also applied in [4] for design space identification. The methodology has been verified against an antibody production process with constraints on antibody concentration and glycosylation.

1.1 Active Subspaces

Consider a nonlinear numerical model $f(x) \in L_1$, $x \in R^d$ distributed according to pdf $p(x)$. Compute the symmetric positive semidefinite matrix $A = E[\nabla f \nabla f^T]$ and its eigenvalue decomposition: $A = W \Lambda W^T$, where

$$\Lambda = diag(\lambda_1, \ldots \lambda_d), \lambda_1 \geq \lambda_2 \geq \ldots \lambda_d \qquad (1)$$

are eigenvalues, and W is the orthogonal matrix of the corresponding eigenvectors forming the basis of R^d. The main idea of the ASM is to find a partition $W = W_1 + W_2$, where W_1 is formed by the eigenvectors of the top k eigenvalues (where $k \ll d$), such that $f(x) \approx g(y)$, where $y = W_1^T x$ and $y \in R^k$. The span of the top k eigenvectors of C is called the "active subspace". It can be shown that optimal (in a certain sense) k corresponds to the largest gap in the spectrum of A [3]. Notice that unlike the global sensitivity analysis techniques, where subsets of input parameters that can be neglected are identified, the active subspaces approach seeks to find important linear combinations of all input parameters, which span the active subspace. Perturbing the inputs along the important directions causes greater change in the prediction, on average, than perturbing along the unimportant ones. The active variables are the linear combinations of the input parameters with weights from the important eigenvectors.

When applied to high-dimensional problems, the ASM allows for the construction of metamodels in low-dimensional active subspaces, reducing computational costs. These metamodels are capable of accurately computing variance-based Sobol' indices for all inputs at a reduced cost as shown in [9]. To build a metamodel $g(y)$ a set of $\{x_i, f(x_i)\}$ and corresponding $\{y_i, f(x_i)\}$, where $y_i = W_1^T x_i$ are computed and used to build $g(y)$ with the aid of metamodeling methods.

1.2 Design Space

A regulatory agency can authorize the production of pharmaceutical products if there is adequate proof of their safety and effectiveness. This entails establishing quality controls within the manufacturing process, which includes a defined set of acceptable operating conditions referred to as the Design Space [3]. These controls ensure that the manufacturing process produces consistent quality products, accounting for variations in materials and processes.

Consider a model $f(x; \theta)$, and the vector of constraint functions $g(x; \theta)$, where x is a vector of process parameters defined in the box domain H^d, θ is a vector of uncertain model parameters defined in m-dimensional real space R^m with a given pdf $\varphi(\theta)$. Define probability $p(x)$ of occurrence of an undesirable (reliability estimation analysis) or desirable (pharmaceutical/chemical engineering) event as

$$p(x) = P(g(x; \theta) \geq g^*) = \int_{\Omega(x)} \varphi(\theta) d\theta. \tag{2}$$

Here g^* is a vector of given thresholds, domain $\Omega(x)$ is defined as $\Omega(x) = (\theta : g(x; \theta) \geq g^*)$. Probability (2) can be presented in a convenient for computation form as:

$$p(x) = \int_{R^m} I(g(x; \theta) \geq g^*) \phi(\theta) d\theta = E_\theta[I(g(x; \theta) \geq g^*)], \tag{3}$$

where I is an indicator function. Define the DS as

$$DS(x; p(x) \geq p^*). \tag{4}$$

Here p^* is a critical (acceptable) value of probability of meeting constraints.

Identification of probabilistic DS is a demanding task and for a typical practical problem the traditional approach based on exhaustive sampling requires costly computations [2]. A novel theoretical and numerical framework for determining probabilistic DS using metamodelling and adaptive sampling was proposed in [7]. It is based on the multi-step adaptive technique using a metamodel for a probability map as an acceptance-rejection criterion to optimize sampling for identification of the DS.

We noted that the concept of DS intersects with the global sensitivity analysis approach developed in [5] for dealing with problems that feature inequality constraints (cGSA). In such cases, the impact of variations in input variables on the model outputs needs to be measured within a non-rectangular, multidimensional domain that is implicitly determined by the constraints imposed on the model.

2   Results

We define the synthetic DS as

$$DS(x; f(x) \geq c), \tag{5}$$

where f acts as "probability $p(x)$", and constant c is a critical value. We assume that function $f(x)$ is known explicitly. It can be achieved, for example by building a metamodel as demonstrated in [7].

Consider a linear model $f(x) = c + v^T x$, $v^T \in R^d$. Its gradient is $\nabla f(x) = v$ and

$$A = \int \nabla f(x) \nabla f(x)^T dx = \int vv^T dx = vv^T = W\lambda W^T. \tag{6}$$

Here $\lambda = \|v\|^2$, $W_1^T = v/\|v\|$. Other eigenvalues are equal to 0. Clearly, $f(x) = g(y) = c + \|v\|y$. That is a reduction from d to 1 dimension. $DS(y, g(y) \geq c)$ is defined by $y \geq 0$, hence DS is defined by inequality $W_1^T x \geq 0$.

Further we consider the general function $f(x)$, such that its low-dimensional metamodel in the active subspace has a quadratic form:

$$g(y) = a^2 y + bx + c. \tag{7}$$

In this case $DS(y, g(y) \geq c)$ is defined by inequalities $y \leq y^*, y \geq y^{**}$, where $y^* = -b/a$, $y^{**} = 0$ (assuming $a > 0$). Hence DS is defined by inequalities $W_1^T x \leq y^*$, $W_1^T x \geq y^{**}$.

A challenge remains in the visualization of DS needed to communicate higher dimensionality probability maps and the inclusion of this information into the formal documentation required for the approval of DS by a government agency. For more than three parameters process parameters visualization of DS is a challenging task. Guidance for Industry, Q8(R2) Pharmaceutical Development [8] has the following recommendations: "When multiple parameters are involved, the design space can be presented for two parameters, at different values (e.g., high, middle, low) within the range of the third parameter, the fourth parameter, and so on." Considering DS in the d-dimensional process parameters space, a graphical presentation of DS via 2D plots prescribed in [8] would result in $N_P = \frac{d(d-1)}{2} 3^{d-2}$ 2D plots.

cGSA can be applied for visualization of DS. Considering a d-dimensional DS problem, if cGSA indicates that a parameter x_i is insignificant, then any 2D projections based on different values of x_i will yield identical results. Consequently, only one value of x_i would be necessary for graphical representation of this parameter. In the case of s non-important parameters N_P will be reduced to $[(d - s)(d + 5s - 1) + 9s(s - 1)]3^{d-s-2}/2$.

For illustration consider a 3D Sobol' "g-function":

$$f(x) = \prod_{i=1}^{3} \frac{|4x_i - 2| + a_i}{1 + a_i} \tag{8}$$

with three different sets of parameters a_i and x_i uniformly distributed in the unit hypercube H^3.

A synthetic DS is defined as $DS(x; f(x) \leq 1)$. The DS shape for different sets of parameters a_i is shown in Fig. 1. In the first case (Fig. 1a) all three 2D projections are important and the prescription given above requires plotting $N_P = \frac{d(d-1)}{2} 3^{d-2} = 9$ 2D projections. cGSA reveals that in this case all three input parameters are important. In the second case (Fig. 1b) although all 2D projections onto the plains (x_1, x_2), (x_1, x_3) and (x_2, x_3) are important, 2D projections on the plane (x_1, x_2) are independent of parameter x_3. cGSA

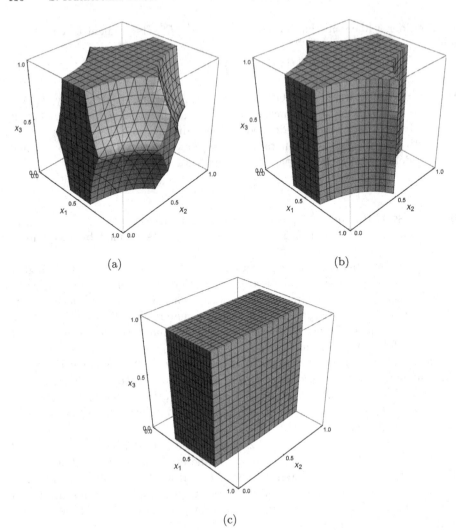

Fig. 1. 3D DS shape. (a) a ={0; 1; 2} ; SI ={ 0.706; 0.0818; 0.0234}, SIT ={ 0.856; 0.229; 0.104}; (b) a ={0; 1; 99}; SI ={0.761; 0.0957; 0.002}, SIT ={0. 887; 0.238; 0.003}; (c) a ={0; 49; 99}; SI ={0.996; 0.0004; 0.0005}, SIT ={0.996; 0.0005; 0.0006}. SI and SIT are the values of the main effect and total Sobol' sensitivity indices respectively.

confirms that in this case parameter x_3 is not important as both the values of the main effect and total Sobol' indices are close to 0.0. Small values of Sobol' indices reflect the fact that a corresponding parameter does not contribute to the description of the DS. Only $N_P = \frac{(d+4)(d-1)}{2} 3^{d-3} = 7$ 2D projections are required for the visualization of DS in this case. In the third case (Fig. 1c) only one 2D projection to the x_1 axis is important while for two other parameters the DS shape is defined in the whole domain of their definition and hence no other

2D projections are of any interest. GSA shows that in this case only parameter x_1 is important.

The ASM applied to the g-function in the third case revealed that there is one-dimensional active subspace with W_1^T=(1,0,0). Metamodel $g(y)$ has a parabola-like form, hence the same approach (see formula (6) and discussion afterwards) can be applied to find two bounds y^*, y^{**}, with the DS defined by $y^* \leq y \leq y^{**}$. Hence $y^* \leq W_1^T x \leq y^{**}$, $y^* \leq x_1 \leq y^{**}$, where $y^* = 0.25$, $y^{**} = 0.75$ in full agreement with Fig. 1c (in Fig. 2 an active variable y is renormalized from [0,1] to [-1,1] interval). We note that polynomials of high even degrees were needed to accurately approximate $g(y)$. The SobolGSA software [6] was used for building a metamodel in the active subspace.

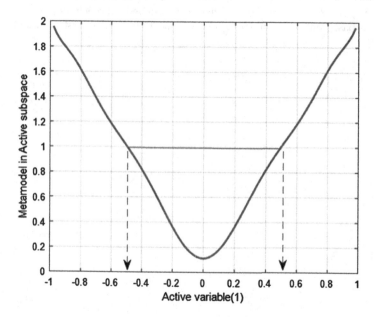

Fig. 2. Metamodel $g(y)$ in the active subspace. G-function with parameters a ={0; 49; 99}. DS is located within the two arrows.

Acknowledgements. We acknowledge the financial support of Eli Lilly and Company and the EPSRC Programme Grant EP/T005556/1.

References

1. Constantine, P.G.: Active Subspaces: Emerging Ideas for Dimension Reduction in Parameter Studies. SIAM (2015)
2. Garcia-Munoz, S., Luciani, C.V., Vaidyaraman, S., Seibert, K.D.: Definition of design spaces using mechanistic models and geometric projections of probability maps. Org. Process Res. Dev. **19**(8), 1012–1023 (2015)

3. International Council for Harmonisation of Technical Requirements for Pharmaceuticals for Human Use: Q8 Guidance for industry. US Department of Health and Human Services and Food and Drug Administration and others (2006)

4. Kotidis, P., et al.: Constrained global sensitivity analysis for bioprocess design space identification. Comput. Chem. Eng. **125**, 558–568 (2019)

5. Kucherenko, S., Klymenko, O.V., Shah, N.: Sobol'indices for problems defined in non-rectangular domains. Reliab. Eng. Syst. Saf. **167**, 218–231 (2017)

6. Kucherenko, S., Zaccheus, O.: SobolGSA Software. Imperial College London, UK (2023)

7. Kucherenko, S., Giamalakis, D., Shah, N., García-Muñoz, S.: Computationally efficient identification of probabilistic design spaces through application of metamodeling and adaptive sampling. Comput. Chem. Eng. **132**, 106608 (2020)

8. Guidance for industry: Q8 (r2) pharmaceutical development (2009)

9. Zhou, C., Shi, Z., Kucherenko, S., Zhao, H.: A unified approach for global sensitivity analysis based on active subspace and kriging. Reliab. Eng. Syst. Saf. **217**, 108080 (2022)

EOCSim: A CloudSim-Based Simulator for Earth Observation Data Processing in Clouds

Arthur Lalayan[1,2]([✉]) [ID], Hrachya Astsatryan[1] [ID], and Gregory Giuliani[3] [ID]

[1] Institute for Informatics and Automation Problems National Academy of Sciences of Armenia, 1 Paruyr Sevak, 0014 Yerevan, Armenia
arthurlalayan97@gmail.com, hrach@sci.am
[2] National Polytechnic University of Armenia, 105 Teryan, 0009 Yerevan, Armenia
[3] GRID-Geneva, Institute for Environmental Sciences, University of Geneva, Bd Carl-Vogt 66, 1211 Geneva, Switzerland
Gregory.Giuliani@unige.ch

Abstract. Advanced computing resources are necessary to efficiently and accurately process Earth Observation data for environmental monitoring. However, selecting the optimal infrastructure for distributed data processing can be costly and time-consuming due to the required historical simulation dataset. The article proposes the Earth Observation Cloud Simulator (EOCSim) model to address the challenge of utilizing the CloudSim environment to optimize performance and reduce resource usage costs. The model's accuracy is demonstrated by its ability to emulate the actual experiments' behavior, achieving a high level of accuracy in the weekly NDVI for the territory of Armenia.

Keywords: Earth observation · CloudSim · distributed computing · performance-cost optimization · Dask · Kubernetes

1 Introduction

Earth observation (EO) data obtained from satellites play a vital role in monitoring the environment at a global scale [1]. However, with the advent of Big EO data, processing these large datasets poses a significant challenge to the scientific community. To handle these data processing workflows, High-Performance Computing (HPC) resources are necessary to parallelize tasks across multiple processors [13]. Distributed data processing frameworks process large-scale EO data across multiple nodes in a distributed computing environment. The frameworks partition the data into smaller pieces based on chunk size, spread across numerous computational nodes, thereby improving the processing performance of large datasets. Optimizing distributed data processing frameworks depends on infrastructure, data characteristics, workload, and other features. Optimizing requires carefully considering the trade-offs between different parameters and features of the data frameworks, such as the number of nodes, the size of the data chunks, the communication overhead, and the processing capacity of nodes.

© The Author(s), under exclusive license to Springer Nature Switzerland AG 2024
I. Lirkov and S. Margenov (Eds.): LSSC 2023, LNCS 13952, pp. 419–426, 2024.
https://doi.org/10.1007/978-3-031-56208-2_43

Multi-parametric Pareto-based optimization [12] is a popular approach for optimizing distributed data processing frameworks by considering multiple objectives, such as performance, cost, or energy consumption. A set of possible configurations for the distributed data processing framework constructs a Pareto frontier to demonstrate the optimal trade-off between the different objectives [18]. Therefore, it enables the optimization of distributed data processing frameworks more efficiently and effectively than traditional single-objective optimization approaches. Finding an optimal framework configuration that balances various objectives fulfilling the requirements of both resource providers and clients, can be daunting, as it is needed to carry out a massive amount of simulations. One approach is to employ modeling and simulation tools to enhance resource provisioning accuracy by considering many parameters. However, conducting experiments that involve fluctuating loads, performance, or system sizes can be costly, whereas simulation tools are better suited for general applications.

The article introduces a user-friendly Earth Observation Cloud Simulator (EOCSim) emulator model for EO data processing workflows based on the CloudSim environment [8]. The primary objective is improving performance and reducing the average resource utilization costs. The EOCSim emulator model identifies the optimal configuration that balances performance and cost. The evaluation shows the suggested approach's high accuracy compared to actual experiments.

2   Related Work

Processing complex and large-scale EO workflows require significant resources, specialized technology, and computational power [5]. Several studies have proposed enhanced frameworks and solutions. For example, [17] offers various regression models to optimize Spark's configuration parameters, while [16] introduces a scheduling strategy to minimize execution time by determining optimal task partitioning and assignments. Many recent studies have focused on developing multi-objective scheduling algorithms to handle conflicting objectives. For instance, the authors [15] suggest a multi-objective scheduling algorithm to produce Pareto-optimal solutions, balancing computational and energy efficiency. Moreover, all methods involve estimating and predicting task execution time, implying that the execution time of EO data processing is typically proportional to the input data size. The limitation of such approaches is the accuracy of task execution time estimation, which can be costly considering computational resource usage.

Therefore, the article suggests using simulators for modeling the behavior of the EO data distributed processing to improve the accuracy of execution time estimation. The simulators, such as SimGrid [9] or CloudSim, can evaluate the performance of distributed applications and systems in various configurations, workloads, and system settings. Different simulators may enhance CloudSim's functionality, such as GridSim [7] or NetworkCloudSim [11]. The mentioned simulators are generic and require customization to handle particular issues. Therefore, the article suggests creating a simulator specifically designed to meet the needs of EO data processing workflows by expanding the capabilities of the CloudSim simulator.

3 Methodology

The methodology section of this study outlines the CloudSim architecture, simulation metrics, and generation scenarios employed in the research.

3.1 CloudSim and Simulation Metrics

The CloudSim simulation tool evaluates specific system issues related to cloud computing environments without requiring low-level knowledge of the underlying infrastructure and services. CloudSim provides a comprehensive and flexible API for modeling and simulating various aspects of cloud computing infrastructures and services. The layered architecture of CloudSim consists of the simulation engine, cloud services, and source code. The critical components of CloudSim include the data center, hosts, virtual machines (VMs), and Cloudlets. The data center represents the physical infrastructure, while hosts represent the physical machines within the data center. Several hosts comprise a data center to set up various resource allocation, scheduling, and power management strategies. A host is a physical computer that offers one or more VMs resources, including CPU, memory, and storage. VMs run on hosts and provide an isolated environment for executing user applications. Cloudlets represent the user tasks performed on VMs.

Simulation metrics provide quantitative measurements of various system characteristics, such as resource utilization or response time. Evaluating the metrics enables insights into the effectiveness of different resource allocation and scheduling policies and identifies potential bottlenecks in the system. The simulation time and cost metrics are used in the emulator to measure the simulation time and evaluate the cost-effectiveness.

Handling large amounts of EO data requires a significant amount of time, making performance measures crucial. Nevertheless, processing massive amounts of data demands considerable computing resources, incurring fees for consumers who rely on global cloud services. Therefore, it is essential to consider performance and cost parameters to improve data processing.

3.2 Scenario

By analyzing changes in environmental variables over time, scientists can develop models to forecast future changes and plan for potential impacts, while decision-makers can develop effective strategies to address environmental problems and reduce the effects of human activities. For instance, air temperature forecasting is a crucial aspect of weather forecasting, helping predict the temperature changes in a particular location over a specific period [3]. Our experiments focused on processing EO data to calculate the Normalized Difference Vegetation Index

(NDVI) [4], which provides information on the amount and density of green vegetation and the photosynthetic activity of the plants. The calculation of the NDVI depends on the near-infrared and red bands. The time series analysis of NDVI detects changes in vegetation cover and identifies areas at risk of land degradation and desertification. We emphases on weekly, monthly, and seasonal time series analysis to understand the trends and patterns of NDVI for different periods.

4   EOCSim Emulator Model Architecture and Evaluation

This section describes the architecture of the EO data processing EOCSim emulator model and its evaluation.

4.1   Architecture

The EOCSim is a modular emulator model to employ the CloudSim simulator. The model has been implemented in Java and is accessible to clients through its REST (Representational State Transfer) API, with its implementation available on GitHub[1]. The packages have been designed flexibly, making it straightforward for future modifications and extensions to be made. Figure 1 illustrates the structure of the EOCSim emulator model consisting of a resource manager, Dask configuration generator, Data size estimator, EO simulator, and Pareto-optimal configuration provider modules.

The client can access the EOCSim emulator model through an HTTP call to a REST API to get the optimal cluster configurations. The client applies to the resource manager by providing the parameters of a data center (number of hosts, description of each host, particularly the number of cores, RAM, and frequency), function (pre-defined functions, like NDVI), area of interest (geographic region), and timeframe for EO data processing. The client can use multi-objective optimization by supplying each machine's hourly cost; by default, the model emphasizes performance optimization.

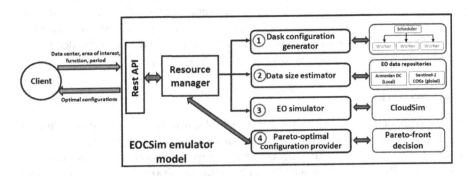

Fig. 1. EOCSim emulator model architecture.

---

[1] https://github.com/ArmHPC/EOCSim.

EO community often uses the Dask parallel Python library [14] to simplify data division into smaller chunks, distributed processing across multiple computer nodes, and aggregate processing results. Using the Dask distribution package age on Kubernetes effectively creates a Dask cluster remotely with the necessary number of nodes, CPU, and RAM resources. The Dask configuration generator module uses input data to create a list of potential cluster configurations in the specified data center.

We consider various client inputs, such as the selected area and timeframe, to estimate the input data size. We use two EO data repositories: the global repository, the Sentinel-2 Cloud-Optimized GeoTIFFs repository[2], and the local repository, the Armenian Datacube [2]. The Sentinel-2 Cloud-Optimized GeoTIFFs collection provides data acquired by the Sentinel-2 satellite, while the Armenian Datacube is a compilation of satellite images covering Armenia, sourced from both Landsat and Sentinel satellites.

After the possible cluster configurations set have been generated and the input data size has been estimated, simulations are carried out using the API provided by the CloudSim, and the described metrics are stored in the simulation result dataset.

The EOCSim emulator model utilizes task execution time as the primary performance metric to identify the optimal configurations for a given task. However, in many cases, clients may have to balance multiple objectives, such as optimizing both performance and cost. The EOCSim emulator model can apply multi-objective Pareto optimization to the simulation data in these situations. By filtering the data in this way, clients can identify the configurations that offer the best trade-offs between performance and cost. Additionally, the Pareto optimization approach can be adapted to other performance metrics, allowing clients to fine-tune their strategy based on various criteria.

4.2 Evaluation

To evaluate the EOCSim, a data center is defined as a physical infrastructure with host machines with specific characteristics. Kubernetes and Dask worker nodes have been used to simulate hosts and VMs in CloudSim. This approach generated multiple Dask clusters with varying numbers of computational nodes, CPU and RAM capacities, and simulated EO data processing using CloudSim. The properties listed in Table 1 were used to create entities in CloudSim. By configuring these entities with the appropriate properties, clients can fine-tune the simulation to match their needs and requirements. This flexibility allows clients to explore various configurations and performance metrics, enabling them to identify the most efficient and effective settings for their system.

---

[2] https://registry.opendata.aws/sentinel-2-l2a-cogs/.

Table 1. CloudSim parameters to generate entities.

Entity	Parameters description
Data center	List of hosts
Host	List of processing elements (PEs) by provided MIPS, RAM of the host, the storage bandwidth capacity and size
VM	Number of PEs, MIPS of each core, RAM, storage bandwidth capacity and size
Cloudlet	The length or size (in MI) of this Cloudlet to be executed in a VM, required PEs, and utilization model

In the subsequent stage, the number of cores, RAM, and Million Instructions Per Second (MIPS) for each host's configuration are provided to the EOCSim simulator. Then identical VMs are created, using the same number of computational nodes and properties for each cluster type (CPU, RAM) from the generated set. During the creation of the Cloudlet, the estimated size of the input data and the chosen processing function are considered. Each VM receives a Cloudlet of the same length to process, which means the total number of Cloudlets equals the number of VMs. The Cloudlet length is calculated by dividing the estimated size of the input data by the number of VMs and then multiplying it by a factor that represents the processing time required for each data unit. This factor is determined based on the chosen processing function and is used to estimate the total execution time of the Cloudlet. The Cloudlet component also receives a parameter known as the utilization model, which establishes the level of computing resources like CPU, RAM, and Network needed during the simulation. The average resource usage is calculated for various input data sizes to select the best utilization model for every type of Dask cluster, each with a different worker count and set of computational characteristics.

The computing resources of CloudLab [10] and the Armenian cloud infrastructure [6] have been used for the experiments to evaluate the EOCSim emulator model. The data center has eight hosts with AMD EPYC 7302P CPU, 16 cores, and 128 GB RAM. Five categories of nodes have been studied, ranging from 1 CPU and 2 GB RAM to 16 CPUs and 32 GB RAM. The weekly NDVI in real Dask clusters has been calculated as an EO data processing task. The input data of the NDVI for the weekly period for the territory of Armenia is approximately 80 GB. Figure 2 compares the real simulations with the suggested EOCSim emulator model.

Figure 2 shows a remarkable level of agreement between the actual simulations and the cluster configurations. Specifically, the plot reveals the sole cluster configuration set that delivers an execution time of less than 1000 s. The quantitative analysis further demonstrates the robustness of the model with an R^2 value of 0.88 and a Root Mean Square Error (RMSE) of 78, indicating a high degree of accuracy in predicting the execution time of the simulations. These findings emphasize the effectiveness of the chosen cluster configurations and validate the overall model's ability to forecast the execution time of the simulations accurately.

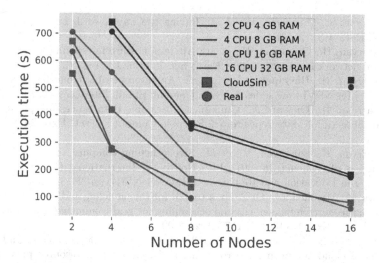

Fig. 2. Comparision of EOCSim emulator model and realistic simulations for weekly NDVI.

5 Conclusion

The article proposes the EOCSim emulator model for EO data processing workflows based on the CloudSim simulator. To improve performance and cut costs, the model aids in selecting the best cluster configuration for distributed computation of EO data. The evaluation results show high accuracy between the experiments and the EOCSim emulator model results. The proposed EOCSim emulator model is valuable to EO data processing, enabling the selection of an optimal Dask cluster configuration to achieve high performance and reduce costs. The evaluation results demonstrate the high accuracy of the EOCSim emulator model in comparison to actual experiments, particularly in predicting the weekly NDVI for Armenian territory with an R^2 value of 0.88 and an $RMSE = 78$. In future studies, it is planned to improve the accuracy of the suggested method and add energy consumption as an optimizable objective.

Acknowledgements. The research was supported by the Science Committee of the Republic of Armenia and the University of Geneva Leading House by the projects entitled "Scalable data processing platform for EO data repositories" (Nr. 22AA-1B015), "Self-organized Swarm of UAVs Smart Cloud Platform Equipped with Multi-agent Algorithms and Systems" (Nr. 21AG-1B052), "Remote sensing data processing methods using neural networks and deep learning to predict changes in weather phenomena" (Nr. 21SC-BRFFR-1B009), and "ADC4SD: Armenian Data Cube for Sustainable Development".

References

1. Anderson, K., Ryan, B., Sonntag, W., Kavvada, A., Friedl, L.: Earth observation in service of the 2030 agenda for sustainable development. Geo-Spat. Inf. Sci. **20**(2), 77–96 (2017)

2. Asmaryan, S., Muradyan, V., et al.: Paving the way towards an Armenian data cube. Data **4**(3) (2019)
3. Astsatryan, H., Grigoryan, H., et al.: Air temperature forecasting using artificial neural network for Ararat valley. Earth Sci. Inform. **14**, 1–12 (2021)
4. Astsatryan, H., Hayrapetyan, A., et al.: An interoperable cloud-based scientific gateway for NDVI time series analysis. Comput. Stand. Interfaces **41**, 79–84 (2015)
5. Astsatryan, H., Lalayan, A., Giuliani, G.: Scalable data processing platform for earth observation data repositories. Scalable Comput.: Pract. Exp. **24**(1), 35–44 (2023)
6. Astsatryan, H., Sahakyan, V., et al.: Strengthening compute and data intensive capacities of Armenia. In: 2015 14th RoEduNet International Conference - Networking in Education and Research (RoEduNet NER), pp. 28–33 (2015)
7. Buyya, R., Murshed, M.: GridSim: a toolkit for the modeling and simulation of distributed resource management and scheduling for grid computing. Concurr. Comput.: Pract. Exp. **14**(13–15), 1175–1220 (2002)
8. Calheiros, R.N., Ranjan, R., et al.: CloudSim: a toolkit for modeling and simulation of cloud computing environments and evaluation of resource provisioning algorithms. Softw.: Pract. Exp. **41**(1), 23–50 (2011)
9. Casanova, H., Legrand, A., Quinson, M.: SimGrid: a generic framework for large-scale distributed experiments. In: Tenth International Conference on Computer Modeling and Simulation (UKSim 2008), pp. 126–131 (2008)
10. Duplyakin, D., Ricci, R., et al.: The design and operation of CloudLab. In: Proceedings of the USENIX Annual Technical Conference (ATC), pp. 1–14 (2019)
11. Garg, S.K., Buyya, R.: NetworkCloudSim: modelling parallel applications in cloud simulations. In: 2011 Fourth IEEE International Conference on Utility and Cloud Computing, pp. 105–113 (2011)
12. Guo, Z., Wong, W., Li, Z., Ren, P.: Modeling and pareto optimization of multi-objective order scheduling problems in production planning. Comput. Industr. Eng. **64**(4), 972–986 (2013)
13. Lee, C.A., Gasster, S.D., Plaza, A., Chang, C.I., Huang, B.: Recent developments in high performance computing for remote sensing: a review. IEEE J. Sel. Top. Appl. Earth Observ. Remote Sens. **4**(3), 508–527 (2011)
14. Rocklin, M.: Dask: parallel computation with blocked algorithms and task scheduling. In: Proceedings of the 14th Python in Science Conference, vol. 130, p. 136. SciPy Austin, TX (2015)
15. Sun, J., Li, H., et al.: Multi-objective task scheduling for energy-efficient cloud implementation of hyperspectral image classification. IEEE J. Sel. Top. Appl. Earth Observ. Remote Sens. **4**, 587–600 (2020)
16. Sun, J., Zhang, Y., et al.: An efficient and scalable framework for processing remotely sensed big data in cloud computing environments. IEEE Trans. Geosci. Remote Sens. **57**(7), 4294–4308 (2019)
17. Yu, Z., Wang, Z., et al.: Parameter optimization on spark for particulate matter estimation. In: 2021 Workshop on Algorithm and Big Data, WABD 2021, pp. 9–13. Association for Computing Machinery, New York (2021)
18. Lalayan, A., Astsatryan, H., Giuliani, G.: A multi-objective optimization service for enhancing performance and cost efficiency in earth observation data processing workflows. Baltic J. Mod. Comput. **11**(3) (2023)

About Methods of Vector Addition over Finite Fields Using Extended Vector Registers

Maria Pashinska-Gadzheva[✉][iD] and Iliya Bouyukliev[iD]

Institute of Mathmatics and Informatics, Bulgarian Academy of Sciences,
Sofia, Bulgaria
{mariqpashinska,iliyab}@math.bas.bg

Abstract. We present optimized algorithm for vector addition over finite prime fields using the extended vector registers of modern central processing units (CPU) and the corresponding extended Intel instruction sets SSE, AVX and AVX512. The presented algorithm is based on representation of the elements of the fields using unsigned 8-bit packed integer, thus allowing for computations over prime fields with up to 127 elements. The efficiency of the presented method is demonstrated in an algorithm for calculating the weight distribution of a linear code over the finite field which is known to be an NP-complete problem. An optimized approach for computing the weight of a vector is also given. The experimental results show faster execution times compared to the corresponding algorithms in the Magma and GUAVA package for GAP packages for finite fields larger than 3.

Keywords: Vector addition over prime fields · Extended registers · Linear codes

1 Introduction

Finite field arithmetic is a basis of many scientific computations connected to mathematical foundations of informatics and more precisely cryptography and coding theory among others. The vector operations over finite fields are connected to many theoretical and practical tasks. Therefore, effective algorithms for vector arithmetic in addition to hardware optimizations make many problems in this area solvable. In this paper, we present some of our work on vector

The research of the first author is partially supported by the Bulgarian National Science Fund under Contract No KP-06-H62/2/13.12.2022. The work of the second author is partially supported by the Bulgarian Ministry of Education and Science, grant no. DO1-325/01.12.2023 for NCHDC. The authors acknowledge also the access to the e-infrastructure provided by the Grant No. D01-325/01.12.2023 "National Centre for High Performance and Distributed Computing" of the Ministry of Education and Science of Bulgaria.

I. Lirkov and S. Margenov (Eds.): LSSC 2023, LNCS 13952, pp. 427–434, 2024.
https://doi.org/10.1007/978-3-031-56208-2_44

addition over prime fields with at most 127 elements. The optimizations of the addition operation are based on vectorization using extended registers present in modern CPUs and a set of specific instructions developed by Intel [1]. This research is an extension of our work on a library [2] for the calculation of some weight characteristics of linear codes over finite fields with $p^m \leq 64$ elements for a prime p. Our goal is by using the same resources - field representation and size of the registers - to implement vector addition over two times larger fields. We compare the efficiency of different instruction sets on the same hardware. We also evaluate the effectiveness of the proposed methods compared to different mathematical commercial and open source software. We use an algorithm for calculation of the weight distribution of a linear code for this purpose.

The extended registers present in modern CPUs are intended to introduce another level of SIMD parallelism that is different from multi-core computations. The presentation of uniform data (in many cases arrays) as vectors in these specialized registers allow to execute operation over multiple elements at the same time as a single CPU instruction. This process is also known as vecotrization. Intel has developed different sets of instructions for the extended registers generalized as SSE, AVX and AVX512 [1] based on the register length. They implement some operations over different data types including addition of integer and real numbers, bitwise and logical operations, etc. For our research purposes, we need the implementation of modulo operation that gives us the remainder after division presented in C/C++ by the '%' operator. This operation is at the base of the arithmetic over finite fields and does not have direct implementation for the extended registers due to its complexity. Therefore, our algorithm for vector addition over prime fields is based on the replacement this specific operations by a set existing functions for the different extended registers. This implementation is more efficient than the existing software packages such as MAGMA [3] and GUAVA package for GAP [4].

The following Sect. 2 gives some preliminaries on linear codes. Section 3 describes in detail the proposed methods for vector addition and the calculation of the weight of a vector. Section 4 presents some experimental results on the efficiency of the proposed algorithms. We give some conclusionary remarks on the use of extended vector registers in Sect. 5.

2   Preliminaries

Without a loss of generality we can consider the elements of finite prime field F_p as the residuals modulo p [5]. Thus, the elements of F_p can be represented by the set of integer numbers $\{0, 1, \ldots, p - 1\}$. Let us consider the n dimensional vector space F_p^n. The (*Hamming*) weight of a vector $v \in F_p^n$ is defined as the number of its non-zero coordinates. A linear code is any k-dimensional subspace of the vector space F_p^n. The elements of a linear code are called *codewords* and the parameters k and n are called *dimension* and *length* of the code respectively. A $k \times n$ matrix G whose rows form a basis of the code is called *generator matrix*. Therefore, all codewords can be generated as linear combination of the rows of a generator matrix. The weight distribution of a linear code C is the sequence $(A_0, A_1, \ldots, A_n)$, where A_i is the number of codewords with weight $i = 0, 1, \ldots n$.

One universal approach to calculate the weight distribution of a linear code is to generate all $(q^k - 1)/(q - 1)$ non proportional codewords as linear combinations of the rows of a generator matrix. For prime felds it can be achieved by using addition of only two vectors [6]. This problem is known to be NP-complete [7] and the optimizations are expected to integrate parallelization. Another approach to optimize the execution time of such algorithm is to accelerate the other two main parts of the computations - vector addition and the calculation of the weight of a vector.

The primary purpose of this work is to present implementation of the vector addition function and calculation of the weight of the vector by using the different instructions form SSE, AVX and AVX512 instruction sets. The arithmetic for finite fields F_2, F_3 and F_4 can be implemented using bitwise representation [8,9]. Therefore, here we consider only fields F_p, where $p > 3$. The elements of the field can be represented by packed 8-bit integers. The implementation of vector addition is different for the signed and unsigned data types. When using signed integer numbers there is further limitation on p, primarily $p < 64$. However, here we use a smaller number of instructions. The unsigned data type allows for the computations to be executed for fields with larger number of elements ($p < 128$). The number of instructions that are used is larger but it does not affect the performance significantly. Therefore, we will present in more detail an implementation using unsigned data type.

The second main function needed for the computations calculates the weight of a vector in the field F_p. The main idea is to find the number of zero elements w_0 in a vector. Therefore, the weight of the vector can be easily obtained using a formula that depends on the length of the used register. The value of w_0 is calculated using instructions for the extended register that mark the zero elements and a the CPU instruction *popcnt*. It returns the number of non-zero bits in a 32-bit or 64-bit computer word depending on the architecture. Its latency is greater than some of the instructions for the extended registers, hence we minimize its use. There is also an alternative for calculation of the non-zero bits in a computer word that uses masks [8].

3 Implementation

In this section we present different implementations of the vector addition over prime finite field F_p with $p > 3$ elements. Let $a, b, c \in F_p^n$ and $wt(c)$ is the weight of vector c. We want to calculate $c = a + b \mod p$ and the value of $wt(c)$. The addition of elements over F_p can be computed by subtracting p when the result is greater than p. Algorithm 1 shows the implementation using this method and the calculation of the weight of the resulting vector. The parallelized versions with extended registers is based on Algorithm 1. The following subsections describe in detail the method for these calculations with different instruction sets.

Algorithm 1. Vector addition and weight computation using subtraction

1: **function** ADD_WEIGHT(a,b,c)
2:     int wt ← 0;
3:     for(int i = 1; i¡=n; i++) do{
4:         c[i] = (a[i] + b[i]);
5:         if (c[i]≥ p) then c[i] = c[i] - p;
6:         if(c[i]≠0) then wt++;}
7: **return** wt;
8: **end function**

3.1   Vector Addition Using Extended Vector Registers

In this subsection we present a method for computing $C = A + B$ mod p using SSE, AVX and AVX512 instruction sets, where A, B and C are extended registers of appropriate length. The elements of F_p are represented by 8-bit integers. We can store 16, 32 or 64 elements into registers with 128, 256 and 512 bits respectively. For simplicity the following descriptions are for vectors with length 16, 32, and 64 for the three types of registers. The different implementations of vector addition over F_p also use at least one of following constant registers $ZERO = (0, 0, \ldots, 0)$ and $P = (p, p, \ldots, p)$ with the appropriate length for the different register lengths.

There are two main approaches for the implementation based on whether the elements are presented by signed or unsigned integers. The implementation integrating signed data types has three main steps - addition of the two vectors, subtraction of vector P and an appropriate blending of the results of the previous steps. Here we use an instruction that allows to blend two vectors (e.g. _mm_blendv_epi8 for 128-bit register). The resulting register gets one component of two registers based on the value of the corresponding component in a third (mask) register. This algorithm works correctly for $p < 64$ since the addition of two elements greater than 64 can result in a negative integer (data overflow).

Algorithm 2 shows a detailed implementation using unsigned numbers. There are a few differences when working with unsigned data type. Firstly, Algorithm 2 allows for computations to be executed using 8-bit integers for finite fields with up to 127 elements without encountering data overflow. The instructions for unsigned integers integrate saturation. When an instruction uses saturation the result is either 0 or FF whenever it gets out of range for the given data type. This implementation uses more instructions. As can be seen, the supplementary operations generate a mask register, marking which elements of the addition result are to be chosen for vector C. An advantage here is the integration of bitwise operations, executed over the full registers.

Let us now consider the implementation using 128, 256 and 512 bit registers. Most operations in the algorithm have a corresponding functions for the 128 and 256-bit registers. One important part is the comparison in the case of unsigned packed data types. The SSE and AVX instructions do not include such comparison functions. Therefore, we compare the data as signed integer only for 'equal'

Algorithm 2. Vector addition for $p < 128$

1: **function** ADD(A,B,C)
2: const P ← $(p, p, \ldots, p)$, ZERO ← $(0, 0, \ldots, 0)$
3: r1 ← A + B // component sum of vectors using saturation
4: r2 ← r1 - P; // component subtraction using saturation
5: m1: if $r1[i] == P$ then $m1[i] \leftarrow FF$ else $m1[i] \leftarrow 0$ //component comparison
6: m2: if $r2[i] == 0$ then $m2[i] \leftarrow FF$ else $m2[i] \leftarrow 0$ //component comparison
7: r3 ← m1 XOR m2 // bitwise XOR
8: r4 ← r3 AND r1 // bitwise AND
9: c ← r4 OR r2 //bitwise OR
10: **end function**

---

operation guaranteeing exact match of the value regardless of the its representation. Example 1 presents a C/C++ implementation of Algorithm 2 using SSE instructions, where the comments show a partial example. The version that uses 256-bit registers is analogous and therefore is not presented here.

Example 1. //a=(1,0,0,100,70,70,...), b=(0,1,0,1,70,20,...), p=101

```
void add_128_u(__m128i  a, __m128i  b, __m128i  c){
__m128i r1, r2, r3, r4, m1, m2, m3;
r1 = _mm_adds_epu8(a,b);         //( 1, 1, 0,101,140, 90,...)
r2 = _mm_subs_epu8(r1,P);        //( 0, 0, 0,  0, 39,  0,...)
m1 = _mm_cmpeq_epi8(r1,P);       //( 0, 0, 0, FF,  0,  0,...)
m2 = _mm_cmpeq_epi8(r2, ZERO);   //(FF,FF,FF, FF,  0, FF,...)
r3 = _mm_xor_si128(m1, m2);      //(FF,FF,FF,  0,  0, FF,...)
r4 = _mm_and_si128(r1,r3);       //( 1, 1, 0,  0,  0, 90,...)
c = _mm_or_si128(r4, r2);}       //( 1, 1, 0,  0, 39, 90,...)
```

The instruction set AVX512 has a function for 'less than' comparison of unsigned data types. Thus, when working with 512-bit registers we can compare the addition result for 'less than' P storing the result in a single *mask*. Here we will also have additional steps since the comparison functions return as result a mask register. Its length depends on the number of elements in the register of the corresponding data type (in our case 64 bits). It has non-zero bit i if the comparison is true for the i-th element. We execute supplementary command to transform this mask into a larger register needed for the rest of the computations using an all one bit register f and a blend function. Example 2 shows implementation using 512-bit registers and AVX512 instruction set.

Example 2.

```
void add_512_u(__m512i  a, __m512i  b, __m512i  c){
__mmask64 mask; __m512i r1, r2, r3, r4, f = _mm512_set1_epi8(-1);
r1 = _mm512_adds_epu8(a, b);
r2 = _mm512_subs_epu8(r1, P);
mask = _mm512_cmplt_epu8_mask(r1,P);
r3 = _mm512_mask_blend_epi8(mask, ZERO,f);
r4 = _mm512_and_si512(r3, r1);
c = _mm512_or_si512(r4,r2);}
```

3.2 Calculation of the Weight of a Vector

We can calculate the weight of a given vector $c \in F_p^n$ using the formula $n - w_0$, where w_0 is the number of the zero coordinates in c. Let us consider the implementation using 128-bit registers. For simplicity we assume $n = 16$ since the register can contain up to 16 elements ($n = 32$ for 256-bit registers and $n = 64$ for 512-bit registers). We want to construct a register $R = (R_1, R_2)$ that has exactly w_0 non-zero bits in its lower 64 bits (R_2). Therefore, the weight of vector c will be $16 - popcnt(R_2)$. For the construction of R we first execute component comparison of c with zero. The result is a register with value FF at the positions of the zero elements. We need each of those positions to be represented by a single bit and to accumulate the 128-bit data into a 64-bit computer word. The mask register $h = (2,2,2,2,2,2,2,2,1,1,1,1,1,1,1,1)$ and a bitwise AND operation are used for this purpose. The right shift of 64-bits and bitwise OR operation accumulate the result into the described form of R as shown in Example 3. The implementation with 256-bit registers needs to execute two more operations - a right shift of 128 bits and another OR operation. However, a shift of full 128-bit lane is not currently possible. Therefore, we consider the 256-bit register as 2 registers of length 128 bits. The OR operation is executed for those two 128-bit registers. This can be implemented with two auxiliary instructions - a compiler functions that does not translate into a CPU instruction and an extraction function for storing the lower 128-bits.

```
Example 3. h = (2,  2,  2,  2,  2,  2,  2,  2,  1,   1, 1, 1, 1, 1, 1, 1)
a + b = c      => (1,  1,  0,  0,39,90, 0,16, 9,100, 0,53,25, 0, 2, 0)
m: m[i]=c[i]>0=> (0,  0,FF,FF, 0, 0,FF, 0, 0,  0,FF, 0, 0,FF, 0,FF)
r1 = m AND h   => (0,  0,  2,  2, 0, 0, 2, 0, 0,  0, 1, 0, 0, 1, 0, 1)
r2 = r1 >> 64  => (0,  0,  0,  0, 0, 0, 0, 0, 0,  0, 2, 2, 0, 0, 2, 0)
R = r1 OR r2   => (0,  0,  2,  2, 0, 0, 2, 0, 0,  0, 3, 2, 0, 1, 2, 1)
R_2 = (0,   0,  3,  2,  0,  1,  2,  1) => w0 = popcnt(R_2)
```

The implementation using AVX512 is conceptually different than the 128-bit version. We calculate w_0 only by comparing c to zero. As previously mentioned the compare functions for the 512-bit registers return as a result a mask register with the appropriate length depending on the data type. Therefore, the calculation of $wt(c)$ can be executed using a single popcnt instruction over the resulting mask of the comparison (64 - popcnt(_mm512_cmpeq_epi8_mask(c, ZERO))).

4   Experimental Results

In this section we present some experimental results for the efficiency of the described methods for vector addition and calculation of the weight of a vector. All presented computations were executed on a single core. We compare the execution times of an algorithm that calculates the weight distribution of a linear code over finite field that contains the two main functions presented in this paper. Table 1 shows the computation times in seconds for the versions

Table 1. Comparison for different vector length

Parameters	128-bit register	256-bit register	512-bit register
n = 40 k = 8 p = 17	5.57	5.21	3.86
n = 120 k = 8 p = 17	9.83	7.25	4.62
n = 240 k = 8 p = 17	15.36	10.58	6.10
n = 40 k = 5 p = 101	1.25	1.14	0.82
n = 120 k = 5 p = 101	2.23	1.69	1.01
n = 240 k = 5 p = 101	3.74	2.48	1.39
n = 40 k = 5 p = 127	3.13	2.85	1.99
n = 120 k = 5 p = 127	5.80	4.22	2.47
n = 240 k = 5 p = 127	9.10	6.19	3.44

integrating the 128, 256 and 512-bit registers. Computations are executed on a Intel Xeon Gold CPU with 24 cores, 48 threads, 2.3 GHz clock frequency and Linux OS. The first column shows the parameters of the codes. The next 3 columns give the execution times in seconds using different register lengths. As can be seen form Table 1 the 128-bit version is approximately 5% slower than the 256-bit version for $n = 40$ and approximately 50% slower for $n = 240$. It is also about 2 times slower than the 512-bit implementation. When we compare 256 and 512-bit version we have an average decrease of 59%. As expected the resulting speedup with the larger registers is greater for bigger lengths.

Table 2 shows a comparison of the presented 128 and 512-bit versions and two major software packages for and mathematical computations - Magma and GAP. The first column again presents the code parameters. Columns 2 and 3 show the execution times in seconds for the calculation of the weight distribution of the code using Magma and GAP packages respectively. Columns 4 and 5 show the execution times for 128 and 512-bit version. The Magma calculations were executed online Magma Calculator running in a virtual machine on an Intel Xeon Processor E3-1220, 3.10 GHz. As can be seen form the experimental results both 128 and 512 bit version are a few times faster than Magma calculator and many

Table 2. Comparison for with different software packages

Parameters	Magma	GAP	128-bit register	512-bit register
n = 40 k = 8 p = 17	24.64	840.56	5.57	3.86
n = 80 k = 8 p = 17	51.91	1667.72	7.34	4.64
n = 40 k = 6 p = 31	1.75	101.11	0.37	0.24
n = 80 k = 6 p = 31	3.63	198.84	0.53	0.29
n = 40 k = 5 p = 101	6.45	1188.84	1.25	0.82
n = 80 k = 5 p = 101	12.33	2022.77	1.67	1.00

times faster than the open source Guava package for GAP. Depending on the length of the code the presented 512-bit version gives speedup between 7 and 12 times compared to the Magma calculator. When comparing to the GAP package the presented methods can be between 150 and 2000 times faster.

5 Conclusionary Remarks

Vector addition over prime finite fields and the computation of the weight of a vector can be used in different algorithms connected to many problems in coding theory and other research areas. The presented methods and the experimental results show that using extended vector registers can result in an improvement of the execution time compared to other packages. The comparison of the implementations using different instruction sets show two main points - the presented method allows for the computations not to be affected by the value of p and that the speedup with larger registers is not necessary proportional to the length increase. Furthermore, integration of AVX512 with other parallel application programming interfaces (APIs) may result in unexpected problem and reduced CPU work frequency [1,10]. Thus, the choice of instruction set for any given application should be carefully considered.

References

1. Intel64 and IA-32 Architectures Optimization Reference Manual. https://www.intel.com/content/www/us/en/developer/articles/technical/intel-sdm.html. Accessed 21 Feb 2023
2. Pashinska-Gadzheva, M., Bouyukliev, I.:LinCodeWeightInv: Library for Computing the Weight Distribution of Linear Codes Over Finite Fields, in preparation
3. Bosma, W., Cannon, J., Playoust, C.: The Magma algebra system I: the user language. J. Symbolic Comput. **24**, 235–265 (1997)
4. GAP package GUAVA. https://www.gap-system.org/Packages/guava.html. Accessed 21 Feb 2023
5. Lidl, R., Niederreiter, H.: Finite Fields, 2nd edn. Cambridge University Press, Cambridge (1997)
6. Bouyukliev, I., Bakoev, V.: A method for efficiently computing the number of codewords of fixed weights in linear codes. Discret. Appl. Math. **156**, 2986–3004 (2008)
7. Berlekamp, E.R., McEliece, R.J., van Tilborg, H.C.: On the inherent intractability of certain coding problems. IEEE Trans. Inform. Theory **24**, 384–386 (1978)
8. Coolsaet, K.: Fast vector arithmetic over F3. Bull. Belg. Math. Soc. Simon Stevin **20**, 329–344 (2013)
9. Bouyukliev, I., Bakoev, V.: Efficient computing of some vector operations over GF(3) and GF(4). Serdica J. Comput. **2**(2), 101–108 (2008)
10. Gottschlag, M., Brantsch, P., Bellosa F.: Automatic core specialization for AVX-512 applications. In: Proceedings of the 13th ACM International Systems and Storage Conference (SYSTOR 2020), pp. 25–35. Association for Computing Machinery (2020)

Grid Search Optimization of Novel SNN-ESN Classifier on a Supercomputer Platform

Dimitar Penkov, Petia Koprinkova-Hristova(✉), Nikola Kasabov,
Simona Nedelcheva, Sofiya Ivanovska, and Svetlozar Yordanov

Institute of Information and Communication Technologies,
Bulgarian Academy of Sciences, Sofia, Bulgaria
{dimitar.penkov,petia.koprinkova,nikola.kasabov,
simona.nedelcheva}@iict.bas.bg, {sofia,svetlozar}@parallel.bas.bg

Abstract. This work is demonstrating the use of a supercomputer platform to optimise hyper-parameters of a proposed by the team novel SNN-ESN computational model, that combines a brain template of spiking neurons in a spiking neural network (SNN) for feature extraction and an Echo State Network (ESN) for dynamic data series classification. A case study problem and data are used to illustrate the functionalities of the SNN-ESN. The overall SNN-ESN classifier has several hyper-parameters that are subject to refinement, such as: spiking threshold, duration of the refractory period and STDP learning rate for the SNN part; reservoir size, spectral radius of the connectivity matrix and leaking rate for the ESN part. In order to find the optimal hyper-parameter values exhaustive search over all possible combinations within reasonable intervals was performed using supercomputer Avitohol. The resulted optimal parameters led to improved classification accuracy. This work demonstrates the importance of model parameter optimisation using a supercomputer platform, which improves the usability of the proposed SNN-ESN for real-time applications on complex spatio-temporal data.

Keywords: Spikking Neural Network · Echo State Network · Classification

1 Introduction

Electroencephalography (EEG) is a method to record the electrical activity of the brain. It is typically non-invasive, with the EEG electrodes placed along the

Supported by the HORIZON-EIC action under the project "Auto-adaptive Neuromorphic Brain Machine Interface: toward fully embedded neuroprosthetics (NEMO-BMI)", No 101070891/01.10.2022. S. Nedelcheva was partially supported by the Bulgarian Ministry of Education and Science under the National Research Programme "Young scientists and postdoctoral students-2" approved by DCM 206/07.04.2022. S. Ivanovska was partially supported by a grant by CAF America. We acknowledge the provided access to the e-infrastructure of the NCHDC - part of the Bulgarian National Roadmap on RIs, with the financial support by the Grant No D01-168/28.07.2022.

I. Lirkov and S. Margenov (Eds.): LSSC 2023, LNCS 13952, pp. 435–443, 2024.
https://doi.org/10.1007/978-3-031-56208-2_45

scalp. The measured in this way signals represent the postsynaptic potentials of pyramidal neurons in the cortex. Since the electrical activity in the brain surface originates in the deeper brain areas that do not contribute directly to an EEG recording, their influence could be assessed accounting for the electrodes orientation and distance to the source of the activity. By far EEG has been applied to numerous domains from brain-computer interface [1–3], emotion recognition [4–8], control of movements [9,10], diagnostic of brain diseases [11] etc. Through the years the EEG data processing methodology has evolved from simple methods, such as mean and amplitude comparison to complicated methods, such as connectivity topology and deep learning [12,13]. In particular, deep learning exhibits better performance in EEG classification in comparison with the conventional methods. Nevertheless, accurate on-line classification and explanation of dynamic spatio-temporal brain data, such as EEG is still an open problem. While there are many methods introduced for brain data classification, most of them lack explainability in relation to the measured brain functions as spatio-temporal patterns.

In order to exploit EEG data for analysis the first step traditionally is to "decode" them [10,14] or to extract a range of signal properties referred to as "features" which are then utilized for detection or classification purposes [15]. The analysis outcome is largely influenced by the quality of extracted features. A recent trend in EEG features extraction and processing exploits recurrent neural networks [11] and especially a member of reservoir family - fast on-line trainable Echo state networks (ESN) [6–9,16,17]. The reservoir computing advantage is in its training algorithms. They usually need a single training epoch by Least Squares (LS) algorithm or on-line training via the Recursive Least Squares (RLS) method. However, their randomly initialized pool of neurons does not account for brain structure, so reservoir computing lacks explainability. Accounting for spatial brain structure in the design of RNN decoders or feature extractor became a natural direction of work nowadays [4]. Recently developed brain-inspired spiking neural network (SNN) models, such as NeuCube [1–3,5,18], demonstrated a good classification accuracy and excellent explanation of the spatio-temporal patterns learned from spatio-temporal brain data, such as EEG and fMRI [19].

In [24] we proposed another combination between fast trainable reservoir computing and brain-inspired architectures - the integration of the SNN module with an ESN classifier, aiming at improved classification accuracy in an on-line learning mode. The overall SNN-ESN classifier has multiple hyper-parameters that are subject to refinement, such as spiking threshold, duration of the refractory period and STDP learning rate for the SNN part; reservoir size, spectral radius of the connectivity matrix, leaking rate and scalling parameters for the ESN part. Since all possible combinations within reasonable intervals of these parameters are too much, their refinement using the exhaustive search on a desktop computer would take enormous time. That is why it was performed on the supercomputer Avitohol.

Further the paper is organized as follows: Sect. 2 presents briefly NeuCube and ESN structures and the proposed hierarchical architecture SNN-ESN for

EEG classification; Sect. 3 presents the optimization procedure and the obtained best hyper-parameters values; finally the concluding remarks summarize the main achievements in the presented work and points the directions for future work.

2   SNN-ESN Architecture

2.1   SNN Structure

The NeuCube architecture is an open one, allowing for new algorithms to be explored for encoding, learning, classification, regression [21]. It consists of three parts [19]: data encoding part, where input streaming data is encoded into spike sequences using a suitable algorithm [22]; a 3D Cube structure of spiking neurons, where every neuron has a 3D spatial coordinates defined through the use of brain-template, such as Talairach or MNI. Initial connections were generated randomly based on the distances between each two neurons; SNN classifiers of evolving spiking neuron networks (eSNN) or dynamic evolving spiking neuron networks (deSNN) [23] are used to separate the outputs of NeuCube into classes. After encoding of the spatio-temporal EEG data into sequences of spikes (spike-time information), the Cube, structured according to a brain template, receives as input EEG recordings at neurons corresponding to the electrodes' positions on the scull. As a result all neurons in the Cube generate spike trains whose dynamics depends on the input signal as well as on both the connectivity within the Cube (small world connectivity at the beginning) and on the spike time dependent plasticity (STDP) of the connections (synapses). Finally, the output classifier takes the Cube spike trains as classification features. The hyper-parameters of the SNN are the parameters of neurons such as threshold of the membrane potential V_{th}, refractory period t_{ref} during with the membrane potential returns to its base value as well as STDP learning rate λ.

2.2   ESN Structure

Echo state networks (ESN) belong to a novel and rapidly developing family of reservoir computing approaches [25–27] whose aim was development of fast trainable recurrent neural network (RNN) architectures able to approximate nonlinear time series dependencies. It incorporates a pool of neurons with sigmoid activation function (usually the hyperbolic tangent) that has randomly generated recurrent connection weights. The only trainable parameters of ESN are the output weights. In case of identity output function the least squares method is applied to train the ESN in a single iteration. For the purpose of online training the recursive version of least squares (RLS) can be applied too [25]. The ESN hyper parameters that are subject of manual tuning, usually via grid search, are the reservoir size (number of neurons), reservoir connection matrix sparsity and spectral radius and leaking rate. Additionally input and output scaling could be included.

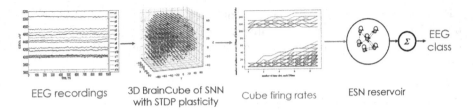

EEG recordings   3D BrainCube of SNN   Cube firing rates   ESN reservoir
                      with STDP plasticity

Fig. 1. Proposed NeuCube - ESN structure.

2.3   SNN-ESN Brain Data Classifier

The proposed novel brain data classifier called SNN-ESN is a hierarchical RNN
composed by two recurrent architectures - a brain inspired NeuCube, that is a
spatio-temporal structure of SNN neurons and a fast trainable ESN as a nonlinear time series classifier. The overall structure is shown on Fig. 1.

In contrast to NeuCube approach, here the EEG data is scaled and fed into
the Cube as generating currents into neurons corresponding to the electrodes
positions. A 3D SNN Cube is initialised by defining the size of the Cube of neurons and their 3D locations, including positions of EEG electordes. The Cube is
designed according to the scalable Talairach atlas [28]. Small-world connectivity
method is used to derive the initial connections in the Cube, where the closer
two neurons are in the 3D space, the higher the probability of them to be connected is. All synapses are plastic, i.e. they adapt their weights via Spike Timing
Dependent Plasticity (STDP) rule that is a kind of unsupervised learning. Thus
the achieved after feeding the input signal to the Cube connectivity will reflect
the EEG data characteristics. Spiking activity of all neurons was recorded and
for a given time window the spiking frequency of all neurons in the Cube is
calculated. Thus a sample EEG record a new time series of Cube firing rates is
extracted as feature vector as shown on Fig. 1. The output classifier is an ESN
reservoir. It receives generated by Cube time series feature per given EEG sample. The achieved reservoir state after presentation of each EEG sample feature
vector was send to its readout and the output weights were adjusted to predict
the correct EEG class. The training was done via RLS in on-line mode.

3   Grid Search Optimization

The SNN Cube is implemented in NEST Simulator, version 3.3 [29], using leaky
integrate-and-fire neuron model. The ESN was implemented in Python.

The benchmark data used here is taken from [20]. The EEG data of 14
channels *Emotiv* measuring device were collected for 1000ms with sampling
frequency of 128 Hz. The test subject is asked to perform three different types
of wrist movement - up, down and straight - that are separated into three EEG
classes and 20 examples per class are collected, making all number of samples

60. The location of the input neurons, corresponding to the used in this case 14 EEG channels, is defined following the 10–20 EEG location system. The Cube consists of 1471 neurons with initial randomly generated connectivity having 80% positive and 20% negative values.

The parallel implementation of the ESN module in Python was done using mpi4py library[1]. The NEST Simulator has its own MPI that distributes the SNN module simulation among the specified number of threads. However it does not allow to run it as a part of another parallel simulation. That is why we run both modules separately.

Two optimization experiments were performed:

1. Complete SNN-ESN model exploration running consecutively each one of them
2. Exploration of the ESN module that receives as input directly EEG data

Hyper-parameters of both modules that has to be refined and their values are given in Table 1. The number of all possible combinations of the SNN module parameters is $3 \times 6 \times 3 = 54$. The ESN module parameters yield $5 \times 5 \times 5 \times 6 \times 6 = 4500$ combinations. Thus the grid search should evaluate accuracy of totally 243000 variants of SNN-ESN model and 4500 for only ESN module respectively.

Table 1. Hyper-parameters subject to optimization

Parameter	Values
SNN	
threshold of membrane potential V_{th}	$-60, -55, -50$
refractory time t_{ref}	$0, 1, 2, 3, 4, 5$
STDP learning rate λ	$0.1, 0.01, 0.001$
ESN	
leaking rate	$0.4, 0.5, 0.6, 0.7, 0.8$
reservoir size	$3000, 3500, 4000, 4500, 5000$
reservoir sparsity	$0.4, 0.5, 0.6, 0.7, 0.8$
scale in	$0.00001, 0.0001, 0.001, 0.01, 0.1, 1$
scale out	$0.00001, 0.0001, 0.001, 0.01, 0.1, 1$

We compare time needed for grid search on an HPC facility of our institute - the supercomputer Avitohol[2] - with estimated time needed to run the same task on a desktop computer architecture. The desktop configuration has 2.60 GHz Intel(R) Core(TM) i7-6500U CPU with 2 cores and 16.0 GB RAM. The HPC System Avitohol consists of 150 servers ProLiant SL250s Gen8 each with dual Xeon CPU E5-2650 v2 at 2.60 GHz and dualXeon Phi 7120P accelerator cards.

---

[1] https://mpi4py.readthedocs.io/en/stable/index.html.
[2] http://www.hpc.acad.bg/system-1/.

In total it has 9600 GB RAM accessible by the regular CPUs and 4800 GB RAM on the accelerator cards. The operating system on the servers is Red Hat Enterprise Linux. The exact version on the servers that were deployed during our tests was 6.7. Table 2 summarizes the estimated simulation time of both optimization tasks on the desktop and supercomputer architectures.

Table 2. Estimated optimization times on desktop and HPC architectures in days.

Task	Module	Time	Processes	Nodes (Cores)
Desktop				
1	SNN	0.868	1	1(2)
1	ESN	112.808	2	1(2)
2	ESN	1.981	2	1(2)
Supercomputer				
1	SNN	0.622	32	1 (16)
1	ESN	8	54	4 (64)
2	ESN	0.124	30	2 (32)

It is obvious that HPC allows to decrease significantly the time needed to solve the optimization time, especially with respect to the Python module. The best SNN parameters are: threshold of membrane potential $V_{th} = -60$, refractory time $t_{ref} = 0$, STDP learning rate $\lambda = 0.001$. The obtained optimal ESN hyperparameter values for all considered reservoir sizes and corresponding test error of the classifier are given in Table 3 for task 1 and in Table 4 for task 2 respectively.

Table 3. Minimum test error for each size of the ESN reservoir and SNN parameters $V_{th} = -60$, $t_{ref} = 0$, $\lambda = 0.001$ (task 1)

Reservoir size	Reservoir sparsity	Leaking rate	Scale in	Scale out	Test error
3000	0.7	0.4	1e−5	1e−5	1.589e−05
3500	0.7	0.6	1e−3	1e−2	1.358e−05
4000	0.6	0.4	1e−4	1e−2	1.590e−05
4500	0.7	0.4	1e−3	1e−5	1.360e−05
5000	0.6	0.4	1e−5	1e−2	1.292e−05

While for task 1 the smallest test error was obtained using the highest tested ESN reservoir size (5000 neurons), for task 2 the best results were obtained with ESN having smaller neurons number (4000). However, the test error in that case is much higher for all considered reservoir sizes. So the features extracted by the SNN module improved significantly the classification accuracy.

Table 4. Minimum test error for each size of the ESN reservoir without the SNN pre-processing (task 2)

Reservoir size	Reservoir sparsity	Leaking rate	Scale in	Scale out	Test error
3000	0.4	0.7	1e−5	1e−2	0.013
3500	0.4	0.5	1e−5	1e−1	0.015
4000	0.4	0.5	1e−5	1e−5	0.011
4500	0.7	0.5	1e−5	1e−2	0.013
5000	0.7	0.4	1e−3	1e−2	0.018

4 Conclusions

The carried out simulation investigation confirmed the significance of HPC facility in solving hard optimization tasks such as presented here grid search of hierarchical RNN model parameters. The decrease of simulation time was about 14–16 times in comparison with a given desktop configuration.

References

1. Padfield, N., Camilleri, K., Camilleri, T., Fabri, S., Bugeja, M.: A comprehensive review of endogenous EEG-based BCIs for dynamic device control. Sensors **22**(15), Article no. 5802 (2022)
2. Ieracitano, C., Mammone, N., Hussain, A., Morabito, F.C.: A novel explainable machine learning approach for EEG-based brain-computer interface systems. Neural Comput. Appl. **34**(14), 11347–11360 (2022)
3. Singanamalla, S.K.R., Lin, C.-T.: Spike-representation of EEG signals for performance enhancement of brain-computer interfaces. Front. Neurosci. **16**, Article no. 792318 (2022)
4. Zhou, J., Zhao, T., Xie, Y., Xiao, F., Sun, L.: Emotion recognition based on brain connectivity reservoir and valence lateralization for cyber-physical-social systems. Pattern Recogn. Lett. **161**, 154–160 (2022)
5. Luo, Y., et al.: EEG-based emotion classification using spiking neural networks. IEEE Access **8**, 46007–46016 (2020)
6. Fourati, R., Ammar, B., Sanchez-Medina, J., Alimi, A.M.: Unsupervised learning in reservoir computing for EEG-based emotion recognition. IEEE Trans. Affect. Comput. **13**(2), 972–984 (2022)
7. Bozhkov, L., Koprinkova-Hristova, P., Georgieva, P.: Learning to decode human emotions with Echo State Networks. Neural Netw. **78**, 112–119 (2016)
8. Bozhkov, L., Koprinkova-Hristova, P., Georgieva, P.: Reservoir computing for emotion valence discrimination from EEG signals. Neurocomputing **231**, 28–40 (2017)
9. Khan, Z.H., Hussain, N., Tiwana, M.I.: Classification of EEG signals for wrist and grip movements using echo state network. Biomed. Res. (India) **28**(3), 1095–1102 (2017)
10. Kim, H., Kim, J.S., Chung, C.K.: Identification of cerebral cortices processing acceleration, velocity, and position during directional reaching movement with deep neural network and explainable AI. NeuroImage **266**, Article no. 119783 (2023)

11. Ruffini, G., Ibañez, D., Castellano, M., Dunne, S., Soria-Frisch, A.: EEG-driven RNN classification for prognosis of neurodegeneration in at-risk patients. In: Villa, A.E.P., Masulli, P., Pons Rivero, A.J. (eds.) ICANN 2016. LNCS, vol. 9886, pp. 306–313. Springer, Cham (2016). https://doi.org/10.1007/978-3-319-44778-0_36

12. Gong, S., Xing, K., Cichocki, A., Li, J.: Deep learning in EEG: advance of the last ten-year critical period. IEEE Trans. Cogn. Dev. Syst. **14**(2), 348–365 (2022)

13. Nakagome, S., Craik, A., Ravindran, A.S., He, Y., Cruz-Garza, J.G., Contreras-Vidal, J.L.: Deep learning methods for EEG neural classification. In: Thakor, N.V. (ed.) Handbook of Neuroengineering, pp. 1–39. Springer, Singapore (2023). https://doi.org/10.1007/978-981-15-2848-4_78-1

14. Phadikar, S., Sinha, N., Ghosh, R.: Unsupervised feature extraction with autoencoders for EEG based multiclass motor imagery BCI. Expert Syst. Appl. **213**, Article no. 118901 (2023)

15. Yuvaraj, R., Thagavel, P., Thomas, J., Fogarty, J., Ali, F.: Comprehensive analysis of feature extraction methods for emotion recognition from multichannel EEG recordings. Sensors **23**(2), Article no. 915 (2023)

16. Jeong, D.-H., Jeong, J.: In-ear EEG based attention state classification using echo state network. Brain Sci. **10**(6), Article no. 321 (2020)

17. Sun, L., Jin, B., Yang, H., Tong, J., Liu, C., Xiong, H.: Unsupervised EEG feature extraction based on echo state network. Inf. Sci. **475**, 1–17 (2019)

18. Kasabov, N.K.: NeuCube: a spiking neural network architecture for mapping, learning and understanding of spatio-temporal brain data. Neural Netw. **52**, 62–76 (2014)

19. Kasabov, N.K.: Time-Space, Spiking Neural Networks and Brain-Inspired Artificial Intelligence. Springer, Cham (2019)

20. Hu, J., Hou, Z.-G., Chen, Y.-X., Kasabov, N., Scott, N.: EEG-based classification of upper-limb ADL using SNN for active robotic rehabilitation. In: 5th IEEE RAS/EMBS International Conference on Biomedical Robotics and Biomechatronics, Sao Paulo, Brazil, pp. 409–414 (2014)

21. NeuCube development environment. https://kedri.aut.ac.nz/neucube

22. Petro, B., Kasabov, N., Kiss, R.: Selection and optimisation of spike encoding methods for spiking neural networks, algorithms. IEEE Trans. Neural Netw. Learn. Syst. **31**(2), 358–370 (2019)

23. Kasabov, N.K., Dhoble, K., Nuntalid, N., Indiveri, G.: Dynamic evolving spiking neural networks for online spatio- and spectro-temporal pattern recognition. Neural Netw. **41**, 188–201 (2013)

24. Koprinkova-Hristova, P., Kasabov, N., Nedelcheva, S., Yordanov, S., Penkov, D.: On-line learning, classification and interpretation of brain signals using 3D SNN and ESN. IJCNN 2023 (accepted)

25. Jaeger, H.: Tutorial on training recurrent neural networks, covering BPPT, RTRL, EKF and the "echo state network" approach. GMD Report 159, German National Research Center for Information Technology (2002)

26. Gallicchio, C., Lukosevicius, M., Scardapane, S.: Frontiers in reservoir computing. In: Proceedings of 28th European Symposium on Artificial Neural Networks, Computational Intelligence and Machine Learning, ESANN 2020, Belgium, pp. 559–566 (2020)

27. Lukosevicius, M., Jaeger, H.: Reservoir computing approaches to recurrent neural network training. Comput. Sci. Rev. **3**, 127–149 (2009)

28. Talairach daemon. http://www.talairach.org

29. Spreizer, S., et al.: NEST 3.3 (3.3) (2022). Zenodo. https://doi.org/10.5281/zenodo.6368024

Parallelisms of $PG(3,4)$ with a Great Number of Regular Spreads

Svetlana Topalova[ID] and Stela Zhelezova[✉][ID]

Institute of Mathematics and Informatics, Bulgarian Academy of Sciences,
Sofia, Bulgaria
{svetlana,stela}@math.bas.bg

Abstract. A *spread* in the finite projective space $PG(n,q)$ is a set of
lines which partition the point set. A *parallelism* is a partition of the set
of lines by spreads. Parallelisms with many regular spreads are of particular interest. All parallelisms of $PG(3,4)$ which are invariant under
automorphisms of orders greater than 2 and some of the parallelisms
with automorphisms of order 2 are known. Among them there are no
parallelisms with more than 13 regular spreads (out of all 21 spreads).
To establish whether a parallelism of $PG(3,4)$ with more than 13 regular spreads exists was an open problem before the present work. We
construct all parallelisms with automorphisms of order 2 and at least 13
regular spreads and succeed to improve the previous result by finding
out that there exist parallelisms of $PG(3,4)$ with 16 regular spreads.

Keywords: finite projective space · spread · parallelism · regular ·
automorphisms

1 Introduction

1.1 Spreads and Parallelisms

Let $PG(n,q)$ be the n-dimensional projective space over the finite field $GF(q)$. A
spread in $PG(n,q)$ is a set of lines which partition the point set. A *parallelism* is
a partition of the set of lines by spreads. More details on spreads and parallelisms
can be found in [9,11,17].

Research on spreads and parallelisms is motivated by their various relations
and applications. Their relation to translation planes is the best known one
[6,9,11], and the recently most investigated relation is that to subspace codes
[8] because of their application in random network coding [12].

The research of the first author is partially supported by the MES, Grant No. DO1-
325/01.12.2023 for NCHDC - part of the Bulgarian National Roadmap on RIs and
of the second by the Bulgarian National Science Fund under Contract No KP-06-
H62/2/13.12.2022.

I. Lirkov and S. Margenov (Eds.): LSSC 2023, LNCS 13952, pp. 444–452, 2024.
https://doi.org/10.1007/978-3-031-56208-2_46

1.2   Regular Parallelisms

A *regulus* of PG(3, q) is a set of $q + 1$ mutually skew lines such that any line intersecting three elements of the regulus intersects all its elements. A spread of PG(3, q) is *regular* if it contains the unique regulus determined by any three of its elements. The regular spread is unique up to isomorphism. In PG(3, 4) there are 3 spreads up to isomorphism - regular, *aregular* (containing none of the reguli determined by any three of its elements) and *subregular*.

A parallelism is *regular* if all its spreads are regular. Regular parallelisms are particularly interesting because of their connection to translation planes. Lunardon [13] and independently Walker [22] proved that every regular parallelism in $PG(3, q)$ leads to a spread in $PG(7, q)$, and determines a translation plane of order q^4. And vice versa such translation planes produce regular parallelisms [10].

Pentilla and Williams have constructed two regular cyclic parallelisms of $PG(3, q)$ for each $q \equiv 2 \pmod 3$ [14]. Among them are the two well known parallelisms of $PG(3, 2)$, which are regular because every spread of this space is regular. The class of Pentilla and Williams includes the regular parallelisms of $PG(3, 5)$ (constructed by Prince [16]) and of $PG(3, 8)$ (first found by Denniston [7]). Outside this class only 6 regular parallelisms of $PG(3, 5)$ are presently known [20]. They are not cyclic and have a full automorphism group of order 3.

The nonexistence of regular parallelisms in PG(3, 3) was first proved by Prince [15], and later confirmed by Betten's classification of all parallelisms of PG(3, 3) [2]. The nonexistence of regular parallelisms in $PG(3, 4)$ was announced by Bamberg [1] in 2012.

1.3   The Problem and Our Motivation

If regular parallelisms do not exist in some projective space PG(3, q), the next question in consideration is about the existence of parallelisms with $q^2 + q + 1 - m$ regular spreads (out of all $q^2 + q + 1$ spreads in the parallelism). A parallelism in $PG(3, q)$ with all but m regular spreads can be used to construct a maximal partial spread in PG(7, q) [11, Theorem 65] and has additional properties and relations if $m \leq 2$ [11, Theorem 67]. That is why small values of m are of particular interest.

From the classification of parallelisms of PG(3, 3) [2] one can see that the smallest value of m is 2. The smallest known value of m in PG(3, 4) is 8 (13 regular spreads) [5, 21]. We show here that there exist parallelisms of PG(3, 4) with $m = 5$.

Parallelisms of PG(3, 4) invariant under automorphisms of orders greater than 2 have been classified [3, 4, 18, 19, 21], but the problem of the classification of parallelisms with full automorphism groups of order at most 2 remains open and is presently infeasible. It might be solvable, however, under certain additional restrictions [4]. We set here the restriction that the parallelisms should have at least 13 regular spreads. This way we succeed to construct parallelisms of PG(3, 4) with a full automorphism group of order 2 and 16 regular spreads ($m = 5$).

We use our own software written in C++. Part of it is based on MPI. Part of the results are obtained on PC, and part on the HPC system Avitohol (http://nchdc.acad.bg/en/resources/iict/avitohol/) with Intel Xeon E5-2650 v2 @2.60 GHz processors.

2 Construction of the Parallelisms

2.1 PG(3, 4) and the Considered Automorphism Group G_2

Consider the finite field $GF(4)$ with generating polynomial $x^2 + x + 1 = 0$ and primitive element ω. To construct $PG(3,4)$ we use $V(4,4)$ - the 4-dimensional vector space over $GF(4)$. The points of $PG(3,4)$ are then all 4-dimensional vectors (v_1, v_2, v_3, v_4) over $GF(4)$ such that $v_i = 1$ if i is the maximum index for which $v_i \neq 0$. We sort these 85 vectors in ascending lexicographic order and then assign them numbers such that $(1, 0, 0, 0)$ is number 1, and $(\omega^2, \omega^2, \omega^2, 1)$ number 85. There are 357 lines (1-dimensional subspaces) in $PG(3,4)$. We sort them in ascending lexicographic order defined on the numbers of the points they contain and assign to each line a number according to this order. The first line contains points $1, 2, 3, 4, 5$ and the last one $21, 37, 38, 64, 75$.

Denote by G the full automorphism group of PG(3, 4). It is conjugate to $P\Gamma L(4, 4)$. Each invertible matrix $(a_{i,j})_{4\times4}$ over $GF(4)$ defines an automorphism of this projective space by the map $v'_i = \sum_j a_{i,j} v_j$.

The parallelisms of PG(3, 4) which have not been classified by now, have either no nontrivial automorphisms, or a full automorphism group G_2 of order 2 [5]. The group G_2 can be defined by the next invertible matrix:

$$\begin{bmatrix} 1 & 0 & 0 & 0 \\ 1 & 1 & 0 & 0 \\ 0 & 0 & 1 & 0 \\ 0 & 0 & 1 & 1 \end{bmatrix}$$

This group fixes one line pointwise and partitions the lines to 189 orbits, 21 of them trivial. The normalizer $N_G(G_2)$ of G_2 in G is defined as $N_G(G_2) = \{g \in G \mid g G_2 g^{-1} = G_2\}$ and is of order 30720.

2.2 The Structure of a Parallelism Invariant Under G_2

A parallelism of PG(3, 4) is a partition of its 357 lines in 21 spreads each comprising 17 mutually disjoint lines. Lemma 1 of [4] gives the structure of a parallelism of PG(3, 4) invariant under G_2. It has 13 spread orbits - 5 of length 1 (fixed spreads) and 8 of length 2. For what follows the constructed parallelisms are considered as a union of two parts: fixed and nonfixed part. The fixed part includes the five fixed spreads, while the nonfixed part includes the 8 spread orbits of length 2 under G_2. One of the fixed spreads has one fixed line and this is exactly the pointwise-fixed line, while each of the other fixed spreads has 5 fixed lines (see Fig. 1).

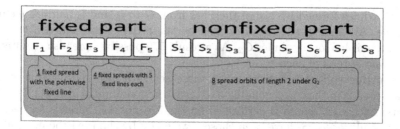

Fig. 1. The structure of a parallelism invariant under G_2.

It is shown in [4] that there are exactly 776928 nonisomorphic fixed parts invariant under G_2, and the types of the spreads in the obtained fixed parts are also presented. There are 7 fixed parts comprising five regular spreads and they have already been extended to parallelisms. The obtained in this case parallelisms contain at most 11 regular spreads. Our aim is to find parallelisms with as many as possible regular spreads. Therefore we are interested in the number of regular spreads in the fixed parts. This information is given in the first and second column of Table 1.

2.3   Construction of the Parallelisms

All the 776921 fixed parts under G_2 have to be extended to parallelisms with at least 13 regular spreads, namely 8 spread orbits $(S_1, S_2, \ldots S_8)$ of length 2 have to be added to each fixed part, and at least $\lceil (13 - a)/2 \rceil$ of them must be regular, where a denotes the number of regular spreads in the fixed part. Each point must be in exactly one line of each spread of the parallelism. That is why without loss of generality we assume that each spread contains a definite line with the first point in it. Hence the first line of an S_i spread is known. We obtain most of the results by different methods and different software based on back-track search techniques.

2.4   Construction Method 1

At first all possible spreads for each S_i are constructed by backtrack search on the lines of the projective space. The spread lines should be from different line orbits under G_2. According to the chosen lexicographic order the first element in each S_i is a line l_1 containing point 1. Then the next two lines l_2 and l_3 are added to form a disjoint triple. The regulus $\rho(l_1, l_2, l_3)$ has two more lines l_4 and l_5 which are added to the spread if the five lines of the regulus are from different line orbits under G_2 (or another l_3 is chosen if they are not). The next spread line l_6 is chosen among the lines which:

- have not been added to the spread yet;
- contain the first missing point in the spread up to the moment;
- are not from the line orbit under G_2 of any of the already chosen five lines.

Then all triples formed by two of the lines from l_1 to l_5 and l_6 as third element are made. All lines of the reguli defined by these triples are added if they are from different line orbits under G_2. As a result 576 possible spreads are obtained for the first spread of each S_i.

We further split the problem to six sub-cases with respect to the number of regular spreads in the corresponding fixed part (first column of Table 1). For each sub-case backtrack search is performed on the 576 possible spreads for each S_i. If all possibilities for S_i have been chosen, we go on by choosing the possibilities for S_{i+1}. This way we find partial parallelisms with regular spread orbits of length 2 (columns 3 and 4 of Table 1). We further consider their extendability to parallelisms (column 5 of Table 1).

Table 1. Parallelisms which are invariant under G_2 and have at least 13 regular spreads

# regular spreads in the fixed part	number of fixed parts	maximal # of regular orbits S_i in partial solutions	maximal # of regular spreads in partial solutions	# of parallelisms with at least 13 regular spreads
5	7	5	15	0
4	55	5	14	0
3	810	5	13	0
2	13731	6	14	0
1	147353	6	13	8
0	614972	8	16	2

2.5   Construction Method 2

As a partial check of the correctness of the obtained results, we have also constructed all parallelisms with G_2 and more than 13 regular spreads by using a different algorithm for extending the fixed parts. As a result the two parallelisms with 16 regular spreads have been found, namely the same results are obtained.

By this extension method we do not construct all the possible regular spreads first. We consider the parallelism as a set of lines grouped in spreads, and use backtrack search to choose the lines of the current spread from the set L of lexicographically sorted lines which do not take part in the current partial solution and contain the smallest point which is not in any line of the current spread. Thus the lines of the constructed spreads are lexicographically sorted.

Since we are interested in parallelisms with at most 7 spreads which are not regular, when there are already 6 or 7 nonregular spreads in the partial solution, we continue by adding only regular spread orbits of length 2. This imposes additional restrictions on the set L. Each three of the lines that have already been added to the current spread define a regulus. We find the smallest line l_m

which is contained in these reguli and therefore must be in the spread, but has not been added to it by now. Since we add the lines in lexicographic order, lines that are bigger than l_m should not be considered at the current step, i.e. we choose only among the lines of L which are lexicographically not bigger than l_m.

2.6   Parallel Implementation

The construction of all the parallelisms with the considered automorphism of order 2 and more than 13 regular spreads by method 2 takes much more time than by method 1 (about 10 min are needed on a 3 GHz PC to extend a fixed part with 2 or 3 regular spreads), but it was worth doing it in order to partially check the obtained results in a very different way. We made this using the high performance computer system Avitohol and our own MPI-based software written in C++. What a PC will have to do in about 4000 h, was completed in 3 days with the help of HPCS Avitohol. The parallel implementation is simple, but very effective and does not imply any communication among the processes. Our software lets each fixed part to be extended by only one process, namely if there are d processes, then the process with number p extends the i-th fixed part only if $i \equiv p \pmod{d}$.

2.7   Isomorphism of the Constructed Parallelisms

We apply a minimality test to all the constructed parallelisms. It checks if there exists an automorphism from the normalizer $N_G(G_2)$ which takes the parallelism to a lexicographically smaller one that has already been constructed. If such an automorphism exists, the current parallelism is rejected. After having constructed all parallelisms we test them for isomorphism once again using suitable invariants.

3   Results and Open Problems

The results are summarized in Table 1. We obtain many partial parallelisms with more than 13 regular spreads that cannot be extended to parallelisms.

One of the fixed parts with 2 subregular, 3 aregular and no regular spreads gives rise to two nonisomorphic parallelisms of PG(3, 4) with 16 regular spreads (row in bold in Table 1). The parallelisms are given explicitly in Table 2, where each row is a spread with 17 lines, and each line is presented by its number. One spread is given from each of the 13 spread orbits under G_2. The full automorphism group of these parallelisms is of order 2 and they can be considered as the first two examples of deficiency 5 regular parallelisms of PG(3, 4).

Only a few of the obtained partial parallelisms with at least 13 regular spreads can be extended to parallelisms (the 5-th column of Table 1). We do not find parallelisms with 14 or 15 regular spreads. We construct 8 parallelisms with 13 regular spreads. Four of them are not new. They are invariant under an elementary abelian group of order 4 and have been constructed in [5]. The other four are new and their full automorphism group is G_2. All parallelisms constructed in this work are available online at http://www.moi.math.bas.bg/moiuser/~stela.

The problem of the existence of parallelisms of PG(3, 4) with more than 16 regular spreads is open. Our present results show that such a parallelism must have no nontrivial automorphisms.

Table 2. Parallelisms of PG(3, 4) with 16 regular spreads and a full authomorphism group of order 2

	1	102	122	143	163	171	183	210	222	244	256	269	281	305	325	328	348
	2	26	52	74	100	167	187	208	228	232	252	273	293	301	313	339	351
fixed	3	27	53	75	101	110	130	137	157	245	257	266	278	300	312	340	352
part	4	29	50	77	98	115	127	138	150	174	194	200	220	308	320	331	343
	5	28	51	76	99	105	125	142	162	175	195	199	219	234	246	277	289
S_1	6	36	42	80	92	124	146	159	177	212	218	235	261	288	302	323	333
S_2	8	30	57	78	84	103	133	156	168	203	226	237	260	262	314	329	355
S_3	10	24	58	67	96	116	129	155	166	191	229	243	250	280	297	316	350
S_4	12	35	59	64	89	108	119	164	184	205	225	242	267	287	309	318	326
S_5	14	38	45	68	91	112	131	140	169	196	221	231	270	283	310	337	354
S_6	16	39	49	72	82	118	149	154	173	193	211	238	264	292	304	311	347
S_7	18	32	56	62	88	126	141	165	170	192	214	233	259	271	295	332	356
S_8	20	23	46	70	95	109	148	151	178	207	216	236	265	286	307	330	353

	1	102	122	143	163	171	183	210	222	244	256	269	281	305	325	328	348
	2	26	52	74	100	167	187	208	228	232	252	273	293	301	313	339	351
fixed	3	27	53	75	101	110	130	137	157	245	257	266	278	300	312	340	352
part	4	29	50	77	98	115	127	138	150	174	194	200	220	308	320	331	343
	5	28	51	76	99	105	125	142	162	175	195	199	219	234	246	277	289
S_1	6	40	42	72	96	128	139	165	178	204	223	241	250	292	307	317	334
S_2	8	35	59	73	84	109	118	164	172	213	215	242	248	267	323	330	345
S_3	10	24	55	81	88	108	133	159	179	182	224	233	260	282	302	311	349
S_4	12	41	46	64	94	103	131	154	189	207	217	238	274	284	296	324	337
S_5	14	30	45	79	89	107	126	146	180	188	214	237	263	287	321	341	344
S_6	16	32	58	67	82	132	140	153	181	186	209	243	271	280	294	315	350
S_7	18	36	48	62	92	124	149	160	177	185	218	239	247	276	306	326	347
S_8	20	23	57	66	93	117	136	155	176	201	226	231	270	285	298	335	356

Acknowledgements. The authors acknowledge the provided access to the e-infrastructure of the NCHDC - part of the Bulgarian National Roadmap on RIs, with the financial support by the Grant No DO1-325/01.12.2023.

References

1. Bamberg, J.: There are no regular packings of PG(3, 3) or PG(3, 4). https://symomega.wordpress.com/2012/12/01/. Accessed 5 Jan 2023
2. Betten, A.: The packings of $PG(3, 3)$. Des. Codes Cryptogr. **79**(3), 583–595 (2016). https://doi.org/10.1007/s10623-015-0074-6
3. Betten, A., Topalova, S., Zhelezova, S.: Parallelisms of $PG(3, 4)$ invariant under cyclic groups of order 4. In: Ciric, M., Droste, M., Pin, Jean-Eric (eds.) 8-th International Conference. CAI 2019. LNCS, vol. 11545, pp. 88–99. Springer, Heidelberg (2019). https://doi.org/10.1007/978-3-030-21363-3_8
4. Betten, A., Topalova, S., Zhelezova, S.: New uniform subregular parallelisms of PG(3, 4) invariant under an automorphism of order 2. Cybern. Inf. Technol. **20**(6), 18–27 (2020). https://doi.org/10.2478/cait-2020-0057
5. Betten, A., Topalova, S., Zhelezova, S.: Parallelisms of PG(3, 4) invariant under an elementary abelian group of order 4. AAECC **33**, 649–674 (2022). https://doi.org/10.1007/s00200-022-00562-7
6. Bruck, R.H, Bose, R.C: The construction of translation planes from projective spaces. J. Algebra **1**(1), 85–102 (1964). https://doi.org/10.1016/0021-8693(64)90010-9
7. Denniston, R. H. F.: Cyclic packings of the projective space of order 8. Atti Accad. Naz. Lincei Rend. Cl. Sci. Fis. Mat. Natur. **54**, 373–377 (1973)
8. Etzion, T., Storme, L.: Galois geometries and coding theory. Des. Codes Cryptogr. **78**(1), 311–350 (2016). https://doi.org/10.1007/s10623-015-0156-5
9. Hirschfeld, J.W.P.: Projective Geometries Over Finite Fields, 2nd edn. Clarendon Press, Oxford (1979)
10. Jha, V., Johnson, N.: Regular parallelisms from translation planes. Discrete Math. **59**, 91–97 (1986). https://doi.org/10.1016/0012-365X(86)90072-5
11. Johnson, N. L.: Combinatorics of Spreads and Parallelisms. Series: Chapman & Hall Pure and Applied Mathematics, 295, CRC Press, Boca Raton (2010)
12. Koetter, R., Kschischang, F.R.: Coding for errors and erasures in random network coding. IEEE Trans. Inf. Theory **54**, 3579–3591 (2008). https://doi.org/10.1109/ISIT.2007.4557321
13. Lunardon, G.: On regular parallelisms in $PG(3, q)$. Discrete Math. **51**, 229–335 (1984). https://doi.org/10.1016/0012-365X(84)90003-7
14. Penttila, T., Williams, B.: Regular packings of $PG(3, q)$. Eur. J. Combin. **19**(6), 713–720 (1998). https://doi.org/10.1006/eujc.1998.0239
15. Prince, A. R.: Uniform Parallelisms of PG(3,3). In: Geometry, Combinatorial Designs and Related Structures : Proceedings of the First Pythagorean Conference, no. 245, pp. 193–200. London Mathematical Society Lecture Note Series (1997)
16. Prince, A.R.: The cyclic parallelisms of $PG(3, 5)$. Eur. J. Combin. **19**(5), 613–616 (1998). https://doi.org/10.1006/eujc.1997.0204
17. Storme, L.: Finite Geometry. In: Colbourn, C., Dinitz, J., Rosen, K. (eds.) Handbook of Combinatorial Designs. 2nd ed. Discrete mathematics and its applications, pp. 702–729. CRC Press, Boca Raton (2007)
18. Topalova, S., Zhelezova, S.: On transitive parallelisms of $PG(3, 4)$. Appl. Algebra Eng. Comm. Comput. **24**(3–4), 159–164 (2013). https://doi.org/10.1007/s00200-013-0194-z
19. Topalova, S., Zhelezova, S.: On point-transitive and transitive deficiency one parallelisms of $PG(3, 4)$. Des. Codes Cryptogr. **75**(1), 9–19 (2015). https://doi.org/10.1007/s10623-013-9887-3

20. Topalova, S., Zhelezova, S.: New Regular Parallelisms of PG(3, 5). J. Combin. Des. **24**, 473–482 (2016). https://doi.org/10.1002/jcd.21526
21. Topalova, S., Zhelezova, S.: New parallelisms of $PG(3, 4)$. Electr. Notes Discrete Math. **57**, 193–198 (2017). https://doi.org/10.1016/j.endm.2017.02.032
22. Walker, M.: Spreads covered by derivable partial spreads. J. Comb. Theory A **38**(2), 113–130 (1985). https://doi.org/10.1016/0097-3165(85)90063-9

Contributed Papers

On Some Quadratic Eigenvalue Problems

Andrey B. Andreev[1,2](✉)[ID] and Milena R. Racheva[2][ID]

[1] Institute of Information and Communication Technologies, Acad. G., Bonchev St., Block 25A, 1113 Sofia, Bulgaria
[2] Department of Mathematics, Informatics and Natural Sciences, Technical University of Gabrovo, 5300 Gabrovo, Bulgaria
{andreev,milena}@tugab.bg

Abstract. In a number of second- and fourth-order boundary value problems the spectral parameter λ appears quadratically as well as linearly. The paper deals with eigenvalue problems of this type. Here, an approach for linearization is proposed for both second- and fourth-order problems. By suitable substitution, a variational system is obtained in which spectral parameter λ appears only linearly but not quadratically.

Theoretical and computational aspects are considered. Combining linearization and the mixed method for fourth-order problems is also discussed. Finally, numerical results are presented.

Keywords: Eigenvalue problem · Quadratic problem · Second-order · Fourth-order

1 Introduction and Preliminaries

General form of the one-dimensional quadratic eigenvalue problem can be written as follows:

$$Ay(x) - \lambda^2 By(x) = \lambda y(x),$$
$$0 < x < l, \tag{1}$$

where A and B are linear ordinary differential operators and A is of higher order (see [1,2]). Also, the operators A and B are positive definite. Various boundary conditions are related to the Eq. (1).

The most useful and important are the second- and fourth-order boundary problems with eigenvalue parameter which appears quadratically [3]. It is precisely such tasks that are the subject of research in this paper. The simultaneous occurence of first and second degree of the eigenvalue parameter λ can be explained, for example, as follows: Some boundary value problems for partial differential equations contain derivatives (usually with respect to time variable) of both first and second order. After separation of variables, a corresponding eigenvalue equation of type (1) is obtained. Then the differential operators A and B

This work is partially supported by the Technical University of Gabrovo under grant 2206C/2023.

I. Lirkov and S. Margenov (Eds.): LSSC 2023, LNCS 13952, pp. 455–462, 2024.
https://doi.org/10.1007/978-3-031-56208-2_47

refer to the spatial variable. More precise analysis for the relation between the partial differential equation and the spectrum of elliptic differential operators could be find in the book [4].

Most of the results of the quadratic eigenvalue theory, concern a linearization, performed on the corresponding algebraic system (see [2,5] and the references therein). Methods for converting the quadratic matrix equation into a linear one are proposed and the issues of symmetry and stability are investigated. A principal difference between these approaches and ours is that we propose a method for linearization of the differential problem. By introducing an additional function, the goal is to obtain a suitable system for purpose to represent in variational form. Thus, a mixed numerical method could be also effectively applied.

For illustration we employ model quadratic eigenvalue problems of second- and fourth-order. In addition, we will show that the method is also applicable under non-homogeneous boundary conditions.

First, let us consider the following Sturm-Liouville problem:

$$-(p(x)y'(x))' + q(x)y(x) - \lambda s(x)y(x) = \lambda^2 y(x), \tag{2}$$
$$\lambda \in \mathbf{C}; \; y(x) \neq 0; \; x \in (0,l),$$

with boundary conditions

$$y'(0)\sin\alpha = y(0)\cos\alpha, \;\; y'(l)\sin\beta = y(l)\cos\beta, \tag{3}$$

where $p > 0$; $q, \dfrac{1}{p} \in L_1(0,l)$, $s \in L_\infty(0,l)$, $\alpha, \beta \in [0,\pi)$.

Quadratic Sturm-Liouville problems of type (2), (3) arise in hydrodynamics, geophysics, also appear in the theory of wave propagation (see [2,6]). In this work the interval $(0,l)$ is considered, but from an application point of view it is natural the considered interval is $(0,\pi)$.

Now, consider a fourth-order quadratic eigenvalue problem. For purpose of illustration we apply a model describing the motion of a beam simply supported at both ends and damped at the midpoint:

$$EIv^{IV}(x) + \lambda^2 \rho Av(x) + \lambda Cv(x) = 0, \tag{4}$$
$$0 < x < l,$$

with homogeneous boundary conditions

$$v(0) = v(l) = 0, \;\; v''(0) = v''(l) = 0. \tag{5}$$

The parameters in (4) include:

- $EI > 0$ is the module of linear deformation (flexural rigidity);
- $\rho A > 0$ is the bending module of rigidity (torsional rigidity);
- $C \geq 0$ represents the rigidity of the elastic base.

For both problems presented above the use of mixed variational methods are very appropriate. In the mixed formulations (see [7,8]), one has two unknown fields to be approximated by Galerkin procedure and by finite element methods in particular. The goal here is to find representations such that the presence of λ^2 is avoided for both problems (2), (3) and (4), (5).

2   Main Results

Let $H^m(0,l)$ be the Sobolev space of order $m \geq 0$ with norm $\| \cdot \|_{m,(0,l)}$. Consequently, the eigenvalues constitute a sequence of complex numbers $\lambda_1, \lambda_2, \lambda_3, \ldots$ and the eigenfunctions are normalized with respect to $\| \cdot \|_{0,(0,l)}$. First, we will present mixed variational formulations of (2), (3) and (4), (5), respectively.

Theorem 1. *Both quadratic eigenvalue problems (2), (3) and (4), (5) admit symmetric variational form with linear eigenvalue parameter.*

Proof. The two cases will be considerd separately.

Case 1 Second-order eigenvalue problem (2), (3). From here on we will drop the argument x. Let us introduce a new function $z = \lambda y$, and denote $Ly = -(py')' + qy$. Then we obtain the system:

$$\left| \begin{array}{l} Lz = \lambda Ly \\ Ly - sz = \lambda z. \end{array} \right.$$

We introduce two functional spaces such that $y \in H^1(0,l) = V_1$ and $z \in H^1(0,l) = V_2$. Multiplying the first equation of the last system by $y_1 \in V_1$ and the second one by $z_1 \in V_2$, we get

$$\left| \begin{array}{l} (Lz, y_1) = \lambda(Ly, y_1) \\ (Ly, z_1) - (sz, z_1) = \lambda(z, z_1), \quad \forall(y_1, z_1) \in V_1 \times V_2. \end{array} \right. \tag{6}$$

We integrate by parts the terms in (6) when $\alpha \neq 0$, $\beta \neq 0$ and get:

$$(Lz, y_1) = -p \, \cot \beta \, z(l)y_1(l) + p \, \cot \alpha \, z(0)y_1(0) + \int_0^l p \, z'y_1' \, dx + \int_0^l q \, zy_1 \, dx;$$

$$(Ly, z_1) = -p \, \cot \beta \, y(l)z_1(l) + p \, \cot \alpha \, y(0)z_1(0) + \int_0^l p \, y'z_1' \, dx + \int_0^l q \, yz_1 \, dx;$$

$$(Ly, y_1) = -p \, \cot \beta \, y(l)y_1(l) + p \, \cot \alpha \, y(0)y_1(0) + \int_0^l p \, y'y_1' \, dx + \int_0^l q \, yy_1 \, dx.$$

Let us introduce the following symmetric bilinear forms:

$$a(y, y_1) = \int_0^l p \, y'y_1' \, dx + \int_0^l q \, yy_1 \, dx - p \, \cot \beta \, y(l)y_1(l) + p \, \cot \alpha \, y(0)y_1(0);$$

$$b(y, y_1) = \int_0^l yy_1 \, dx; \quad c(y, y_1) = \int_0^l s \, yy_1 \, dx.$$

Then, the system (6) transforms into the following variational system: find $(\lambda, y, z) \in \mathbf{C} \times V_1 \times V_2$ such that

$$\left| \begin{array}{l} a(z, y_1) = \lambda a(y, y_1) \\ a(y, z_1) - c(z, z_1) = \lambda b(z, z_1), \quad \forall(y_1, z_1) \in V_1 \times V_2. \end{array} \right. \tag{7}$$

Obviously, the bilinear forms in (7) are symmetric and continuous. In conclusion, the mixed variational formulation is symmetric and this presentation is not unique [7]. It should be noted that λ appears linearly in (7), which is the main advantage of the proposed method.

Remark: The case $\alpha = \pi/2$ and/or $\beta = \pi/2$ is much easier from computational point of view . The authors studied the Sturm-Liouville problem in [8], which corresponds to $\alpha = \beta = 0$.

Case 2 Fourth-order eigenvalue problem (4), (5). In the same manner, for the problem (4), (5) we introduce a new function $z = \lambda v$. Then we obtain the system:

$$\left| \begin{array}{l} EIz^{IV} = \lambda EIv^{IV} \\ EIv^{IV} + \lambda \rho Az + Cz = 0 \end{array} \right. \tag{8}$$

with boundary conditions (5).

We consider two variational functional spaces: $V = H_0^2(0, l)$ and $\Sigma = H_0^2(0, l)$.

Multiplying the first equation of (8) by $v_1 \in V$ and the second one by $z_1 \in \Sigma$ and integrating by parts, we get:

$$\int_0^l EIz^{IV} v_1 \, dx = \int_0^l EIz'' v_1'' \, dx; \quad \lambda \int_0^l EIv^{IV} v_1 \, dx = \lambda \int_0^l EIv'' v_1'' \, dx;$$

$$\int_0^l EIv^{IV} z_1 \, dx = \int_0^l EIv'' z_1'' \, dx.$$

Let us denote the following bilinear forms for any functions $u, v \in H_0^2(0, l)$:

$$\widetilde{a}(u, v) = \int_0^j EIu'' v'' \, dx; \quad \widetilde{b}(u, v) = \int_0^l \rho Auv \, dx; \quad \widetilde{c}(u, v) = \int_0^l Cuv \, dx.$$

Thus we can write the system (8) in the following variational form:

$$\left| \begin{array}{l} \widetilde{a}(z, v_1) = \lambda \widetilde{a}(v, v_1) \\ \widetilde{a}(v, z_1) + \widetilde{c}(z, z_1) = -\lambda \widetilde{b}(z, z_1), \quad \forall (v_1, z_1) \in V \times \Sigma. \end{array} \right. \tag{9}$$

This formulation is symmetric. It should be noted that λ appears linearly in (9), which is the main advantage of the proposed method.

For both quadratic eigenvalue problems we can use finite element method. Therefore, discrete finite element spaces should be introduced. First, for the problem (2), (3) consider coupled finite element spaces $V_{1,h}$ and $V_{2,h}$, such that $V_{1,h} \times V_{2,h} \subset V_1 \times V_2$ and h is the standard discretization parameter. Then, the mixed finite element system corresponding to (7) is: find $(\lambda_h, y_h, z_h) \in \mathbf{C} \times V_{1,h} \times V_{2,h}$ such that

$$\left| \begin{array}{l} a(z_h, y_1) = \lambda_h a(y_h, y_1) \\ a(y_h, z_1) - c(z_h, z_1) = \lambda_h b(z_h, z_1), \quad \forall (y_1, z_1) \in V_{1,h} \times V_{2,h}. \end{array} \right. \tag{10}$$

The equations in (10) could be unified by the equality:

$$a(z_h, y_1) + a(y_h, z_1) - c(z_h, z_1) = \lambda_h \left[a(y_h, y_1) + b(z_h, z_1) \right], \tag{11}$$
$$\forall (y_1, z_1) \in V_{1,h} \times V_{2,h}.$$

Similarly, for the fourth-order quadratic eigenvalue problem (4), (5) we have the finite element spaces $V_h \times \Sigma_h \subset V \times \Sigma$ and from (9) we can fomulate: find $(\lambda_h, v_h, z_h) \in \mathbf{C} \times V_h \times \Sigma_h$ such that

$$\left| \begin{array}{l} \widetilde{a}(z_h, v_1) = \lambda_h \widetilde{a}(v_h, v_1) \\ \widetilde{a}(v_h, z_1) + \widetilde{c}(z_h, z_1) = -\lambda_h \widetilde{b}(z_h, z_1), \quad \forall (v_1, z_1) \in V_h \times \Sigma_h. \end{array} \right. \tag{12}$$

The two equations in the mixed FE system (12) could be combined as follows:

$$\widetilde{a}(z_h, v_1) + \widetilde{a}(v_h, z_1) + \widetilde{c}(z_h, z_1) = \lambda_h \left[\widetilde{a}(v_h, v_1) - \widetilde{b}(z_h, z_1) \right],$$
$$\forall (v_1, z_1) \in V_h \times \Sigma_h. \tag{13}$$

From discrete systems (10), (11) and (12), (13) we can obtain linearized matrix equations corresponding to the second- and fourth-order quadratic eigenvalue problems, respectively.

Concerning the fourth-order problem, let us outline an approach allowing the use of mixed finite element method. The aim is to reduce the smoothness requirement of variational functions. Thus, appart from $z = \lambda v$ we set $-v'' = \sigma$. The system corresponding to (8) is:

$$\left| \begin{array}{l} -EIv'' = EI\sigma \\ -\rho A z = -\lambda \rho A v \\ -EI\sigma'' + \lambda \rho A z + \lambda c v = 0, \end{array} \right. \tag{14}$$

with homogeneous boundary conditions $v(0) = v(l) = 0$; $\sigma(0) = \sigma(l) = 0$ and $z(0) = z(l) = 0$.

Multiplying the three equations from (14) by $\sigma_1 \in \widetilde{\Sigma}$, $z_1 \in \Sigma_1$ and $v_1 \in \widetilde{V}$ respectively, where $\widetilde{\Sigma}$ and $\widetilde{V}$ are subspaces of $H_0^1(0, l)$. After integration we get:

$$\left| \begin{array}{ll} -(EIv'', \sigma_1) = (EI\sigma, \sigma_1), & \forall \sigma_1 \in \widetilde{\Sigma} \\ -(\rho A z, z_1) = -\lambda(\rho A v, z_1), & \forall z_1 \in \Sigma_1 \\ -(EI\sigma'', v_1) = -\lambda(\rho A z, v_1) - \lambda(cv, v_1), & \forall v_1 \in \widetilde{V}. \end{array} \right. \tag{15}$$

Consider the following bilinear forms for any $u, v \in H_0^1(0, l)$:

$$a_1(u, v) = \int_0^l EIu'v' \, dx; \quad b_1(u, v) = \int_0^l \rho A u v \, dx;$$

$$b_2(u, v) = \int_0^l EIuv \, dx; \quad c_1(u, v) = \int_0^l cuv \, dx.$$

Obviously, integrating by parts, the system (15) takes the form

$$\left| \begin{array}{l} a_1(v, \sigma_1) = b_2(\sigma, \sigma_1), \\ -b_1(z, z_1) = -\lambda b_1(v, z_1), \\ a_1(\sigma, v_1) = -\lambda b_1(z, v_1) - \lambda c_1(v, v_1), \quad \forall (v_1, z_1, \sigma_1) \in \widetilde{V} \times \Sigma_1 \times \widetilde{\Sigma}. \end{array} \right. \tag{16}$$

Finally, we are looking for a number $\lambda \in \mathbf{C}$ and nontrivial finctions $(v, z, \sigma) \in \widetilde{V} \times \Sigma_1 \times \widetilde{\Sigma}$ such that

$$
\begin{aligned}
& a_1(v, \sigma_1) + a_1(\sigma, v_1) - b_1(z, z_1) - b_2(\sigma, \sigma_1) \\
& = -\lambda \left[b_1(v, z_1) + b_1(z, v_1) + c(v, v_1) \right] \quad \forall (v_1, z_1, \sigma_1) \in \widetilde{V} \times \Sigma_1 \times \widetilde{\Sigma}.
\end{aligned}
\tag{17}
$$

The Eqs. (16) and (17) have symmetrical form. Another important advantage is that variational spaces $\widetilde{V}$, Σ_1 and $\widetilde{\Sigma}$ are subspaces of $H_0^1(0, l)$. Taking this fact into consideration we define mixed finite element problems corresponding to discrete Eqs. (12), (13).

Let $\widetilde{V}$, Σ_1 and $\widetilde{\Sigma}$ be subspaces of $H_0^1(0, l)$. Consider finite element spaces $\widetilde{V}_h \subset \widetilde{V}$, $\Sigma_{1,h} \subset \Sigma_1$ and $\widetilde{\Sigma}_h \subset \widetilde{\Sigma}$. Then, the mixed finite element problem is: find $(\lambda_h, v_h, z_h, \sigma_h) \in \mathbf{C} \times \widetilde{V}_h \times \Sigma_h \times \widetilde{\Sigma}_h$ such that

$$
\left|
\begin{aligned}
& a_1(v_h, \sigma_1) = b_2(\sigma_h, \sigma_1), \\
& -b_1(z_h, z_1) = -\lambda_h b_1(v_h, z_1), \\
& a_1(\sigma_h, v_1) = -\lambda_h b_1(z_h, v_1) - \lambda_h c_1(v_h, v_1), \quad \forall (v_1, z_1, \sigma_1) \in \widetilde{V}_h \times \Sigma_{1,h} \times \widetilde{\Sigma}_h.
\end{aligned}
\right.
\tag{18}
$$

Also, we can combine the equations from (18) into the form:

$$
\begin{aligned}
& a_1(v_h, \sigma_1) + a_1(\sigma_h, v_1) - b_1(z_h, z_1) - b_2(\sigma_h, \sigma_1) \\
& = -\lambda_h \left[b_1(v_h, z_1) + b_1(z_h, v_1) + c(v_h, v_1) \right] \quad \forall (v_1, z_1, \sigma_1) \in \widetilde{V}_h \times \Sigma_{1,h} \times \widetilde{\Sigma}_h.
\end{aligned}
\tag{19}
$$

Finally, discrete presentation (18), (19) could be applied instead of (12), (13). Let us emphasize that in this case (see [9]) even linear finite elements are applicable for the fourth order nonlinear problem under consideration.

3   Matrix Representations of the Mixed Variational Equations

Let $\varphi_i(x)$, $i = 1, 2, \ldots, n_1$, $n_1 \in \mathbf{N}$, be the shape functions of the finite element spaces V_h and Σ_h. Then y_h and z_h have the representation

$$
y_h(x) = \sum_{i=1}^{n_1} y_i \varphi_i(x) \quad \text{and} \quad \sigma_h(x) = \sum_{j=1}^{n_1} s_j \varphi_j(x), \quad x \in [0, l].
$$

We denote $\mathcal{Y} = (y_1, y_2, \ldots, y_{n_1})^T$ and $\mathcal{S} = (s_1, s_2, \ldots, s_{n_1})^T$ and then from (11) we obtain matrix equation which could be written as

$$
\begin{pmatrix} -\mathcal{C} & \mathcal{A}^T \\ \mathcal{A} & \mathcal{O} \end{pmatrix} \begin{pmatrix} \mathcal{S} \\ \mathcal{Y} \end{pmatrix} = \lambda_h \begin{pmatrix} \mathcal{B} & \mathcal{O} \\ \mathcal{O} & \mathcal{A} \end{pmatrix} \begin{pmatrix} \mathcal{S} \\ \mathcal{Y} \end{pmatrix}.
\tag{20}
$$

where the matrices $\mathcal{A}$, $\mathcal{B}$ and $\mathcal{C}$ correspond to the bilinear forms $a(\cdot, \cdot)$, $b(\cdot, \cdot)$ and $c(\cdot, \cdot)$, respectively, and $\mathcal{O}$ is square null matrix with dimensions $n_1 \times n_1$.

The matrices $\mathcal{A}$, $\mathcal{B}$ and $\mathcal{C}$ are symmetric, consequently the global block matrices in (20) are symmetric matrices of type $2n_1 \times 2n_1$.

Similarly, if $\varphi_i(x)$, $i = 1, 2, \ldots, n_2$, $n_1 \in \mathbf{N}$ are the shape functions of the finite element spaces $\widetilde{V}_h, \Sigma_{1,h}$ and $\widetilde{\Sigma}_h$, then v_h, z_h and σ_h are presented as

$$v_h(x) = \sum_{i=1}^{n_2} v_i \varphi_i(x), \quad z_h(x) = \sum_{i=1}^{n_2} z_i \varphi_i(x) \quad \text{and} \quad \sigma_h(x) = \sum_{j=1}^{n_2} z_j \varphi_j(x), \quad x \in [0, l].$$

We denote $\mathcal{V} = (v_1, v_2, \ldots, v_{n_2})^T$, $\mathcal{Z} = (z_1, z_2, \ldots, z_{n_2})^T$ and $\mathcal{S} = (s_1, s_2, \ldots, s_{n_2})^T$ and then from (19) we obtain matrix equation which could be written as

$$\begin{pmatrix} \mathcal{O} & \mathcal{O} & \mathcal{A}_1^T \\ \mathcal{O} & -\mathcal{B}_1 & \mathcal{O} \\ \mathcal{A}_1 & \mathcal{O} & -\mathcal{B}_2 \end{pmatrix} \begin{pmatrix} \mathcal{V} \\ \mathcal{Z} \\ \mathcal{S} \end{pmatrix} = \lambda_h \begin{pmatrix} -\mathcal{C}_1 & -\mathcal{B}_1^T & \mathcal{O} \\ -\mathcal{B}_1 & \mathcal{O} & \mathcal{O} \\ \mathcal{O} & \mathcal{O} & \mathcal{O} \end{pmatrix} \begin{pmatrix} \mathcal{V} \\ \mathcal{Z} \\ \mathcal{S} \end{pmatrix}$$

or

$$\begin{pmatrix} -\mathcal{B}_2 & \mathcal{O} & \mathcal{A}_1^T \\ \mathcal{O} & -\mathcal{B}_1 & \mathcal{O} \\ \mathcal{A}_1 & \mathcal{O} & \mathcal{O} \end{pmatrix} \begin{pmatrix} \mathcal{S} \\ \mathcal{Z} \\ \mathcal{V} \end{pmatrix} = \lambda_h \begin{pmatrix} -\mathcal{O} & \mathcal{O} & \mathcal{O} \\ \mathcal{O} & \mathcal{O} & -\mathcal{B}_1^T \\ \mathcal{O} & -\mathcal{B}_1 & -\mathcal{C}_1 \end{pmatrix} \begin{pmatrix} \mathcal{S} \\ \mathcal{Z} \\ \mathcal{V} \end{pmatrix},$$

where the matrices $\mathcal{A}_1$, $\mathcal{B}_1$, $\mathcal{B}_2$ and $\mathcal{C}_1$ correspond to the bilinear forms $a_1(\cdot, \cdot)$, $b_1(\cdot, \cdot)$, $b_2(\cdot, \cdot)$ and $c_1(\cdot, \cdot)$, respectively, and $\mathcal{O}$ is square null matrix with dimensions $n_2 \times n_2$.

4 Numerical Example

Consider the problem (2), (3) when $p(x) \equiv 1$, $q(x) \equiv 0$, $s(x) \equiv 1$ and $\alpha = \beta = 0$ on the interval $[0, 1]$.

The choice of this extremely simplified example is motivated by the fact that in this case the exact eigenpairs are known. The results confirm the proposed method and illustrate how does the method works. The exact eigenfunctions are computed as $y(x) = \sin k\pi x$, $k = 1, 2, \ldots$ and the exact eigenvalues are

$$\lambda^{+,-} = \frac{-1 \pm \sqrt{1 + 4k^2\pi^2}}{ch12}.$$

The interval $(0, 1)$ is uniformly divided into N subintervals and thus the mesh parameter h is equal to $1/N$.

We solve the Eq. (11) using the matrix representation (20) by means of linear finite elements. This is an illustration that eigenvalue problem in which the eigenvalue parameter λ appears quadratically as well as linearly could be successfully solved even if simplest finite elements are used.

In Table 1 approximations of $\lambda_1^{\pm}$, $\lambda_2^{\pm}$ and λ_3^{+} are presented. The trends are given and, as it is seen, they depend on the sign of the approximated eigenvalue.

It is well-known that standard conforming finite element methods give approximations of the eigenvalues from above, while for any finite element method of mixed type this is not clear in general. From Table 1 it is seen that for the concrete numerical example under consideration the demonstrated method in fact approximates the absolute values of the exact eigenvalues asymptotically from above. The numerical results demonstrate also that the method works even in case of course meshes.

Table 1. Approximations $\lambda_{1,h}^+, \lambda_{1,h}^-, \lambda_{2,h}^+, \lambda_{2,h}^-, \lambda_{3,h}^+$ of the exact eigenvalues λ_1^+, λ_1^-, $\lambda_2^+, \lambda_2^-, \lambda_3^+$ obtained after finite element implementation of the proposed method by means of linear FEs

N	$\lambda_{1,h}^+$	$\lambda_{1,h}^-$	$\lambda_{2,h}^+$	$\lambda_{2,h}^-$	$\lambda_{3,h}^+$
8	2.7011062	−3.7011062	5.9650265	−6.9650265	9.4868931
16	2.6861189	−3.6823787	5.8433684	−6.8433675	9.0345715
32	2.6823787	−3.6823787	5.8131145	−6.8131145	8.9682250
64	2.6814441	−3.6814441	5.8055640	−6.8055639	8.9465378
128	2.6812104	−3.6812104	5.8036771	−6.8036771	8.9401578
256	2.6811520	−3.6811520	5.8032055	−6.8032055	8.9385631
512	2.6811374	−3.6811374	5.8030876	−6.8030876	8.9380186
Trend	↘	↗	↘	↗	↘
Exact	2.6811326	−3.6811326	5.8030483	−6.8030483	8.9380316

References

1. Eisenfeld, J.: Quadratic eigenvalue problems. J. Math. Anal. Appl. **23**(1), 58–70 (1968)
2. Tisseur, F., Meerbergen, K.: The quadratic eigenvalue problem. SIAM Rev. **43**(2), 235–286 (2001)
3. Collatz, L.: Eigenwertaufgaben mit Technischen Anwendungen, Leipzig, Acad. Verlag (1963)
4. ch1Raviart, P.-A., Thomas, J. M.: Introduction à l'analyse numérique des équation aux dérivées partielles. Collection Math. Appliquees pour la Maîtrise. Masson, Paris, p. 224 (1983)
5. Fan, H.Y., Lin, W.W., Van Dooren, P.: Normwise scaling of second order polynomial matrices. SIAM J. Matrix Anal. Appl. **26**(1), 252–256 (2004)
6. Al-Gwaiz, M.A.: Sturm-Liouville Theory and its Applications, vol. 264. Springer, London (2008)
7. Babuška, I., Osborn, J.: Eigenvalue problems. In: Lions, P.G., Ciarlet, P.G. (eds.) Handbook of Numerical Analysis, Vol. II, , Finite Element Methods (Part 1) North-Holland, Amsterdam, pp. 641-787 (1991)
8. Andreev, A.B., Racheva, M.R.: Finite element approximation for the Sturm-Liouville problem with quadratic eigenvalue parameter. In: Dimov, I., Fidanova, S. (eds.) Advances in High Performance Computing. Studies in Computational Intelligence, vol. 902, pp. 368–375. Springer, Cham (2021). https://doi.org/10.1007/978-3-030-55347-0_31
9. Ishihara, K.: A mixed finite element approximation for the buckling of plates. Numer. Math. **33**, 195–210 (1979)

Numerical Determination of Time-Dependent Volatility for American Option When the Optimal Exercise Boundary Is Known

Miglena N. Koleva[✉] and Lubin G. Vulkov

University of Ruse, 8 Studentska str., 7017 Ruse, Bulgaria
{mkoleva,lvalkov}@uni-ruse.bg

Abstract. The paper is devoted on developing a robust free-boundary based method for reconstruction time-dependent volatility of American put options. This problem is posed as an inverse problem: given the optimal exercise boundary, find the volatility function. We propose a linearization refereed to the semi-time layers algorithm of decomposition of the approximate solution for which the transition to the new time level is carried out by two ODE problems. The correctness of the method is discussed. We test the efficiency of the approach for synthetic and close to deal market data.

1 Introduction

American option is a financial contract that gives the holder to exercise its option at any time until a prescribed future date. American options are the most used derivatives in the financial markets. The American option pricing problem can be studied either as a linear complimentary problem (LCP) or as a free boundary value problem (FBVP), see [8] for more details.

In this paper, the American option problem is considered as a specific free boundary problems in partial differential equations (PDEs) related to financial markets. It aims to describe the price distribution.

Here, we work with the front fixing method to find the volatility if the exercise boundary is known.

In the Black-Scholes-Merton model of option pricing [3,8–10,12,14] is supposed that the volatility is a constant. However, with this strong assumption the Black-Scholes model cannot capture features of empirically observed option prices. The deterministic local volatility model is very popular and attractive in financial practice, but the funding of the local or time-dependent volatility function is a very challenging task. Results in this direction are obtained in [2,4,5,16].

The *direct* European and American option problems are known to be well-posed, i.e. it possesses a unique solution, which depends continuously on the data, see e.g. [1,3,7,8]. In contrast to the *direct* European and American option

© The Author(s), under exclusive license to Springer Nature Switzerland AG 2024
I. Lirkov and S. Margenov (Eds.): LSSC 2023, LNCS 13952, pp. 463–471, 2024.
https://doi.org/10.1007/978-3-031-56208-2_48

problem, only few numerical results are available for the related *inverse* problem, see e.g. [2,4,5,16]. Let us note that concerning known free boundary, in [6] is considered one-dimensional one phase Stefan problem for recovering a boundary flux condition for the heat equation.

The paper is composed as follows. Section 2 provides an introduction to the direct American put option with time-dependent volatility. The inverse problem is formulated in the next section. Section 4 is devoted to the full description of our numerical method for solving the inverse problem. Numerical test examples are discussed in Sect. 5 and the paper is finalized by some conclusions.

2   The American Put Option

The American put-option problems for the Black-Sholes equation consists in determining of a function $P(S,t)$ for $(S,t) \in [0,+\infty) \times [0,T]$, where t is a time variable, S is the underlying asset price, and P is the option price [8,15]. The definition domain of P is subdivided by an a priori unknown internal boundary $S(t)$, $t \in [0,T]$ into two parts $\Omega_1 = \{(S,t) : S \geq s(t)\}$ and $\Omega_2 = \{(S,t) : S < s(t)\}$. Here Ω_1 is the continuation region, Ω_2 is the stopping region, and $s(t)$ is the optimal exercise boundary that must be determined simultaneously with $P(S,t)$. In Ω_1, the function $P(S,t)$ is a solution to the parabolic equation

$$\frac{\partial P}{\partial t} + \frac{1}{2}\sigma^2(t)S^2\frac{\partial^2 P}{\partial S^2} + \left(r(t) - d(S,t)\right)S\frac{\partial P}{\partial S} - r(t)P = 0, \quad (S,t) \in \Omega_1, \quad (1)$$

with terminal condition

$$P(S,T) = 0, \quad S \in (K,+\infty), \tag{2}$$

and boundary conditions

$$P(s(t),t) = K - s(t), \quad \lim_{S \to \infty} P(S,t) = 0, \quad t \in (0,T). \tag{3}$$

Here σ^2 is the volatility, r is the risk-free interest rate, d is the dividend rate, and K is the option strike price. In Ω_2, this function has a simple form: $P(S,t) = K - S$, $(S,t) \in \Omega_2$, where $K - S$ is called pay off function. The position of the internal boundary $s(t)$ at time $t = T$ is $s(T) = \min\{Kr(T)/d(T), K\}$ [8] and at other times is determined by the equality

$$\frac{\partial P}{\partial S}(s(t),t) = -1, \quad t \in [0,T), \tag{4}$$

providing continuity of the solution derivative with respect to S. Thus, the behavior of $P(S,t)$ in Ω_2 is of little interest and our aim is to construct approximate solution of the problem (1)-(4).

Let us discuss some features of the problem. First, we have an unknown free boundary $s(t)$ on the left. Second, the domain is semi-infinite in S. Usually, $S = S_{max}$ is taken as the right boundary for sufficiently large S_{max} and $P(S_{max},t) =$

0 is assumed instead of the second condition in (3). Since the truncated domain must be large enough, this results in a large cost. There are number of papers devoted to this problem, see e.g., [7,13,14].

Third, at the point $(S(T), T)$ the first-order derivative of a solution with respect to S is discontinuous. On the one hand, from the terminal condition (2), we have $\frac{\partial P}{\partial S}(S(T), T) = 0$. From (4) it follows $\frac{\partial P}{\partial S}(S(T), T) = -1$. Besides, the second-order derivative of a solution with respect to time increases infinitely as t approaches T. This leads to serious difficulties in the numerical treatment of the American option problems.

3   Formulation of the Inverse Problems

It is well-known that the most of the options in markets are American options and often real data are available for the moving the expire day. Also, since the American option is a Stefan-type problem, it is naturally the inverse volatility problem at known exercise boundary to be studied.

By using changes of variables [1,15] we reduce the free boundary problem into a fixed fixed boundary problem

$$\tau = T - t, \quad p(x, \tau) = \frac{P(S, t)}{K}, \quad \widetilde{s}(\tau) = \frac{s(t)}{K}, \quad x = \ln \frac{S}{\widetilde{s}(\tau)}. \tag{5}$$

In these variables, the equation (1) takes the form

$$\frac{\partial p}{\partial \tau} - \frac{1}{2}\sigma^2(\tau)\frac{\partial^2 p}{\partial x^2} - \left(r(\tau) - d(\tau) - \frac{1}{2}\sigma^2(\tau) + \frac{\widetilde{s}'(\tau)}{\widetilde{s}(\tau)}\right)\frac{\partial p}{\partial x} + r(\tau)v = 0, \tag{6}$$

for $x > 0$, $0 < \tau \le T$ and $\widetilde{s}'(\tau)$ is the derivative of $\widetilde{s}(\tau)$ with respect to τ.

Terminal condition (2), which becomes initial condition, and boundary conditions (3), (4) transform to

$$p(x, 0) = 0, \quad x \ge 0, \tag{7}$$

$$p(0, \tau) = 1 - \widetilde{s}(\tau), \tag{8}$$

$$\frac{\partial p}{\partial x}(0, \tau) = -\widetilde{s}(\tau) \tag{9}$$

$$\lim_{x \to +\infty} p(x, \tau) = 0, \tag{10}$$

$$\widetilde{s}(0) = \min\{r(0)/d(0), 1\} \tag{11}$$

For the direct American put option problem (6)-(11), we have to determine the functions $p(y, t)$ and $\widetilde{s}(t)$. However, if $\widetilde{s}(t)$ is known it is possible one or more of the coefficients of the equation to be identified.

In this paper, we are concentrated on numerically solving inverse problem for identifying the volatility $\sigma^2(t)$ and solution $P(S, t)$ i.e. $\sigma^2(\tau)$ and $p(x, t)$, respectively, through the substitution (5), for known free boundary $s(t)$ $(\widetilde{s}(\tau))$.

4   Numerical Method for Solving Inverse Problem

We introduce large enough $x_{\max}$ and study the problem in the truncated finite domain $[0, T] \times [0, x_{\max}]$. Values of $x_{\max}$ can be chosen according to the criterion in [10]. We define uniform meshes $\overline{\omega}_h = \{x_i = ih, \ i = 0, 1, \ldots, N, \ h = x_{\max}/N\}$, $\overline{\omega}_\tau = \{\tau_n = n\triangle\tau, \ n = 0, 1, \ldots, M, \ \triangle\tau = T/M\}$ and denote by p_i^n the approximate value of $p(x, \tau)$ at mesh point (x_i, τ_n).

First we consider the problem (6)-(10), assuming that $\widetilde{s}$ is known at each time layer. For the numerical solving the inverse problem the boundary conditions (8), (9) can be used in different fashion, for example

- taking (9) as boundary condition and a point observation (8);
- using (6) and both (8), (9) to define natural boundary condition [1] and a point observation given by (8);
- taking (8) as boundary condition and the observation defined by (9).

Further, in order to minimize the influence of the noisy data, and to ensure the correctness of the method, we subtract the equations (8) and (9) to derive the boundary condition at $x = 0$

$$-\frac{\partial p}{\partial x}(0, \tau) + p(0, \tau) = 1. \tag{12}$$

The second-order fully implicit finite difference approximation of (6), (7), (10), (12) is

$$\frac{p_i^{n+1} - p_i^n}{\triangle\tau} - \frac{1}{2}(\sigma^{n+1})^2 p_{\overline{x}x,i}^{n+1} - \left(S_i^{n+1} - \frac{1}{2}(\sigma^{n+1})^2\right) p_{\mathring{x},i}^{n+1} + r^{n+1}p_i^{n+1} = 0,$$

$$S_i^{n+1} = r^{n+1} - d_i^{n+1} + \frac{\widetilde{s}^{n+1} - \widetilde{s}^n}{\triangle\tau\widetilde{s}^{n+1}}, \quad i = 1, 2, \ldots, N-1,$$

$$\frac{p_0^{n+1} - p_0^n}{\triangle\tau} - (\sigma^{n+1})^2\frac{p_{x,0}}{h} - \left(S_0^{n+1} - (\sigma^{n+1})^2\left(\frac{1}{h} + \frac{1}{2}\right) - r^{n+1}\right)p_0^{n+1} \tag{13}$$

$$-(\sigma^{n+1})^2\left(\frac{1}{h} + \frac{1}{2}\right) = -S_0^{n+1},$$

$$p_N^{n+1} = 0, \quad p_i^0 = p(x, 0), \quad i = 0, 1, \ldots, N,$$

where $p_{\mathring{x},i} = (p_{x,i} + p_{\overline{x},i})/2$, $p_{\overline{x}x,i} = (p_{x,i} - p_{\overline{x},i})/h$, $p_{x,i} = (p_{i+1} - p_i)/h$, $p_{\overline{x},i} = p_{x,i-1}$.

To solve the non-linear system (13), first we apply the linearization [4] $(\sigma^{n+1})^2p^{n+1} = (\sigma^{n+1})^2p^n + (\sigma^n)^2p^{n+1} - (\sigma^n)^2p^n + O(\triangle\tau^2)$ and then decomposition of the solution at each time level

$$p^{n+1} = u^{n+1} + (\sigma^{n+1})^2v^{n+1}. \tag{14}$$

Substituting (14) in (13), we get

$$\frac{u_0^{n+1}}{\Delta \tau} - \frac{1}{h}(\sigma^{n+1})^2 u_{x,0}^{n+1} + \left((\sigma^{n+1})^2 \frac{2+h}{2h} + r^{n+1} - \mathcal{S}_0^{n+1} \right) u_0^{n+1}$$
$$= \frac{p_0^n}{\Delta \tau} - \frac{(\sigma^n)^2}{h} \left(p_{x,0}^n - \frac{2+h}{2} p_0^n \right) - \mathcal{S}_0^{n+1},$$

$$\frac{u_i^{n+1}}{\Delta \tau} - \frac{1}{2}(\sigma^{n+1})^2 u_{\bar{x}x,i}^{n+1} - \left(\mathcal{S}_i^{n+1} - \frac{1}{2}(\sigma^{n+1})^2 \right) u_{\mathring{x},i} + r^{n+1} u_i^{n+1}$$
$$= \frac{p_i^n}{\Delta \tau} - \frac{1}{2}(\sigma^n)^2 \left(u_{\bar{x}x,i} - u_{\mathring{x},i} \right), \quad i = 1, \ldots, N-1, \qquad u_N^{n+1} = 0$$

(15)

$$\frac{v_0^{n+1}}{\Delta \tau} - \frac{1}{h}(\sigma^{n+1})^2 v_{x,0}^{n+1} + \left((\sigma^{n+1})^2 \frac{2+h}{2h} + r^{n+1} - \mathcal{S}_0^{n+1} \right) v_0^{n+1}$$
$$= \frac{p_{x,0}^n}{h} + \frac{2+h}{2h}(1 - p_0^n),$$

$$\frac{v_i^{n+1}}{\Delta \tau} - \frac{1}{2}(\sigma^{n+1})^2 v_{\bar{x}x,i}^{n+1} - \left(\mathcal{S}_i^{n+1} - \frac{1}{2}(\sigma^{n+1})^2 \right) v_{\mathring{x},i}^{n+1} + r^{n+1} v_i^{n+1}$$
$$= \frac{1}{2} \left(p_{\bar{x}x,i}^n - p_{\mathring{x},i}^n \right), \quad i = 1, 2, \ldots, N-1, \qquad v_N^{n+1} = 0.$$

(16)

Solving (15) and (16) and in view of (14) and (8), we find

$$(\sigma^n)^2 = \frac{1 - \tilde{s}^n - u_0^n}{v_0^n}, \quad n = 1, 2, \ldots, M.$$

(17)

Then, we obtain the solution p_i^n, $i = 0, 1, \ldots, N-1$ from (14) for $(\sigma^n)^2$, known from (17).

Theorem 1. *The numerical approach (15)-(17) is correct, if*

$$h < \frac{2(\sigma^n)^2}{(\sigma^n)^2 + 4\|\mathcal{S}^{n+1}\|}, \quad n = 1, 2, \ldots, M-1, \quad \|v\| = \max_{0 \le i \le N} |v|.$$

(18)

Proof (outline). We have to prove that $v_0^n \ne 0$ and $\sigma^n > 0$, $n = 1, 2, \ldots, M$. The condition (18) ensure that the maximum principle for problems (13), (15) and (16) hold. From (13), since $\tilde{s}(\tau)$ is positive, monotonically decreasing [11], we get $p^n \ge 0$, $n = 0, 1, \ldots, M$. Then, multiplying $(i-1)$-th, i-th and $(i+1)$-th equations of (13) by coefficients $1/h^2, -2/h^2, 1/h^2$, respectively and summing up, we derive $p_{\bar{x}x}^n \ge 0$, $n = 1, 2 \ldots, M$, since $p_{\bar{x}x}^0 \ge 0$. In the same manner, as $p_{\mathring{x}}^0 \le 0$, we get $p_{\mathring{x}}^n \le 0$, $n = 1, 2 \ldots, M$. Therefore, the right-hand side of (16) for $i = 1, 2, \ldots, N-1$ is nonnegative. For $i = 0$, in view of (12), we derive $\left(p_{x,0}^n + (2+h)(1 - p_0^n)/2 \right)/h = (p_0'' + \tilde{s}(\tau))/2 > 0$. Therefore, $v_0^n > 0$, $n = 1, 2, \ldots, M$. $\square$

5   Numerical Examples

In this section we test the efficiency of the proposed numerical approach.

Example 1. We verify the accuracy of the proposed method (15)-(17), $x_{\max} = 1$, $T = 1$ for the test example with exact solution $p(x,t) = (e^x - 1)((x - 0.25)^4 - x + 1)$. We take $r(\tau) = 0.02 + 0.04t$, $d(\tau) = 0.02 + 0.01\sin(2\pi t)$, $\widetilde{s}(\tau) = (1 - 0.7023 - ((t - 0.5) - (t - 0.45)^2))$ and add residual function in the right-hand side of (6), (7)-(10), such that to obtain the exact solution.

Errors and order of convergence are defined by $\mathcal{E}_N = \|p(x,T) - p_i^N\|$, $\varepsilon_N = \max_{0 \leq n \leq M} |\sigma^2(t) - (\sigma_i^n)^2|$, $CR = \log_2(\mathcal{E}_N/\mathcal{E}_{2N})$, $CR_\sigma = \log_2(\varepsilon_N/\varepsilon_{2N})$.

In Table 1 we give results for the identified volatility σ, computed on spatial mesh with N grid nodes and price p. The ratio $\tau/h^2 = 1$ is fixed. We observe second order spatial convergence of the numerical price. The convergence of the recovered volatility is also effected by this accuracy, due to (17). In Table 2, setting $\tau = h$, we illustrate the first order and close to first order temporal order of convergence for option price and volatility, respectively.

Table 1. Errors of recovered p and σ, $\tau = h^2$, Example 1

	$N = 20$	$N = 40$	$N = 80$	$N = 160$	$N = 320$	$N = 640$	$N = 1280$
$\mathcal{E}_N$	1.505e-3	3.795e-4	9.503e-5	2.377e-5	5.944e-6	1.486e-6	3.715e-7
CR	1.987	1.998	1.999	2.000	2.000	2.000	
ε_N	2.987e-3	1.036e-3	3.584e-4	1.236e-4	4.245e-5	1.449e-5	4.921e-6
CR_σ	1.528	1.531	1.536	1.542	1.550	1.551	

Table 2. Errors of recovered p and σ, $\tau = h$, Example 1

	$M = 40$	$M = 80$	$M = 160$	$M = 320$	$M = 640$	$M = 1280$	$M = 2560$
$\mathcal{E}_M$	1.141e-2	5.666e-3	2.848e-3	1.440e-3	7.263e-4	3.652e-4	1.832e-4
CR	1.010	0.992	0.984	0.987	0.992	0.995	
ε_M	2.483e-2	1.718e-2	1.042e-2	5.920e-3	3.291e-3	1.830e-3	1.022e-3
CR_σ	0.532	0.722	0.815	0.847	0.847	0.840	

Example 2. Now, we take the values of the free boundary, solving direct problem (13) with the method proposed in [1] for $r = 0.02$, $d = 0$, $x_{\max} = 2$, $\tau = h^2$, $T = 1$ and

$$(A)\ \ \sigma^2(\tau) = 0.02 + 0.2t^3, \qquad (B)\ \ \sigma^2(\tau) = \begin{cases} 0.05, \ t \in [0, T/4], \\ e^{-2t}, \ t \in (T/4, 3T/4], \\ 0.36, \ t \in (3T/4, T]. \end{cases}$$

In Figs. 1, 2 (left) we depict the exact and restored volatility (A) and (B), respectively versus time to maturity and in Figs. 1, 2 (right) we plot the corresponding free boundaries, computed by direct method [1]. We observe a very good fitting of the recovered volatility to the exact one. The bigger errors arise at points, where the function $\widetilde{s}(\tau)$ is non-smooth. We also compare the numerical solution p, computed by inverse method (15)-(17) with the one, obtained by direct method [1], see Fig. 3. The magnitude of the biggest error arising close to the initial time, is also good enough.

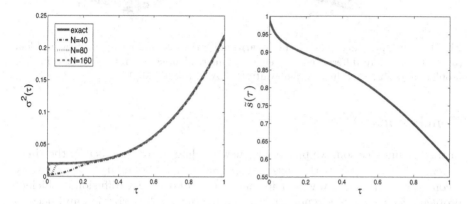

Fig. 1. Restored volatility (A) (*left*) and the corresponding free boundary, Example 2.

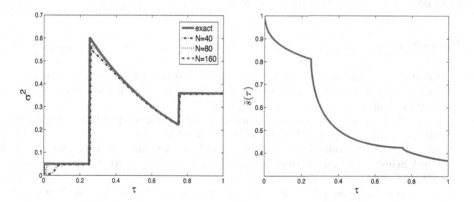

Fig. 2. Restored volatility (B) (*left*) and the corresponding free boundary, Example 2.

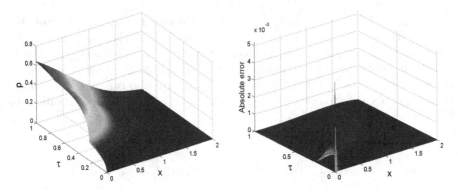

Fig. 3. Time-space graphics of the logarithmic price, computed by inverse problem (*left*) and absolute difference between the numerical price, computed by the inverse problem and direct method [1], volatility (B), $N = 160$, Example 2.

Conclusions

In this communication, we proposed a new method for reconstruction the time-dependent volatility from the observed optimal exercise boundary. This volatility reconstruction problem was first formulated as a non-linear diffusion coefficient problem. To overcome the non-linearity, we perform a linearization and decomposition with respect to the unknown coefficient. Numerical results verified the efficiency of the method.

Acknowledgements. This research is supported by the Bulgarian National Science Fund under Project KP-06-N 62/3 from 2022.

References

1. Company, R., Egorova, V.N., Jódar, L.: Solving American option pricing models by the front fixing method: numerical analysis and computing. In: Abstract and Applied Analysis (2014). Article ID 146745
2. Deng, Z.C., Hon, Y.C., Isakov, V.: Recovery of the time-dependent volatility in option pricing model. Inv. Probl. **32**(11), 115010 (2016)
3. Ehrhardt, M., Mickens, R.: A fast, stable and accurate numerical method for the Black-Scholes equation of American options. Int. J. Theoret. Appl. Finance **11**(5), 471–501 (2008)
4. Georgiev, S., Vulkov, L.: Numerical determination of time-dependent implied volatility by a point observation. In: Georgiev, I., Kostadinov, H., Lilkova, E. (eds.) Advanced Computing in Industrial Mathematics. Studies in Computational Intelligence, vol. 961, pp. 99–109. Springer, Cham (2021)
5. Georgiev, S.G., Vulkov, L.G.: Fast reconstruction of time-dependent market volatility for European options. Comp. Appl. Math. **40**, 30 (2021)
6. Ghanmi, C., Aouadi, S.M., Triki, F.: Identification of a boundary influx condition in a one-phase Stefan problem. Appl. Anal. **101**(18), 6573–6595 (2022)

7. Gyulov, T.B., Valkov, R.L.: American option pricing problem transformed on finite interval. Int. J. Comput. Math. **93**, 821–836 (2016)
8. Jiang, L.: Mathematical Modeling and Methods of Option Pricing, p. 344. World Scientific, Singapore (2005)
9. Kandilarov, J.D., Valkov, R.L.: A numerical approach for the American call option pricing model. In: Dimov, I., Dimova, S., Kolkovska, N. (eds.) Numerical Methods and Applications. Lecture Notes in Computer Science, vol. 6046, pp. 453–460. Springer, Berlin (2010)
10. Kangro, R., Nicolaides, R.: Far field boundary conditions for Black-Scholes equations. SIAM J. Numer. Anal. **38**(4), 1357–1368 (2000)
11. Samuelson, P.A.: Rational Theory of Warrant Pricing. Ind. Manage. Rev. **6**(2), 13 (1965)
12. Ševčovič, D.: Analysis of the free boundary for the pricing of an American call option. Eur. J. Appl. Math. **12**(1), 25–37 (2001)
13. Valkov, R.: Fitted finite volume method for a generalized Black-Scholes equation transformed on finite interval. Numer. Algor. **65**, 195–220 (2014)
14. Windcliff, H., Forsyth, P., Vetzal, K.R.: Analysis of the stability of the linear boundary condition for the Black-Scholes equation. J. Comput. Finan. **8**, 65–92 (2004)
15. Wu, L., Kwok, Y.K.: A front-fixing method for the valuation of American option. J. Finan. Eng. **6**(2), 83–97 (1997)
16. Zhang, K., Teo, K.L.: A penalty-based method from reconstructing smooth local volatility surface from American options. J. Ind. Manage. Optim. **11**(2), 631–644 (2015)

Exploring the Global Solution Space of a Simple Schrödinger-Poisson Problem

Robert Kosik$^{(\boxtimes)}$, Johann Cervenka , Dominic Waldhör , Felipe Ribeiro ,
and Hans Kosina

Institute for Microelectronics, Vienna, TU Wien, Austria
{kosik,cervenka,waldhoer,ribeiro,kosina}@iue.tuwien.ac.at
http://www.iue.tuwien.ac.at

Abstract. Quantum electron transport in nanostructures can be
described by a coupled Schrödinger-Poisson system. For the simulation of
IV-characteristics we need to solve the stationary system self-consistently
with the bias as a varying parameter. We use a full Newton method and
monitor the number of negative eigenvalues in the Jacobian in order to
detect path switching during bias-stepping. Selected simulation examples
demonstrate numerically that the solution to the Schrödinger-Poisson
model at a fixed bias is in general not unique and the global solution
structure of the continuation problem is non-trivial.

Keywords: Schrödinger-Poisson · Wigner-Poisson · continuation
problem · bifurcation · multiple solutions · resonance

1 Introduction

This work started out with the task of implementing a solver for the Wigner-
Poisson respectively sigma-Poisson [3] systems intended for the simulation of
stationary quantum transport in resonant tunneling diodes. However, simulation
attempts for IV-curves always ran into bifurcation points at a low bias and we
were not able to reproduce results in the literature.

In order to gain a better understanding and to exclude implementation errors
a corresponding Schrödinger-Poisson solver was implemented. The Schrödinger-
Poisson system presents better opportunities for parallelization and this allows
us to ensure we are near the numerical limit and calculate an accurate solution.

2 Schrödinger-Poisson Model

We consider one-dimensional electron quantum transport through a structure
embedded between two electrodes. The electrodes have the same material (effective mass $m_L = m_R$) and are both doped with a constant concentration N_d.

Financial support by the Austrian Science Fund (FWF): P33151 is gratefully acknowledged. Some of the computational results presented here have been achieved using the
Vienna Scientific Cluster (VSC5).

I. Lirkov and S. Margenov (Eds.): LSSC 2023, LNCS 13952, pp. 472–480, 2024.
https://doi.org/10.1007/978-3-031-56208-2_49

Particles are injected from the left and the right boundaries into the electrodes with a Fermi-Dirac distribution whose parameters are calculated from the doping N_d of the electrodes and the condition of charge neutrality in equilibrium (bulk at zero bias). This simple model for the distribution of incoming particles was introduced in [2] in the context of Wigner function simulations. A variety of Schrödinger-Poisson models used in device simulation are discussed in [4]. Incoming waves are assumed to be plane waves as in the quantum transmitting boundary method [1,5].

A particle injected from the left electrode with wave vector k_L fulfills a Schrödinger equation

$$-\frac{\hbar^2}{2}\frac{d}{dx}\left(\frac{1}{m(x)}\frac{d}{dx}\right)\psi_{k_L}(x) + (V(x) - E_L)\psi_{k_L}(x) = 0 \tag{1}$$

Here $m(x)$ presents the effective mass, a material-dependent parameter. The energy eigenvalue is $E_L = \frac{(\hbar k_L)^2}{2m_L}$. The potential energy V is given as

$$V(x) = qU(x) + V_{\text{off}}(x) \tag{2}$$

where U is the electrostatic potential, V_{off} is the offset from the band gap (material-dependent) and $q = -e$ is the electron charge.

The Schödinger modes (injected from the left) fulfill the boundary conditions

$$\psi_L' + \imath k_L \psi_L = 2\imath k_L \tag{3}$$

$$\psi_R' = \imath k_R \psi_R \tag{4}$$

where the subscripts L, R denote the value of ψ_{k_L} and the spatial derivative ψ_{k_L}' on the left and right boundary respectively.

By conservation of energy we have

$$k_R = \sqrt{k_L^2 - 2m_R V_R/\hbar^2} \tag{5}$$

where V_R denotes the difference in potential energy between the right and left boundary.

A single mode produces a particle density $n_{k_L}(x) = \psi_{k_L}(x)\psi_{k_L}^*(x)$. The total density $n(x)$ is then given by the sum of densities produced from particles injected from the left and particles injected from the right weighted with the Fermi-Dirac distribution.

The set of Schrödinger equations represented by Eq. (1) has to be solved self-consistently with the Poisson equation for the electrostatic potential U:

$$\frac{d}{dx}\left(\frac{1}{\epsilon(x)}\frac{d}{dx}\right)U(x) = -\rho(x). \tag{6}$$

Here $\epsilon(x)$ is the permittivity (material-dependent) and ρ is the electrostatic charge density given as

$$\rho(x) = q(n(x) - n_D(x)) \tag{7}$$

where n is the particle density (as calculated from solutions of Eq. (1)) and $n_D(x)$ is the given doping density. Dirichlet boundary conditions are used in the Poisson equation ($U_L = 0$, and U_R is given by the bias).

3   Discretization

The examples presented in this article will assume space-independent material. In particular, effective mass and electric permittivity are constant, and there is no offset from the bandgap. Furthermore, the doping $n_D(x)$ will be assumed piecewise constant.

3.1   Poisson Equation

We assume the density $n(x)$ on the right hand side is presented through piecewise cubic splines, i.e.,

$$n(x) = \sum_{i=0}^{3} a(i)x^i = \sum_{i=0}^{3} b(i)p_i(x) \qquad \text{(on an interval)} \tag{8}$$

where the p_i denote numerically suitable polynomial basis functions which we have adapted from the literature on finite element methods, see [7].

For calculation of the electrostatic potential $U(x)$ we make an ansatz of piecewise quintic splines. Details will not be discussed here but the method is analogous to the solution method for the Schrödinger equation discussed next.

3.2   Schrödinger Equation

For the Schrödinger equation we make again an ansatz based on spline functions $sp(x)$, seventh order in this case. We impose continuity conditions for ψ and ψ' on the inner nodes. Two boundary conditions are given for the Schrödinger equation. We then need 6 equations per interval to get an equal number of equations and unknowns. These equations are formed by plugging the spline ansatz $sp(x)$ into the Schrödinger equation, multiplying with a basic spline function q_j and integrating over the interval:

$$\int_{x_i}^{x_{i+1}} \left(-\frac{\hbar^2}{2} \frac{d}{dx} \left(\frac{1}{m(x)} \frac{d}{dx} \right) sp(x) + (V(x) - E_L) sp(x) \right) q_j(x)dx = 0. \tag{9}$$

Here the q_j denote the first six chosen basis functions for the spline space (polynomials with order < 6).

This method is similar to a finite element method. An advantage of our variant is that for piecewise constant $m(x)$ the continuity condition for ψ' is replaced by a jumping condition which can be imposed exactly on a conforming mesh.

For summation of the densities n_k we use Gauss-Legendre quadrature of degree 4 on each interval in k-space which correctly integrates polynomials up to order 7. Points with $k_R = 0$ or $k_L = 0$ are always in the initial mesh which is then refined by interval subdivision.

3.3   Newton Method

To solve the coupled Schrödinger-Poisson system we take the total particle density $n(x)$ as the fundamental unknown n_1. We calculate the electrostatic potential U by solving the Poisson equation (6) where n_1 enters on the right hand side. The electrostatic potential enters the set of Schrödinger equations (1) via the potential energy V as a coefficient of ψ. Densities n_k produced by the modes are summed up weighted with the Fermi-Dirac distribution which gives a new density n_2. Then $n_1 = n_2(n_1)$ is our form of the coupled Schrödinger-Poisson system.

We denote by Z the Jacobian of n_2 as a function of n_1

$$Z = \frac{Dn_2}{Dn_1} \tag{10}$$

and the linearized form of the Schrödinger-Poisson equation (as used in the Newton method) becomes

$$n_1 + \triangle n_1 = n_2(n_1) + Z\triangle n_1 \tag{11}$$

which gives

$$(I - Z)\triangle n_1 = n_2 - n_1 \tag{12}$$

where I denotes the identity matrix. Hence

$$J = I - Z \tag{13}$$

is the Jacobian matrix entering in the Newton method.

The Jacobian was calculated by fully symbolic differentiation of the discretized system. This leads to inhomogeneous Schrödinger type equations which have to be solved numerically for the calculation of the Jacobian. We employ a simple Newton method combined with line search, i.e., the optimal damping is calculated at each step in the Newton method. Assembly of the Jacobian is parallelized using MPI on the cluster and CUDA for GPUs.

For the simulation of IV-characteristics we need to solve the Schrödinger-Poisson system with the bias as a varying parameter. We solve the continuation problem by stepping with the bias starting from the equilibrium solution. The bias is increased by a chosen small quantity and the Newton iteration uses the solution from the previous bias point as an initial guess. This solution path can be extended as long as the Jacobian is not singular.

A noticeable numerical observation is that for self-consistent solutions almost all eigenvalues in the Jacobian are larger than 1. If eigenvalues of J are larger than 1, this implies that the eigenvalues of Z (Eq. (10)) are negative which is a manifestation of the repulsive Coulomb interaction between electrons.

4   Bulk

In applications bulk denotes the case with constant material and constant doping $n_D(x) = N_d$. The Schrödinger-Poisson system for bulk and zero bias has an analytical solution $n(x) = N$ with self-consistent potential energy $V(x) = 0$. This is an anomalous potential in the sense of [6]. The temperature assumed for the Fermi-Dirac distribution is 300 K. Electrode doping N_d is 2×10^{18} cm^{-3}. The material is GaAs. Numerically, the corresponding Jacobian appears to have one negative eigenvalue, all other eigenvalues are larger than 1.

When we make a small random bias step (e.g., 0.01 V) and start the Newton scheme with the analytical equilibrium solution as initial guess then the iteration will converge in a few steps. Calculating the Jacobian for the solution we find that the number of negative eigenvalues in the Jacobian is zero, i.e., we have switched path. The two solutions (one at bias zero, one at the bias of 0.01 V) are either not connected or in case they are connected the solution path must have passed through a bifurcation point.

Further numerical experiments show that the former is the case. There exist two disjoint solution paths (IV-curves). One with no negative eigenvalues, the other has exactly one negative eigenvalue in the Jacobian. These two paths can be traced out by bias-stepping and monitoring the number of negative eigenvalues. If the number of negative eigenvalues changes we reduce the bias step in order to avoid path switching.

Exemplary solutions are depicted in Fig. 1. We could not satisfactorily resolve what happens at zero bias for the solution with no negative eigenvalues. There could hypothetically be a bifurcation point, but we suspect that there are two different solutions at zero bias so close to each other that they cannot be distinguished by our numerical method.

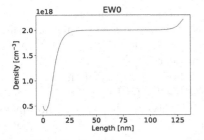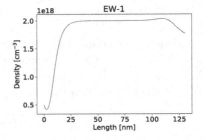

(a) Self-consistent solution with no negative eigenvalue in the Jacobian.

(b) Self-consistent solution with one negative eigenvalue in the Jacobian.

Fig. 1. Density distribution of the two solutions for bulk at 0.195 V are plotted. The solutions show different boundary effects near the right boundary.

5   Simulation of NIN-Structures

NIN-structures consist of constantly doped n-type electrodes with an undoped (intrinsic) region in the center of the simulation domain. A typical nin-structure is depicted in Fig. 2. We introduce the doping parameter f which describes the constant doping N_i in the center region (width 10. nm) with $N_i = fN_d$.

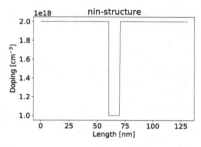

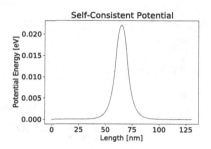

(a) Doping of a nin-structure for $f = 0.5$.   (b) Self-consistent potential.

Fig. 2. Doping of a nin-structure with doping parameter $f = 0.5$. The equilibrium solution is calculated by stepping with the parameter f. The doping of the nin-structure results in a single-barrier like self-consistent potential.

The Newton method needs a good initial guess in order to ensure convergence. For bias zero a good initial guess is to put $n(x) = n_D(x)$. However, for low values of the doping parameter f this initial guess will not converge. In these cases one can use stepping with the doping parameter f (while keeping the bias fixed at zero) to calculate the equilibrium solution for a nin-structure.

5.1 Bifurcation

Using doping-stepping we find an equilibrium solution for $f = 0$ with no negative eigenvalues. Three more equilibrium solutions are depicted in Fig. 3. These 4 solutions are not independent but lie on 4 solution segments connected through bifurcation points. They are found by bias-stepping till running in a bifurcation point and then stepping (with reversed direction) on the emerging second path till running in a bifurcation point again and so on (going full circle).

All bifurcation points lie near $\pm 0,004$ V and solution paths starting from equilibrium cannot be extended beyond this domain. However, it is possible to find solutions at a higher bias by starting from the bulk solution at a given bias and stepping with the doping parameter f while keeping the bias fixed. Subsequently, bias-stepping gives additional solution segments.

5.2 Resonances in NIN-Structures

The particle number contributed by a mode $\psi_k(x)$ is defined as the integral of $n_k(x)$ over the simulation domain. The type of resonances discussed here carry no current but contribute such a high particle number that they dominate the overall solution (blowup modes). We present an illustrative example for a nin-structure with $f = 0.1$ at a bias of -0.04 V in Fig. 4. The potential is shown in Fig. 4a.

In Fig. 4b we plot the contributed particle number as a function of the energy of the injected mode in a logarithmic scale. Here particles are injected with

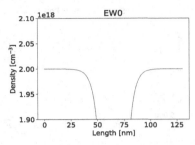

(a) Equilibrium solution with no negative eigenvalues in the Jacobian.

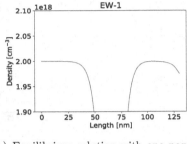

(b) Equilibrium solution with one negative eigenvalue in the Jacobian.

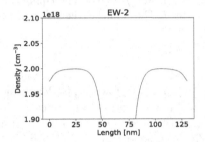

(c) Equilibrium solution with two negative eigenvalues in the Jacobian.

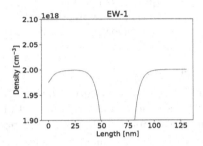

(d) Equilibrium solution with one negative eigenvalue in the Jacobian (mirror).

Fig. 3. Density distribution of four equilibrium solutions for a nin-structure with $f = 0$ exhibiting different effects near the boundary. The solutions are connected by bifurcation points discovered through bias-stepping.

amplitude 1 from the left. The energy levels of the three largest resonances are indicated in Fig. 4a relative to the potential. They are evanescent modes where particles get trapped in a potential well.

We plot the spatial density distribution $n_{k_L}(x)$ for the two largest resonant modes in Fig. 4c and Fig. 4d which shows the typical pattern reminiscent of eigenstates in a potential well. When injected with amplitude 1 the local density of the largest resonance reaches a maximum above 10^6. Full width at half maximum for this sharp spike is 3×10^{-8} eV. To detect the resonances a large initial k-grid is necessary which is then adaptively refined. Massive parallelization is needed to make all this computationally tractable.

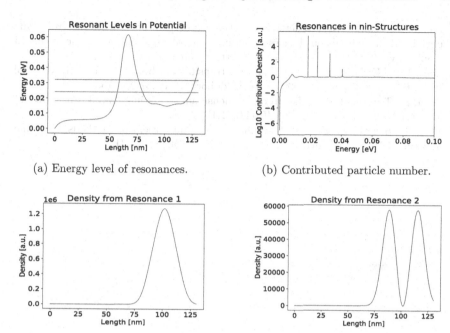

(a) Energy level of resonances.

(b) Contributed particle number.

(c) Resonance at $E_{k_L} = 0.018\,\mathrm{eV}$.

(d) Resonance at $E_{k_L} = 0.024\,\mathrm{eV}$.

Fig. 4. Resonances from bound states in a nin-structure with $f = 0.1$ at a bias of $-0.04\,\mathrm{V}$.

6   Conclusion

The existence of bifurcation points in IV-curves originally detected with a sigma-Poisson solver was confirmed with a full Newton Schrödinger-Poisson implementation. The Schrödinger-Poisson solver can use an adaptive k-grid needed to resolve sharp resonances and a much more reliable and accurate solution is possible. Numerical results in this work sketch a complex picture of the global structure of the solution space which so far has not been described in the literature.

References

1. Aktosun, T.: On the Schrödinger equation with steplike potentials. J. Math. Phys. **40**(11), 5289–5305 (1999). https://doi.org/10.1063/1.533032
2. Frensley, W.R.: Boundary conditions for open quantum systems driven far from equilibrium. Rev. Mod. Phys. **62**(3), 745–791 (1990). https://doi.org/10.1103/revmodphys.62.745
3. Kosik, R., Cervenka, J., Kosina, H.: Numerical constraints and non-spatial open boundary conditions for the Wigner equation. J. Comput. Electron. **20**(6), 2052–2061 (2021). https://doi.org/10.1007/s10825-021-01800-w

4. Laux, S.E., Kumar, A., Fischetti, M.V.: Analysis of quantum ballistic electron transport in ultrasmall silicon devices including space-charge and geometric effects. J. Appl. Phys. **95**(10), 5545–5582 (2004). https://doi.org/10.1063/1.1695597
5. Lent, C.S., Kirkner, D.J.: The quantum transmitting boundary method. J. Appl. Phys. **67**(10), 6353–6359 (1990). https://doi.org/10.1063/1.345156
6. Senn, P.: Threshold anomalies in one-dimensional scattering. Am. J. Phys. **56**(10), 916–921 (1988). https://doi.org/10.1119/1.15359
7. Solin, P., Segeth, K., Dolezel, I.: Higher-Order Finite Element Methods. Chapman and Hall/CRC (2003). https://doi.org/10.1201/9780203488041

Author Index

I. Lirkov and S. Margenov (Eds.): LSSC 2023, LNCS 13952, pp. 481–482, 2024.
https://doi.org/10.1007/978-3-031-56208-2